Carbohydrate Chemistry

Volume 42

A Specialist Periodical Report

Carbohydrate Chemistry
Volume 42

Editors
Amelia Pilar Rauter, *Universidade de Lisboa, Portugal*
Thisbe K. Lindhorst, *Christiana Albertina University of Kiel, Germany*
Yves Queneau, *Université de Lyon, France*

Authors
Mohammed Ahmar, *University of Lyon, France*
Ana Ardá, *CIC bioGUNE, Derio, Spain*
Jacqueline Bezençon, *University of Basel, Switzerland*
Roman Bielski, *Wilkes University, Wilkes-Barre, PA, USA*
F. Javier Cañada, *CIB-CSIC, Madrid, Spain*
Stéphane Chambert, *University of Lyon, France*
Helena Coelho, *CIC bioGUNE, Derio, Spain*
Stephen J. Cowling, *The University of York, UK*
Gour Chand Daskhan, *Indian Institute of Science, Bangalore, India*
Simone Dedola, *Iceni Diagnostics Ltd, Norwich, UK*
Deniz Eris, *University of Basel, Switzerland*
Beat Ernst, *University of Basel, Switzerland*
Beatriz Fernández de Toro, *CIB-CSIC, Madrid, Spain*
Robert A. Field, *John Innes Centre and Iceni Diagnostics Ltd, Norwich, UK*
Silvia Galante, *CIB-CSIC, Madrid, Spain*
Ana Gimeno, *CIC bioGUNE, Derio, Spain*
Ana M. Gómez, *Instituto de Química Orgánica General, Madrid, Spain*
John W. Goodby, *The University of York, UK*
Sławomir Jarosz, *Institute of Organic Chemistry, Polish Academy of Sciences, Warsaw, Poland*
N. Jayaraman, *Indian Institute of Science, Bangalore, India*
Jesús Jiménez-Barbero, *CIC bioGUNE, Derio, Spain and Ikerbasque, Bilbao, Spain*
Simon Kleeb, *University of Basel, Switzerland*
Paul Kosma, *University of Natural Resources and Life Sciences-Vienna, Austria*
Pavol Kováč, *NIDDK, LBC, National Institutes of Health, Bethesda, MD, USA*
J. Cristobal López, *Instituto de Química Orgánica General, Madrid, Spain*
Grahame Mackenzie, *University of Hull, UK*
Krishnagopal Maiti, *Indian Institute of Science, Bangalore, India*

Michał Malik, *Institute of Organic Chemistry, Polish Academy of Sciences, Warsaw, Poland*
Silvia Miranda, *Instituto de Química Orgánica General, Madrid, Spain*
Lijuan Pang, *University of Basel, Switzerland*
Giulia Pergolizzi, *John Innes Centre, Norwich, UK*
Yves Queneau, *University of Lyon, France and University of Hull, UK*
Ana Poveda, *CIC bioGUNE, Derio, Spain*
Said Rabbani, *University of Basel, Switzerland*
Christoph P. Sager, *University of Basel, Switzerland*
Gopal Kumar Samanta, *Indian Institute of Science, Bangalore, India*
Javier Sastre, *CIB-CSIC, Madrid, Spain*
Oliver Schwardt, *University of Basel, Switzerland*
Anja Sigl, *University of Basel, Switzerland*
Martin Smiesko, *University of Basel, Switzerland*
Laurent Soulère, *University of Lyon, France*
Luca Unione, *CIC bioGUNE, Derio, Spain*
Pablo Valverde, *CIC bioGUNE, Derio, Spain*
Zbigniew J. Witczak, *Wilkes University, Wilkes-Barre, PA, USA*
Peng Xu, *NIDDK, LBC, National Institutes of Health, Bethesda, MD, USA*
Rui Xu, *University of Lyon, France*
Zonglong Yang, *University of Lyon, France*

ROYAL SOCIETY OF CHEMISTRY

THE QUEEN'S AWARDS FOR ENTERPRISE: INTERNATIONAL TRADE 2013

ISBN: 978-1-78262-538-4
PDF eISBN: 978-1-78262-665-7
EPUB eISBN: 978-1-78262-974-0
ISSN: 0306-0713
DOI: 10.1039/9781782626657

A catalogue record for this book is available from the British Library

Published by The Royal Society of Chemistry,
Thomas Graham House, Science Park, Milton Road,
Cambridge CB4 0WF, UK

Registered Charity Number 207890

For further information see our web site at www.rsc.org

Printed in the United Kingdom by CPI Group (UK) Ltd, Croydon,
CR0 4YY

Preface

DOI: 10.1039/9781782626657-FP007

This volume is dedicated to the memory of Derek Horton, a scientist that has shown to the world the importance of the glycosciences for societal and industrial innovation. With more than 500 publications and numerous patents, he was also well known in the carbohydrate community for his editorial activities: he was a founder and editor of the leading journal Carbohydrate Research, and the editor of Advances in Carbohydrate Chemistry and Biochemistry, formerly known as Advances in Carbohydrate Chemistry, a book series containing peer reviewed contributions covering all areas of glycosciences.

In our memory remains his enthusiasm to find new topics and new authors, encouraging researchers, including the younger ones, to submit their contributions to the Advances, in recognition of their excellence and novelty. He loved challenges, and another of his passions was the nomenclature of carbohydrates. Indeed, he was a reference in this domain, and nomenclature experts always wanted to hear his voice on the matter! He was always first when giving a compound name, and a more complex structure was a bigger challenge and a joy for him! He chaired the American Chemical Society Committee for Carbohydrate Nomenclature and belonged to the expert panel of the Recommendations 1996 for the Nomenclature of Carbohydrates. Since then, as a member of the IUPAC Joint Commission of Biochemical Nomenclature and as a task group member of the IUPAC project dedicated to the revision of the Nomenclature of Carbohydrates, he honored us with his outstanding contributions to carbohydrate nomenclature.

Derek Horton was a Fellow of both the American Society of Chemistry (ACS) and the Royal Society of Chemistry (RSC), and received numerous awards, to name the Haworth Medal from the RSC, the Wolfrom Award and the Claude S. Hudson Award from the ACS. He held the Isbell Chair of Carbohydrate and Natural Product Chemistry at American University (1993–2009) and his love for teaching and research can be demonstrated by the number of students that have learnt how to study and how to investigate with him. This book contains contributions of some of his friends, who wish to dedicate this work to him and to the friendship they chaired in life.

Glycoconjugation and glycoside chemistry are the major topics of this volume. Chapter 1, authored by Rob Field, provides a critical survey of glycoconjugation methodologies, relevant to access homogenous natural glycoproteins and analogues thereof, emphasizing recent progress in the area. In this chapter, chemical and chemo-enzymatic glycoconjugation methods are described, along with illustrations of typical applications. These methods are unique tools to access glycoproteins for understanding the role of each glycan in glycoprotein structure–function relationships.

In Chapter 2, the application of solution NMR to the study of conformation, dynamics of molecules embodying sugars in their structure (oligo and polysaccharides, glycopeptides and glycomimetics) and to the investigation of glycan-related molecular recognition events is authored by Jesús Jiménez-Barbero and coworkers (Chapter 4). They present recent selected examples for analysing saccharide conformation, and for describing key structural features of glycan molecular recognition events with different natural and synthetic receptors.

In Chapter 3 the evolution of methodologies for the conjugation of single amino group-containing carbohydrates to protein carriers is described by Pavol Kováč and his coworker. They present the squaric acid-based method as the most efficient one for converting proteins to glycoconjugates with predictable carbohydrate-carrier ratios, and give examples of proven protocols for the conjugation of low- and high-molecular-weight carbohydrates.

Paul Kosma contribution describes the latest findings in Kdo glycosidation, a demanding research area for the development of glycoarrays, immunoreagents and vaccines that can be accessed by reading Chapter 4.

Cyclic oligosaccharides have attracted the interest of glycochemists as a result of their immense industrial importance arising from their molecular and supramolecular properties. As opposed to the production of cyclic oligosaccharides biosynthetically, their chemical synthesis remains a challenge. In this Chapter 5, authored by N. Jarayaman and coworkers, advances in cyclic oligosaccharide synthesis are compiled, particularly those described over the last decade, with emphasis on modifications of glycosidic bonds or individual sugar moieties.

The developments in the Ferrier rearrangement reported in the last three years are the subject of the critical review authored by Ana Gómez and Cristobal López (Chapter 6), who demonstrate the usefulness of glycals to the synthesis of glycosides bearing a 2,3-unsaturated glycosyl group and the various reaction conditions described for stereoselectivity control. In addition, *S*-, *N*- and *C*-glycosylation starting from glycals is also presented and discussed.

In Chapter 7, authored by Beat Ernst and coworker, they report their original work on how to improve solubility, metabolic stability as well as passive permeability of mannoside antagonists of the mannose-binding adhesin FimH, by modifying the substitution pattern and by introducing heteroatoms at the aglycone. These molecules may give an important contribution to an efficient non-antibiotic strategy to overcome antibiotic resistance in urinary tract infections.

An overview of steroid glycoside synthetic methodologies is given in Chapter 8, authored by Yves Queneau and coworkers, who highlight their nature as amphiphilic systems and their ability to interact with other biomolecules within cell membranes and to contribute to organized molecular and supramolecular systems.

Recent advances in the synthesis of imino sugars, comprising the preparation of five- and six-membered heterocycles containing an endocyclic nitrogen atom, authored by Slawomir Jarosz and coworker (Chapter 9) and amino sugar synthesis, with emphasis on diastereo- and enantioselectivity (Chapter 10) authored by Zbigniew Witczak and Roman

Bielski complete this volume 42 of the Specialist Periodical Reports – Carbohydrate Chemistry: Chemical and Biological Approaches.

We hope the readers enjoy this selection of contributions that focus on the progress of synthetic methodologies for small molecule amino and imino sugars, glycoconjugation and glycosylation of a variety of acceptors leading to molecules with a diversity of medicinal properties. Insights onto molecular recognition studies by NMR also enrich the content and multidisciplinarity of this volume.

Amelia P. Rauter
Thisbe K. Lindhorst
Yves Queneau

CONTENTS

Cover

Tetrahydropyran-enclosed ball-and-stick depiction of a glucose molecule, and (in the background) part of an α-glycosyl-(1→4)-D-glucose oligosaccharide and a glycosidase, all representative of the topics covered in Carbohydrate Chemistry Chemical and Biological Approaches.
Cover prepared by R. G. dos Santos.

Contemporary glycoconjugation chemistry

Giulia Pergolizzi,[a] Simone Dedola[b] and Robert A. Field*[a,b]
DOI: 10.1039/9781782626657-00001

Glycoproteins are typically expressed in nature with heterogeneous glycan components; understanding glycoprotein structure–function relationships is therefore challenging. This has been the driving force for increasing investigation for glycoconjugation techniques, in order to access homogenous natural glycoproteins and analogues thereof. These tools will be invaluable in efforts to unveil the precise role of each glycan structure. This mini-review provides a critical survey of glycoconjugation methodologies, with an emphasis on more recent progress in the field. Chemical and chemo-enzymatic glycoconjugation methods will be described, along with illustrations of typical applications.

1 Introduction

It is now almost redundant to state that carbohydrates are crucial molecules in nature, taking part in a wide variety of biological process:[1,2] prevalent on cell surfaces, carbohydrates are involved in cell–cell signalling events and regulate the interaction with extracellular matrix; oligosaccharides modulate glycoprotein folding in the ER and are subsequently modified to achieve fully functional mature glycoproteins;[3,4] recognition of carbohydrates by lectins forms the basis for inflammatory and immune responses, as well as for the adhesion and infection processes of pathogenic microorganisms. Surprisingly therefore, the importance of glycan structure in biology has been underestimated in the past and this class of molecule is only now gaining recognition for its significant role on the crowded stage of scientific understanding (for context, see Fig. 1).

The level of diversity[5] present in glycan structures can be overwhelming due to the intrinsic nature of carbohydrates. A detailed theoretical examination of all the possible monosaccharide combinations that can give rise to disaccharides through hexasaccharides estimated an astonishing 10^{12} possible structures.[6] Based on biosynthesis considerations, refinement of this number arrived at a projected number of mammalian glycans of just over than 3000, to which may be added other 4000 estimated glycosaminoglycans.[7] So, a projected ~7000 mammalian glycans overall, requiring overall the action of ~700 enzymes for their biosynthesis.[7] The study of such a complex set of carbohydrates, known as the "glycome", and its relation to biological processes is referred to as "glycomics";[8] the latter has seen increasing attention in recent years thanks to the development of glycan microarrays,[9–12] nanoparticle[13–15] and biosensors,[13,16–19] and improvements in glycoanalysis techniques.[20,21]

[a]*Department of Biological Chemistry, John Innes Centre, Norwich NR4 7UH, UK. E-mail: rob.field@jic.ac.uk*
[b]*Iceni Diagnostics Ltd, Norwich Research Park, Norwich NR4 7UH, UK*

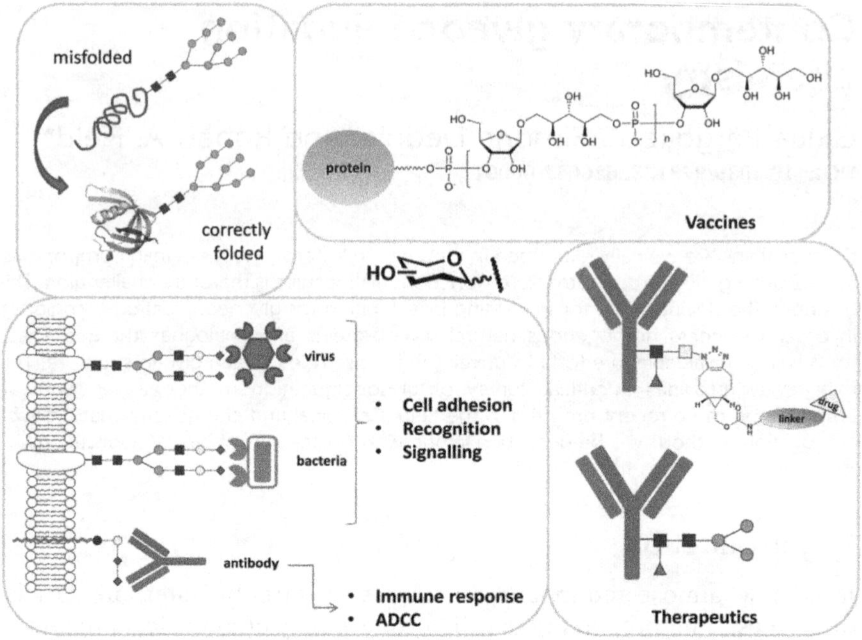

Fig. 1 Glycoconjugates and their biological relevance.

However, in the absence of the biological context – conjugated to protein or lipids, for instance – the true function of glycans may be lost. There is therefore a pressing need for efficient methods to prepare natural, homogenous glycoproteins, or structural mimetics thereof.

An ideal glycoconjugation method should fulfil a series of requirements. Very often either the protein or the carbohydrate moieties are not available in large quantities, so conjugation methods should be highly efficient to avoid the loss of valuable material. Related to this issue, the possibility of recovering the starting materials in a usable form is very much desirable. For glycoconjugation applied to glycoprotein synthesis the reaction conditions should ideally involve neutral pH, performed at room temperature, and generally mild so as not to disrupt the protein structure/activity. Most of the glycoconjugation methods used to date to generate non-native glycan–protein conjugates have relied on the random modification of one or more of the amino acid side chains of lysine, cysteine or tyrosine.

It is clear that glycoconjugates have a crucial role in fundamental biology, with scope for application as therapeutics.[17,22,23] Given the complexity and variety of carbohydrates, and the difficulties in obtaining homogeneous forms of glycans and glycoproteins from natural sources, efficient methods for the preparation of homogeneous glycoconjugates are key. With regard to natural glycoproteins,[2] glycans may be connected to polypeptides *via* the amide bond of the side chain of Asn, (*N*-glycans),[24,25] *via* glycosidic bonds to the side chain of Ser or Thr (*O*-glycans),[26–28] or *via* an ethanolamine phosphate linkage that bridges a protein C-terminus to a glycolipid, as in glycosylphosphatidylinositol (GPI) anchors.[29] Recent progress in the field chemical synthesis of glycoproteins has been extensively reported,[30–32] highlighting powerful approaches to accurately control the site and nature of glycosylation. Recent example of homogeneous glycoproteins obtained by chemical means are, among the others, the total synthesis of EPO,[33,34] interferon β,[35] MCP-3,[36] homogeneous antifreeze glycoprotein,[37] the α and β subunit of human glycoprotein hormone.[38,39] In term of non-natural glycoconjugates, linkers that permit glycan coupling to the reactive side chains lysine and cysteine have traditionally been popular.[30,40] In addition, where chemical synthesis fails to afford reasonable yields, the synthesis of natural and unnatural oligosaccharides and their conjugation to proteins can be achieved by chemo-enzymatic methods, which have become a valuable approach to tackle difficult syntheses of less straightforward oligosaccharide conjugates.[41]

As summarised above, it is evident that the structural diversity and potential heterogeneity of natural glycoproteins makes deciphering structure–function relationships a challenging task – one that may be overcome by the preparation of homogeneous glycoproteins by means of chemical or biological methods, or more often these days by a combination of both. Likewise, efficient conjugation methods that are tolerant of diverse glycan and protein structures will open new avenues for conjugate vaccines.[42] The conjugation of carbohydrates to proteins, in particular, may be achieved by a range of approaches that have been extensively reviewed in the past.[30,43,44] Herein we illustrate a range of newer methodologies and their applications, with reference to some of the classic methods that provided the inspiration for continued method development.

2 Established chemical methodologies

The definition of a "glycoconjugate" is rather broad and comprises classes of molecules where a carbohydrate unit is covalently linked to another molecule, generally a protein. Conjugation of carbohydrates to proteins has been a challenging task, tackled in classic work performed by Pauly, which dates back to the beginning of the twentieth century, where he was able to couple glucose and galactose to serum globulin.[45] His seminal work inspired later work by Avery and Goebel in the 1930s,[46] which described the synthesis of diazonium salt derivatives of glucose and galactose and their conjugation to serum globulin for use in a series of immunological studies. The reaction of diazonium sugar derivatives with proteins is non-specific resulting in indiscriminate modification of tyrosine, histidine, lysine and, if used in large excess, with tryptophan and arginine.[47–50] Since then, efforts at improving the production of glycoconjugates have been unceasing.

Synthetic carbohydrates generally have the reducing end available for conjugation; on the other hand, carbohydrates from natural sources may have the reducing end available or not, depending on the nature of the glycan concerned. There are a variety of well-established reducing end conjugations that exploit the reactivity of the exposed aldehyde of ring-open reducing sugars towards amines.[51] A classic approach involves the reductive amination of a reducing sugar with the amino group of lysine with cyanoborohydride,[11,52,53] giving secondary amines that are stable to hydrolysis. The aldehyde form of reducing sugars can also react with hydrazines, hydrazides or aminooxy-functionalised proteins to give hydrazones or oximes[11] (Fig. 2a). Where the reducing end of a glycan is not available for coupling, internal sugar units or the non-reducing terminus may be oxidised with periodate to generate aldehydes that can undergo reductive amination (Fig. 2b).

Alternatively, the reducing end or specific functional groups, such as carboxylic acid in uronic acids or amine in amino sugars, can be selectively modified by the use of heterobifunctional cross-linkers introducing a variety of functional groups[11,51] (Fig. 2c), such as amine, hydrazine or thiol, which can react with *N*-hydroxy-succinimide (NHS) ester- or maleimide-functionalised proteins. In the opposite sense, the carbohydrate can be functionalised with linkers carrying NHS-ester, *para*-nitrophenyl ester, squarate,[54] maleimide, isocyanate, isothiocyanate or acyl azide[55] (Fig. 2c–d) to react with lysine and cysteine of the protein, or with amine, thiol, or hydrazine-functionalised proteins.[51] The hydroxyl groups of carbohydrate moiety can also be traditionally activated by cyanogen bromide (CNBr)[56] or carbonyldiimidazole (CDI)[57] to react with amino groups of the proteins and generate isourea and carbamate bonds, respectively (Fig. 2e–f). If the carboxylic groups on the sugar or proteins are free and not activated by NHS-ester, they can be activated *in situ* with water soluble carbodiimide,[58] forming an *O*-acylisourea active ester that can be attacked by amino groups forming an amide bond.

Historically, some of the most important glycan–protein conjugates investigated have been non-natural materials, created as potential antimicrobial vaccines. Vaccination has had a major impact in improving human health, eliminating various diseases, such as pneumonia, influenza, typhoid fever, pertussis, measles and smallpox, which have caused millions of deaths in the past.[59,60] Glycoconjugate vaccines are particularly effective in this context; while the *Haemophilus influenza* type B (Hib) capsular polysaccharide alone is poor at inducing an immune response, immunogenicity was triggered by conjugation of the polysaccharide to a carrier protein.[61] In this instance, protein–glycan conjugation was achieved through the bifunctional linker adipic acid dihydrazide (ADH), which provides access to a protein acyl hydrazide by EDC coupling with glutamate or aspartate through one end; CNBr activation of the hydroxyl groups of the polysaccharide then allows isourea bond formation through the other end of the adipic acid linker, as illustrated in Fig. 3.

The Hib vaccines have been further improved by Verez-Bencomo *et al.*[62] using a synthetically prepared capsular polysaccharide, obtained with a controlled degree of polymerization, which enabled the production of a homogeneous glycoconjugate in a cost efficient manner. This example paved the way for the development of glycoconjugate vaccines against *Neisseria meningitides*, *Salmonella typhi* and *Streptococcus pneumoniae*.[63]

Pozsgay *et al.*[64] have applied well-known oxime formation to obtain glycoconjugates. The approach consists of a two-step functionalization of BSA (Fig. 4): first, the protein is treated with the commercially available succinimidyl 3-(bromoacetamido)propionate **2.1**. The conjugation yield

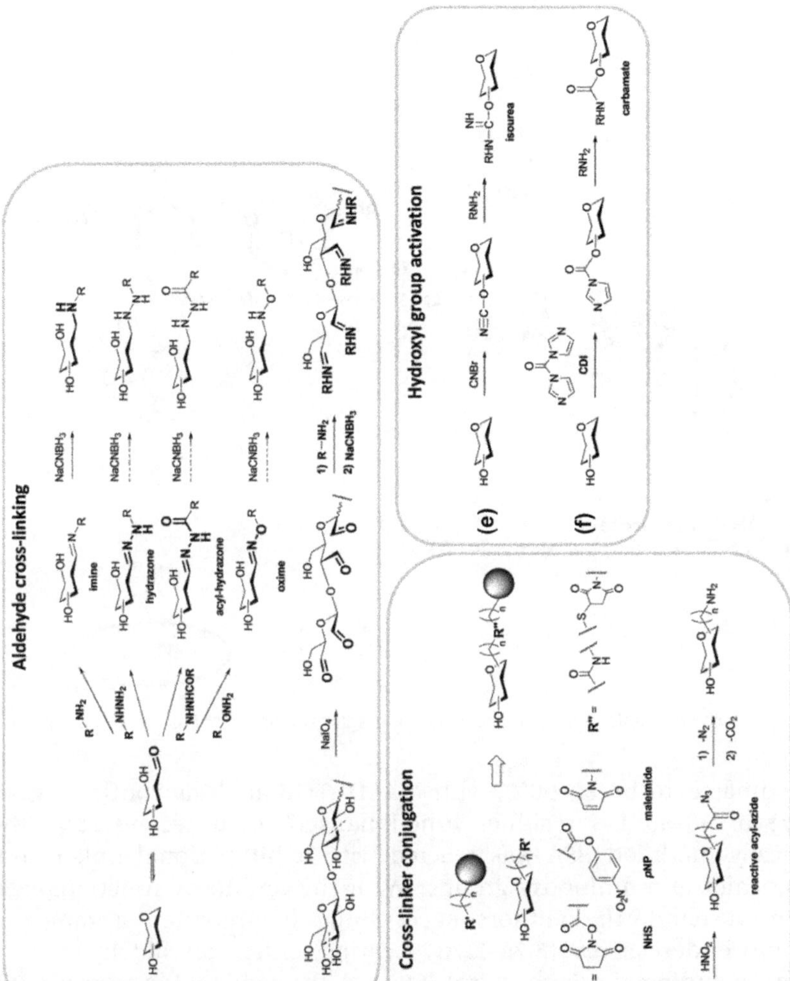

Fig. 2 Overview of established chemical methodologies for glycoconjugation.

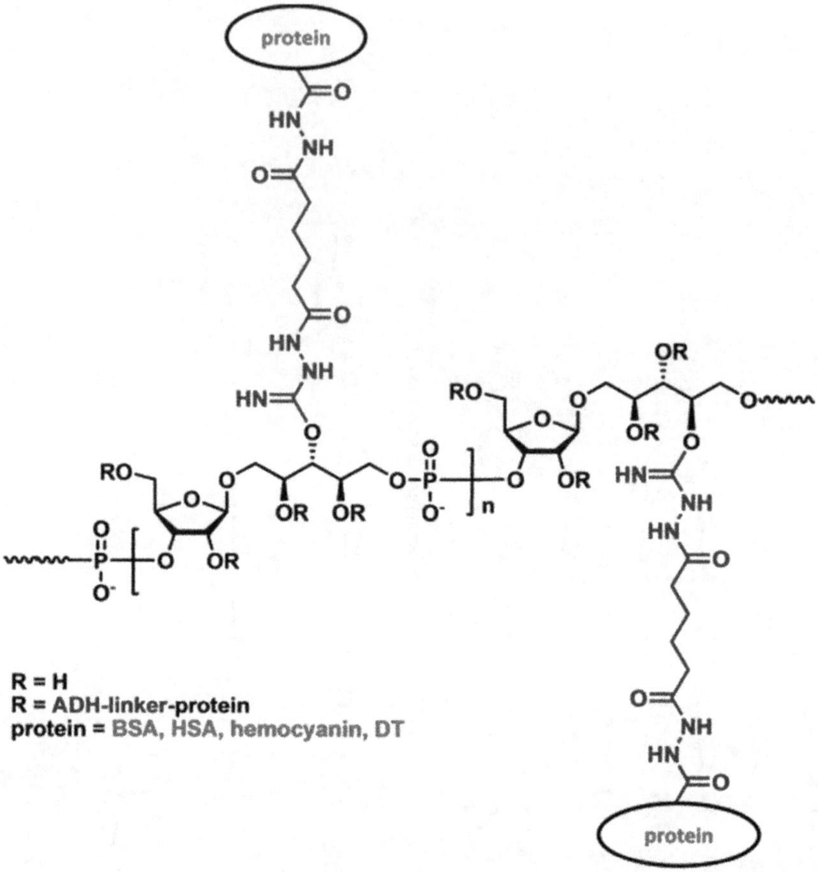

Fig. 3 First fully synthetic glycoconjugate vaccine against *Haemophilus influenza* type B.[61]

was estimated to be 80–90%, with MALDI-TOF analysis confirming an average of 30–35 Lys residues functionalised. In a second step, the chemically modified BSA was reacted with a bifunctional linker containing thiol and aminooxy groups. With the aminooxy functionalized protein in hand, the authors successfully incorporated a range of L-rhamnose derivatives (**2.7a–2.7d**) bearing either an aldehyde or a ketone, including the keto-ribitol **2.8** and the keto-tetrasaccharide **2.5** (Fig. 4), which yielded the BSA glycoconjugate **2.6**.

Lysine on BSA was also targeted for glycoconjugation by Kovac *et al.* by using a series of squarate derivative installed on *Vibrio cholerae* polysaccharide fragments,[65] of which an example (**2.10**) is shown in Fig. 5. Also in this case, the advantage of using squarate derivatives as **2.9** is that it can be used in excess and the unreacted material can be recovered.

A remarkable example of classic methodology has been recently reported by the Davis group, which relates to a new approach in antibacterial vaccines design.[66] The Hep_2KDO_2 tetrasaccharide in Fig. 6 was obtained by a new synthetic strategy starting from mannose; it was then coupled *via* the well-established isothiocyanate methodology to a

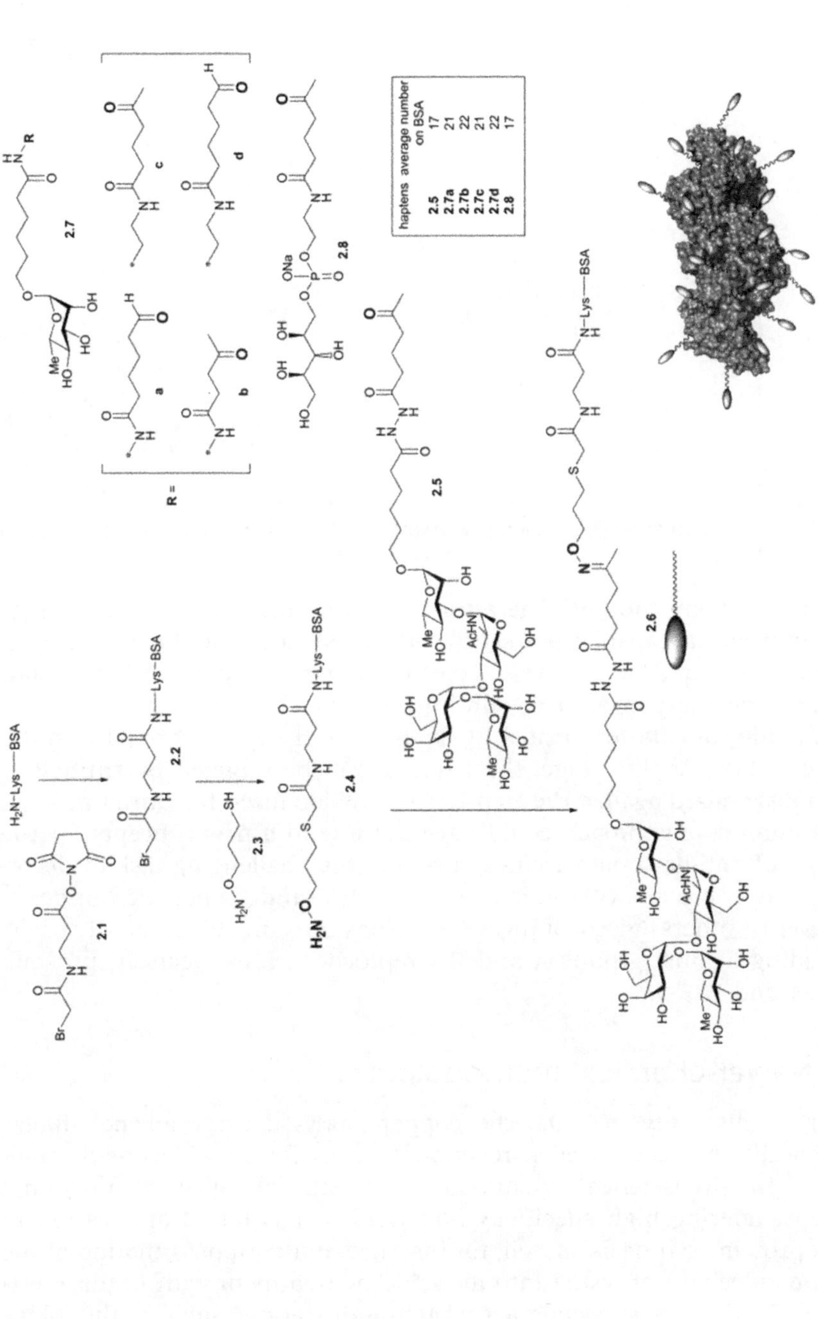

haptens	average number on BSA
2.5	17
2.7a	21
2.7b	22
2.7c	21
2.7d	22
2.8	17

Fig. 4 Preparation of aminooxylated BSA and conjugation with keto-tetrasaccharide **2.5** *via* oxime formation.[64]

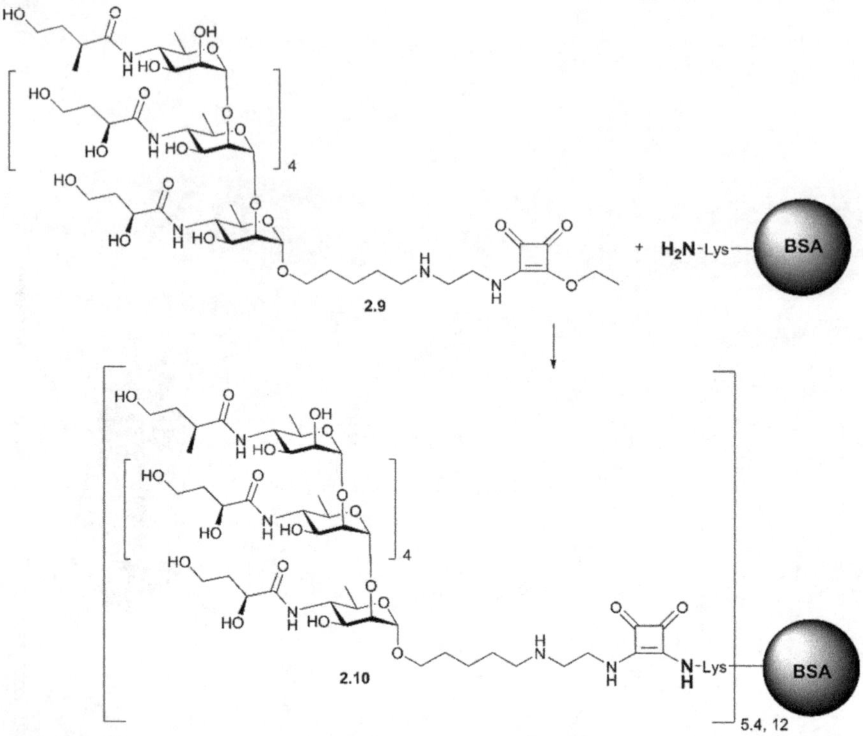

Fig. 5 Conjugation of hexasaccharide fragment **2.9** of *Vibrio cholerae* polysaccharide to BSA protein carrier.[65]

diphtheria toxin mutant. The tetrasaccharide mimics a common inner core of bacterial capsular polysaccharide (CPS). Access to the inner core is usually prevented by the presence of the outer core polysaccharide, but using γ-cyclodextrin as a co-administered inhibitor of the outer polysaccharide transporter, results in exposure of the inner core part on the bacteria LPS. At this stage, the bacteria can be targeted by antibodies previously raised against the Hep_2KDO_2-functionalized DT-carrier protein. This innovative approach could pave the way to a new concepts for the design of anti-bacterial vaccines, avoiding the challenging task of decorating carrier protein with limited accessibility *O*-antigen polysaccharides.[67]

Several others glycoconjugate vaccines are in Phase I–III trials, including against groups A and B streptococci, breast cancer, prostate cancer and HIV-1.[63]

3 Newer chemical methodologies

After its discovery in 2002, the copper-catalysed azide–alkyne dipolar cycloaddition (CuAAC) click reaction[68,69] has found widespread application in glycoscience[70–73] due to its broad tolerance of functional groups, offering high specificity and yield. It has found application in glycoproteins synthesis as well, for instance, in the transformation of the amine sidechain of lysine into an azide by treatment with imidazole-1-sulfonyl azide for subsequent CuAAC modification. Such methodology

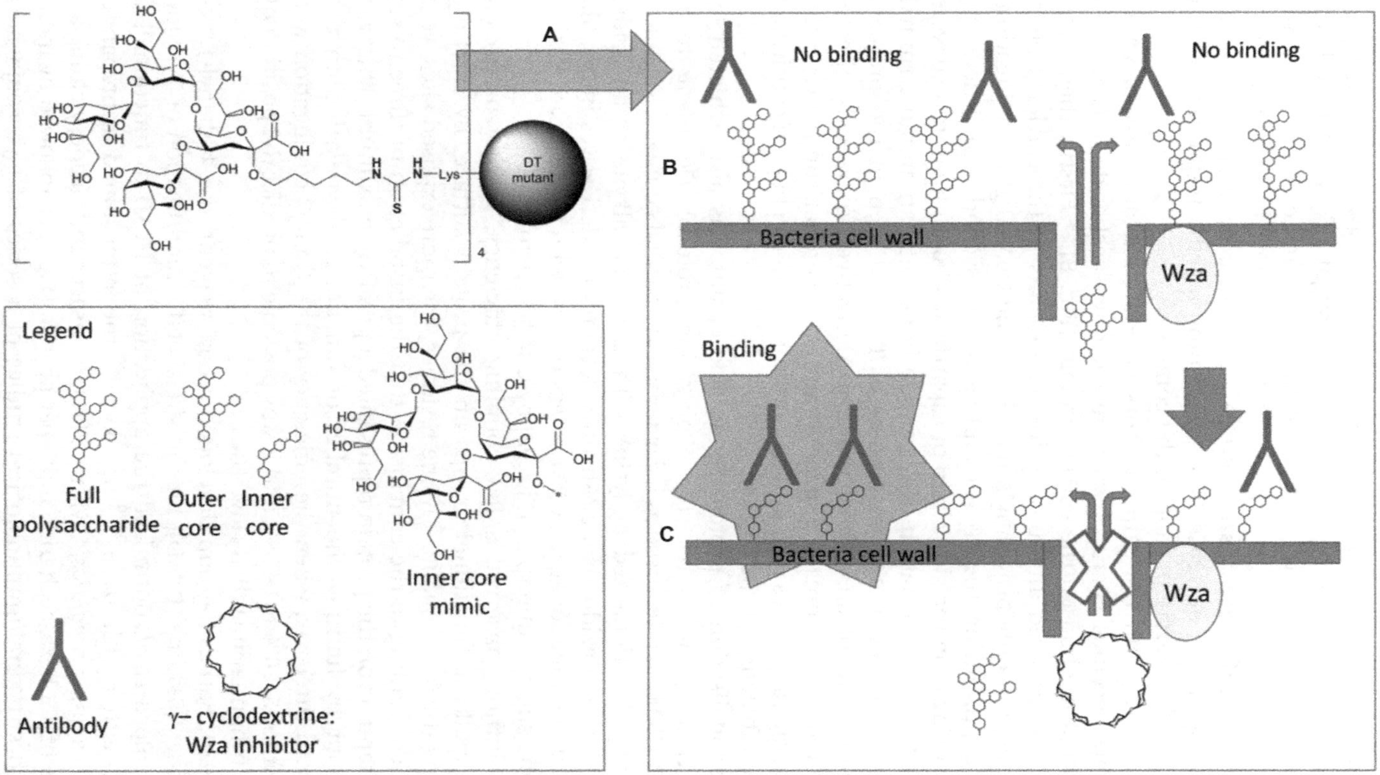

Fig. 6 An initial treatment **(A)** with DT-mutant decorated with the inner core mimic Hep_2KDO_2 will raise antibody **(B)** against the inner-core carbohydrate portion of bacterial polysaccharide. Successively, γ-cyclodextrin are used to inhibit Wza, which is the outer core polysaccharide transporter, exposing the inner core to the antibody and generating the immune response **(C)** to kill the bacteria.[66,67]

have been applied by Lipinski *et al.*[74] *en route* to the preparation of a vaccine against *Candida albicans*. Of the 40 surface amino groups present on the chicken serum albumin (CSA) surface, about half were converted into azide form (Fig. 7). Subsequent CuAAC reaction gave an average of 17 and 19 oligosaccharides per protein from sugar alkynes **3.1** and **3.2**, respectively. In a first attempt towards the generation of the vaccines, the authors analysed these two different linkers construction in order to compare the influence of an unstructured *versus* a structured linker: the former appears to decrease undesired immunogenic responses against the non-carbohydrate portion of the conjugate.

Later work from the same authors generated tetanus toxoid glycoconjugate vaccines bearing two different oligosaccharides installed in orthogonal method on the carrier protein (Fig. 8).[75] Firstly, they transformed the exposed amine of lysine side chains of the carrier protein into azide, not only providing a handle for the CuAAC glycoconjugation, but also preventing the cross-coupling in the second step, where the carboxylic acid on the side chain of aspartic acid and glutamic acid were activated for the formation of an amide bond with an amino terminal functionalized β-mannan trisaccharide. The last step of the protocol was the CuAAC of an alkyne-laminarin derivative, followed by the re-conversion of the unreacted azide to amine. A comparison between the two vaccine candidates with and without the laminarin hapten confirmed a better efficiency of the former.

Functionalisation of protein lysine side chains was also achieved by Crotti *et al.*[76] with a two-step conjugation protocol. *Diphtheria* toxin mutant, CRM_{197}, often used as a carrier in commercial conjugate vaccines, was firstly functionalized at lysine with either an alkyne or an azide, installed *via* the azido/alkyne *N*-hydroxysuccinimide derivatives in Fig. 9b. Secondly, the functionalized CRM_{197} was reacted with the respective azido or alkyne sugar counterpart (Fig. 9a) by CuAAC chemistry.

The authors observed a low coupling efficiency, leaving unreacted azide or alkyne functional groups on the protein surface. By carefully mapping the functionalized lysine residues, the more exposed sites were identified, leading to the conclusion that the installation of three to six sugar haptens on the protein represented a practical maximum. This was borne out by limiting the number of functional alkyne/azide on the protein to three–six, when the efficiency of the CuAAC conjugation was raised from 5–20% to >95%, and it was possible to predict the position of the each hapten installed (Fig. 10).

In 2010, Barbas[77] group developed a new reagent, the 4-phenyl-1,2,4-triazoline-3,5-dione (PTAD), for the site-specific modification of Tyr side chains. Berti *et al.* optimized the application of PTAD to functionalize CRM_{197} either with azide[78,79] or alkyne;[80] the latter aimed to prepare a vaccines against *candidiasis*, with a β-(1,3) hexasaccharide azide, yielding an average of 3.5 hexasaccharide units per protein (Fig. 11), followed by more in-depth structural/immunological studies of a series of glycoconjugates obtained by defined conjugation (PTAD) and random conjugation.[81]

The great potential of CuAAC for glycoconjugate synthesis was enhanced by Shoda's method[82] for the direct preparation of sugar anomeric azides

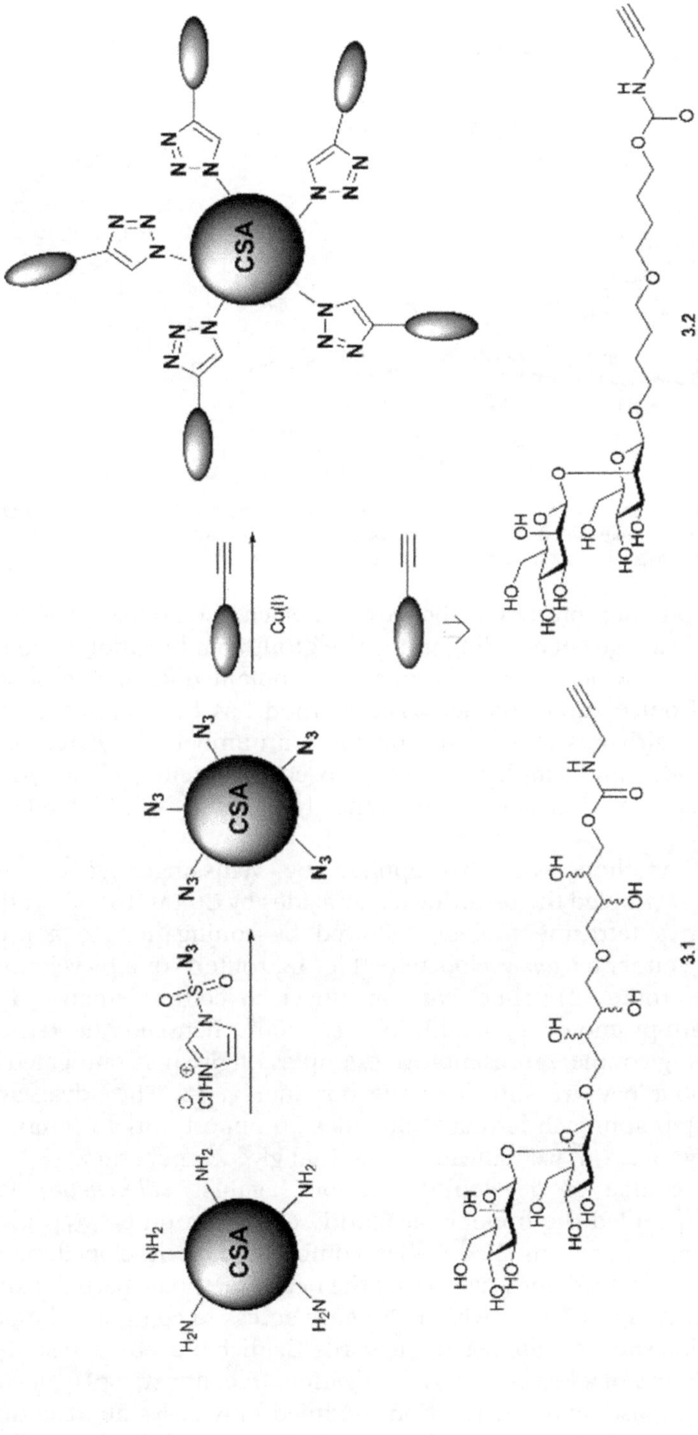

Fig. 7 Conversion of side chain amine groups into azide followed by CuAAC reaction with the stereo-diversified glycoconjugate **3.1** and the structurally defined glycoconjugate **3.2**.[74]

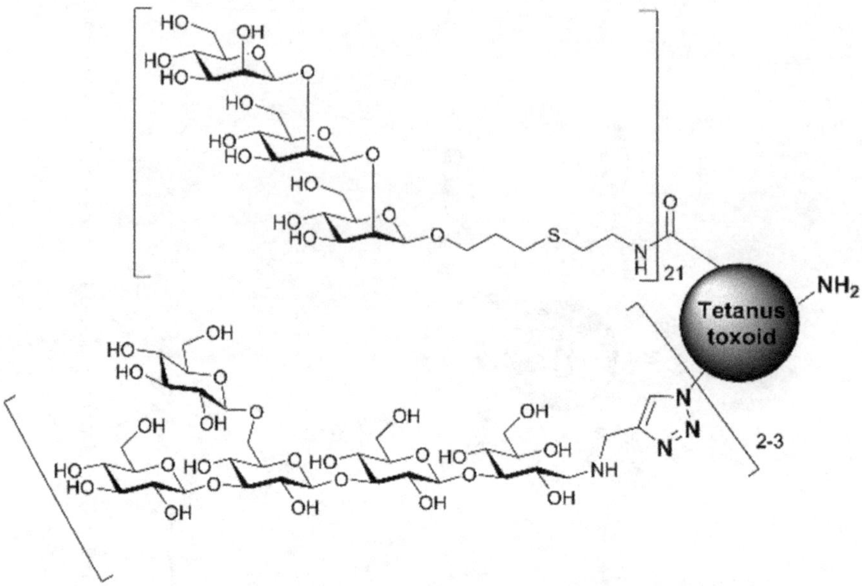

Fig. 8 Bifunctionalized tetanus toxoid carrier protein bearing β-mannan trisaccharide and laminarin installed in a two-step protocol showed better binding to Dectin-1 on dendritic cells than the conjugate with only the β-mannan.[75]

directly from unprotected carbohydrates. Such methodology offers the obvious advantages of avoiding protective group manipulation of complex carbohydrates, which are often expensive or obtained in small amount by isolation from natural products. The method has been applied, among others, by Fairbanks *et al.*[83] who prepared a number of different carbohydrate azides and employed CuAAC to couple them with a synthetic peptide fragment of MUC1 obtaining the glycopeptide with high efficiency (Fig. 12).

Shoda's method was also applied by Winssinger *et al.*[84] who successively coupled the resulting sugar azides by CuAAC to a short linker containing a tetrazine moiety, followed by conjugation to a protein containing either a *trans*-cyclooctene (Fig. 13, route 1) or a bicyclononyne (Fig. 13, route 2) ring *via* an inverse-electron-demand Diels–Alder strain-promoted cycloaddition reaction[85] between the tetrazine/strained-ring couple, representative examples of glycans conjugated with this methodology are shown in the box in Fig. 13. The advantage of this two-step approach is that it provides an opportunity to remove the potentially toxic Cu^{I} well ahead of the final glycoconjugation.

In the course of developing vaccines against *Salmonella*, Micoli *et al.*[86,87] installed the lipopolysaccharide *O*-antigen on CRM_{197} protein carrier using different methodologies, combining new developed methods with well-established one (Fig. 14). Of the newer methods, partial disulfide bond reduction of CRM_{197} with TCEP gave access to co-localised cysteine modification sites that are reactive towards 1,2-dichloracetone, resulting in the installation of a ketone for oxime ligation. In contrast, a pH-controlled transglutaminase-catalysed reaction modified only at lysine, resulting in transfer of Cbz-Gln-Gly-Peg3-azide on one/two lysine residue depending on

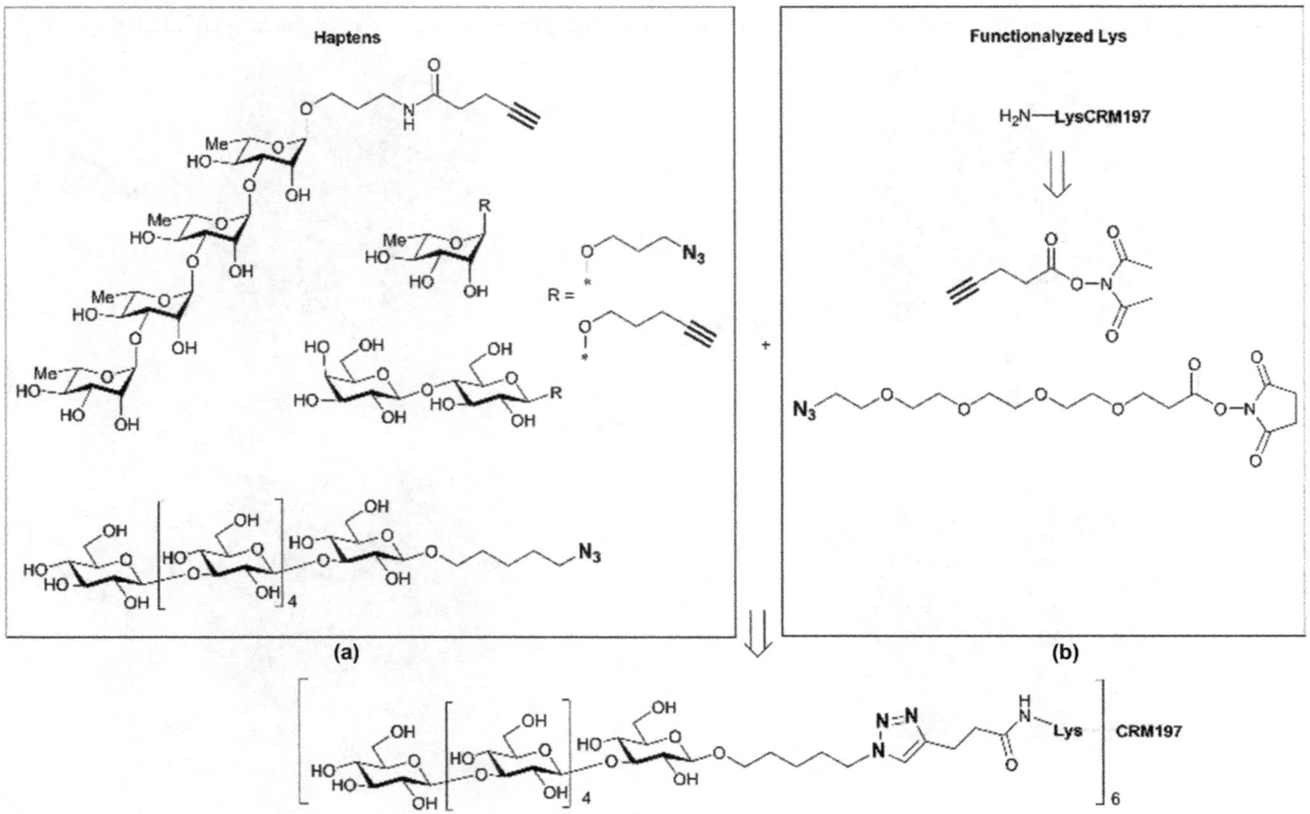

Fig. 9 Functionalization of terminal amines of CRM_{197} and azide/alkyne sugar derivative building blocks for the subsequent CuAAC reaction with an example of final glycoconjugate product.[76]

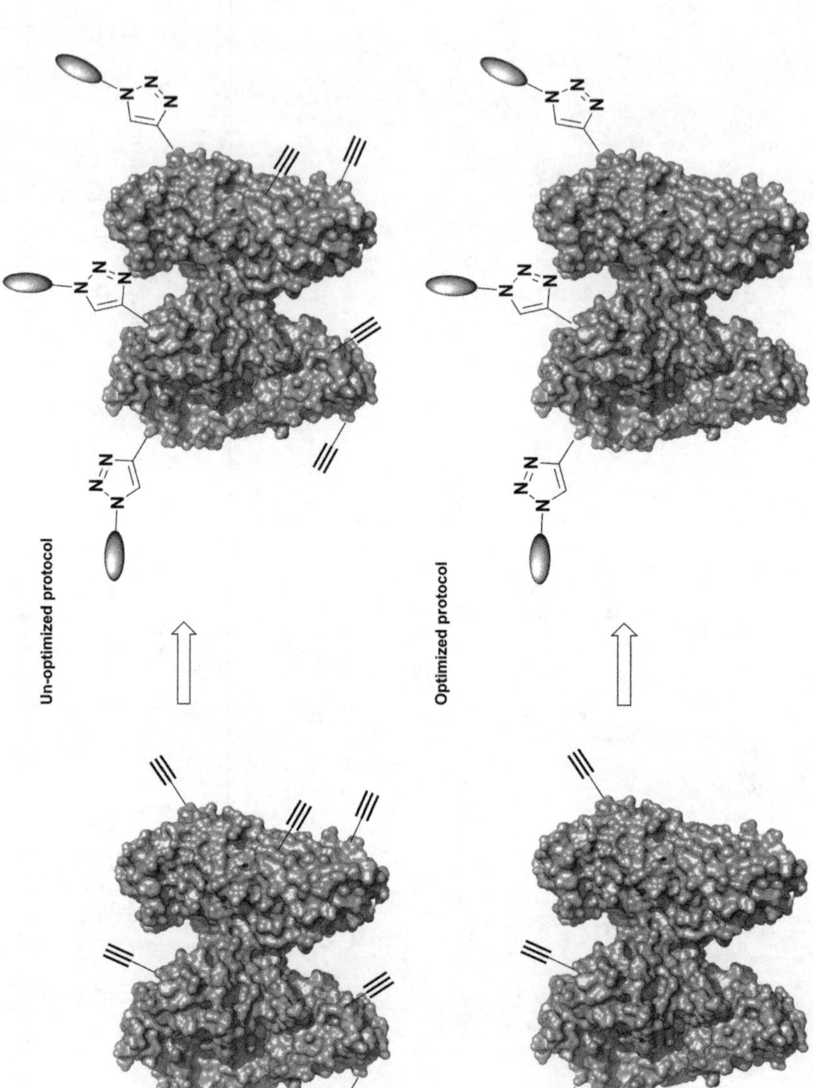

Fig. 10 Cartoon representation of the optimized functionalization protocol for CuAAC conjugation of CRM_{197} alkyne derivative.[76]

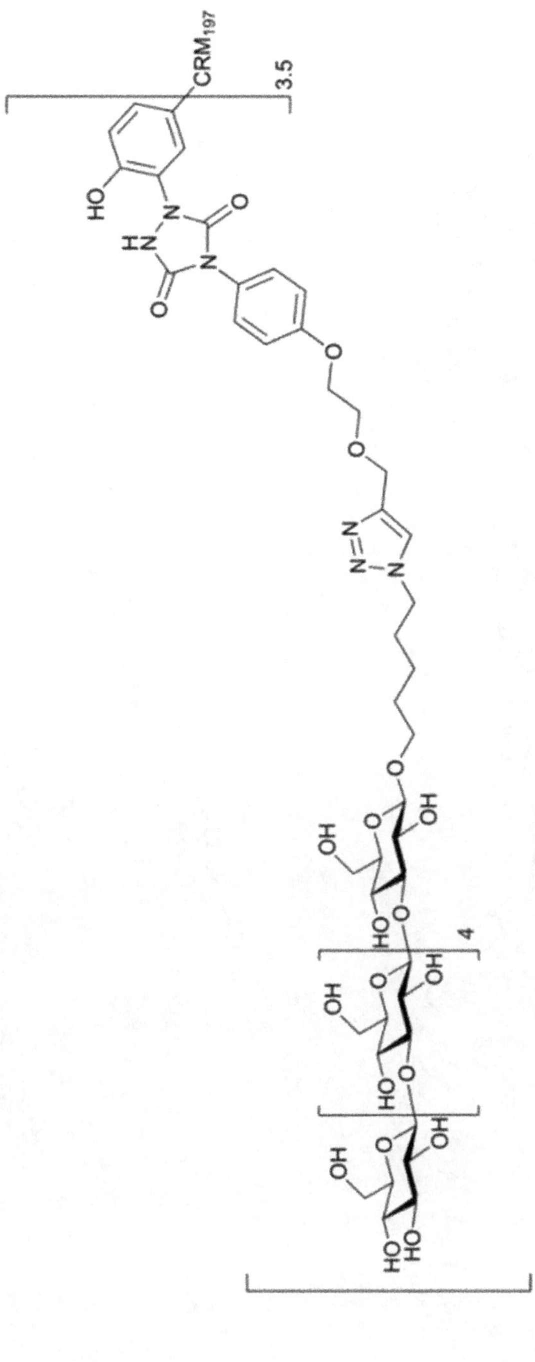

Fig. 11 Glycoconjugation product originating from the PTAD-alkyne decorated CRM_{197} and successive CuAAC reaction.[80]

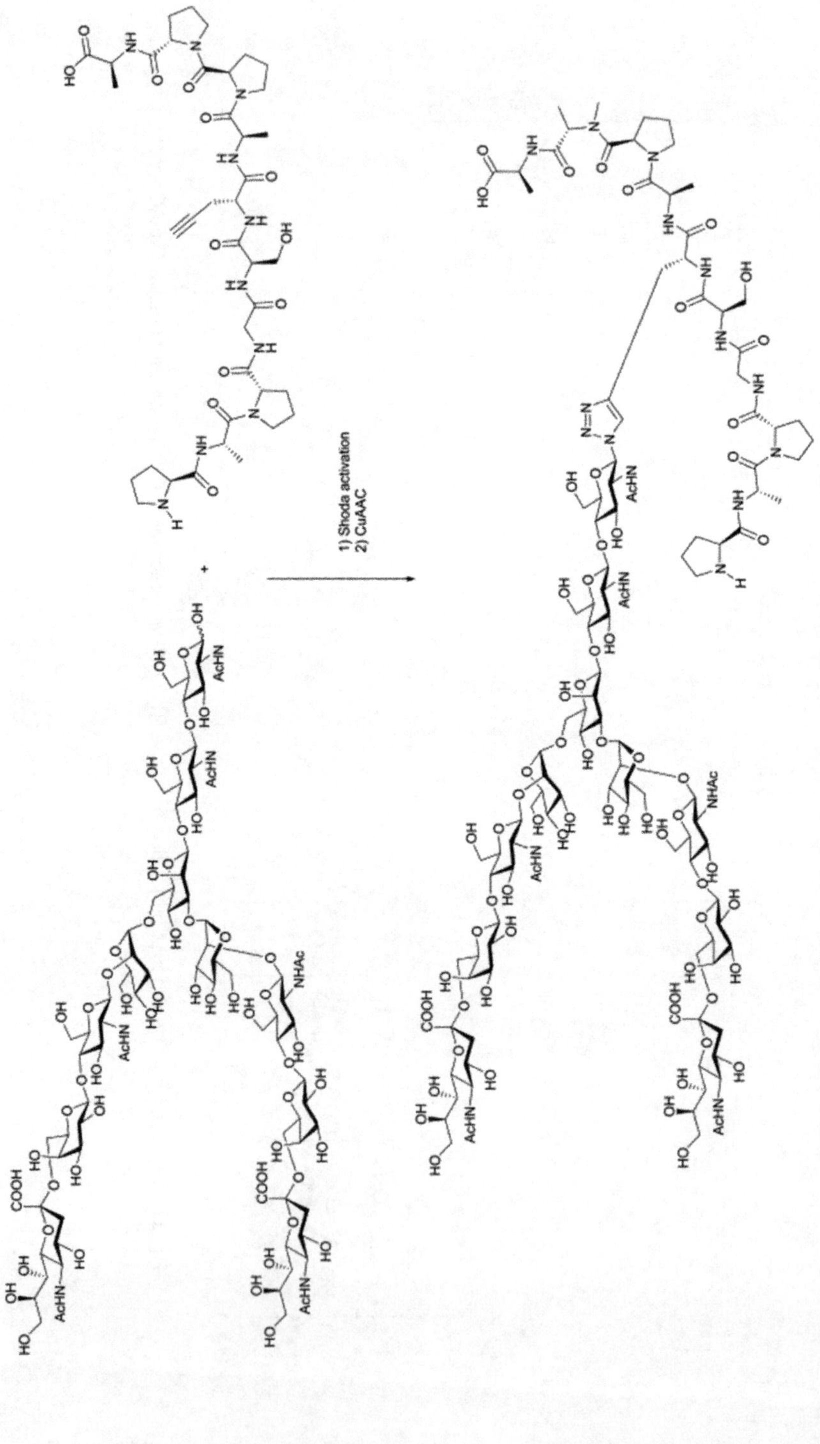

Fig. 12 Straightforward preparation of glycopeptide fragment of MUC1 by a stepwise azide formation/CuAAC protocol.[83]

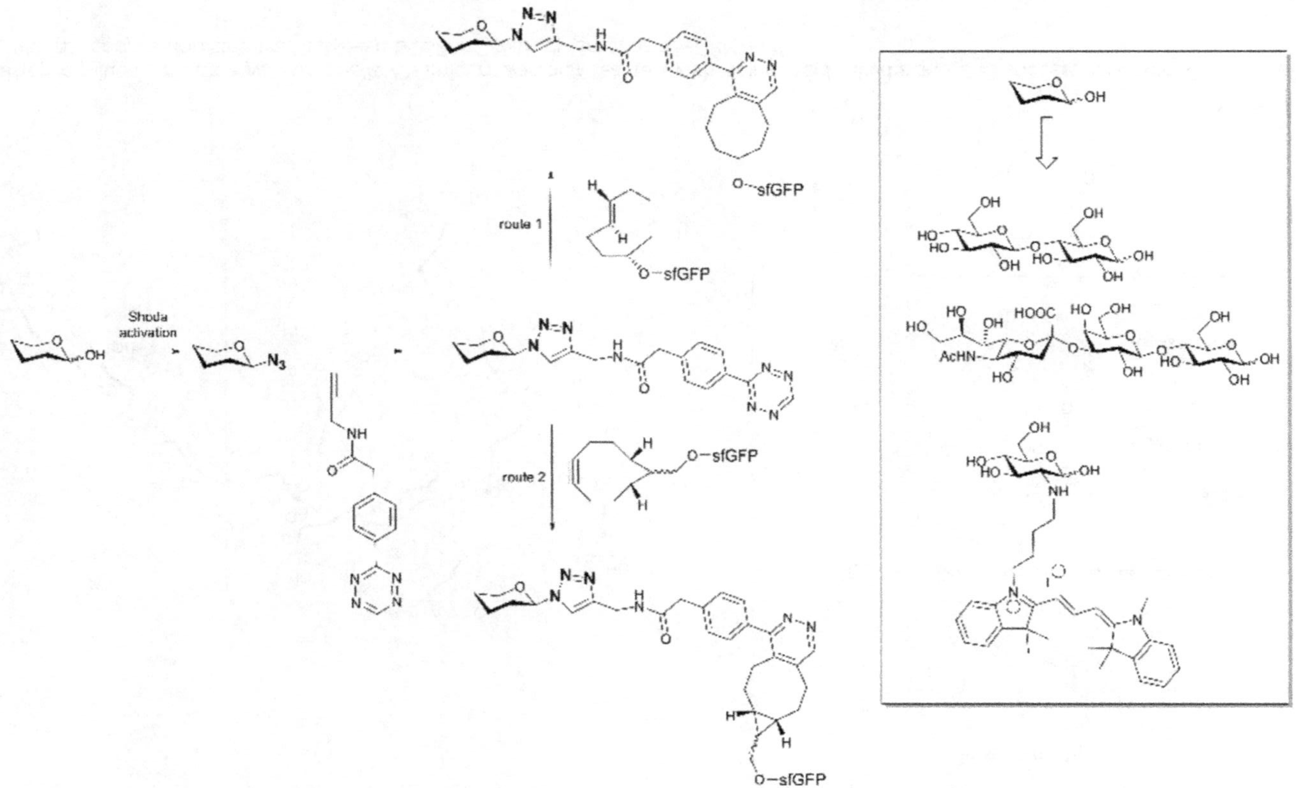

Fig. 13 Scheme for the stepwise cycloaddition reaction. Shoda's activation protocol followed by a CuAAC reaction with a tetrazine-alkyne bi-functional linker, which allowed the following copper free strain promoted final glycoconjugation.[84]

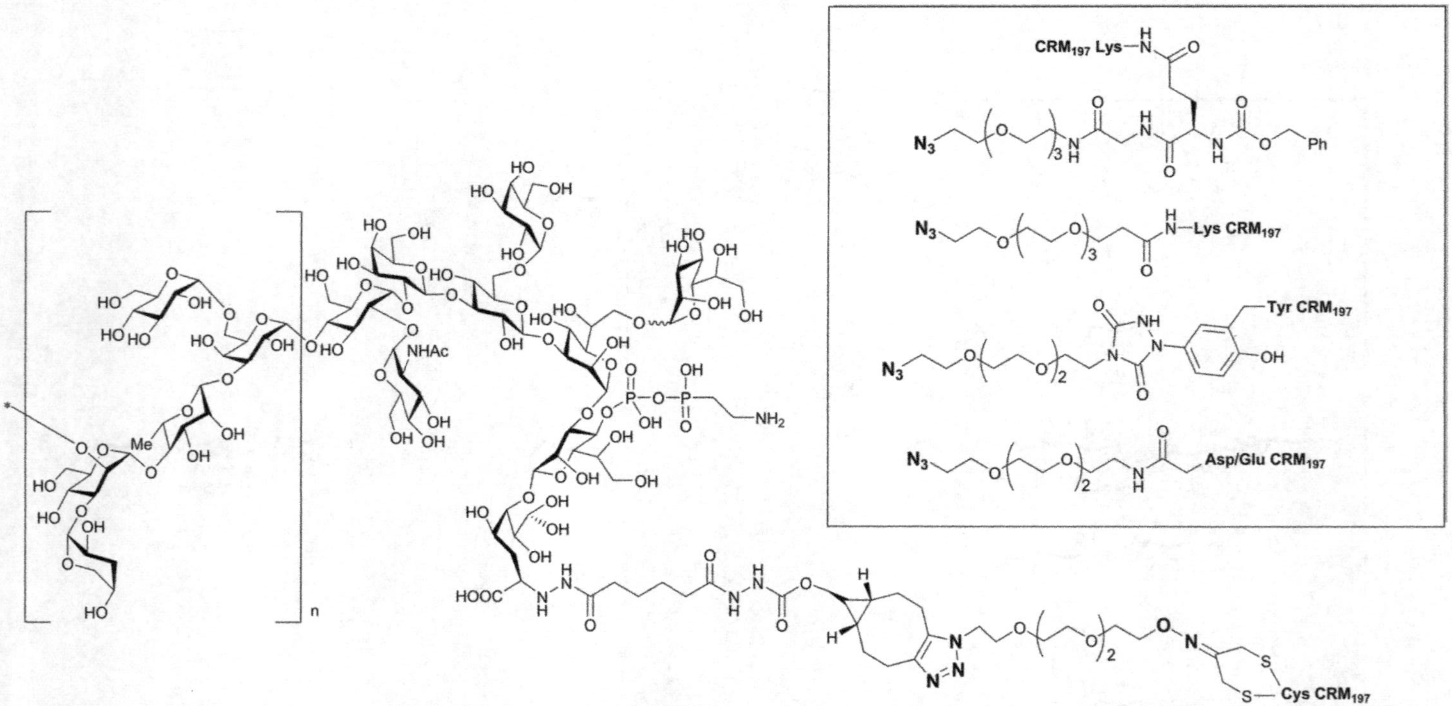

Fig. 14 Multistep conjugation strategy: *Salmonella O*-antigen was first reacted with adipic acid dihydrazide and successively functionalized with a strained cyclooctyne; the final conjugation step was achieved by SPAAC with different azido-CRM_{197}.[86]

the pH employed. Functionalisation of an average of 7 of the 39 lysine residues available on CRM_{197} was achievable in this manner. Further, triazolidinone-ene-Peg4-azide was used to functionalize an average of 1.5–3.8 of the eighteen tyrosine residues in the protein. Random functionalization of glutamates (38 total residues) and aspartate (28 total residues) by conventional activation of their side chain carboxyl groups resulted in an average of 4.8 amino acids modified. The library of glycoconjugate was then subjected to systematic assessment by immunization of mice. Interestingly the highest immune response was obtained with the CRM_{197} conjugates containing the highest number of antigens, regardless the amino acid connectivity.

Most of the effort in new glycoconjugation methodologies has been directed to the formation of *O*-linked glycosides at Ser, Thr and Tyr residue or *N*-glycosylation at Asn. The recent discovery of an unusual arginine GlcNAc transferase activity triggered the effort towards the generation and biological role of this type of modification.[88] Ornithine was used as a precursor for the GlcNAc-Arg and combined with chemical conjugation of a GlcNAc-based *S*-alkyl-isothiourea residue furnished the required adduct, which was used to generate antibodies that can specifically recognize this novel *N*-GlcNAc modification (Fig. 15).

Extensive work on the 'tag-and-modify' approach has been carried out by the Davis group,[40] spanning from the modification of a cysteine tag to the inclusion of non-proteinogenic amino acids (*e.g.* dehydroalanine,

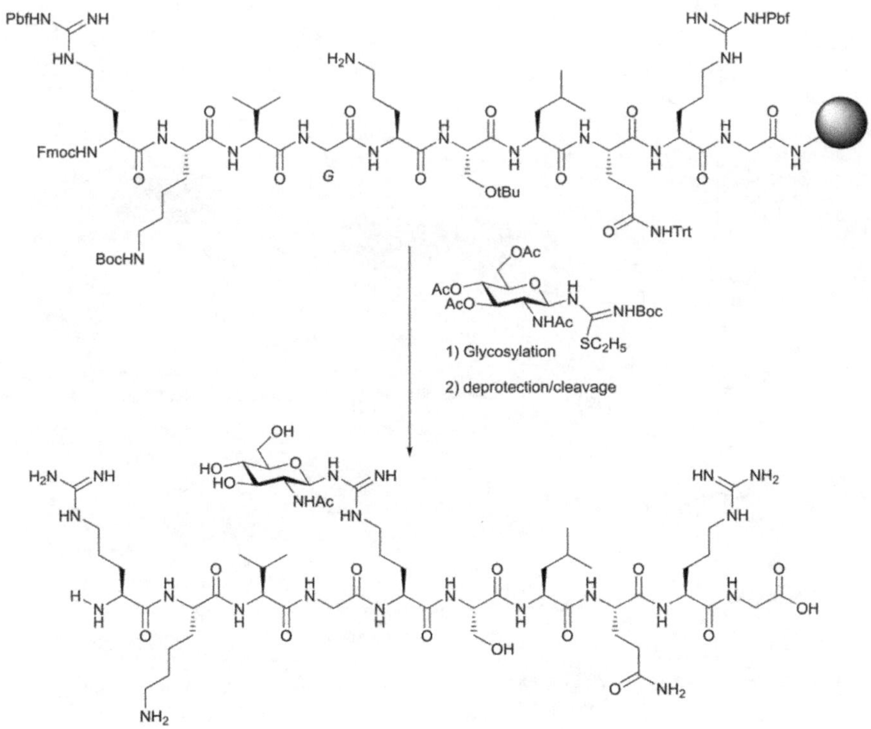

Fig. 15 Ornithine side chain glycosylation afforded the desired GlcNAc-Arg glycopeptide.[88]

homoallylglycine, *S*-allylcysteine, azidohomoalanine) in a protein sequence. Starting from free sugars of different complexity and chain length, Davis *et al.* optimized the formation of a glycosyl thiol by using the Lawesson reagent (LR),[89] offering the advantage of avoiding the need for protecting group manipulation (Fig. 16a). The sugar thiol can be used as such as nucleophile to react with an electrophilic cysteine (phenylselenyl) sulfide (Fig. 16b); this principle can also be reversed, exploiting the nucleophilicity of a free cysteine thiol on a protein in reaction with an (phenylselenyl) sulfide thioglycoside (Fig. 16c–d).

By using the thioglycoside **3.3–3.9** (Fig. 17) and selected mutants of the model proteins, such as *Bacillus lentus* subtilisin SBL-S156C or *Sulfolobus solfataricus* SSbG-Cys344Cys432, a number of single and multiply glycosylated proteins were successfully obtained. Notably, all sugar thiols **3.3–3.9** were prepared directly from free sugars with LR, reacted with the cysteine (phenylselenyl)sulfide of SBL-S156C,[89] with a conversion yield exceeding 95% for all substrates, illustrating the potential of the methodology. Thioglycosides **3.3**, **3.7**, **3.10** and **3.11** (Fig. 17) were instead coupled by a selenylsulfide-mediated conjugation to SBL-SSePh (Fig. 16b) again with >95% conversion.[94]

Exploiting the incorporation of the unnatural amino acid L-homoallyglycine and using a radical hydrothiolation reaction (Fig. 16e), Davis *et al.*[92] also developed a mild glycoconjugation reaction to install several thioglycoside in model proteins, such as β-glycosidase from *Sulfolobus solfataricus* SsbG-Hag43, *Nostoc punctiforme* "cuboid" protein Np276-Hag61

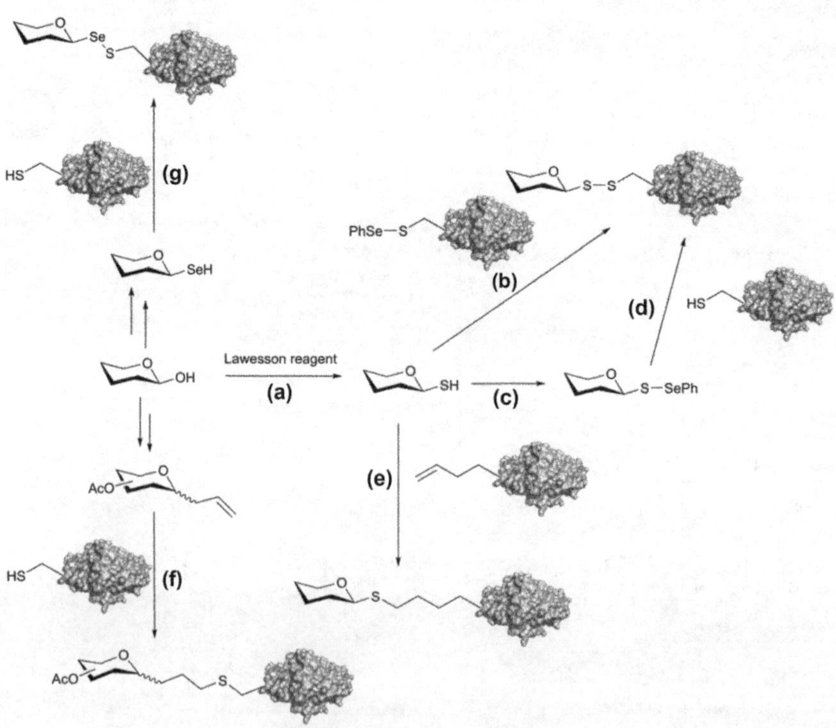

Fig. 16 General scheme for the "Tag and Modify" approach.[89–94]

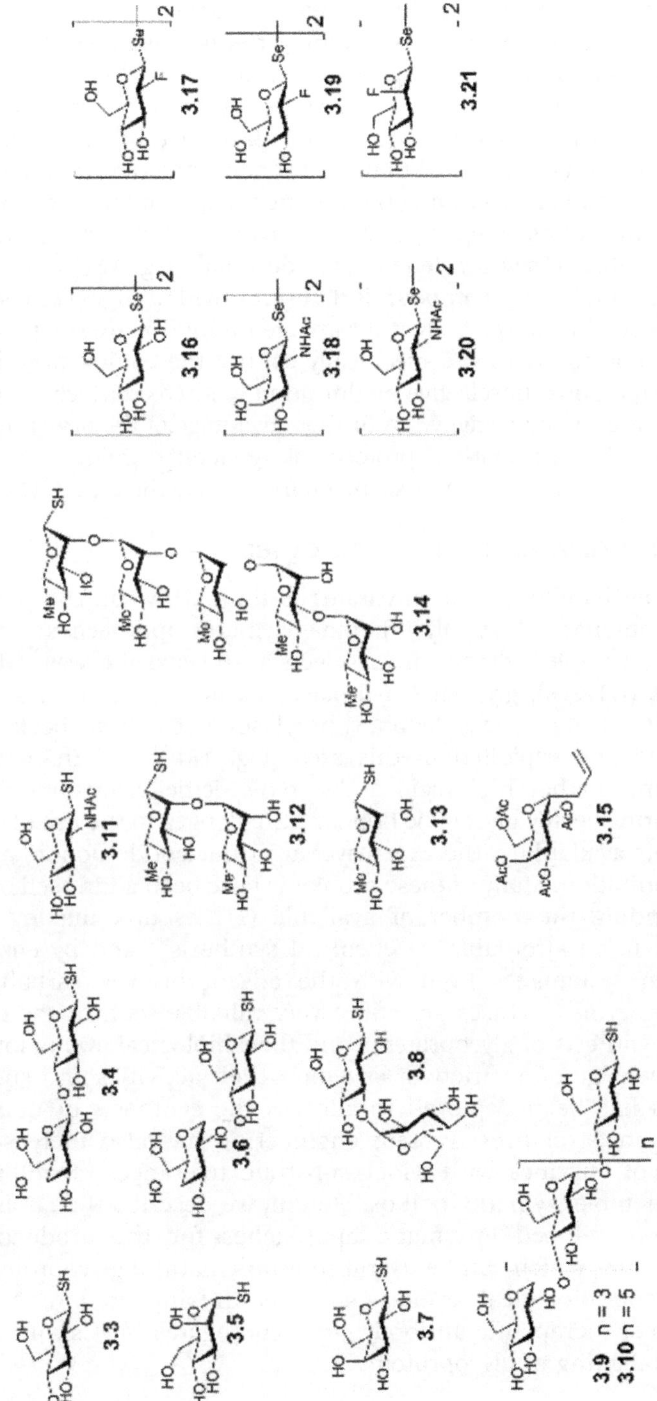

Fig. 17 Thio- and seleno-glycoside derivatives used for the "Tag and Modify" approach.[89–94]

and on virus like particle Qβ-$(Hag16)_{180}$; also in this case, with few exceptions, the conjugation yield was >95%. For instance, using this methodology, Qb-Hag16 multimer was conjugated with a series of rhamnosyl derivatives (**3.12**, **3.13**) including the pentasaccharide, **3.14**, an *O*-antigen polysaccharide, creating an impressive multi-glycosylated carrier bearing 180 glycan chains per protein. The same principle was applied by Dondoni *et al.* who reported a complementary method where the allyl *C*-glycoside **3.15** was conjugated to the side chain thiol of cysteine residues contained in native BSA following the generic scheme of Fig. 16f.[90] A recent update by Davis *et al.*[91] reported the conjugation of the free cysteine of subtisilin SBL-S156C *via* a selenenyl sulfide bond (Fig. 16g) obtained by reacting the protein with compounds **3.16–3.21**, with a highly efficiency of conjugation yield (>95%). A great advantage offered by these approaches lies in the opportunity to site-specifically control the conjugation, in contrast to the traditional unselective modification methods that rely on natural amino acid side chain reactivity.[93] A further advantage of the newer methods is that the glycoconjugation protocol is generally mild, leaving the protein function intact, and it is toxic metal-free, unlike the CuAAC chemistry.

4 Newer enzymatic methodologies

Where the availability of carbohydrates is limited by difficult synthesis or inadequate natural supply, chemo-enzymatic approaches can help synthesising complex glycans in a selective and specific way. Glycosyltransferases (GTases), glycoside hydrolases (GHs or glycosidases), glycosynthases (GSs) and glycoside phosphorylases (GPs) have been largely characterised and exploited in this area (Fig. 18).[41,95–97] In particular, GTases, thanks to their high regio- and stereo-specificity, are often the only choice to form specific glycosidic bonds.[98] Challenges in the use of GTases include their availability, the expensive sugar-nucleotide donors and the feedback inhibition. Many of these problems have been addressed over the years expanding the number of available GTases and making sugar-nucleotides more affordable by chemical synthesis[99] and by enzymatic regeneration systems.[100] Even with the advent of new carbohydrate-modifying enzymes, GTases are still a very valuable tool for the chemo-enzymatic synthesis of glycoproteins and their biological evaluation.

New knowledge of a variety of enzymes, their activities and substrate preferences has been deployed to address the synthesis of numerous glycoconjugate structures. Protein engineering has also increased the availability of enzymes with wider substrate tolerance, expanding the available chemo-enzymatic toolbox. Herein, we describe the application of recently developed enzymatic approaches for the production of glycoconjugates, which can be useful to mimic natural glycoproteins for structural and biological studies, such as GPI-anchored or *N*-linked glycoproteins, therapeutic antibody–drug conjugates (ADCs), or chemo-enzymatic labelling of glycoproteins.

4.1 Sortase enzymes

Sortase enzymes were first isolated in the late '90s from *Staphylococcus aureus.* These enzymes were found to be responsible for anchoring

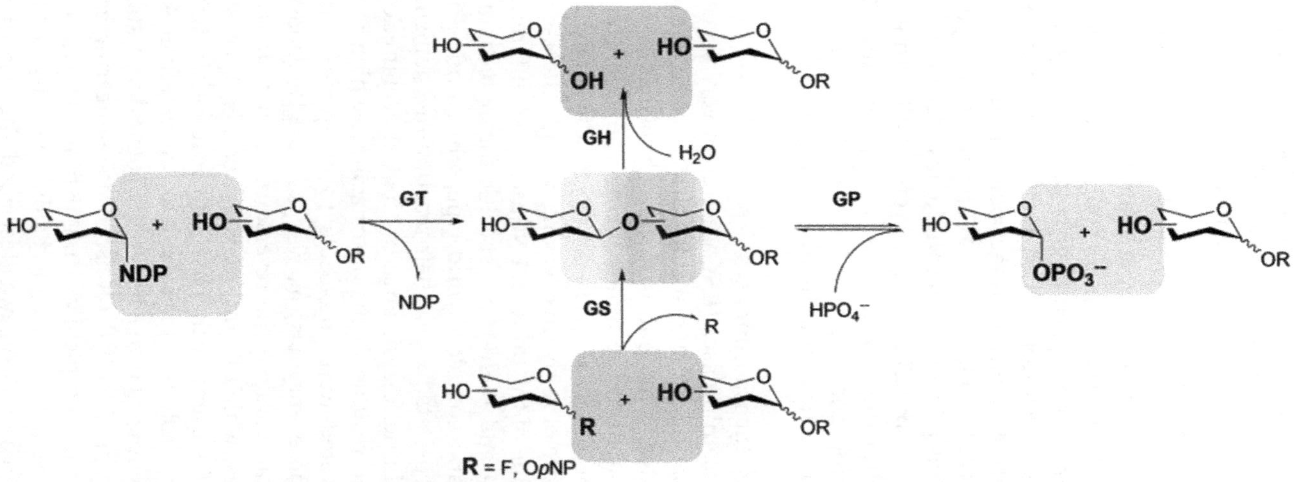

Fig. 18 Enzymes involved in glycosidic bond formation/cleavage. GT = glycosyltransferase, GH = glycoside hydrolase, GS = glycosynthases, GP = glycoside phosphorylase, NDP = nucleotide diphosphate.

surface proteins to the peptidoglycan cell wall of Gram-positive bacteria.[101,102] These enzymes are transpeptidases, cleaving the peptide bond between threonine and glycine of a conserved LPXTG peptide sequence to form an activated thioester. The thioester in turn reacts with the amino group of a terminal glycine of peptidoglycans, but can be used to react with the phosphoethanolamine group of GPI anchors. This enzyme has been extensively used for protein engineering and, more recently, has found applications in the synthesis of glycoproteins, in particular GPI-anchored proteins. The Guo group has studied the chemo-enzymatic synthesis of GPI-anchored proteins using the sortase reaction, initially for the synthesis of small GPI analogue–peptide conjugates,[103] moving on to more complex GPI derivatives conjugated to glycopeptides (Fig. 19).[104,105]

Finally, the full conserved core of GPI anchors was successfully coupled to intact GFP protein (**4.8**), and to intact polypeptide sequences of CD52 (**4.9**) and CD24 (**4.10**),[106] (Fig. 20).[107] While the total synthesis of GPI glycolipid anchors has advanced a lot in the last decade,[108] the synthesis of full GPI anchored proteins is still very challenging. The chemo-enzymatic approach using sortases is very promising, although the impact of introducing the unnatural LPXTGG sequence between the protein and the GPI anchor remains to be established.

4.2 *Endo*-β-*N*-acetylglucosaminidases

Endo-β-*N*-acetylglucosaminidases (ENGases) are a very successful class of *endo*-glycosidases that have found use for the chemo-enzymatic synthesis of *N*-glycan conjugates.[41] In contrast to *exo*-glycosidases, which cleave a single terminal sugar from the non-reducing end of an oligosaccharide, ENGases cleave the internal β-(1→4) glycosidic bond of the *N,N′*-diacetylchitobiose core of *N*-glycans, leaving a single GlcNAc unit bound to the asparagine residue of the protein. ENGases have different specificities; for example, Endo-A and Endo-H are specific for high mannose, Endo-M for high mannose and complex/hybrid type, while Endo-S is specific for human IgG with fucosylated core. Their reaction mechanism is substrate-assisted, where anchimeric assistance by the C-2 acetamido group of the penultimate HexNAc residue forms an oxazoline, which is subsequently opened by an activated water molecule to complete hydrolysis of the glycosidic bond. These enzymes, as well as *exo*-glycosidases and other *endo*-glycosidases, can catalyse the reverse reaction of *trans*-glycosylation and act as synthases, which form glycosidic bonds. Excess sugar donor, pH or a mixture of solvents can push the reaction in favour of *trans*-glycosylation over hydrolysis. However, a more powerful way to exploit these enzymes for synthesis is to activate the sugar donor substrate and to avoid the competing product hydrolysis. Glycoengineering can modify key residues in the enzyme, completely extinguishing the hydrolytic activity.[41]

The first use of the native *trans*-glycosidase activity of ENGases dates back to early ‘90s, when Takegawa described the use of Endo-A from *Arthrobacter protophormiae* for the synthesis of homogeneous $Man_6GlcNAc_2$ ribonuclease B (RNase B) (Fig. 21, A).[109] Heterogeneous RNase B was treated with Endo-A to give homogeneous GlcNAc-RNase B (**4.11**); then, $Man_6GlcNAc_2$-Asn (**4.12**) was added together with Endo-A,

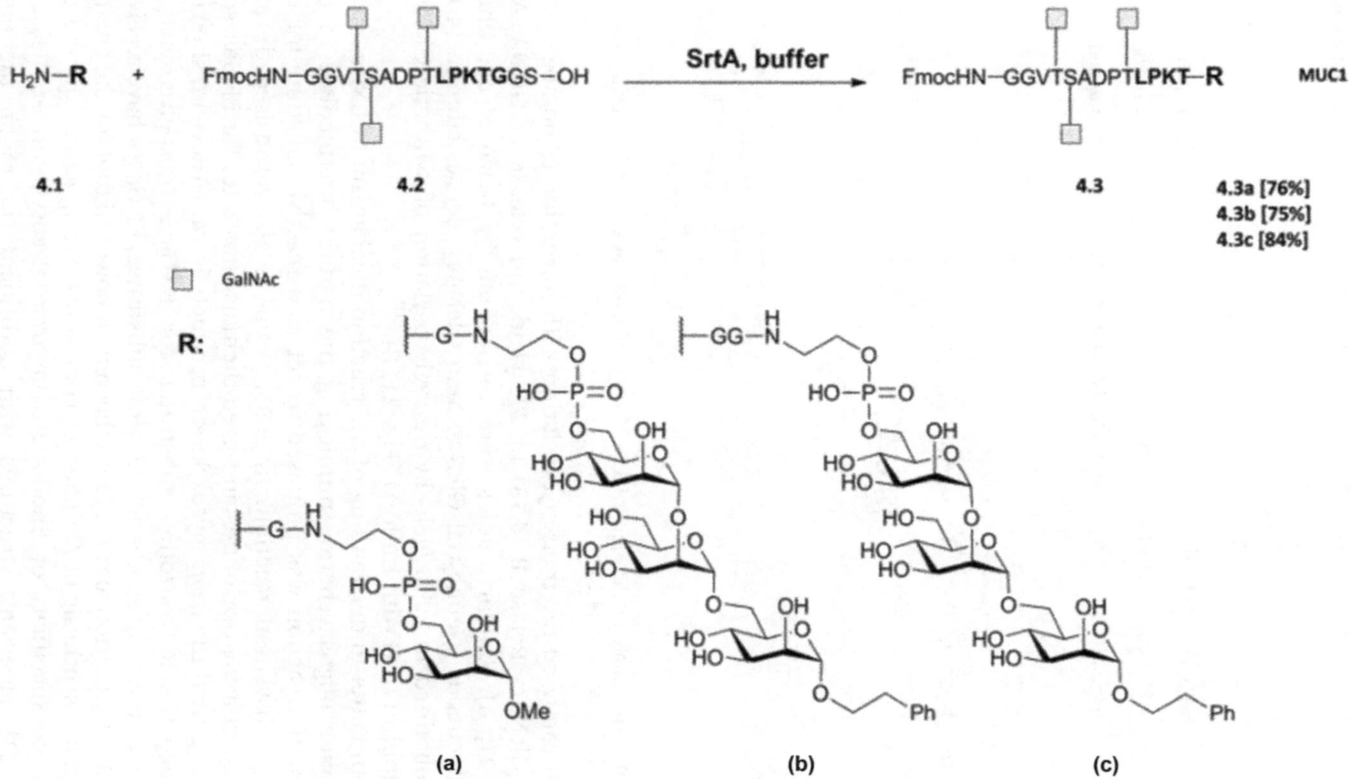

Fig. 19 SrtA-catalysed ligation of GPI–glycopeptide analogues (4.3a–c).[104]

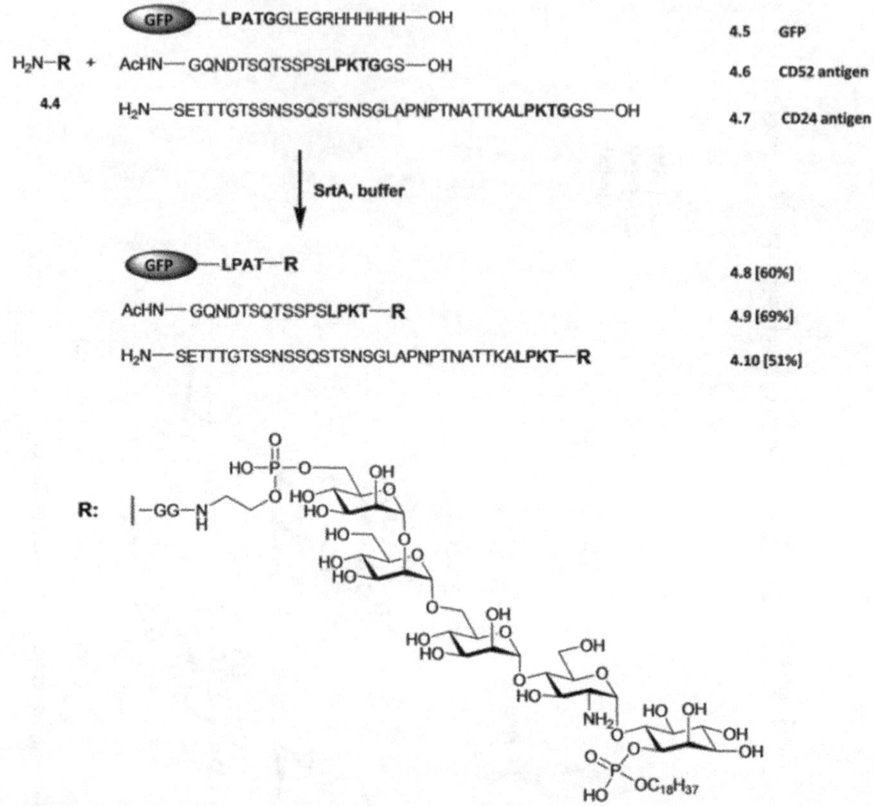

Fig. 20 SrtA-catalysed ligation of GPI–GFP conjugate (4.8),[107] GPI–CD52 (4.9) and GPI–CD24 analogues (4.10).[106]

which catalysed the *trans*-glycosylation with formation of homogeneous $Man_6GlcNAc_2$ RNase B (**4.13**) in 2% yield and release of GlcNAc-Asn (Fig. 21, A). Similar studies were carried out by Inazu *et al.*, where homogeneous $NeuAc_2Gal_2GlcNAc_2Man_3GlcNAc_2$ eel calcitonin (**4.16**) was obtained in 8.5% yield by *trans*-glycosylation of synthetic GlcNAc-calcitonin (**4.14**) with Endo-M (Fig. 21, B).[110]

With better understanding of the reaction mechanism of ENGases, an oxazoline oligosaccharide mimicking the natural intermediate in the enzymatic reaction was proposed to act as a reactive substrate for the *trans*-glycosylation reaction. In 2005, Wang *et al.* synthesised di- and tetra-saccharide oxazolines and used them as substrates for ENGases;[111] they obtained the *trans*-glycosylation products in *ca.* 75% yield, showing the benefits of oxazoline substrates for ENGase *trans*-glycosylation. With the advent of synthetic oxazoline substrates, ENGases have received increased attention and they have become a powerful tool for the chemo-enzymatic synthesis of *N*-linked glycoproteins. Wang *et al.* explored the substrate specificity of Endo-A in the *trans*-glycosylation of RNase B (Fig. 22), accessing structures with additional sugars on the outer mannoses of the $Man_3GlcNAc_2$ core (**4.18**), such as Gal (**4.19**) or lactose (**4.20**) units, and also changing the core itself, such as in $Man_4GlcNAc_2$ (**4.21**) or introducing azide group in C-6 of mannose unit (**4.22**).[112]

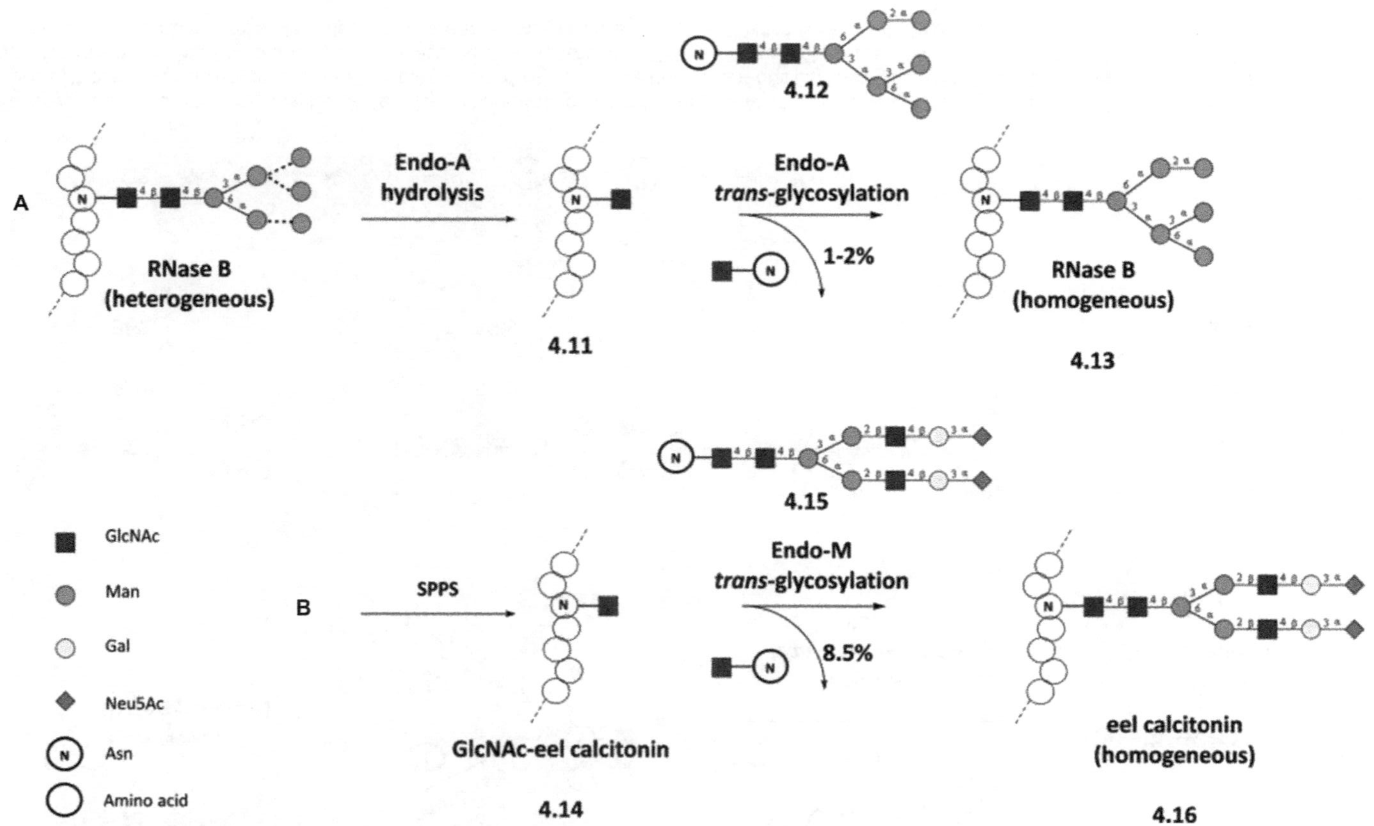

Fig. 21 ENGase-catalysed *trans*-glycosylation for the production of homogeneous *N*-glycan proteins using natural oligosaccharide-Asn donor. (A) Endo-A *trans*-glycosylation for the production of homogeneous $Man_6GlcNAc_2$ RNase B (**4.13**).[109] (B) Endo-M *trans*-glycosylation for the production of homogeneous $NeuAc_2Gal_2GlcNAc_2Man_3GlcNAc_2$ eel calcitonin (**4.16**).[110]

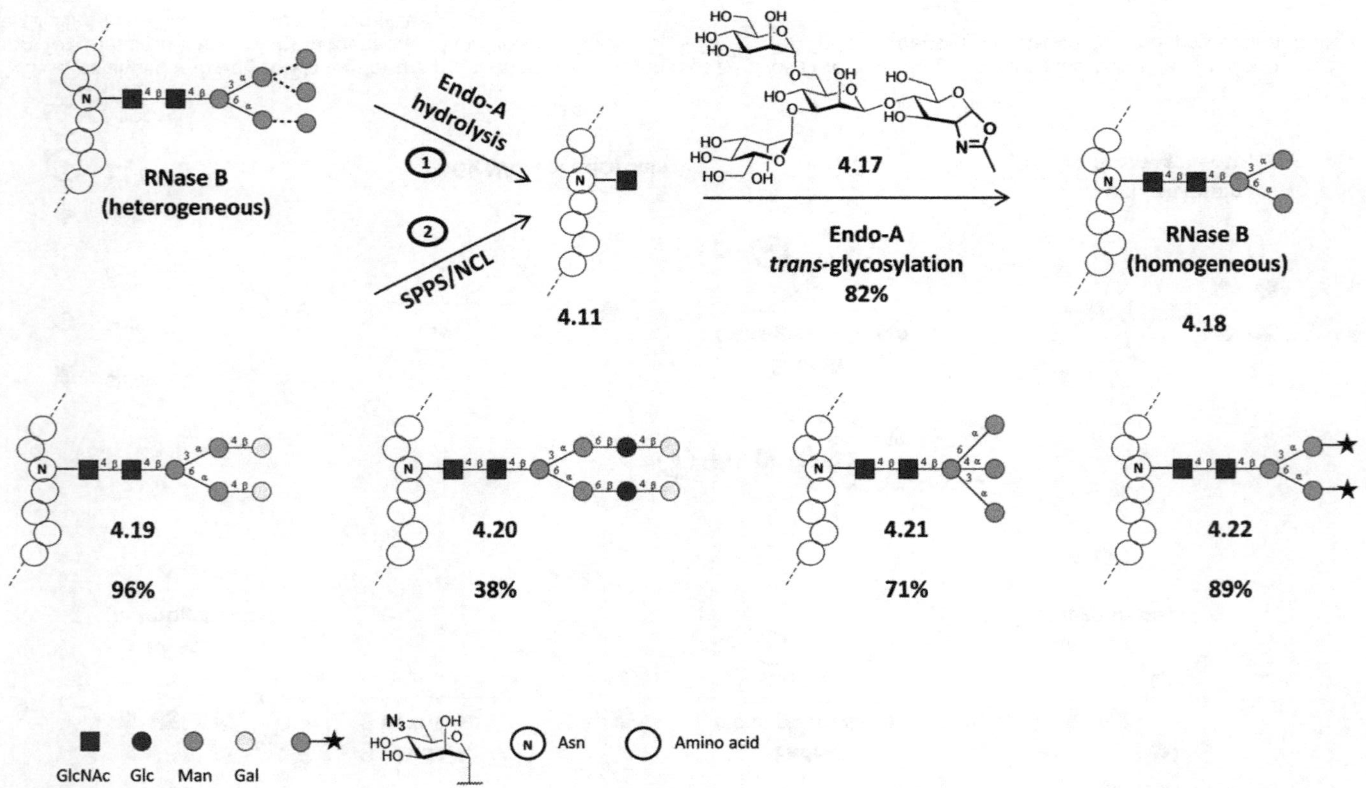

Fig. 22 Endo-A-catalysed *trans*-glycosylation for the production of homogeneous *N*-glycan RNase B using synthetic oligosaccharide oxazoline donors. (1) Heterogeneous RNase B was trimmed by Endo-A-catalysed hydrolysis to give homogeneous GlcNAc-RNase B (**4.11**). (2) Homogeneous GlcNAc-RNase B (**4.11**) was chemically synthesised by a combination of solid-phase peptide synthesis (SPPS) and native chemical ligation (NCL). Then, Endo-A-catalysed *trans*-glycosylation with a variety of oligosaccharide oxazoline donors afforded homogeneous *N*-glycan RNase B (**4.18–4.22**) in high yields.[112]

Endo-A showed very relaxed substrate specificity, making it suitable for production of a large variety of different *N*-linked proteins. Moreover, Endo-A was found useful for glycosylation of complex proteins, such as the hydrophobic saposin C.[113] In 2008, Wang *et al.* engineered *Pichia pastoris* to produce human IgG-Fc in high yield, which was subsequently deglycosylated by Endo-H to obtain homogeneous GlcNAc-IgG-Fc. This intermediate was then successfully glycosylated by Endo-A with sugar oxazoline donors to obtain homogeneous glycosylated IgG-Fc.[114] This experiment showed that Endo-A was able to access GlcNAc on IgG-Fc without the need to denature the protein, which made *N*-glycosylation more straightforward and quick to perform.

Davis *et al.* investigated the use of unnatural linkers between GlcNAc and the peptide/protein to enhance the stability of the *N*-glycan conjugates toward chemical and enzymatic hydrolysis (Fig. 23).[115]

First, they compared the natural GlcNAc-Asn linkage (**4.23a**) with a series of GlcNAc-linker-amino acids (**4.23b–f**); they also tested the ability of Endo-A to *trans*-glycosylate these materials using sugar oxazoline donor **4.24** (Fig. 23, A). Surprisingly, the natural GlcNAc-Asn linkage gave the poorest yield toward *trans*-glycosylation compared to triazole, disulfide, selenylsulfide and thiol GlcNAc linkages. When the protein SBL-S156C was modified with GlcNAc-disulfide (**4.26a**), selenylsulfide (**4.26b**) and thiol-linkages (**4.26c**), the GlcNAc-thiol SBL-S156C gave the best *trans*-glycosylation yield (Fig. 23, B). However, when the GlcNAc-thiol was in a hindered position on the protein, the *trans*-glycosylation yield dropped, suggesting that the GlcNAc position is critical for enzyme access to the glycosylation site.

In 2012, Davis *et al.* identified the gene of a new ENGase, Endo-S, from the genome sequence of *Streptococcus pyogenes* and expressed it in *E. coli*. This enzyme was found to be active on *N*-glycans with a fucosylated core, in contrast to Endo-A and Endo-H which were completely inactive on such substrates.[116] Endo-S allowed remodelling of the glycans present on human IgG-Fc (Fig. 24), giving homogeneous glycoforms for use in antibody-dependent cell-mediated cytotoxicity (ADCC) studies.

The enzymatic remodelling of *N*-glycans on mammalian IgG-Fc was also used to install an azide for subsequent coupling by strain-promoted azide–alkyne cycloaddition (SPAAC) with a drug in order to generate antibody–drug conjugates (ADCs) (Fig. 25).[117] In this approach, the heterogeneous *N*-glycans were fully trimmed with Endo-S2, followed by transfer of azide-modified GalNAc (GalNAz, F_2-GalNAz and F_2-GalNBAz), by β-(1→4)-Gal transferase. The final step involved the SPAAC coupling with the drug-linker. A comparative study showed that bicyclo[6.1.0]nonine (BCN) was superior to dibenzoazacyclooctyne (DBCO) for the SPAAC coupling and that the ideal length of PEG linker was between 4 and 8 ethylene units.

4.3 Chemo-enzymatic synthesis with oxime conjugation

Another approach to the synthesis of homogeneous glycoproteins is based on oxime conjugation, wherein aldehydes that are chemically or enzymatically introduced onto proteins or into sugar units can be readily coupled to aminooxy groups to generate oxime-linked products. Herein,

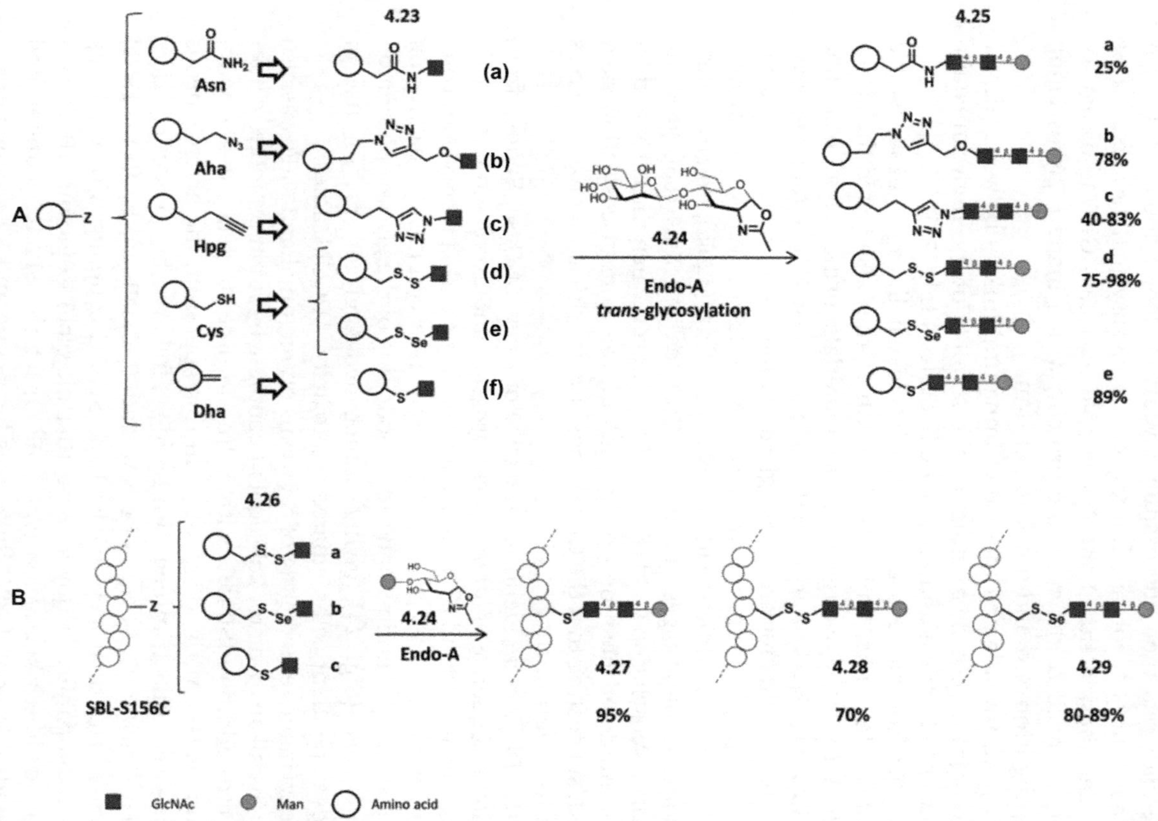

Fig. 23 Effect of a variety of GlcNAc-linked amino acids and peptides on Endo-A-catalysed *trans*-glycosylation for the production of homogeneous glycoproteins using synthetic oligosaccharide oxazoline donors **4.24**. (**A**) Endo-A-catalysed *trans*-glycosylation with **4.24** proved to be more efficient with GlcNAc acceptor linked to amino acids by a variety of linkers (**4.23b–f**) than with natural GlcNAc-Asn (**4.23a**). (**B**) SBL-S156C containing GlcNAc linked by S–S, Se–S and S linker was efficiently glycosylated by Endo-A-catalysed *trans*-glycosylation. *S*-glycoconjugation proved to be the most efficient method.[115]

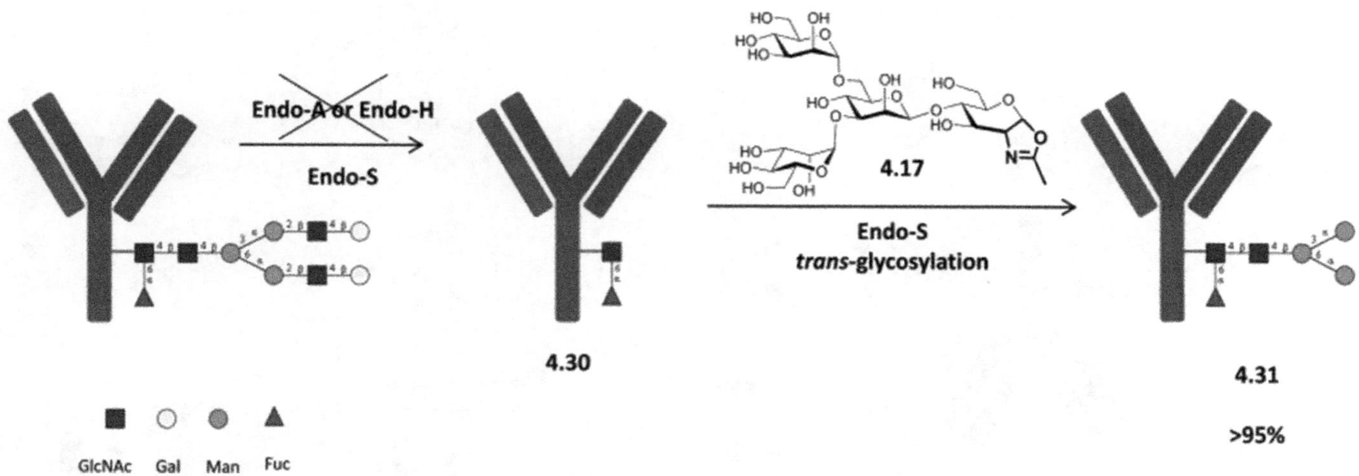

Fig. 24 Endo-S-catalysed *trans*-glycosylation of complex-type *N*-linked glycans with core α-(1,6)-Fuc on human IgG-Fc.[116]

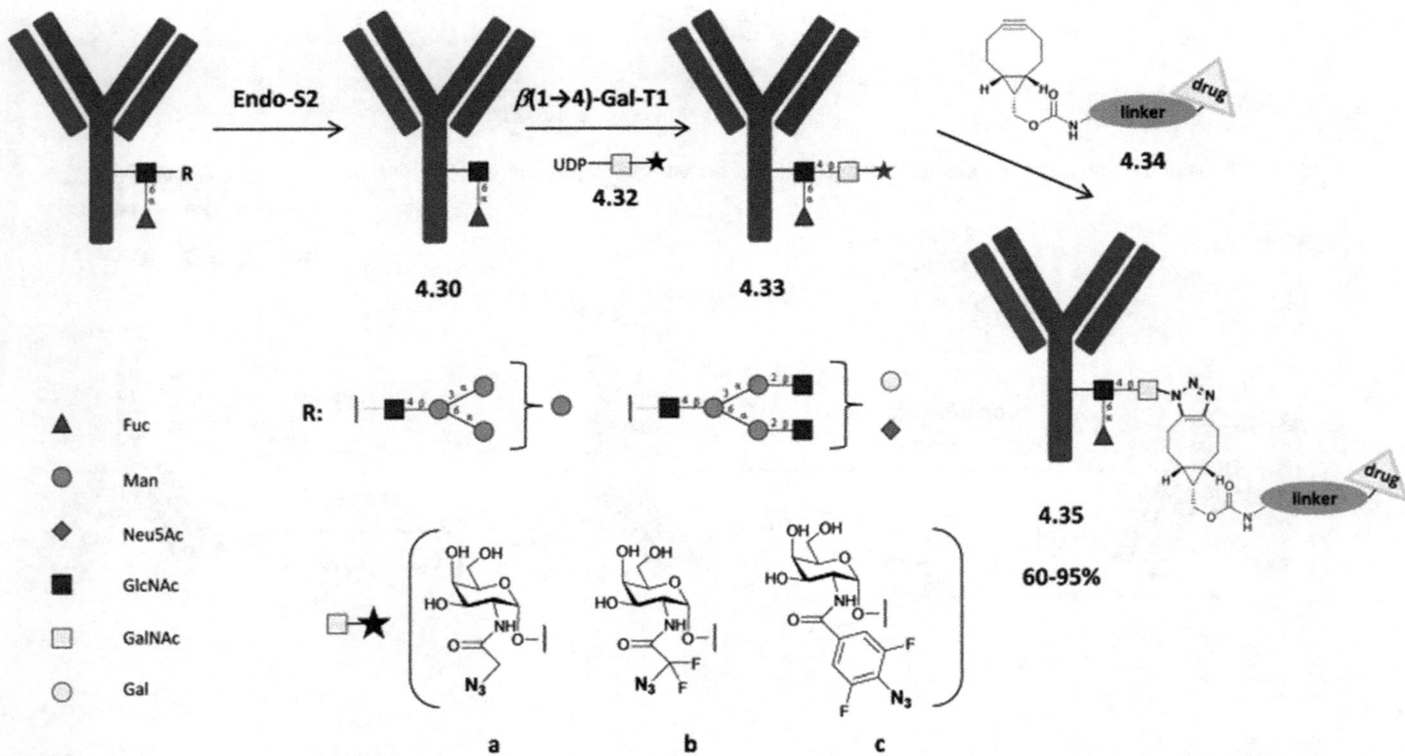

Fig. 25 Chemo-enzymatic conjugation of toxic drugs to homogeneous *N*-glycan of IgG-Fc for the development of efficacious antibody–drug conjugates (ADC).[117]

we present few different examples of this methodology, which is useful to for extending and remodelling the glycans on intact proteins, or for labelling or drug conjugation applications.

Bertozzi *et al.* engineered IgG-Fc (**4.36**) in order to introduce a LCTPSR consensus sequence for the formylglycine generating enzyme (FGE), which converted Cys into formylglycine (fGly) by oxidation (**4.37**), thus introducing an aldehyde functional group for conjugation to aminooxy-modified sugar (Fig. 26).[118] That is, a synthetic aminooxy-GlcNAc (**4.38**) was conjugated to fGly and the glycan subsequently extended by Endo-S-D233Q and sugar oxazoline donor **4.40** to obtain a homogeneous glycan-oxime-linked IgG (**4.41**, Fig. 26).

Zhou *et al.* treated IgG-Fc with galactosyltransferase and sialyltransferase to obtain homogeneous *N*-glycans with a terminal sialic acid (**4.43**), whose C-7/8 were chemically oxidised to aldehyde[119] allowing its conjugation with aminooxy drug-linkers to produce ADCs (Fig. 27). Alternatively, the oxime conjugation could be used for introducing labels or extra sugars.[120]

The enzymatic introduction of synthetic 2-keto-galactose using galactosyltransferase onto the *N*-glycans of IgG-Fc allowed subsequent oxime conjugation chemistry for coupling drugs, labels or extra sugars on the protein skeleton (Fig. 28).[121]

The introduction of an aldehyde onto a sugar for oxime conjugation can also be achieved by galactose oxidase (GO) oxidation, which oxidises C-6 of Gal to an aldehyde group.[122] Engineering of GO has allowed the development of enzyme mutants which are able to oxidise C-6 of Man or GlcNAc, expanding the potential of this class of enzymes for labelling and conjugation.[123]

4.4 Oligosaccharyltransferases

N-Glycosylation is a post-translational protein modification present in all kingdoms of life. The bacterial *N*-glycosylation system has been recently disclosed and the key enzymes have been studied and characterised for their application in the chemo-enzymatic synthesis of *N*-glycoproteins.[124,125] The main bacterial *N*-glycosylation system involves the transfer of an oligosaccharide, synthesised in the cytoplasm, from a lipid anchor to the asparagine side chain of a protein. The transfer, which takes place into the periplasm, is catalysed by a membrane-bound oligosaccharyltransferase (OST) to a consensus peptide sequence, D/EXNXS/T, of the protein. The identification of an oligosaccharyltransferase, PglB, from *Campylobacter jejuni,* and its expression in *E. coli* has opened up new possibilities for glycoengineering.[126–128] Currently, the most popular methodology to produce glycoconjugated vaccines is based on the protein glycan coupling technology (PGCT) (Fig. 29, A).[129] With this technique, glycoconjugates can be efficiently obtained by engineering *E. coli* with three plasmids (1): the first is dedicated to the synthesis of the desired glycan/polysaccharide on a pyrophosphate lipid anchor; a second encodes the carrier protein, CP, containing the necessary D/EYNXS/T consensus sequence for the PglB; and a third plasmid for PglB expression. Once the *N*-glycan is synthesized and extended onto a lipid anchor by the action of glycosyltransferases in the

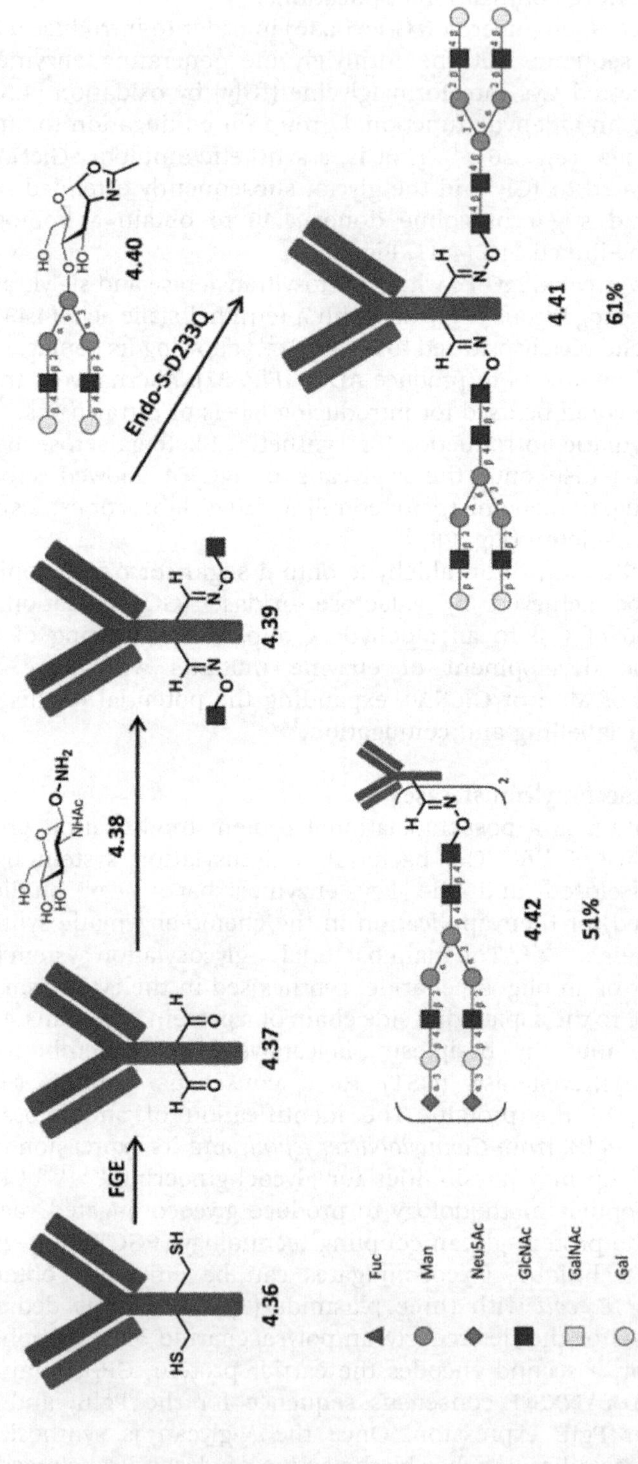

Fig. 26 Chemoenzymatic glycosylation of IgG by engineered aldehyde tags.[118]

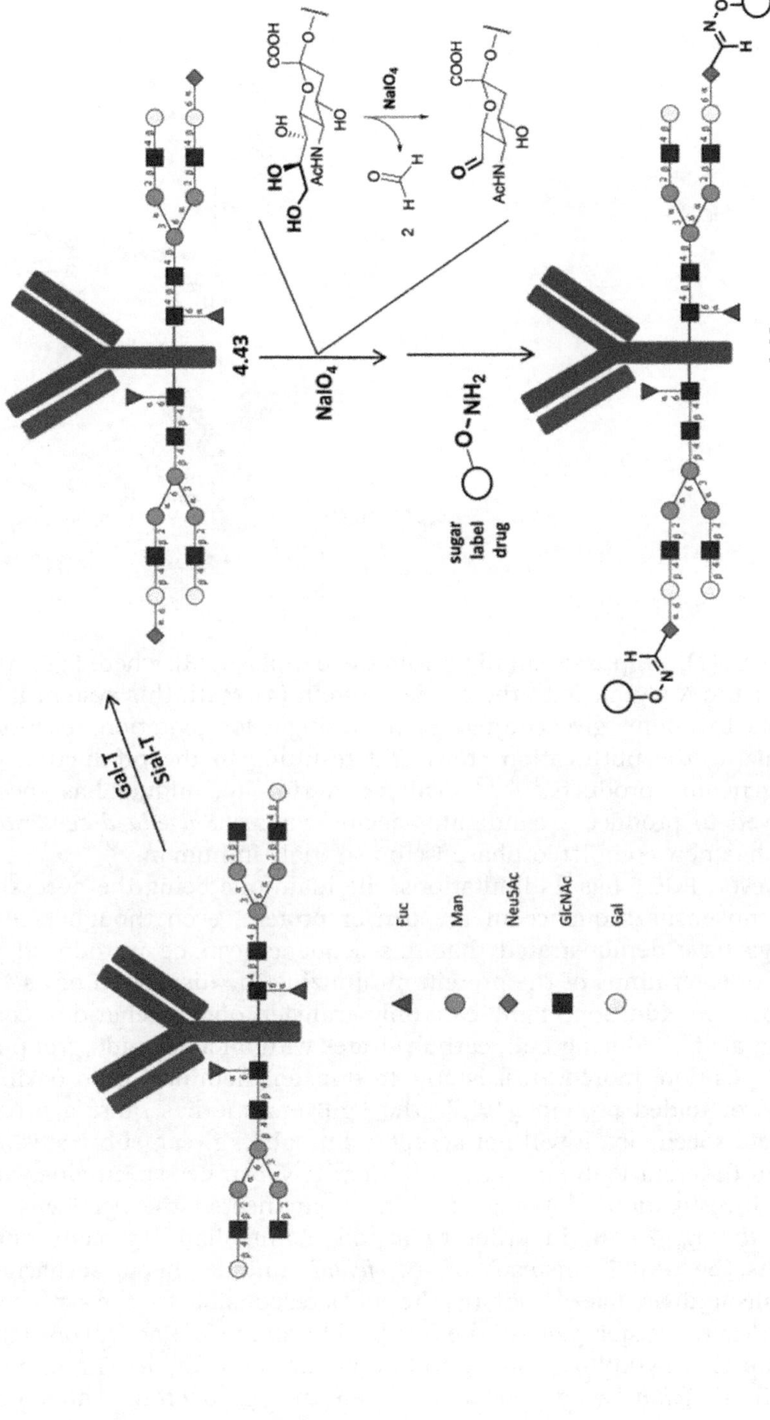

Fig. 27 Enzymatic introduction of sialic acid and C-7/8 chemical oxidation to aldehyde for oxime conjugation.[120]

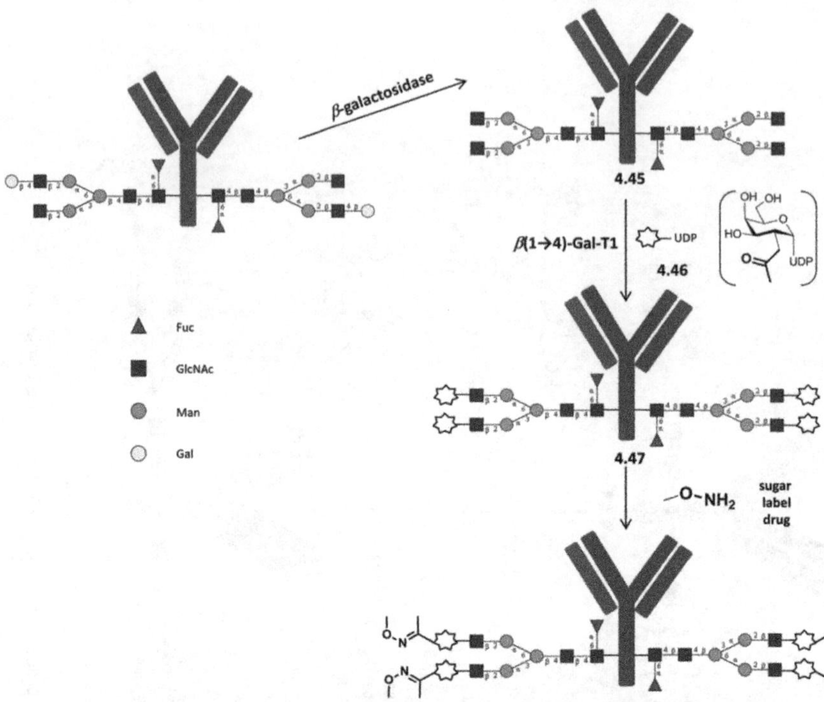

Fig. 28 Enzymatic introduction of ketone for oxime conjugation by glycosyltransferases.[121]

cytoplasm (2), a flippase can flip it into the periplasm (3), where PglB will transfer the *N*-glycan onto the carrier protein (4). With this system it is possible to obtain glycoconjugates in a single fermentation reaction, minimizing the purification steps and resulting in the production of homogeneous products.[130,131] This powerful technique has been employed to produce a candidate vaccine against *Shigella dysenteriae*, which has now completed phase I clinical trials in human.[132]

However, PGCT has its limitations, the main one being the necessity for a consensus sequence in the carrier protein, even though recent findings have demonstrated that this sequence can be introduced at the N- or C-terminus of the protein minimizing the disruption of its 3D structure. In addition, PglB can only transfer oligosaccharides containing, at the reducing end, carbohydrates with an acetamido group at the C2 position; moreover, it is able to transfer them only onto flexible regions of folded proteins. While the PglB enzyme has quite a broad substrate specificity, it will not accept mammalian glycan substrates. At present, different PglB enzymes with broader substrates specificities are under investigation.[133] Wang *et al.* have engineered the *pgl* locus of *C. jejuni* (Fig. 29, B) in order to install mammalian *N*-glycans onto proteins by PGCT instead of *C. jejuni* unique heptasaccharide, $Glc_1GalNAc_5diNAcBac_1$.[134] Firstly, the genes responsible for the synthesis (*pglDEF*) and transfer (*pglC*) of the first bacillosamine (diNAcBac) onto the undecaprenyl pyrophosphate (Und-PP) anchor, and for the Glc branch (*pglI*) in *C. jejuni* were deleted (1) from the *pgl* operon, which was

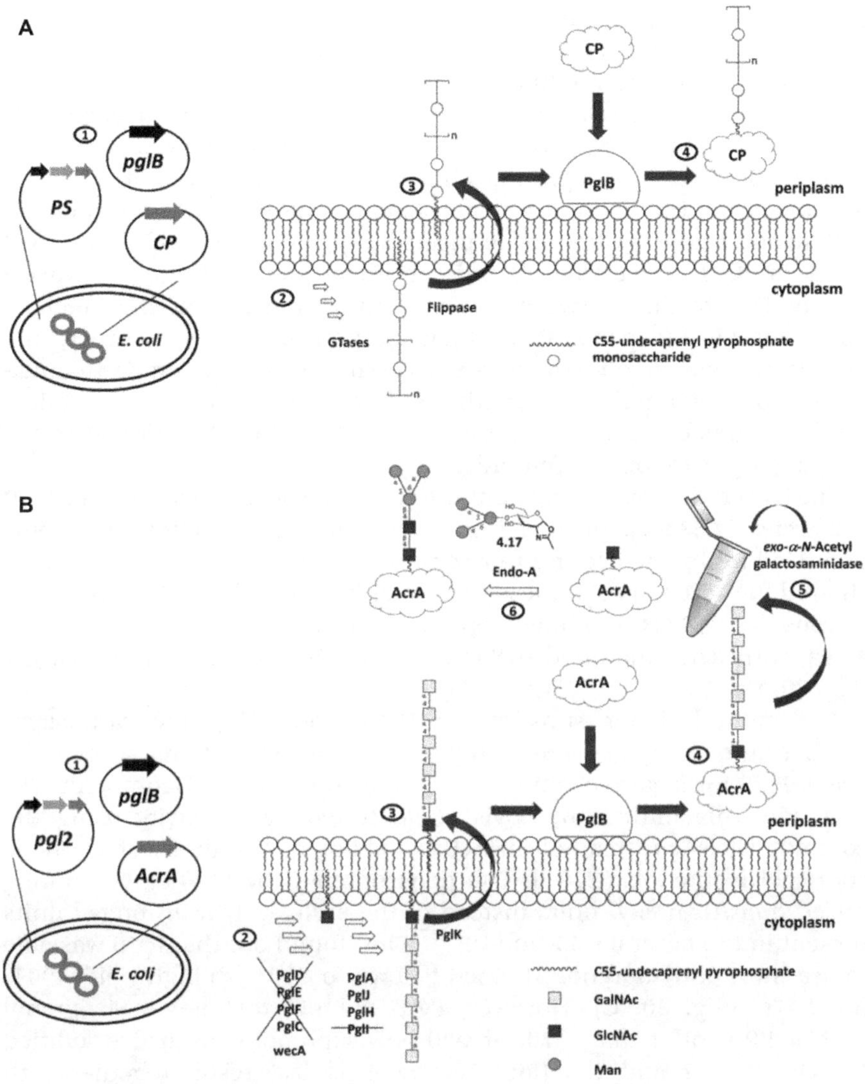

Fig. 29 (A) Protein Glycan Coupling Technology (PGCT).[129] (B) PGCT for the development of a homogeneous glycoprotein carrying a mammalian *N*-glycan.[134]

renamed *pgl2*. In the absence of PglC, *E. coli* WecA transferred GlcNAc-1-P onto the lipid anchor; then, GalNAc units were successively transferred by the remaining PglAJH glycosyltransferases (2). The oligosaccharide was then flipped into the periplasm by PglK flippase (3) and transferred by PglB onto the AcrA protein (4); once isolated, the resulting glycoprotein was trimmed by *exo*-α-*N*-acetylgalactosaminidase to give a homogeneous GlcNAc-Acra (5). Finally, *endo*-α-*N*-acetylglucosaminidase (Endo-A) catalysed the *trans*-glycosylation between GlcNAc-Acra and Man_3GlcNAc oxazoline **4.17** to give a homogeneous glycoprotein carrying a mammalian *N*-glycan (6).

The activity of PglB from *C. jejuni* has been investigated not only *in vivo* with its application in the PGCT, but also *in vitro* for chemo-enzymatic synthesis of glycoproteins using natural or unnatural substrates. A key factor in these studies was the opportunity to explore the enzyme tolerance to sugar or lipid modifications (Fig. 30).

Ito *et al.* have carried out *in vitro* *N*-glycosylation of peptides with chemically synthesised donors to explore the chemo-enzymatic route to glycoproteins (Fig. 30, A).[135] PglB was able to transfer its natural biosynthetic intermediates, Und-PP-$(diNAcBAc)_1$ (**4.50a**), Und-PP-$(GalNAc_2diNAcBac_1)$ (**4.50b**) and Und-PP-$(Glc_1GalNAc_5diNAcBac_1)$ (**4.50c**), to TAMRA-modified peptides (**4.49**) suggesting relaxed specificity towards the carbohydrate moiety. Due to the *in vitro* ability of PglB to transfer any intermediate onto Und-PP independent of their natural biosynthetic order, the experiment also demonstrated the necessity for *in vivo* compartmentalization of oligosaccharide synthesis into the cytoplasm and oligosaccharyltransferase into the periplasm, in order to obtain glycoproteins carrying a precise oligosaccharide.

The Imperiali group carried out the chemo-enzymatic synthesis of PglB substrates (**4.53a–c**), introducing azide-modified diNAcBAc or GalNAc onto Und-PP by the action of glycosyltransferases PglCAJ.[136] Then, they showed the acceptance of these substrates by PglB, which transferred the glycans onto a FITC-modified peptide (**4.52**) with yields increasing from **4.54a**, with azide-modified diNAcBac, to **4.54c**, with the distal GalNAz (Fig. 30, B).

The natural donor substrate of PglB, containing the rare sugar bacillosamine and C55-Und-PP, represents a substantial limitation to the use of PglB for *in vitro* chemo-enzymatic synthesis of *N*-glycoproteins. To overcome this limitation, Davis *et al.* tested PglB activity using the common GlcNAc instead of diNAcBac, anchored to a library of synthetic short polyisoprenols (C10 to C40 polyisoprenol GlcNAc-PP), containing mixed *trans*/*cis*-prenyl units instead of the exclusively *trans*-prenyl units present in the natural C55-Und-PP.[137] They found out that PglB was able to use short polyisoprenol GlcNAc-PP down to C20 with high yields (90% for **4.57c**) (Fig. 30, C). However, PglB did not tolerate polyisoprenol GlcNAc-PP shorter than C20, or C20 polyisoprenol with azide-modified GlcNAc on C-2 and C-6 (Fig. 30, D). This last result together with Imperiali's demonstrates that PglB tolerates azide-modification as long as the structure of the substrate is not too different from the natural one.

Another bacterial *N*-glycosylation pathway has been discovered wherein a cytoplasmic *N*-glycosyltransferase (NGT) transfers Glc from UDP-Glc directly to the asparagine side chain of a **N**XS/T consensus sequence.[138] NGT isolated from *Actinobacillus pleuropneumoniae* is an inverting glycosyltransferase that exhibits a preference for similar consensus sequence to PglB OST from *C. jejuni* but, in contrast to PglB, it uses nucleotide-activated monosaccharides instead of lipid-bound oligosaccharides as donor substrates. While NGT is able to transfer a single Glc onto the protein, an α-(1$\rightarrow$6)glucosyltransferase present in *A. pleuropneumoniae* can extend the nascent oligosaccharide in the cytoplasm. NGT, even if strictly specific for the formation of β-glucosidic linkages (very low activity with Gal), can be useful for the chemo-enzymatic direct introduction of Glc units onto proteins, which can be

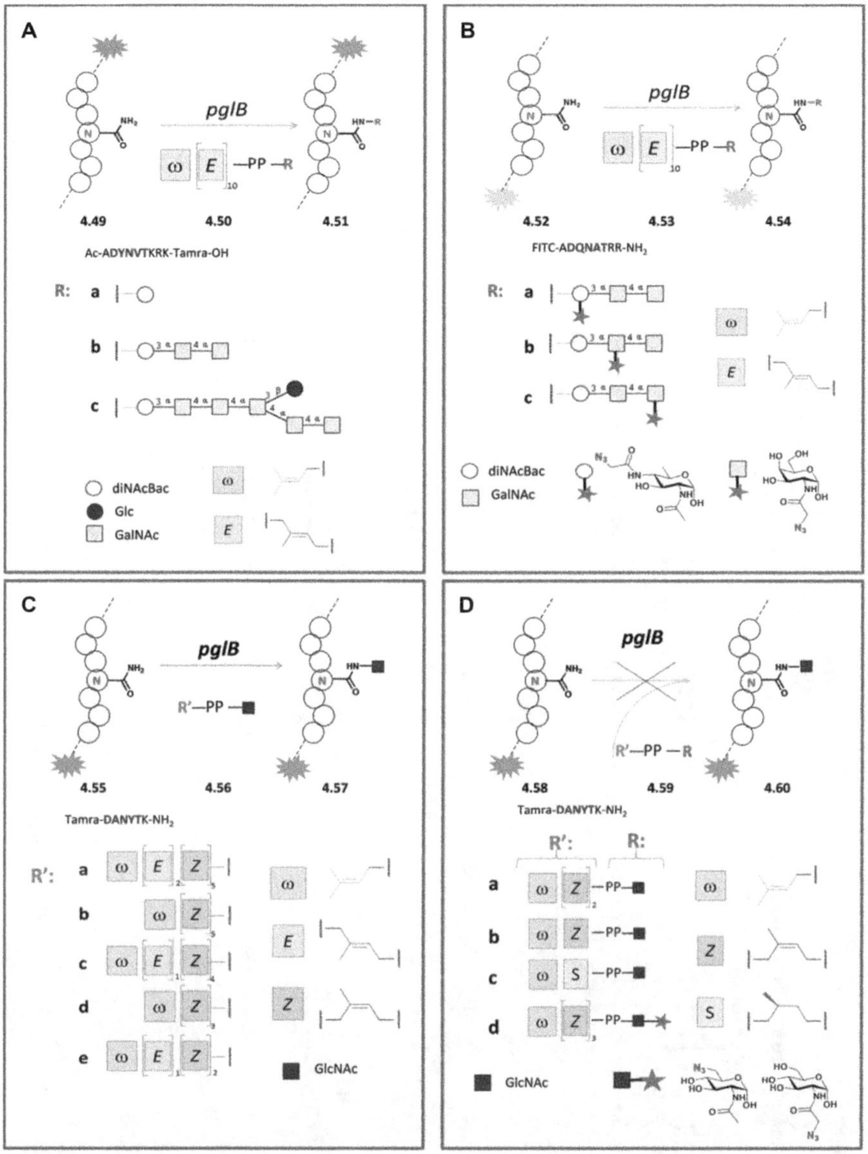

Fig. 30 (A) *In vitro N*-glycosylation of peptides by PglB with synthetic natural oligosaccharides, **4.50a–c**.[135] (B) *In vitro N*-glycosylation of peptides by PglB with synthetic unnatural oligosaccharides, **4.53a–c**.[136] (C) *In vitro N*-glycosylation of peptides by PglB using synthetic unnatural C20 to C40 polyisoprenol GlcNAc-PP substrates, **4.56a–e**.[137] (D) *In vitro N*-glycosylation of peptides by PglB using synthetic unnatural C10 to C20 polyisoprenol GlcNAc-PP substrates, **4.59a–d**.[137]

further decorated with complex oligosaccharides by the use of *endo*-glycosidases and oxazoline donors. Wang *et al.* synthesised a glycoform of polypeptide C34, a potent HIV inhibitor, by *trans*-glycosylation with Endo-M N175A of Glc-C34 **4.63**, obtained by NGT glucosylation of the synthetic peptide **4.61**, or of the fully synthetic GlcNAc-C34 **4.64**, in comparable high yields (Fig. 31).[139] Moreover, the Glc-bound glycoform (**4.66**) was completely resistant to PNGase F hydrolysis and 10 times more

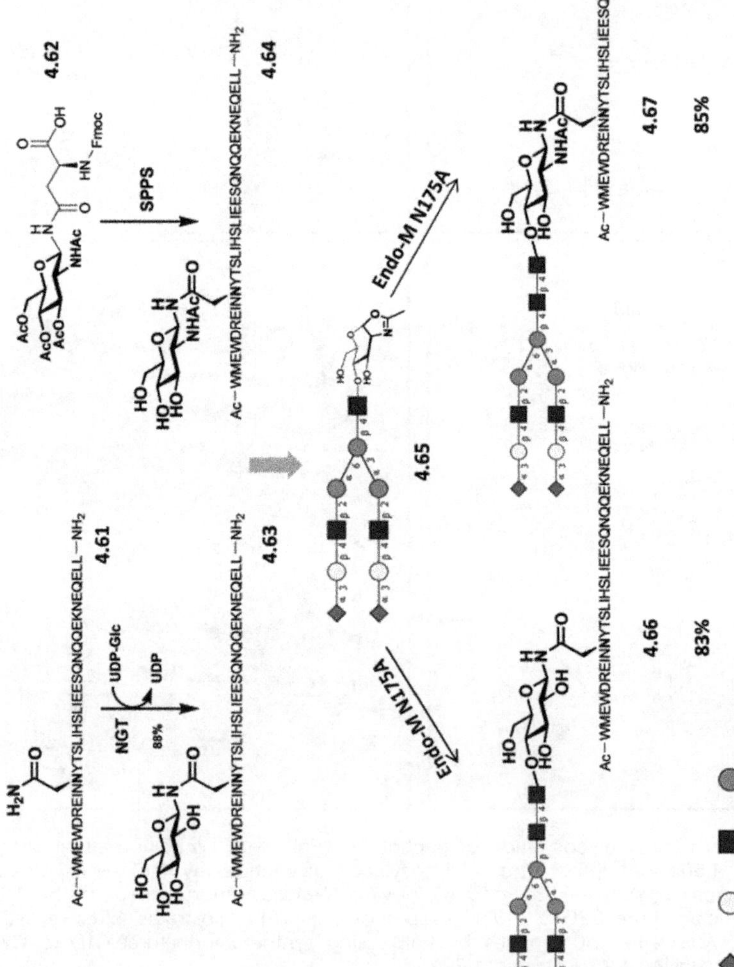

Fig. 31 NGT-catalysed peptide glycosylation.[139]

resistant to Endo-M hydrolysis compared to the GlcNAc-bound glycoform (**4.67**). The introduction of Glc-bound *versus* GlcNAc-bound oligosaccharides onto proteins could be a substantial advantage in the synthesis of inhibitors or therapeutics, which need to resist *in vivo* degradation.

5 Conclusions and future perspective

Rapid progress in biology and biomedicine places increasing demands on methods for the construction of precise and often elaborate glycoconjugate structures with which to explore and explain biological phenomena. In this article, we have provided an overview of the increasing range of chemical and chemo-enzymatic methodologies used for glycoconjugation reactions in recent years. Even for the chemist, biochemistry and molecular biology approaches are becoming routine tools to complement the traditional chemical synthesis toolbox. Indeed, the powerful integration of chemical and chemo-enzymatic methods offers many and varied new ways to address future questions in this vast field that is glycobiology.

Acknowledgements

Studies at the JIC were supported by the UK BBSRC Institute Strategic Programme on Understanding and Exploiting Metabolism (MET) [BB/J004561/1] and the John Innes Foundation.

References

1 R. A. Dwek, *Chem. Rev.*, 1996, **96**, 683–720.
2 A. Varki, R. D. Cummings, J. D. Esko, H. H. Freeze, P. Stanley, C. R. Bertozzi, G. W. Hart and M. E. Etzler, *Essentials of Glycobiology*, Cold Spring Harbor Laboratory Press, Cold Spring Harbor (NY), 2nd edn, 2009.
3 L. Ellgaard and A. Helenius, *Nat. Rev. Mol. Cell Biol.*, 2003, **4**, 181–191.
4 D. N. Hebert and M. Molinari, *Physiol. Rev.*, 2007, **87**, 1377–1408.
5 K. W. Moremen, M. Tiemeyer and A. V. Nairn, *Nat. Rev. Mol. Cell Biol.*, 2012, **13**, 448–462.
6 R. A. Laine, *Glycobiology*, 1994, **4**, 759–767.
7 R. D. Cummings, *Mol. Biosyst.*, 2009, **5**, 1087–1104.
8 J. Hirabayashi and K. Kasai, *Trends Glycosci. Glycotechnol.*, 2000, 1–5.
9 T. Feizi, *Curr. Opin. Struct. Biol.*, 2003, **13**, 637–645.
10 N. Laurent, J. Voglmeir and S. L. Flitsch, *Chem. Commun.*, 2008, 4400–4412.
11 C. D. Rillahan and J. C. Paulson, *Annu. Rev. Biochem.*, 2011, **80**, 797–823.
12 X. Song, J. Heimburg-Molinaro, R. D. Cummings and D. F. Smith, *Curr. Opin. Chem. Biol.*, 2014, **18**, 70–77.
13 S. Cunningham, J. Q. Gerlach, M. Kane and L. Joshi, *Analyst*, 2010, **135**, 2471–2480.
14 Y. P. Kim and H. S. Kim, *ChemBioChem*, 2016, **17**, 275–282.
15 M. J. Marin, C. L. Schofield, R. A. Field and D. A. Russell, *Analyst*, 2015, **140**, 59–70.
16 A. Hushegyi and J. Tkac, *Anal. Methods*, 2014, **6**, 6610–6620.
17 R. Jelinek and S. Kolusheva, *Chem. Rev.*, 2004, **104**, 5987–6015.

18 B. Wang and J.-I. Anzai, *Materials*, 2015, **8**, 8590–8607.
19 X. Zeng, C. A. Andrade, M. D. Oliveira and X. L. Sun, *Anal. Bioanal. Chem.*, 2012, **402**, 3161–3176.
20 Z. Roth, G. Yehezkel and I. Khalaila, *Int. J. Carbohydr. Chem.*, 2012, **2012**, 1–10.
21 S. Yang, A. Rubin, S. T. Eshghi and H. Zhang, *Proteomics*, 2016, **16**, 241–256.
22 L. Cipolla, A. C. Araujo, D. Bini, L. Gabrielli, L. Russo and N. Shaikh, *Expert Opin. Drug Discovery*, 2010, **5**, 721–737.
23 J. E. Hudak and C. R. Bertozzi, *Chem. Biol.*, 2014, **21**, 16–37.
24 N. Lannoo and E. J. Van Damme, *Plant Sci.*, 2015, **239**, 67–83.
25 F. Schwarz and M. Aebi, *Curr. Opin. Struct. Biol.*, 2011, **21**, 576–582.
26 A. Gooley and K. Williams, *Glycobiology*, 1994, **4**, 413–417.
27 J. Hansen, O. Lund, J. Nilsson and K. Rapacki, *Nucleic Acids Res.*, 1998, **26**, 387–389.
28 E. Hounsell, M. Davies and D. Renouf, *Glycoconjugate J.*, 1996, **13**, 19–26.
29 B. Eisenhaber, P. Bork and F. Eisenhaber, *Protein Eng.*, 2001, **14**, 17–25.
30 D. P. Gamblin, E. M. Scanlan and B. G. Davis, *Chem. Rev.*, 2009, **109**, 131–163.
31 L.-X. Wang and M. N. Amin, *Chem. Biol.*, 2014, **21**, 51–66.
32 C. Unverzagt and Y. Kajihara, *Chem. Soc. Rev.*, 2013, **42**, 4408–4420.
33 M. Murakami, T. Kiuchi, M. Nishihara, K. Tezuka, R. Okamoto, M. Izumi and Y. Kajihara, *Sci. Adv.*, 2016, **2**, e1500678.
34 P. Wang, S. Dong, J.-H. Shieh, E. Peguero, R. Hendrickson, M. A. S. Moore and S. J. Danishefsky, *Science*, 2013, **342**, 1357–1360.
35 I. Sakamoto, K. Tezuka, K. Fukae, K. Ishii, K. Taduru, M. Maeda, M. Ouchi, K. Yoshida, Y. Nambu, J. Igarashi, N. Hayashi, T. Tsuji and Y. Kajihara, *J. Am. Chem. Soc.*, 2012, **134**, 5428–5431.
36 N. Yamamoto, Y. Tanabe, R. Okamoto, P. E. Dawson and Y. Kajihara, *J. Am. Chem. Soc.*, 2008, **130**, 501–510.
37 B. L. Wilkinson, R. S. Stone, C. J. Capicciotti, M. Thaysen-Andersen, J. M. Matthews, N. H. Packer, R. N. Ben and R. J. Payne, *Angew. Chem., Int. Ed.*, 2012, **51**, 3606–3610.
38 B. Aussedat, B. Fasching, E. Johnston, N. Sane, P. Nagorny and S. J. Danishefsky, *J. Am. Chem. Soc.*, 2012, **134**, 3532–3541.
39 P. Nagorny, N. Sane, B. Fasching, B. Aussedat and S. J. Danishefsky, *Angew. Chem., Int. Ed.*, 2012, **51**, 975–979.
40 J. M. Chalker, G. J. L. Bernardes and B. G. Davis, *Acc. Chem. Res.*, 2011, **44**, 730–741.
41 R. M. Schmaltz, S. R. Hanson and C.-H. Wong, *Chem. Rev.*, 2011, **111**, 4259–4307.
42 R. Adamo, A. Nilo, B. Castagner, O. Boutureira, F. Berti and G. J. L. Bernardes, *Chem. Sci.*, 2013, **4**, 2995–3008.
43 C. I. C. Crucho, P. Correia-da-Silva, K. T. Petrova and M. T. Barros, *Carbohydr. Res.*, 2015, **402**, 124–132.
44 B. G. Davis, *J. Chem. Soc., Perkin Trans. 1*, 1999, 3215–3237.
45 H. Pauly, *Z. Phys. Chem.*, 1907, **42**, 508–518.
46 O. T. Avery and W. F. Goebel, *J. Exp. Med.*, 1929, **50**, 533–550.
47 E. W. Gelewitz, W. L. Riedeman and I. M. Klotz, *Arch. Biochem. Biophys.*, 1954, **53**, 411–424.
48 A. N. Howard and F. Wild, *Biochem. J.*, 1957, **65**, 651–659.
49 M. Tabachnick and H. Sobotka, *J. Biol. Chem.*, 1960, **235**, 1051–1054.
50 C. M. Reichert, C. E. Hayes and I. J. Goldstein, *Methods in Enzymology*, Academic Press, 1994, vol. 242, pp. 108–116.

51 G. T. Hermanson, *Bioconjugate Techniques*, 2nd edn, 2008.
52 G. R. Gray, *Arch. Biochem. Biophys.*, 1974, **163**, 426–428.
53 R. Roy, E. Katzenellenbogen and H. J. Jennings, *Can. J. Biochem. Cell Biol.*, 1984, **62**, 270–275.
54 R. Saksena, X. Ma and P. Kováč, *Carbohydr. Res.*, 2003, **338**, 2591–2603.
55 R. U. Lemieux, D. R. Bundle and D. A. Baker, *J. Am. Chem. Soc.*, 1975, **97**, 4076–4083.
56 J. Porath, R. Axen and S. Ernback, *Nature*, 1967, **215**, 1491–1492.
57 G. S. Bethell, J. S. Ayers, W. S. Hancock and M. T. Hearn, *J. Biol. Chem.*, 1979, **254**, 2572–2574.
58 N. Nakajima and Y. Ikada, *Bioconjugate Chem.*, 1995, **6**, 123–130.
59 R. Rappuoli, M. Pizza, G. Del Giudice and E. De Gregorio, *Proc. Natl. Acad. Sci. U. S. A.*, 2014, **111**, 12288–12293.
60 W. G. van Panhuis, J. Grefenstette, S. Y. Jung, N. S. Chok, A. Cross, H. Eng, B. Y. Lee, V. Zadorozhny, S. Brown, D. Cummings and D. S. Burke, *N. Engl. J. Med.*, 2013, **369**, 2152–2158.
61 R. Schneerson, O. Barrera, A. Sutton and J. B. Robbins, *J. Exp. Med.*, 1980, **152**, 361–376.
62 V. Verez-Bencomo, V. Fernández-Santana, E. Hardy, M. E. Toledo, M. C. Rodríguez, L. Heynngnezz, A. Rodriguez, A. Baly, L. Herrera, M. Izquierdo, A. Villar, Y. Valdés, K. Cosme, M. L. Deler, M. Montane, E. Garcia, A. Ramos, A. Aguilar, E. Medina, G. Toraño, I. Sosa, I. Hernandez, R. Martínez, A. Muzachio, A. Carmenates, L. Costa, F. Cardoso, C. Campa, M. Diaz and R. Roy, *Science*, 2004, **305**, 522–525.
63 R. D. Astronomo and D. R. Burton, *Nat. Rev. Drug Discovery*, 2010, **9**, 308–324.
64 J. Kubler-Kielb and V. Pozsgay, *J. Org. Chem.*, 2005, **70**, 6987–6990.
65 S. J. Hou, R. Saksena and P. Kovac, *Carbohydr. Res.*, 2008, **343**, 196–210.
66 L. Kong, B. Vijayakrishnan, M. Kowarik, J. Park, A. N. Zakharova, L. Neiwert, A. Faridmoayer and B. G. Davis, *Nat. Chem.*, 2016, **8**, 242–249.
67 D. Bundle, *Nat. Chem.*, 2016, **8**, 201–202.
68 V. V. Rostovtsev, L. G. Green, V. V. Fokin and K. B. Sharpless, *Angew. Chem., Int. Ed.*, 2002, **41**, 2596–2599.
69 C. W. Tornøe, C. Christensen and M. Meldal, *J. Org. Chem.*, 2002, **67**, 3057–3064.
70 S. Dedola, S. A. Nepogodiev and R. A. Field, *Org. Biomol. Chem.*, 2007, **5**, 1006–1017.
71 A. Dondoni, *Chem. – Asian J.*, 2007, **2**, 700–708.
72 M. Meldal and C. W. Tornøe, *Chem. Rev.*, 2008, **108**, 2952–3015.
73 V. K. Tiwari, B. B. Mishra, K. B. Mishra, N. Mishra, A. S. Singh and X. Chen, *Chem. Rev.*, 2016, **116**, 3086–3240.
74 T. Lipinski, T. Luu, P. I. Kitov, A. Szpacenko and D. R. Bundle, *Glycoconjugate J.*, 2011, **28**, 149–164.
75 T. Lipinski, A. Fitieh, J. Pierre, H. L. Ostergaard, D. R. Bundle and N. Touret, *J. Immunol.*, 2013, **190**, 4116–4128.
76 S. Crotti, H. Zhai, J. Zhou, M. Allan, D. Proietti, W. Pansegrau, Q. Y. Hu, F. Berti and R. Adamo, *ChemBioChem*, 2014, **15**, 836–843.
77 H. Ban, J. Gavrilyuk and C. F. Barbas, *J. Am. Chem. Soc.*, 2010, **132**, 1523–1525.
78 A. Nilo, L. Morelli, I. Passalacqua, B. Brogioni, M. Allan, F. Carboni, A. Pezzicoli, F. Zerbini, D. Maione, M. Fabbrini, M. R. Romano, Q.-Y. Hu, I. Margarit, F. Berti and R. Adamo, *ACS Chem. Biol.*, 2015, **10**, 1737–1746.
79 A. Nilo, M. Allan, B. Brogioni, D. Proietti, V. Cattaneo, S. Crotti, S. Sokup, H. Zhai, I. Margarit, F. Berti, Q.-Y. Hu and R. Adamo, *Bioconjugate Chem.*, 2014, **25**, 2105–2111.

80 Q.-Y. Hu, M. Allan, R. Adamo, D. Quinn, H. Zhai, G. Wu, K. Clark, J. Zhou, S. Ortiz, B. Wang, E. Danieli, S. Crotti, M. Tontini, G. Brogioni and F. Berti, *Chem. Sci.*, 2013, **4**, 3827–3832.
81 R. Adamo, Q.-Y. Hu, A. Torosantucci, S. Crotti, G. Brogioni, M. Allan, P. Chiani, C. Bromuro, D. Quinn, M. Tontini and F. Berti, *Chem. Sci.*, 2014, **5**, 4302–4311.
82 T. Tanaka, H. Nagai, M. Noguchi, A. Kobayashi and S.-I. Shoda, *Chem. Commun.*, 2009, **23**, 3378–3379.
83 D. Lim, M. A. Brimble, R. Kowalczyk, A. J. Watson and A. J. Fairbanks, *Angew. Chem., Int. Ed.*, 2014, **53**, 11907–11911.
84 T. Machida, K. Lang, L. Xue, J. W. Chin and N. Winssinger, *Bioconjugate Chem.*, 2015, **26**, 802–806.
85 M. L. Blackman, M. Royzen and J. M. Fox, *J. Am. Chem. Soc.*, 2008, **130**, 13518–13519.
86 G. Stefanetti, Q. Y. Hu, A. Usera, Z. Robinson, M. Allan, A. Singh, H. Imase, J. Cobb, H. Zhai, D. Quinn, M. Lei, A. Saul, R. Adamo, C. A. MacLennan and F. Micoli, *Angew. Chem., Int. Ed.*, 2015, **54**, 13198–13203.
87 G. Stefanetti, A. Saul, C. A. MacLennan and F. Micoli, *Bioconjugate Chem.*, 2015, **26**, 2507–2513.
88 M. Pan, S. Li, X. Li, F. Shao, L. Liu and H. G. Hu, *Angew. Chem., Int. Ed.*, 2014, **53**, 14517–14521.
89 G. J. L. Bernardes, D. P. Gamblin and B. G. Davis, *Angew. Chem., Int. Ed.*, 2006, **45**, 4007–4011.
90 A. Dondoni, A. Massi, P. Nanni and A. Roda, *Chemistry*, 2009, **15**, 11444–11449.
91 O. Boutureira, G. J. Bernardes, M. Fernandez-Gonzalez, D. C. Anthony and B. G. Davis, *Angew. Chem., Int. Ed.*, 2012, **51**, 1432–1436.
92 N. Floyd, B. Vijayakrishnan, J. R. Koeppe and B. G. Davis, *Angew. Chem., Int. Ed.*, 2009, **48**, 7798–7802.
93 E. J. Grayson, G. J. Bernardes, J. M. Chalker, O. Boutureira, J. R. Koeppe and B. G. Davis, *Angew. Chem., Int. Ed.*, 2011, **50**, 4127–4132.
94 D. P. Gamblin, P. Garnier, S. van Kasteren, N. J. Oldham, A. J. Fairbanks and B. G. Davis, *Angew. Chem., Int. Ed.*, 2004, **43**, 828–833.
95 R. A. Field, *Nat. Chem. Biol.*, 2011, **7**, 658–659.
96 E. C. O'Neill and R. A. Field, *Carbohydr. Res.*, 2015, **403**, 23–37.
97 S. Salamone, C. Guerreiro, E. Cambon, I. Andre, M. Remaud-Simeon and L. A. Mulard, *Chem. Commun.*, 2015, **51**, 2581–2584.
98 H. Yu and X. Chen, *Org. Biomol. Chem.*, 2016, **14**, 2809–2818.
99 G. K. Wagner, T. Pesnot and R. A. Field, *Nat. Prod. Rep.*, 2009, **26**, 1172–1194.
100 K. M. Koeller and C.-H. Wong, *Glycobiology*, 2000, **10**, 1157–1169.
101 S. K. Mazmanian, G. Liu, H. Ton-That and O. Schneewind, *Science*, 1999, **285**, 760–763.
102 R. P. Novick, *Trends Microbiol.*, 2000, **8**, 148–151.
103 X. Guo, Q. Wang, B. M. Swarts and Z. Guo, *J. Am. Chem. Soc.*, 2009, **131**, 9878–9879.
104 Z. Wu, X. Guo and Z. Guo, *Chem. Commun.*, 2010, **46**, 5773–5774.
105 Z. Wu, X. Guo, Q. Wang, B. M. Swarts and Z. Guo, *J. Am. Chem. Soc.*, 2010, **132**, 1567–1571.
106 Z. Wu, X. Guo, G. Gu and Z. Guo, *Org. Lett.*, 2013, **15**, 5906–5908.
107 Z. Wu, X. Guo, J. Gao and Z. Guo, *Chem. Commun.*, 2013, **49**, 11689–11691.
108 A. V. Nikolaev and N. Al-Maharik, *Nat. Prod. Rep.*, 2011, **28**, 970–1020.
109 K. Takegawa, M. Tabuchi, S. Yamaguchi, A. Kondo, I. Kato and S. Iwahara, *J. Biol. Chem.*, 1995, **270**, 3094–3099.

110 M. Mizuno, K. Haneda, R. Iguchi, I. Muramoto, T. Kawakami, S. Aimoto, K. Yamamoto and T. Inazu, *J. Am. Chem. Soc.*, 1999, **121**, 284–290.
111 B. Li, Y. Zeng, S. Hauser, H. Song and L.-X. Wang, *J. Am. Chem. Soc.*, 2005, **127**, 9692–9693.
112 H. Ochiai, W. Huang and L.-X. Wang, *J. Am. Chem. Soc.*, 2008, **130**, 13790–13803.
113 H. Hojo, H. Tanaka, M. Hagiwara, Y. Asahina, A. Ueki, H. Katayama, Y. Nakahara, A. Yoneshige, J. Matsuda, Y. Ito and Y. Nakahara, *J. Org. Chem.*, 2012, **77**, 9437–9446.
114 Y. Wei, C. Li, W. Huang, B. Li, S. Strome and L.-X. Wang, *Biochemistry*, 2008, **47**, 10294–10304.
115 M. Fernandez-Gonzalez, O. Boutureira, G. J. L. Bernardes, J. M. Chalker, M. A. Young, J. C. Errey and B. G. Davis, *Chem. Sci.*, 2010, **1**, 709–715.
116 J. J. Goodfellow, K. Baruah, K. Yamamoto, C. Bonomelli, B. Krishna, D. J. Harvey, M. Crispin, C. N. Scanlan and B. G. Davis, *J. Am. Chem. Soc.*, 2012, **134**, 8030–8033.
117 R. van Geel, M. A. Wijdeven, R. Heesbeen, J. M. M. Verkade, A. A. Wasiel, S. S. van Berkel and F. L. van Delft, *Bioconjugate Chem.*, 2015, **26**, 2233–2242.
118 E. L. Smith, J. P. Giddens, A. T. Iavarone, K. Godula, L.-X. Wang and C. R. Bertozzi, *Bioconjugate Chem.*, 2014, **25**, 788–795.
119 G. Reuter, R. Schauer, C. Szeiki, J. P. Kamerling and J. F. Vliegenthart, *Glycoconjugate J.*, 1989, **6**, 35–44.
120 Q. Zhou, J. E. Stefano, C. Manning, J. Kyazike, B. Chen, D. A. Gianolio, A. Park, M. Busch, J. Bird, X. Zheng, H. Simonds-Mannes, J. Kim, R. C. Gregory, R. J. Miller, W. H. Brondyk, P. K. Dhal and C. Q. Pan, *Bioconjugate Chem.*, 2014, **25**, 510–520.
121 P. K. Qasba, *Bioconjugate Chem.*, 2015, **26**, 2170–2175.
122 E. C. Rodriguez, K. A. Winans, D. S. King and C. R. Bertozzi, *J. Am. Chem. Soc.*, 1997, **119**, 9905–9906.
123 J. B. Rannes, A. Ioannou, S. C. Willies, G. Grogan, C. Behrens, S. L. Flitsch and N. J. Turner, *J. Am. Chem. Soc.*, 2011, **133**, 8436–8439.
124 C. M. Szymanski, R. Yao, C. P. Ewing, T. J. Trust and P. Guerry, *Mol. Microbiol.*, 1999, **32**, 1022–1030.
125 H. Nothaft and C. M. Szymanski, *Nat. Rev. Microbiol.*, 2010, **8**, 765–778.
126 M. F. Feldman, M. Wacker, M. Hernandez, P. G. Hitchen, C. L. Marolda, M. Kowarik, H. R. Morris, A. Dell, M. A. Valvano and M. Aebi, *Proc. Natl. Acad. Sci. U. S. A.*, 2005, **102**, 3016–3021.
127 J. L. Baker, E. Çelik and M. P. DeLisa, *Trends Biotechnol.*, 2013, **31**, 313–323.
128 J. Cuccui and B. Wren, *J. Pharm. Pharmacol.*, 2015, **67**, 338–350.
129 V. S. Terra, D. C. Mills, L. E. Yates, S. Abouelhadid, J. Cuccui and B. W. Wren, *J. Med. Microbiol.*, 2012, **61**, 919–926.
130 M. M. Kämpf, M. Braun, D. Sirena, J. Ihssen, L. Thöny-Meyer and Q. Ren, *Microb. Cell Fact.*, 2015, **14**, 1–12.
131 J. Cuccui, R. M. Thomas, M. G. Moule, R. V. D'Elia, T. R. Laws, D. C. Mills, D. Williamson, T. P. Atkins, J. L. Prior and B. W. Wren, *Open Biol.*, 2013, **3**, 130002.
132 C. F. R. Hatz, B. Bally, S. Rohrer, R. Steffen, S. Kramme, C.-A. Siegrist, M. Wacker, C. Alaimo and V. G. Fonck, *Vaccine*, 2015, **33**, 4594–4601.
133 A. A. Ollis, S. Zhang, A. C. Fisher and M. P. DeLisa, *Nat. Chem. Biol.*, 2014, **10**, 816–822.
134 F. Schwarz, W. Huang, C. Li, B. L. Schulz, C. Lizak, A. Palumbo, S. Numao, D. Neri, M. Aebi and L.-X. Wang, *Nat. Chem. Biol.*, 2010, **6**, 264–266.

135 A. Ishiwata, Y. Taguchi, Y. J. Lee, T. Watanabe, D. Kohda and Y. Ito, *ChemBioChem*, 2015, **16**, 731–737.

136 V. Lukose, G. Whitworth, Z. Guan and B. Imperiali, *J. Am. Chem. Soc.*, 2015, **137**, 12446–12449.

137 F. Liu, B. Vijayakrishnan, A. Faridmoayer, T. A. Taylor, T. B. Parsons, G. J. L. Bernardes, M. Kowarik and B. G. Davis, *J. Am. Chem. Soc.*, 2014, **136**, 566–569.

138 F. Schwarz, Y.-Y. Fan, M. Schubert and M. Aebi, *J. Biol. Chem.*, 2011, **286**, 35267–35274.

139 J. V. Lomino, A. Naegeli, J. Orwenyo, M. N. Amin, M. Aebi and L.-X. Wang, *Bioorg. Med. Chem.*, 2013, **21**, 2262–2270.

Recent advances in the application of NMR methods to uncover the conformation and recognition features of glycans

Ana Ardá,[a] Helena Coelho,[a] Beatriz Fernández de Toro,[b] Silvia Galante,[b] Ana Gimeno,[a] Ana Poveda,[a] Javier Sastre,[b] Luca Unione,[a] Pablo Valverde,[a] F. Javier Cañada[b] and Jesús Jiménez-Barbero*[a,c]
DOI: 10.1039/9781782626657-00047

This chapter is dedicated to the presentation of different examples of the application of solution NMR to the study of conformation, dynamics of sugar molecules (oligo and polysaccharides, glycopeptides and glycomimetics) and to the investigation of glycan-related molecular recognition events. Selected examples since 2012 are presented depending on the chemical nature of the sugar molecule, on the environment (free or bound) and on the nature of the receptor.

1 Introduction

NMR methods constitute a powerful tool widely applied in the field of glycosciences[1] to analyse the structure, dynamic and conformational behaviour of saccharides and particularly in the study of sugar–protein interactions.

This report gathers selected recent examples (since 2012) of studies of the conformation, structure, dynamics and binding features of oligosaccharides, glycopeptides and glycomimetics. It has not been our intention to be exhaustive, since many groups are working in this field, making frequent contributions. It has been our aim to present different examples of the application of NMR methods and protocols for analysing saccharide conformation and for describing key structural features of glycan molecular recognition events with different natural and synthetic receptors.

2 Mono-, oligo- and poly-saccharides. Conformation and dynamics in solution

Nowadays, from a technical perspective, the boundaries of NMR for studying large glycans are continuously growing, due to the development of new avenues in methodologies and to the access to magnets which operate at very high field. In any case, when the number of monosaccharide units increases, the global flexibility increases, and the study

[a]*CIC bioGUNE, Bizkaia Technology Park, Building 800, 48160 Derio, Spain. E-mail: jjbarbero@cicbiogune.es*
[b]*Chemical & Physical Biology, CIB-CSIC, Ramiro de Maeztu 9, 28040 Madrid, Spain*
[c]*Ikerbasque, Basque Foundation for Science, Maria Diaz de Haro 13, 48012 Bilbao, Spain*

of the conformational preferences of large oligosaccharides remains a challenging task. The area has benefited from the introduction of paramagnetic probes. These probes induce two main paramagnetic effects on the NMR properties of the nearby nuclei: an enhancement of the relaxation, the paramagnetic relaxation enhancement (PRE), and a change in the chemical shift, the pseudocontact shifts (PCS). The PRE effect has a dependency with $1/r^6$, where r is the distance between the affected nuclei and the paramagnetic centre, and induces broadening of the NMR signals. The PCS effect occurs only when the paramagnetic centre presents a non-isotropic magnetic susceptibility (which is the case of some lanthanide cations), and it is dependent on the relative position of the paramagnetic centre (its magnetic susceptibility tensor, $\Delta\chi$) and the affected nuclei. Furthermore, its dependence with the distance is $1/r^3$, and thus the PCS effect does not decrease as rapidly as the PRE, and its effects can be observed up to at 40 Å distance for certain paramagnetic metals. The conformational information is provided by obtaining the distance (r) and the relative orientation of the individual nuclei with respect to the magnetic susceptibility tensor $\Delta\chi$ as from a minimum number of experimentally obtained PCS. However, for a simpler treatment of the data, the former should be constant, and thus the paramagnetic metal must be attached to the oligosaccharide with the least possible flexibility.[2] Different linkers have been described so far, which have been covalently attached to the reducing end of the sugar.[3] The strategy that was first applied to describe the conformational ensemble of simple sugars like lactose[4] and chitobiose,[5] has been recently applied to more complex glycan structures, like *N*-glycans and ganglioside oligosaccharides. In the case of symmetrically branched *N*-glycans, PCS permitted to magnetically differentiate resonances from the different *N*-glycan arms (Fig. 1), opening the possibility to study their different recognition by specific receptors.[6]

Kato *et al.* also proposed to use experimental PCSs to be compared with theoretical ones computed from large conformational ensembles extracted from extensive Molecular Dynamic simulations (MD), in order to validate the MD-derived conformational space. In such a way the conformational space for GM1 tri-, GM2 tetra- and GM3 pentasaccharide were compared, concluding that the global minimum around the Neu5Acα2-3Gal linkage is modified by the presence of the Galβ1-3GalNAc branch in GM2 and GM3, shifting from $\Phi/\Psi = -90° \pm 11°, 57° \pm 11°$ to $\Phi/\Psi = -174° \pm 11°, -32° \pm 11°$.[7–9] The same protocol was applied to two large triantennary high-mannose-type *N*-glycans: Man9 and its derivative where the terminal mannose at the central branch was missing (Man8D).[10] Both *N*-glycans were produced by genetically engineered yeast cultivated with ^{13}C-glucose as carbon source,[11] producing uniformly ^{13}C-labeled Man9 and Man8D. In this case, in order to obtain a good fit between the theoretical and experimental PCS, replica-exchange molecular dynamics (REMD) were necessary, that allow jumping between different minima, enhancing the sampling of the conformational space. The results show that removal of the single terminal mannose at branch D2 causes that Man8D occupies a broader conformational space than

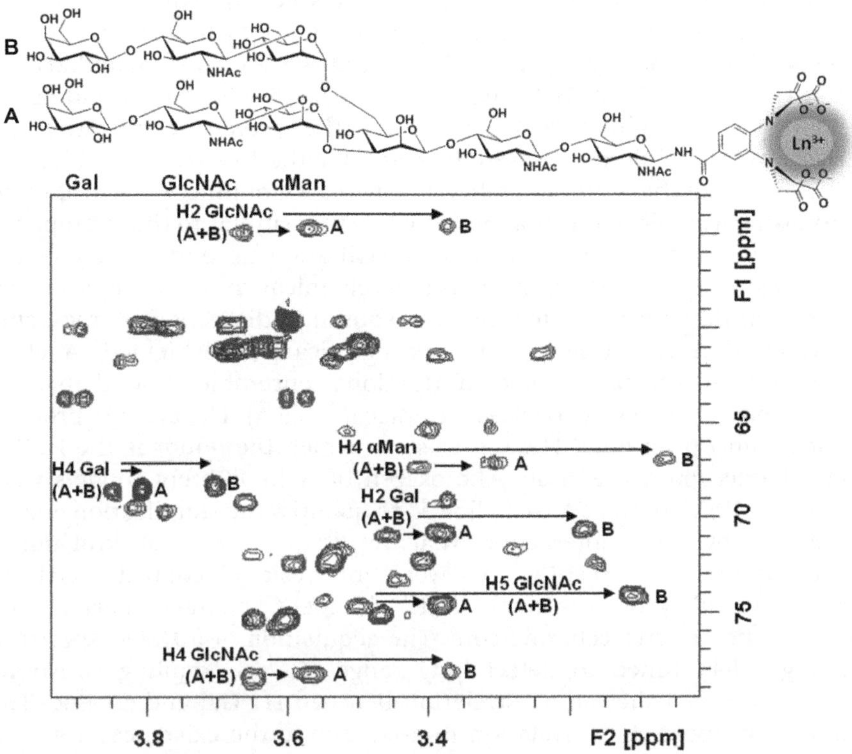

Fig. 1 Superimposition of $^{1}H-^{13}C$ HSQC spectra for the lanthanide-tagged *N*-glycan, in the presence of the diamagnetic reference ion (La^{3+}) and the paramagnetic ion (Dy^{3+}). Key differences of chemical shifts due to PCS from both branches, A and B, are highlighted (adapted from ref. 6, Canales *et al.*).

Man9, for which the presence of the above mentioned terminal Man hinders the back folding of D2 and D3 branches. The paramagnetic relaxation enhanced effect (PRE), produced by the attachment of the paramagnetic relaxation agent TEMPO at the reducing end of Man9, agreed with these findings, showing that the anomeric protons of the mannose residues at the D1 branch suffer less relaxation enhancement that those located at the D2 and D3 branches.[12]

Besides complex oligosaccharides, the analysis of complex mixtures by using NMR usually remains a challenge. The use of paramagnetic substances to filter out the NMR signals of specific molecules has also been a matter of interest. The use of Gd^{3+} salts seems to be a robust method to suppress the signals of anionic molecules.[13] It has been shown that the ^{1}H and even ^{13}C NMR signals of highly negatively charged polysaccharides (hyaluronic acid) can be removed in the presence of neutral polysaccharides (chondroitin sulphate or dextran) in the presence of minute amounts of Gd^{3+}. The performance of these experiments may be improved in the presence of T_2 filters.

In the conformational side, the geometry of the biologically relevant Le^{X} branched oligosaccharide has been re-examined,[14] through a novel approach that consists in covalently linking the small Le^{X} trisaccharide to a $^{13}C/^{15}N$ labeled protein. In such a way, the saccharide displays a longer rotational motion correlation time, which permits to obtain stronger NOEs. Thus, by acquiring 2D ^{13}C F2-filtered-NOESY (in D_2O) and 2D ^{15}N F2-filtered-NOESY (in H_2O) experiments, 28 NOEs corresponding to the oligosaccharide were observed for the protein conjugated Le^X trisaccharide, while only 9 were detected for the free trisaccharide, due to its unfavourable tumbling rate. This increased number of NOE allowed proposing a single conformation, which is very similar to those proposed earlier, but with a much narrower distribution around the glycosidic linkages. The comparison of the proton chemical shifts of the branched trisaccharide with those for the corresponding disaccharide fragments revealed that H5 of Fuc in Le^X is significantly deshielded. A close inspection of the proposed conformational ensemble showed that H5 Fuc is indeed very close to O5Gal (average $d = 2.5$ Å). The authors propose that an unconventional H-bond exist, in which the donor is the H5Fuc and the acceptor, the endocyclic oxygen of Gal. Different models were built and DFT optimized, that allowed to quantify the stabilization energy provided by such interaction that resulted *ca.* 1.7–1.8 kcal mol^{-1}. Interestingly, very recently, this weak interresidual contact has been experimentally proven through NMR[15] for a sLe^X pentasaccharide, containing the Le^X trisaccharide core. The acquisition of a HSQC spectrum at high field tuned to detect long range ^1H–^{13}C coupling constants ($1/4\,J = 125$ ms) showed a correlation between H1 Gal and C5 Fuc. The authors propose the correlation to arise due to the existence of a $^4J_{C,H}$ through the H1Gal-C1Gal-O5Gal···H5Fuc-C5Fuc bonding pathway, thus confirming the existence of an hydrogen bond O5Gal···H5Fuc. The authors also found that specific NMR parameters change with the temperature. In particular, the coupling constant between NH and H2 and the ^{13}C T2* of C1, C2 and C3 of the GlcNAc residue experience a temperature dependence that suggests a conformational motion in the Le^X core in the ms time scale. Indeed, MD simulations find a swing motion of the NHAc group of GlcNAc, which would be establishing alternative hydrogen bonds either with OH2Gal on the one side or with OH2Fuc, on the other side. These studies reveal the Le^X epitope has a fairly special structure. Recently, fucosylated chondroitin sulfate (fCS) has been isolated from the sea cucumber *Holothuria forskali*.[16] fCS has attracted much attention because they have been reported to possess different interesting biological activities such as anticoagulant, antithrombotic, antiinflammatory, anti-HIV and metastasis blocking. The fCS polysaccharides were extracted, purified, and characterized to have the repeating unit→)GalNAcβ4,6*S*(1-4) [FucαX(1-3)]GlcAβ(1→, where X indicates the different sulfation pattern of fucose. Interestingly, NMR studies permitted to propose that the conformation of the repeating unit is reminiscent of that of Le^X core, as described above, where the Fuc residue sits on top of the GalNAc residue, being this ring stacking independent of the sulfation pattern, which permits to speculate about the possible mimicry of these fCS structures with the Le^X epitope.

Interresidual hydrogen bonding in carbohydrates through hydroxyl groups in water is controversial. The experimental evidence of the existence of hydrogen bonds in carbohydrates in water *via* the detection of long range NMR scalar couplings across hydrogen bonded nuclei had been previously reported for the tetrasaccharide (Neu5Acα2-8Neu5Acα2-)$_2$ (SiA)$_4$.[17] α2-8 poly-sialic acid had been previously proposed to adopt a secondary helical structure in solution. In this study, the utilization of a fully labelled ^{13}C, ^{15}N (SiA)$_4$ in H_2O permitted to acquire the CBCANH experiment, typically employed for proteins. The optimization for long range ^{13}C–^{15}N coupling constants permitted to detect an intrarresidual correlation between the N of the acetamido group and C8 of the glycerol chain, but only for the central and reducing end residues. The authors proposed this correlation to arise because of the existence of the coupling constant $^4J_{C8-N(H)}$ that involves the hydrogen bond O8···H(N). This interaction is not expected to happen at the non-reducing end residue, where the glycerol chain would be flexible. Further support for the rigidification of the glycerol chain in central and reducing end residues arose from the observation of the long range coupling constant $^4J_{H7-C2}$ *via* an HSQMBC experiment, as a result of the ***W*** conformation along H7–C2. These two geometrical constraints, together with experimental NOE data (obtained through CNH-NOE and ^{1}H–^{13}C HSQC-NOESY experiments) permitted to propose a 2_4 helix secondary structure for the tetrasaccharide (SiA)$_4$.

In sucrose, there is still some controversy remaining with respect to the persistence of intramolecular hydrogen bonding in solution.[18,19] The observation of HSQC-TOCSY inter-ring correlations was taken as an experimental evidence for the presence of two hydrogen bonds that involve the same hydroxyl groups that alternate between donor and acceptor of the H-bond.[20] Care must be however taken, since in order to obtain such experimental data high concentrations of the sugar have to be used (300 to 800 mM) and some of the observed crosspeaks can arise from inter- instead of intra-molecular interactions due to aggregation, as is demonstrated through DOSY experiments. Hydrogen bond participation has also been explored in heparin oligosaccharides, in particular in Arixtra, a pentasaccharide that was designed in order to mimic the heparin pentasaccharidic fragment responsible for antithrombin-II binding. In this case, the temperature coefficients and the energy barrier for solvent exchange of the sulfamate groups ($NHSO_3^-$),[21] as well as temperature coefficients, proton–water chemical exchange, and chemical shift differences for the hydroxyl protons,[22] suggested the participation of different hydrogen bonds shaping the 3D structure of this important low molecular weight heparin.

As mentioned above, the conformational rigidity found for the Le^X core is not common in carbohydrates, which usually present multiple minima around ψ and ω and/or present shallow and broad energy minima. The production of isotopically labelled glycans, especially ^{13}C labelled, offers attractive possibilities in their structural studies. It facilitates access to parameters like heteronuclear coupling constants, for which specific sequences have been proposed,[23,24] which contain structural information,[25] but furthermore provides the wider chemical shift dispersion of the ^{13}C isotope, especially important when dealing with large oligosaccharides.[11] This strategy, for instance, permitted the NMR structural elucidation of a novel *N*-linked oligosaccharide structure from a sample of only 110 μM, derived from a hyperthermophilic archaeon.[26] Archea and eubacteria show much greater structural variability in their *N*-glycosylated structures than eukaryotes.

Since NMR spectroscopy provides structural information in an indirect way and thus, in order to obtain 3D structures scientists must rely on molecular modelling protocols. One of the most extended ones is molecular dynamics (MD) simulations, for which different force fields (FF) are available for carbohydrates (CHARMm, GLYCAM06, GROMOS, *etc*...). The CHARMm force field has been recently reparametrized for the (1-6) glycosidic linkages, when important deviations between the experimental and the MD-derived $^3J_{H,H}$ for the hydroxymethyl groups for different α- and β-(1-6) linked disaccharides were found.[27] Generally, FFs are parameterized for systems at room temperature. The ability of different FFs to reproduce experimental data at different temperatures was also explored, by studying the conformational equilibrium of the hydroxymethyl group in cellobiose and longer cellooligomers as a function of temperature. Different force fields and solvation models were tested, showing that all of them predict an overestimation of the temperature dependence of the conformational equilibrium around the hydroxymethyl group.[28]

The conformational preferences of different natural oligosaccharide structures with diverse biological relevance have also been studied by a classical combination of NMR parameters and MD simulations. For instance the oligosaccharides derived from pathogens like *Moraxella cattarhalis*[29] or *E. coli* O13,[30] which present interest for vaccine development. Tumour associated carbohydrate antigens (TACA's), as the dimeric Le(x) and Le(x)–Le(a) or [Le^X and Le^X–Le^A], were also investigated, showing that they populate two conformers differing in Ψ of the linkage GlcNAcβ(1-3)Gal (between Le(x)–Le(x) or Le(X)–Le(a)). These structures, overexpressed in certain cancerous tissues, are thus able to present the two different epitopes, which could have important implications in their molecular recognition by specific monoclonal antibodies (mAb).[31]

3 The bound state

In the last few years many studies on conformational and structural features of the interaction between carbohydrates and different type of receptors have taken advantage of NMR methodologies. These techniques are extensively used since they provide information at molecular/atomic level. Thus, NMR experiments have been extensively applied to study the interaction between sugars and their mimetics with biomolecules such as lectins, antibodies, enzymes, nucleic acids as well as synthetic receptors or even for analyzing sugar–sugar interactions. Moreover, these methodologies have been extended to understand in depth the nature of these key interactions. Therefore, according to the paramount importance and the enormous applicability of NMR methods this topic is continuously updated.[32–34]

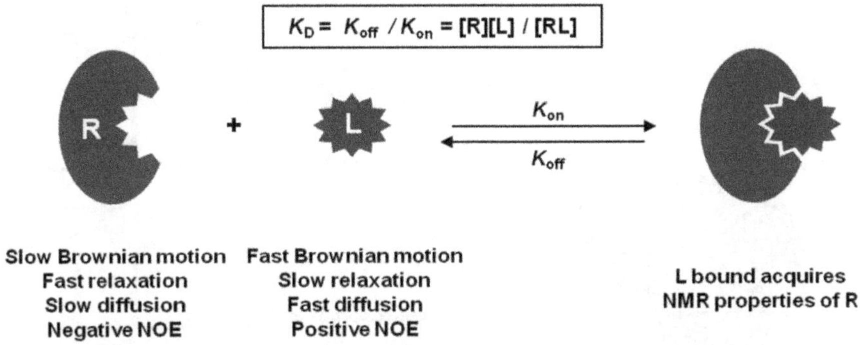

Fig. 2 Scheme of the different NMR-related motional and spectroscopic properties of receptors and ligands, as well as their complexes.

Two main strategies are used to analyse and characterize the interaction between carbohydrates and macromolecules by NMR techniques; they are ligand and receptor-based methods. In one hand, NMR parameters of the receptor can be monitored, as measuring differences in chemical shift or relaxation times between the bound and the free state. On the other hand, more feasible ligand-based methods are focused on changes in the ligand NMR signals as a consequence of the binding process (Fig. 2).

These useful methodologies together with computational modelling protocols provide an accurate picture of the binding process.

4 Oligosaccharide–protein interactions: lectins

Lectins, proteins with exquisite target specificity for glycans, and therefore potential translators of the sugar code, have attracted wide attention.[35] Since they are related with adhesion, transport, endocytosis, cell growth, cell migration, routing and some other important biological processes they are being exploited as biomarkers as well as therapeutic targets or drugs. Among methodologies providing detailed structural knowledge about the interaction between carbohydrates and lectins, NMR data furnish a dynamic viewpoint of the binding event. High-resolution NMR experiments together with molecular modelling protocols give insights into solution structural features and dynamics of the recognition process. The level of specificity of the lectin, affinity constants as well as conformational and structural features of the sugar–protein complexes can be drawn from NMR data.

Despite WaterLOGSY methods seem to be highly sensitive,[36] trNOESY and especially STD experiments (Fig. 3) have proven to be extremely useful ligand-based techniques to detect carbohydrate–lectin interactions and analyze bound ligand features.[37]

The weak interaction between carbohydrates and lectin receptors in the mM–μM range makes them appropriate systems to be analyzed with this type of NMR techniques. STD epitope mapping strategies have also been successfully employed for analyzing the lectin binding abilities of large oligosaccharides. These molecules often display multimodal

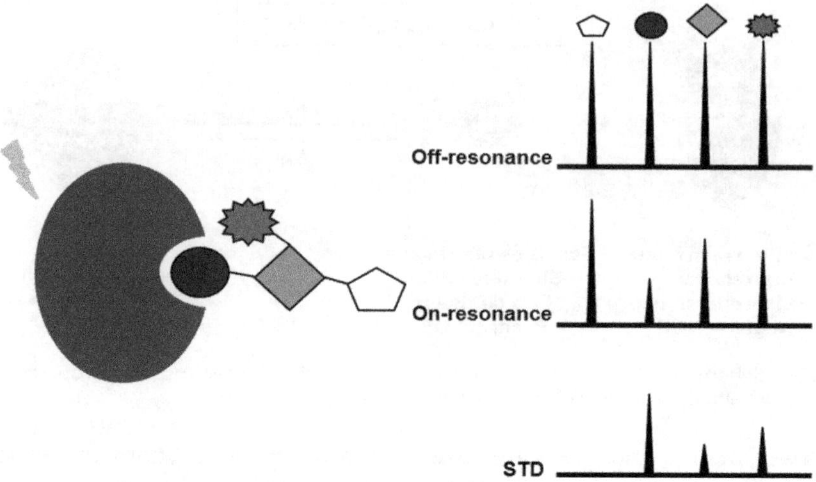

Fig. 3 Schematic representation of the STD experiment. The STD spectrum is obtained by subtraction of the spectrum obtained after irradiation of the protein resonances from that recorded with off-resonance saturation. Epitope mapping can be elucidated through the differences of the visible NMR signal intensities.

binding; then, presentation of the overall structure of a glycan could be considered a determining factor for its recognition by lectins.[38] However, the study of the recognition processes involving large carbohydrates has remained elusive due to their more intricate and overlapped spectra. In the last few years, STD-TOCSY experiments have permitted to deal with the problems derived from the existence of overlapping NMR signals (Fig. 4).

This technique combined with modelling protocols has been successfully applied for the analysis of the binding of complex-type biantennary glycans to WGA.[39] This lectin, with a dual specificity for GlcNAc and Neu5Ac, monosaccharides usually found at opposite ends of *N*-glycans, binds in a specific manner through the GlcNAc end of the large decasaccharide tested. This study evidenced differences in the binding mode between small and large oligosaccharides, in terms of epitope recognition, pointing to the need to evaluate molecular recognition processes at the highest possible resolution with natural glycans.

STD experiments have also been successfully used to screen a set of sialylated glycans for binding with the serum protein complement factor H, involved in innate immunity.[40] The results established Neu5Acα2-3Gal as the minimum essential residue for the interaction. NMR together with X-ray crystallography data permitted to dissect the interactions responsible for the recognition, establishing some structural basis in the relationship between FH mutations and immune related diseases.

These protocols have also been applied to define the binding preferences of different influenza hemagglutinin (HA) subtypes for α2,3-sialyllactose and α2,6-sialyllactose, sugars commonly present on cell surface. These studies proved the ability of influenza virus to switch hosts by subtle modifications in the HA structure.[41]

Receptor-based methods have also been successfully applied to gain insight about the interaction between carbohydrates and their lectin-type

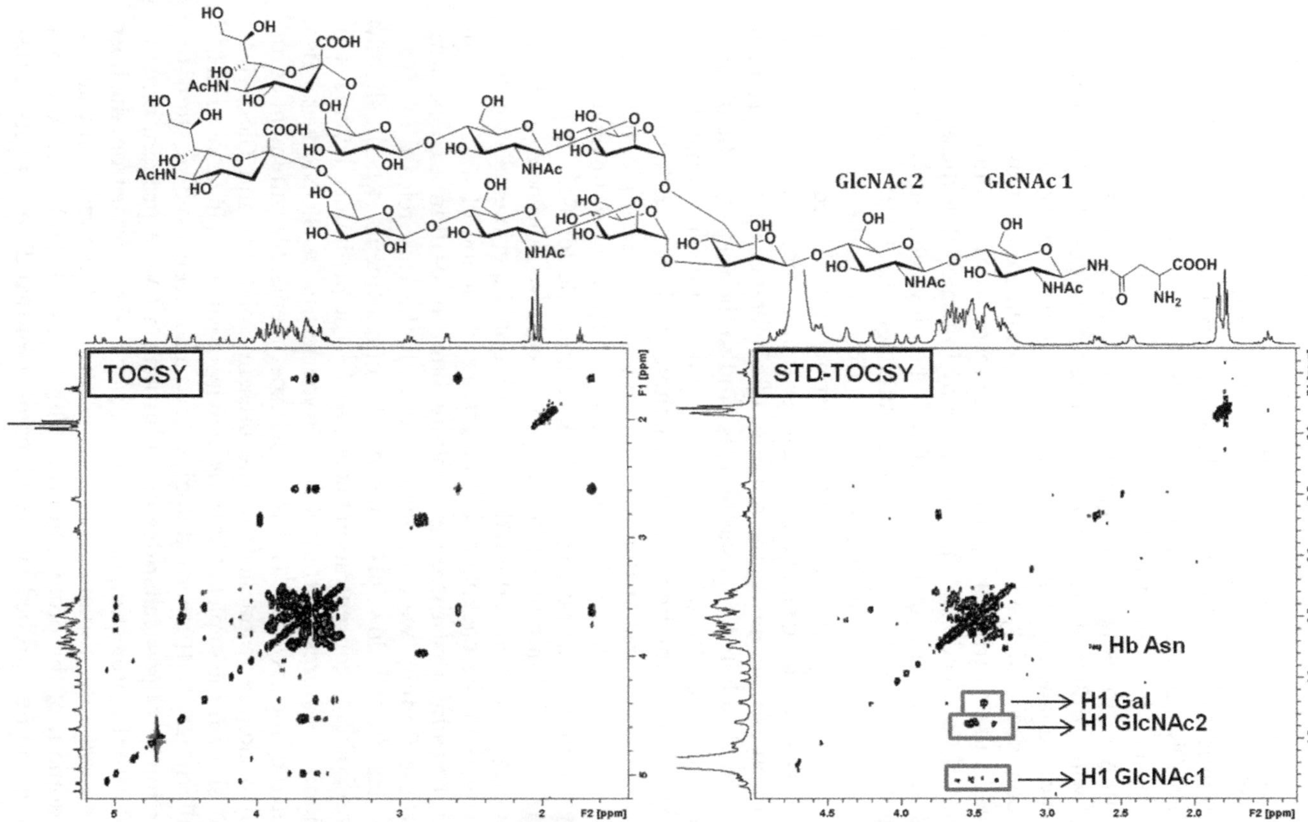

Fig. 4 Illustration of TOCSY and STD-TOCSY experiments of a biantennary sialylated *N*-glycan in absence (left) and presence (right) of a model lectin. STD-TOCSY allows obtaining the ligand recognition features despite the signals overlap in the ^{1}H dimension (adapted with permission from ref. 39 (*J. Am. Chem. Soc.*, 2013, **135**, 2667, Copyright 2013 American Chemical Society).

receptors. Although these methods require ^{15}N and/or ^{13}C isotopically labelled protein, monitoring protein resonances are useful methodologies that provide information about receptor binding epitopes. Titration experiments with ^{15}N-labeled Gal-1, Gal-3 and Gal-7 and *p*-nitrophenyl lactopyranoside indicated that this lactose derivative interacts with the three galectins in the canonical binding site described for galactose, with subtle differences between three galectins.[42] Moreover, comparison of the changes produced by different ligands could serve to identify residues that are selectively affected and characterize binding disparities.

In a similar manner, Angulo *et al.* have performed a high-resolution NMR study to investigate the specificity of Langerin lectin when interacting with different heparin-like oligosaccharides.[43] The binding process was analyzed by trNOESY and STD experiments. Whereas for the larger polysaccharides, a Ca^{2+}-independent binding event was found, for the small sulfated trisaccharides only STD response was obtained in the presence of Ca^{2+} cations. The results clearly indicated that these charged sugars specifically interact with different binding sites in the protein, depending on the size of the ligand. Although the NMR ligand-based methods employed did not delineate the binding site in the protein, for the larger oligosaccharides, cross-linking experiments validated the existence of an interface binding site between Langerin monomers. In contrast, STD competition experiments in the presence of a reference ligand determined that the small sulfated oligosaccharides bind to the Ca^{2+}-specific binding site located at the CRD of the monomeric units composing langerin.

Receptor-based methods are the method of choice to delineate the key structural features of the lectin binding site. As example, the structural features of the molecular recognition processes involving the asialoglycoprotein C-type lectin (ASGP), responsible for Ca^{2+}-dependent binding and following liberation of galactose-derived glycoconjugates in endosomes have been explored. HSQC titration experiments monitored the changes in the backbone resonances as a function of Ca^{2+} concentration, revealing that three Ca^{2+} cations bind to ASGP in a cooperative way. These findings were in agreement with the easy release of glycoconjugates in the low pH endosomal compartments where slightly reduction of Ca^{2+} concentration takes place.[44] The pH-induced changes in the binding features of the related C-type lectin DC-SIGNR were also examined by means of 2D ^{1}H–^{15}N HSQC experiments. Measurement of the protein backbone resonances in the presence of Man- and Glc-derived ligands showed the depleted binding activity of this protein at lower pH.[45]

Similarly, 2D NMR methodologies have also been used to study the interaction of oligosaccharides of chitin with a CVNH protein, process related with the protection of the fungus *Magnaporte oryzae* from the host immune response in plants.[46] ^{1}H–^{15}N HSQC titrations revealed that the LysM domain of the protein interact with $(GlcNAc)_5$ and $(GlcNAc)_6$ in a dynamic manner, probed by the appearance of two different sets of resonances for several residues of the protein. Furthermore, the comparison with the backbone resonances perturbations produced by $(GlcNAc)_4$ with a known binding mode, allowed the characterization of the binding for both $(GlcNAc)_5$ and $(GlcNAc)_6$.

5 Oligosaccharide–protein interactions: antibodies

The most abundant post-translational differentiation mechanism in proteins is glycosylation.[47] Covalently linked glycans play key roles in almost all biological processes, from protein folding, stability and solubility to cell differentiation and proliferation.[48] Even more, glycosylation of protein antigens influences adaptive immune response, being the immune recognition of glycosylated peptides linked to autoimmunity.[49]

Guillain–Barré syndrome is an autoimmune disorder in which the peripheral nervous system is attacked. The human natural killer cell carbohydrate, HNK-1, has an important role in peripheral nerve regeneration and is the target of autoantibodies in polyneuropathies. Tsvetkov *et al.*[50] analysed the specificity of two different monoclonal HNK-1 antibodies, (HNK-1 antibody (IgMk) and HNK-1 412 antibody before known as L2 antibody (IgG2ak)) by SPR and STD-NMR. Although the 3-*O*-sulfated glucuronic acid is the key for the recognition process,[51,52] the NMR data[50] demonstrated that the monosaccharides of the attached *N*-acetyllactosamine sugar also provide interactions with both antibodies. The STD-NMR data also disclosed that the lactose fragment at the "reducing end" of the HNK-1 pentasaccharide remains outside of the binding site.

Lowary and coworkers[53] have described a new generation of the single chain variable fragment (scFv) of the CS-35 antibody, and several mutants. A combination of techniques including SPR, circular dichroism (CD) and STD-NMR, were used to characterize the binding of the terminal hexasaccharide motif (Ara6) in mycobacterial lipoarabinomannan (LAM) with the wild-type scFv and its mutants. The mutants were designed to manipulate the binding site and allowed to investigate the details of the binding for the Ara6 antigen. Although Ser31 is not in direct contact with Ara6, nevertheless its mutation to Ala inhibited the interaction. Thus, it was assumed that this residue is involved in the orientation of other residues that are essential for Ara6 binding. When three aromatic residues (Trp33, Tyr98, and Phe95) where changed to alanine residues, it was demonstrated that CH–π interactions are main contributors to build the affinity and specificity of Ara6–CS-35 binding. STD-NMR data showed that the His35 is the most important residue in the hydrogen-bond network. Although some hydrogen-bonding interactions help to strengthen the binding, the recognition event is tolerant to the manipulation of the hydrogen bonds, having such interactions a secondary role in the process.

The lipopolysaccharide (LPS) of gram-negative bacteria induces a strong response of the immune system. Marchetti *et al.*[54] have used a multidisciplinary approach (organic synthesis, NMR, molecular mechanics and MD simulations) to decipher the interaction of the LPS O-antigen of *Burkholderia anthina* with the 5D8 mAb. Two O-antigen molecules, with either trisaccharide or hexasaccharide repeating units, were synthesized and characterized. STD-NMR data highlighted that the galactose residues (Gal) in both molecules were in closer contacts to the 5D8 mAbs than the two rhamnose residues (Rha). Their conformations in the free and bound states were also evaluated. The data demonstrated that the trisaccharide revealed few changes between free and bound with the antibody. On the other hand, hexasaccharide recognition by 5D8 mAb, takes place through a conformational selection process wherein the bound-conformation is the most abundant in solution.[54] These informations are good starting-points for conceiving novel glycoconjugate–immunogens as vaccine candidates.

Antibodies are also essential elements in vaccine development to immunotherapy for HIV. Indeed, several broadly neutralizing monoclonal antibodies are actually available that bind glycans of the gp120 glycoprotein.[55] Pancera *et al.*[56] have studied the structural basis underlying the interaction of PG9 and PG16 antibodies with the HIV-1 V1–V2 region, using a combination of X-ray crystallography, NMR and mutagenesis. The deduced differences between PG16 and PG9 in the light-chain residues that interact with the sialic acid and the preceding galactose moieties ($Arg94_{LC}$ and $His95A_{LC}$ in PG9 compared to $Thr94_{LC}$ and $Arg95A_{LC}$ in PG16), allowed to create a PG9-16-RSH variant that have the capacity to neutralize HIV-1, with better breadth (PG9) and potency (PG16) than for either antibody alone.

Interestingly, the terminal sialic acid moiety of gp120 is important for the recognition by PG16 and PG9-16-RSH, while the absence of this unit does not affect the binding to PG9. Nevertheless, PG16 and PG9-16-RSH antibodies were able to neutralize HIV-1, thus suggesting that there are additional interactions with the V1–V2 residues. Although it had been previously described that PG9 and PG16 required $Man_5GlcNAc_2$ at position $Asn160_{gp120}$ to neutralize HIV-1,[57] the STD data[56] assessed that PG16 preferentially binds the complex-type *N*-glycan at $Asn173_{gp120}$, and binds to $Man_5GlcNAc_2$ at $Asn160_{gp120}$ with weaker affinity when compared to PG9.

6 Oligosaccharide–protein interactions: enzymes

Carbohydrate metabolism is performed by a great variety of enzymes including those that synthesize (glycosyltransferases) or break (glycosidases) the glycosidic linkage, but also others enzymes that modify functional groups and configurations in saccharide moieties (*e.g.* epimerases or mutases).

Calcium-dependent mannuronan C-5 epimerases are produced by *Azobacter vinelandii* have a key role in the epimerization of the carbohydrate-based acid, β-D-mannuronic acid, to α-L-guluronic acid in alginate polymers. This family of epimerases shows a modular structure composed of catalytic activating modules and reaction modules. Buchinger *et al.*[58] have determined the structure of three individual reaction modules applying standard NMR methodology on a $^{15}N/^{13}C$ labelled enzyme. In addition to the structural elucidation of the enzyme, the authors also characterized the aminoacids present in the active site of the enzyme by titration ^{1}H–^{15}N HSQC experiments. Similar NMR studies have been applied to the study of carbohydrate-binding-modules (CBMs) present in many glycosidases, facilitating substrate identification.[59] For instance, Grondin *et al.*[60] found an unusual mode of galacto-sugars binding by the CBM family 32, focused on CpGH 84A CBM32-1 as a relevant example from *Clostridium perfringens*.

Regarding the enzymatic reaction, NMR can provide real time information of the evolution of the substrates and the products in the enzymatic process.[61,62] In this context, Bunker *et al.*[63] proved the conversion of D-xylulose into xylulose 5-phosphate (Xu5P) by human xylulokinase and inferred the reaction kinetic parameters from the NMR data.[63]

Similarly, Pélissier *et al.*,[64] studying a set of selectively mutated enzymes, demonstrated the key role of Arg43, Asp126 and Lys66 of *N*-acetylmannosamine-6-phosphate 2-epimerase from *Clostridium perfringens* (CpNanE) following the conversion of ManNAc-6P into GlcNAc-6P by using real time ^{1}H-NMR.

New enzyme specificities have been shown employing NMR methodologies. Poulin *et al.*[65] applied STD-NMR experiments complemented with CORCEMA and molecular modelling calculations to characterize essential recognition features of the pyranose-furanose mutase from *Campylobacter jejuni* (cjUNGM), the first bi-functional mutase that recognizes two different substrates: UDP-Gal and UDP-GalNAc. Similar NMR strategies have been used by Bohlmann *et al.* to characterize a novel heparanase from the pathogenic bacteria *Burkholderia pseudomallei*.[66]

Over the last two decades several sugar-nucleotide-dependent biosynthetic enzymes have been characterized using discontinuous HPLC-based assays. Singh *et al.*[67] have shown that NMR is a powerful tool to characterize short live or unstable intermediates and can complement those studies overcoming HPLC drawbacks. Using fully ^{13}C labelled substrate and a set of ^{1}H–^{13}C HSQC spectra the authors were able to characterize four sugar-nucleotide dependent enzymes involved in the synthesis of calicheamicin: CalS3, CalS13, CalS1 and CalS2, as well as the stereospecificity of some of them.[67]

7 Polysaccharide interactions

Polysaccharides may also bind organic molecules. Different reports have addressed the high-field NMR study of the binding of amylopectin to amphiphilic probes, composed by a aliphatic tail and a 8-*O*-alkylated 1,3,6-trisulfonic acid trisodium salt (HPTS-alkyl).[68] This polysaccharide forms parallel double helices in the solid state, with left-turned at the non-reducing ends. In contrast, in aqueous medium with hydrophobic guests, linear α(1-4)-Glc polysaccharides display single left-turned helices. Using a chemical shift perturbation and DOSY experiments, the authors demonstrated the existence of binding to the external ramifications of amylopectin and to internal chains. NOESY experiments demonstrated the helical arrangement of the glucan chains around the aliphatic chain. Additional studies demonstrated that the minimal length of the glucan required for efficient probe binding corresponded to two helix turns, twelve residues and provided quantitative measurements of the binding affinities.[69]

The HPTS-alkyl probe has also been employed as supramolecular chemical shift reagent to analyse complex mixtures of homo-oligosaccharides (between 1–20 residues of α(1-4)-Glc saccharides).[70] These molecules tend to reorient themselves and to form helices around the alkyl chain. This fact, together with the ring current shifts, produces a clear dispersion of the NMR signals in HSQC experiments. The effect is specially evidenced in the NMR signals of the reducing ends of the different saccharides. The binding affinities could also be determined.

Similar experimental conditions have been employed to monitor the dynamic character of the interactions between (1-3/1-4)β-glucans and bile salts.[71]

8 Glycopeptides. The free state

It has been estimated that approximately 7000 pentasaccharide sequences exist in normal tissues, which represent minimal epitopes for binding to their receptors.[72] Consequently, gaining insights into the conformational impact of distal glycosylation on the peptide scaffold is of paramount importance. Nishimura *et al.* have provided evidence of the occurrence of site-specific structural alterations during multiple sialylation at Ser/Thr residues of human MUC1 tandem repeats. The analysis of $^{3}J\alpha N$ coupling constants and NOEs between the peptide core and the inner GalNAc residues revealed that the conformational impact of distal sialylation may greatly influence the secondary structure of the peptide. Consequently, the essential epitope structure observed in anti-KL-6 mAb, a clinically important disease probe, depends on both the peptide site-specific glycosylation and on the length and the chemical structure of the glycosidic motif.[73]

The full characterization of intact glycans attached to a glycoprotein has been so far almost elusive due to the required expensive and damaging methods. Typically glycans in proteins are indirectly analysed by using a combination of sample consuming mass spectrometry and enzymatic digestion or degradation of either glycan or peptide chains. However, even when results are successful, information regarding the stereochemistry or type of glycosidic linkage remains elusive. In fact, in MS-based glycomics, it is almost impossible to distinguish between the different hexoses, although novel avenues have been recently.[74] In any case, the stereochemistry and the linkage point are the reasons behind the immense variability of the sugar code. Recently, an elegant NMR-based method, for detection and analysis of posttranslational modifications of proteins, has been presented.[75] The authors used simple ^{1}H–^{13}C (at natural abundance) and ^{1}H–^{1}H correlation experiments under denaturing conditions to properly observe the glycan NMR signals. Since the signals arising from the saccharides close to the protein are often broadened beyond detection due to a fast transverse relaxation rate, T_2, the use of protein-denaturing conditions ensure the existence of only random coil structures for the polypeptide with the consequent increase in flexibility and faster tumbling motion. In this way, the fingerprint region of a variety of glycoproteins from different kingdoms was conformed. Obviously, information on novel glycosylation patterns could also be provided. Very recently, De Castro and coworkers have described an unprecedented *N*-glycosylation pattern constituted for a unique central core structure that is common to chloroviruses genus, while additional occurring decorations represent a molecular signature for each species.[76] The results open new questions about the differences in the glycosylation machinery in the three domains of life, the variety of glycosylation patterns in nature and how different organisms are able to process their own *N*-glycans.

The expression of glycoproteins from natural sources is normally poor and glycosylation is often inhomogeneous and incomplete.[77] In this context, the functional transfer of the glycosylation machinery of the prokaryote *Campylobacter jejuni* into *E. coli*[78] represented a major advance to structural biology. Using this method, a target glycoprotein can be uniformly $^{13}C/^{15}N$-labelled, enabling its NMR analysis through advanced filtering- and/or editing-NMR techniques. In fact, this method allowed showing that the bacterial *N*-linked glycan of *C. jejuni* forms a well-defined rod-like structure.[79] In contrast with the general assumption that glycans on the protein surface are flexible entities, without a well-defined three-dimensional architecture.

In a similar context, Unverzagt and coworkers have employed chemoselective ligation methods to achieve the semisynthesis of biologically active human cytokine interleukin-6 (IL-6) in two glycoforms, one decorated with a single monosaccharide unit and a second one with a biantennary nonasaccharide.[80] The activity of the two semisynthetic IL-6 glycoforms was found to be equal to the fully recombinant reference IL-6. This, together with many recent advances in chemical protein modification,[81] synthesis,[82] engineered polypeptide and protein glycosylation,[83] as well as the introduction of exogenous NMR active probe on inner or distal glycans on glycoprotein[84,85] are opening fascinating new perspectives on the application of NMR for understanding the molecular function and dynamics of peptide glycosylation.

9 Glycopeptide–antibody interactions

Carbohydrate-mediated interactions are also presents in diverse disease mechanisms, essentially in the immune system.[86] Carbohydrate antigens are valuable as structures for the design of vaccines or inhibitors against several events as well cancer, autoimmune diseases, bacterial and virus infections. Alterations in protein *N*- and *O*-glycosylation is regularly featured in malignant transformation.[87] These changes in the glycosylation pattern may represent an excellent starting point for the development of a carbohydrate-based cancer vaccines, educating our immune system in the generation of antibodies.[88] The Tn antigen (α-*O*-GalNAc-Ser/Thr) is one of the most specific human tumor-associated structures.[89] Glycosylation can alter the structure, even the function of the active peptide by steric and stereoelectronic influences or mediating interactions with glycan-binding proteins.

Knowledge about the recognition process of anti-carbohydrates antibodies is essential to a rational design of glycan-based vaccines. STD-NMR experiments have permitted to identify the binding epitope of the T-antigen to 1E8 ScFv, an anti-T-antigen antibody.[90,91] The galactose unit of T-antigen was primarily recognized. In contrast, Woods's group[92] identified that the anti-T-antigen antibody JAA-F11 mostly binds *via* the methyl group of GalNAc, followed by H2, H3 and H4 of Gal.

Corzana *et al.* used X-ray diffraction data, together with NMR experiments assisted with MD protocols to disclose the non-equivalence of Ser and Thr as *O*-glycosylation points, along with its implication on the bioactive conformation of the peptide.[93] In this work, the authors present a detailed analysis, at atomic resolution, of the interactions between SM3 mAb and two Ser and Thr-containing glycopeptides, pointing out the existence of subtle differences between them. Peregrina *et al.* have also defined the molecular recognition of several glycopeptides bearing the Tn antigen by three lectins, defining the epitope recognition and emphasizing the role of the linking aminoacid in the molecular recognition event.[94] These works demonstrate the privilege for NMR assisted by molecular modelling to uncover the conformation and recognition features of glycopeptides. Actually, the intrinsic flexibility of the carbohydrate antigens may hamper the detailed X-ray analysis,[95] while NMR spectroscopy has proved to be notably useful to obtain structural insights into the structure and dynamics of glycan–protein complexes.[96]

A recent study has combined STD-NMR and microarray technology to unravel the fine-epitope specificity of two cancer-related monoclonal antibodies (anti-MUC1 and anti-Tn) raised against the MUC1 tandem repeated domain. In the tumour-associate state, MUC1 is overexpressed and aberrantly glycosylated, mostly carrying shortened *O*-glycans, (*i.e.*, Tn, T, Sialyl-Tn or Sialyl-T), as well as a highly immunogenic PDTRP non-glycosylated peptide sequence.[97] The anti-MUC1 mAbs recognize naked MUC1-peptides (VU-3C6 and VU-11E2) and only recognize GalNAc when the sugar is in the peptide immunogenic sequence (PDTR). In contrast, the anti-Tn family (14D6 and 8D4) does not recognized naked MUC1 peptides and exclusively binds to the GalNAc moiety. Indeed, the Tn motif in the MUC1 sequence is required for binding. Actually, STD epitope mapping experiments demonstrated that the GalNAc residue is mainly involved in binding, while the peptide backbone is only marginally implicated in intermolecular contacts. Both anti-Tn mAbs showed preference for the Tn-Ser motif respect to the Tn-Thr alternative, demonstrating that the chemical nature of the amino acids plays a key role for anti-Tn antibodies recognition. The reason relies on the additional flexibility of Tn-Ser glycopeptides, which allows the proper complementary conformation in the bound state, without a major entropy penalty.[98]

10 Glycopeptide–lectin interactions

STD binding competition experiments have also been used to evaluate dissociation constants (K_d) and compare the affinity of different glycopeptides for a lectin. Moreover, direct determination of a specific K_d is also possible by means of STD methodologies.[99] Marcelo *et al.* applied these NMR methods to determine the binding affinity of MUC1-derived glycopeptides (GalNAc-α-1-*O*-Ser/Thr) to the macrophage galactose-type lectin (MGL).[100] STD-based experiments established that the sugar moiety plays a main role in the recognition and, further, they allowed the accurate determination of K_d. STD titration experiments based on time dependence of saturation build-up curves were applied to determine the affinity constant for a MGL–ligand pair and the K_d values for the remaining glycopeptides were determined by STD displacement experiments.

The favourable entropy contribution to the binding of biomedical-relevant glycopeptides to endogenous lectins has been discussed by Cudic *et al.* Binding of lactose and of the Thomsen–Friedenreich antigen to galectin-3 showed divergent shifts in the ^{1}H–^{15}N HSQC spectra, fact that pointed to differences in the binding for these two ligands.[101] Combining a protocol that included systematic glycopeptide synthesis, Isothermal Titration Calorimetry (ITC), NMR experiments, and cell-based assays, the authors characterized key thermodynamic, structural and cell biological parameters of different glycopeptides.[102] Specifically, they examined the thermodynamic balance associated to the binding event dissecting the enthalpy and the entropic contribution for four ligands with or without peptide scaffold and different chemical nature of the saccharide antigen. The most fascinating result is that, for peptide-free glycosides, the binding to human galectin-3 (hGal3) is enthalpically driven, as usually observed for glycan–receptor interactions. However, for the MUC1 glycopeptides, a shift from enthalpy-towards an entropy-driven process is observed. Strikingly, HSQC titrations of ^{15}N-hGal3 with the MUC1-glycopeptide ligands only revealed minor structural changes at the sugar binding site. From the ligand point of view, STD showed that only protons of the sugar and the directly linked Thr9 moieties were affected. The authors suggested that the release of water molecules to the bulk could be responsible of the observed entropy gain.

11 Glycomimetics: structural and conformational features

The design and employment of glycomimetics as therapeutic agents remains an attractive area of research. During the last few years, several publications and reviews continue appearing and illustrating the potential of glycomimetics in addressing the drawbacks of natural carbohydrates in terms of synthetic feasibility, molecular size, physiological stability, weakness in affinity or selectivity[103–105] and its comparison or combination with glycocluster effects.[106] NMR studies have contributed to provide detailed descriptions about the conformational preferences of these types of compounds, both free and bound to proteins naturally involved in sugar recognition processes.[107,108]

Two major types of sugar mimetics have been proposed, with the substitution of either the endocyclic or the exocyclic anomeric oxygen atom by another atom (carbon, sulfur, nitrogen).[109] A diversity of strategies for isosteric substitution may take place.[110] As a result, they might be differences in the presence of stereoelectronic effects, in the intramolecular hydrogen-bond patterns, in the amphiphilicity of the sugar ring or in the flexibility of the rings. These features can also influence the interactions of these compounds with the key residues at the binding sites of their possible receptors.

Whereas the replacement of the exocyclic anomeric oxygen atom by a carbon atom leads to *C*-glycosyl compounds, the substitution of the endocyclic one produces carbasugars. In both cases, the stereoelectronic stabilization coming from the anomeric effect is lost. Recently, NMR methods in combination with theoretical calculations have demonstrated that the replacement of the endocyclic oxygen atom by a CF_2 group induces the polarization of the C1–CF_2 bond, because of the high electronegativity of the fluorine atoms. This phenomenon allows restoring the *exo*-anomeric effect in a *gem*-difluoro carbadisaccharide, which displays to adopt the natural glycoside conformation.[111]

On the other hand, the conformational behaviour of diastereomeric sugar sulfoxides at the glycosidic linkage has been analyzed by NMR.[112] The conformational behaviour was shown to depend on the absolute configuration (*R*/*S*) of the stereogenic sulphur atom.

12 Glycomimetic interactions: lectins and enzymes

Among ligand-based methods trNOESY experiments provide helpful information to study carbohydrate–lectin interactions, especially to disclose conformational demands for the bound ligand. Small and large oligosaccharides usually exhibit high flexibility around glycosidic bonds and this intrinsic feature allows them to access to different conformations. In the last years, many efforts have been made to delineate the binding features of galectins,[113–115] involved in a wide range of biological functions as immune and inflammatory responses and tumour development, including cell adhesion, migration, and survival. Molecular recognition processes involving natural galactose-type ligands, as well as glycomimetics has been explored by means of NMR.[116,117] For many years, *C*-glycosyl compounds have been attractive mimetics for glycosides. Recently, Vidal *et al.*[118] have observed different conformational preferences in the free state for the lacto-*N*-biose mimic Galβ(1→3)GlcβOMe and its *C*-glycosyl analogue Galβ-*C*-(1→3)GlcβOMe. However, NMR data also indicated that both molecules are recognized by human galectin 1 (hGal-1) with a similar geometry, *syn*-Φ/*syn*-ψ. The transferred NOESY experiments assisted by docking and molecular dynamics simulations clearly showed the exclusive recognition of *syn*-type conformations of β-1,3/1,4-linked galactosides by hGal-1.[118] In this study, comparisons of the nOe signals from NOESY (free) and trNOESY (bound) were used to detect changes in the conformational preferences of the ligand during the binding process (Fig. 5).

In view of the context-dependent involvement of hGal-1 in disease, as in pancreas carcinoma progression or in glioblastoma invasion[119] these data represent a useful hint for the design of biopharmaceuticals.

The replacement of the interglycosidic oxygen by a sulphur atom leads to thio-oligosaccharides, which have been used as glycosidase inhibitors.[120] Calle *et al.* have addressed the conformational distortion in the molecular recognition process by β-galactosidases from *Escherichia coli* of

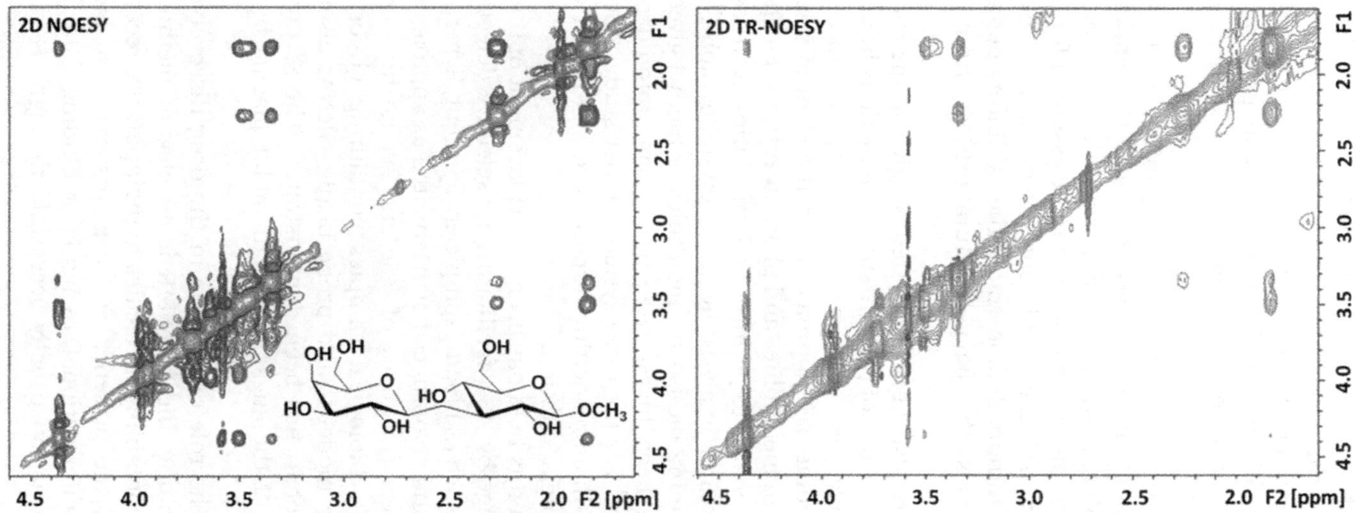

Fig. 5 NOESY and TR-NOESY experiments of a model *C*-glycosyl disaccharide in the absence (left) and presence (right) of human galectin-1. The binding event is pointed out by change of sign of the NOE cross peaks between both experiments. Moreover, changes of the ligand upon the addition of the protein are revealed by the disappearance of some NOE patterns as shown in the TR-NOESY experiment (adapted with permission from ref. 118 (*Chemistry*, 2013, **19**, 14581, Copyright © 2013 WILEY-VCH Verlag GmbH & Co).

two inhibitors based on aglyconic thiodisaccharides. The authors applied STD-NMR and transferred (TR)-NOESY experiments to obtain structural and conformational details of the recognition process. The experimental observations suggested that the two inhibitors display a different binding mode that explains the differences in their inhibition effect on the β-galactosidase. Moreover, the results showed that the bound conformations of the two ligands differed from those that are predominant in the free state showing that the enzyme prefers to bind to the higher energy conformations.[120]

Varela and coworkers used NMR experiments in association with molecular modelling techniques to assess the conformation of methyl β-D-galactofuranosyl-(1→5)-thiofuranosides in solution. The β-D-Galf-(1→5)-D-Galf motif is considered to be essential for the survival or virulence of many pathogenic microorganisms, but is absent in high eukaryotes.[121] The authors demonstrated the inhibitory potential of the thio analogues versus the β-D-galactofuranosidase from *Penicillium fellutanum.*

Seleno sugars has also been suggested as glycomimetics. In fact, a method for the direct study of protein–sugar interactions by using ligand-detected ^{77}Se NMR has been proposed. Selenium represents a component of the human proteome (selenoproteome) in the form of selenocysteine and it is also found in metabolites and pharmaceuticals. Furthermore, its physical properties enable to perform high resolution NMR experiments and accurate structural analysis by X-ray crystallography. The interactions of Se-containing bio-isoster molecules of methyl glycosides with model lectins have been investigated by regular proton-decoupled 1D ^{77}Se NMR spectra.[122] For the three systems under study, ^{77}Se NMR parameters, especially line-broadening, experienced significant changes in response to the protein addition.

A step forward in this field has been the replacement of the O atom in the glycosidic linkage by ^{77}Se, facilitating its selective detection even in complex biofluids, such as blood. Isotopic enrichment, hyperpolarization and use of broadband cryoprobes has been cited as approaches to offset the moderate nuclear sensitivity of ^{77}Se and get no interference from "matrix" signals in biological samples containing glycosylselenides. The binding of the model ligand probe bis(β-D-GlcNAc)diselenide to a GlcNAc-binding protein has been demonstrated. The Se tag permitted the fine detection of the sugar mimic in rabbit plasma by ^{77}Se NMR spectroscopy.[123,124]

In this context, the role of fluorine in the drug discovery process is continuously expanding and ^{19}F NMR is an elegant method to detect binding of the ^{19}F-containing molecules to their putative receptors.[125,126] Last but not least, the presence of ^{19}F atoms may also facilitate the provision of key conformational, dynamic or binding data that can be added to the information usually accessible through ^{1}H or ^{13}C NMR studies.[127,128]

Indeed, the presence of fluorine in carbohydrates allows setting up a fast and efficient screening method of compound mixtures for binding to a given receptor.[129]

Even more, polyfluorination of certain carbohydrate molecules has allowed to access to high-affinity binders. In this context, in addition to specificity and affinity levels, of particular interest is the ability of STD experiments to provide information on the ligand-binding epitope. In fact, STD-based experiments of a tetrafluorinated analogue of UDP-galactofuranose against UDP-galactopyranose mutase (UGM) have been reported. UGM is a therapeutically relevant carbohydrate-processing enzyme involved in the *Mycobacterium tuberculosis* cell wall biosynthesis. The data evidenced an enhancement of affinity for this glycomimetic to both the reduced and oxidized form of UGM, when compared with natural UDP-galactofuranose. This noticeable difference in binding properties is thought to arise from attractive interactions linked to the electronic properties of the innovative perfluorinated system, from a privileged conformation adopted by the furanose ring, from hydrophobic desolvation, or a combination of all these factors. Additionally, the introduction of fluorine atoms allows to extend the NMR methodological repertoire to ^{19}F-NMR.[130]

Development of strategies based on the incorporation of NMR sensitive sensors has improved the prospects of NMR to analyse the binding of complex glycans. For instance, recent examples highlighted the application of ^{19}F to detect molecular binding.[128] The use of difluoroacetamide moiety as a chemical tag to monitor interactions between GlcNAc derived carbohydrates and WGA has been recently reported. The introduction of this group allowed easy detection and characterization of the binding as well as the study of CH–π interactions in this system.[131]

Other sensor-based strategies applied lanthanide-chelating carbohydrates to study sugar–lectin interactions. As a proof of the potential of these methods the interaction between galectin-3 and a lactose mimic loaded with a paramagnetic ion has been recently explored.[132] Paramagnetic relaxation enhancements (PREs) along with pseudo contact shifts (PCSs) effects were effectively transferred from the carbohydrate to the lectin indicating the existence of interaction. PCSs for residues located 42 Å apart from the metal core could be observed, pointing to the ability of these methods to provide structural data about the binding event (Fig. 6).

Chemically simplified synthetic mimics of specific glycan moieties have been studied by NMR in their interactions with biomolecular receptors for screening compounds libraries.[133,134] For example, Ernst and collaborators[133] using a simplified model of sialic acid included in a mimic of sialyl Lewis^{x}, have applied a NMR-based Fragment Base Drug Discovery strategy for testing chemical fragment libraries. The introduction of nitroxide spin labels in the glycan mimic allowed the discovery of potential ligands at secondary binding sites in the E-selectin receptor. On the other side, Nativi and coworkers[135] have synthetized a tricyclic mimetic of the Tn antigen and its different selectivity against known Tn lectin receptors (ECL, MGL and HPA) has been characterized by STD-NMR.

A mixture of glycomimetics has been explored as anti-infective agents against the multidrug-resistant pathogen *Pseudomonas aeruginosa*. Thus, more than 20 substituted cinnamido D-mannose derivatives were

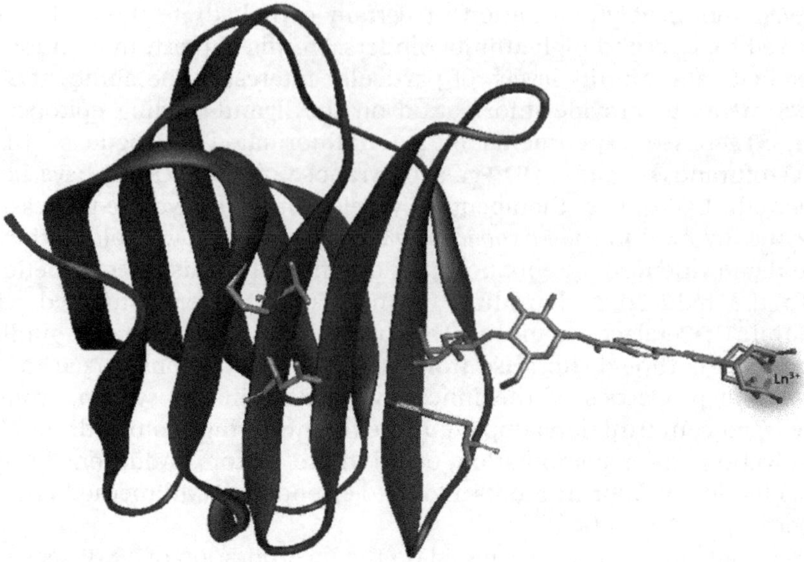

Fig. 6 Structure of human galectin-3 CRD in complex with a lactose moiety bearing a lanthanide binding tag. Significant chemical shift perturbations (PCS) in the protein (residues represented as sticks) can be detected upon ligand binding (modelled from galectin structure PDB 2NN8).

used to obtain structure–activity relationship (SAR) data of their interaction with LecB, one of the lectins involved in host damage and bacterial uptake processes promoted by *Pseudomonas aeruginosa*. The different behaviour of the ^{15}N labelled form of LecB signals detected through the $^{1}H,^{15}N$-TROSY-HSQC NMR experiments upon addition of the ligands were decisive to point out the interesting potential of some of those glycomimetics as inhibitors.[136]

The CRD of DC-SIGN plays a fundamental role in infection processes due to the interaction with glycoproteins present on the cell surface of some pathogens including HIV, SIV and hepatitis C. Many efforts have been made to develop inhibition strategies based on blocking the carbohydrate-binding site of this lectin. Thus, the interaction of different mannose saccharidic structures, including glycomimetics[137,138] and Lewis^{X} derivatives[139] has been studied. A combination of ligand- and protein-based NMR experiments has been applied to obtain a dynamic picture of the interaction in solution between a pseudo-1,2-mannobioside and the C-type lectin DC-SIGN, involved in gp120 interaction in the early stages of HIV infection. This work revealed that, albeit the glycomimetic shares shape and conformational properties with its natural counterpart Manα1–2Man, it exhibits a unique binding mode to the lectin, as opposed to the several binding orientations described for the natural ligand. This unique binding mode results in a high selectivity of this antagonist for DC-SIGN. Indeed, no binding is detected to Langerin. This is a mannose binding C-type lectin expressed on Langerhans cells, also present in the same mucosal environment as interstitial dendritic cells expressing DC-SIGN, but known for its protective action against HIV infection.

In conclusion, continuously reported studies clearly show that NMR techniques are of paramount importance to gain insight about solution structural features of ligand–protein complexes, important step toward the understanding of carbohydrate-receptor recognition processes, which finally are extremely correlated with biomolecular functions. Recently, X-ray crystallography, NMR spectroscopy and glycocluster assays have been fruitful combined to discern the structural requirements for the antiviral and mitogenicity activity of BanLec lectin.[140] Thus, NMR methods can be unambiguously viewed as crucial tools in the multidisciplinary strategy required for the progress in the field of chemical glycobiology.

13 Other interactions. Artificial receptors for carbohydrates

In the last decade, the increasing research about protein–carbohydrate interactions has also led to an arising interest in designing efficient and smaller organic structures, with the purpose of mimicking the binding site of lectins to get better selectivity in carbohydrate recognition events. Very different chemical architectures have been described in order to mimic the molecular recognition process carried out by the carbohydrate binding proteins in Nature. NMR techniques have been extensively employed to study the effectiveness and physicochemical features that make possible these processes, usually based on two kinds of interactions: polar interactions with OH groups and CH–π stacking interactions.

One of the most studied families of artificial receptors in recent years is that of tripodal receptors, consisting of a central or two central cores (aliphatic) to establish stacking interactions, bearing side arms to create polar contacts. Mazik *et al.* tested the interaction capability of a full set of tripodal receptors with trimethyl- and trimethoxybenzene centers, concluding that polar groups in the aromatic moiety can not help to form additional polar interactions and moreover decrease affinity due to intramolecular hydrogen bonds.[141] Additionally, pyrrole and pyrimidine units were proved to be the more suitable five- and six-membered polar rings for side arms,[142] and the same pattern was noticed in free substituents and linkers of macrocyclic receptors.[143] Roelens *et al.* evidenced that the simplest structure, consisting of a triethylbenzene unit with three pyrrolic-based branches, shows the best fit with mannosides: raising the length of side arms or the number of polar groups or arms reduces the affinity.[144] Other side structures, as trihydroxybenzamides, have also shown good recognition properties specific towards α(1-2)-branched mannans,[145] using NMR chemical shift perturbation and line broadening analysis, assisted by UV data and molecular modelling techniques (Fig. 7a).

The groups of Mazik[146] and Rieth[147] have also investigated other acyclic tetrapodal receptors with dimesitylmethane and biphenyl scaffolds, respectively, as central units. The first class of hosts bound disaccharides better than monosaccharides. Interestingly, Rieth described different host–guest ratios and geometries for the conjugates, as well as distinct selectivities for the β-*OctGlc* and β-*OctMan* anomers, when the concentration of the pyranoside was changed.

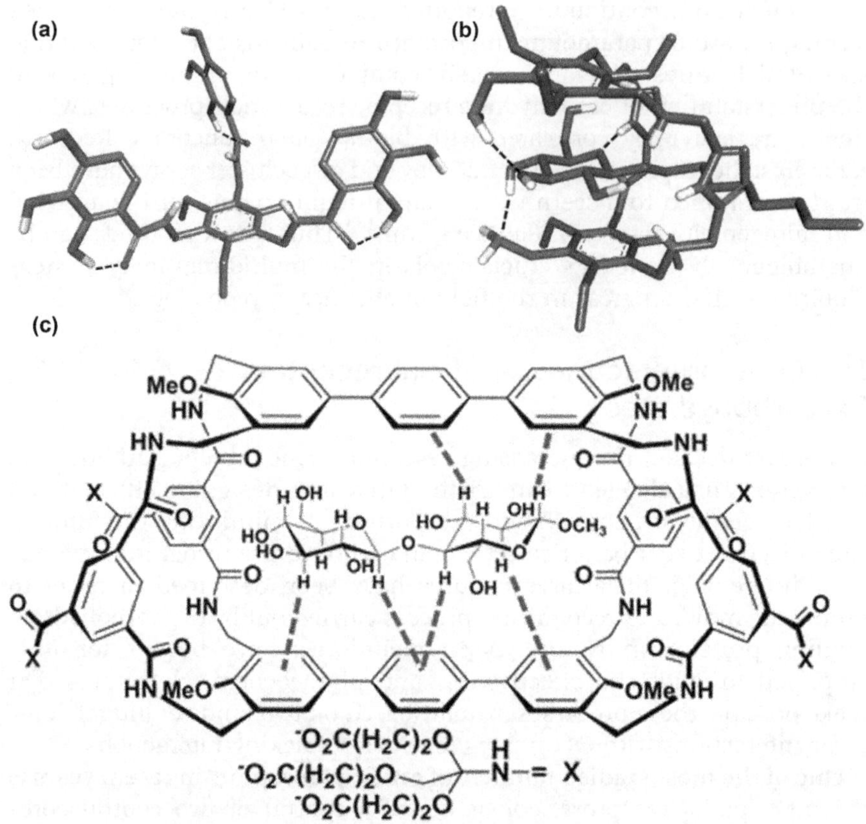

Fig. 7 Examples of lectin mimics where interactions based on hydrogen bonds and CH–π stacking are exploited to promote carbohydrate binding: (a) Intramolecular hydrogen bonds between OH groups and the amide carbonyl group confer stability and preorganize the structure of a tripodal receptor[145] (figure built with Discovery Studio 4.5 program). (b) Intermolecular hydrogen bonds between the amino groups of a tripodal receptor and the oxygen atoms from OctβGlc stabilize the complex[149] (figure built with Discovery Studio 4.5 program). (c) Schematic view of the CH–π stacking interactions between the hydrogen atoms from lactose and the aromatic rings from a synthetic lectin[153] (figure built with ChemDraw Professional 15.0 program).

NMR studies focused on these tripodal receptors have been also applied to study "expanded" structures with a second aromatic center opposite to the first. In bicyclic cages, steric effects and competitive solvents have been described as limitations for achieving proper binding properties. However, their affinity could be considerably improved by the use of small pyrrole-based linkers, instead of pyridine-based ones.[148] The systematic dissection of one aminopyrrolic arm within a bicyclic template allowed highlighting the key importance of amino groups on intermolecular hydrogen bonding, even more than the pyrrole nitrogen atom[149] (Fig. 7b). NMR studies on the effect of the receptor chirality in diverse structures with long and flexible linkers showed that, in all cases, the four "*S*" isomers had a marked preference for mono- and dimannosides.[150] Also, the influence of benzene-, pyridine- and pyrimidine-based branches in macrocyclic and acyclic scaffolds has been investigated.[151] In some cases, the preference of this kind of receptors for all-equatorial saccharides has been described.[151]

Another group of interesting and broadly studied receptors is that of macrocyclic oligoamides, reported by Davis and coworkers. In the last few years, improving of binding properties towards β-glycosides have been pursued through varying or modifying certain structural features.[152] With the purpose of assessing how flexibility can affect the behaviour of these compounds, the upper and lower aromatic bases were extended from two to three benzene units, finding a remarkable preference for disaccharides than for monosaccharides, but reducing the selectivity between the first ones. Nevertheless, very large binding constants, fairly close to those of natural lectins, were achieved[153] (Fig. 7c). More easily available structures, retaining the binding power but simplifying synthetic routes, have also been described.[154,155]

Finally, it is also worth mentioning the studies related to tune the selectivity and other physicochemical properties like solubility, through receptors bearing external dendrimeric systems: negative-charged chains can favour binding of amino sugars by additional electrostatic forces,[156] while steric and entropic factors can modulate inclusion of oligosaccharides depending on their number of residues.[157] Indeed a receptor for β(1-4)-linked Glc-polysaccharides has been proposed by the group of A. P. Davis.[157] The receptor consists on an aromatic cage that displays all equatorial carbohydrate selectivity (Glc) in aqueous medium, through stabilizing CH/π interactions. NMR data demonstrated that the linear glucan chains may pass through the cage resembling the thread through the needle. Interactions with large polysaccharides (cellulose) were demonstrated through AFM experiments.

A completely different family of chiral and artificial receptors has been developed by Schmitt *et al.* These so-called hemicryptophanes are large heteroditopic cages with a cyclotribenzylene moiety (CTB) as aromatic core. It is well known that besides chirality, different substituents in the linker and the CTB moiety can also be used to modify selectivity. For instance, P-type hemicryptophanes (*P-RRR-3 and P-SSS-3*) can distinguish the anomers of glucose, but M-type cages (*M-RRR-3 and M-SSS-3*) were noticed to prefer mannose derivatives.[158] Conformational changes in hemicryptophanes were accomplished by the presence of minor substituents, allowing the observation of significant changes in the association constants between anomers.[159]

Ouchi *et al.* have used NMR titrations to find out details about the recognition mechanism and the stoichiometry of the recognition process of Neu5Ac by fluorescent dyes like lanthanide-macrocyclic polyazacarboxylate complexes,[160] and boronic acid-functionalized squarylium cyanine structures.[161]

An uncharged and water-soluble macrocyclic host, curcubit[7]uril[162] showed effective binding to all kind of aminoglycosides in water, showing upgraded association constants in this competitive medium. Analyses of Nuclear Overhauser effects permitted to explain the larger stabilization of α anomers over the β ones, as well as the electrostatic contribution of the primary amine inside the ring formed by carbonyl groups.

Recently, Rauschenberg *et al.*[163] have investigated the sugar complexes formed by tripeptides with the Cys-X-Cys sequence, attached through disulfide bonds. Aromatic amino acids, mainly histidine and tyrosine, were placed in the "X" position to provide stacking interactions. The excessive flexibility prevented acquisition of good NOE data, but titrations were performed to clarify that a parallel arrangement between the two peptides increased the affinity. Ding and coworkers[164] have synthetized a polymer that displays a synergic effect of the two constituents in the binding process: the recognition unit, a dipeptide, and the phenylthiourea fragment. Indeed, retention of carbohydrates and glycopeptides at the polymer surface was achieved.

14 Cyclodextrins

Carbohydrates not only act as ligands, but they can also provide scaffolds for molecular recognition processes. It is well known that cyclodextrins (CDs) are able to form an inclusion complex with specific guest molecules. In the last years, NMR experiments combining with other techniques have been used to highlight different recognition events. In this context, one of the most recurrent challenges in drug discovery is to confer suitable pharmacokinetic properties to developed pro-drugs, so that these can reach their corresponding therapeutic targets. Complexation of these intermediates with appropriate scaffolds is usually an effective strategy to improve some properties, including solubility. In this context, cyclodextrins (CDs) have been exploited as efficient carriers for its ability to form inclusion complexes with mostly hydrophobic structures and solubilize them in aqueous media. NMR, often combining with other techniques, is a frequent tool used to study interaction mechanisms between these oligosaccharides and guest molecules.

In the last years, there are numerous examples in literature related to study of CD–guest interactions through NMR approaches. The typical changes in chemical shifts of guest and host when they are mixed in the same medium have been employed to prove whether a conjugate is formed between native or modified α- and β-CD and substrates like naproxen,[165] carbendazim[166] or astaxanthin;[167] and subsequently establish its stoichiometry. The complexation mode of curcumin[168] and a difluorinated derivative[169] has been completely described by analyzing changes in chemical shifts and combining analyses from 2D-ROESY spectra with computational modelling. Jahed *et al.* demonstrated that aromatic rings located in the two ends of curcumin are immersed into different β-CD units, so therefore a 2 : 1 host–guest complex is formed. In the same line, Floare *et al.*[170] evidenced that both aromatic rings of flufenamic and tolfenamic acids interact with β-CD inner protons through an inclusion mechanism, but variations in chemical shifts when modified β-CD concentration proved that two distinct 1 : 1 complexes can coexist in solution, instead of the 2 : 1 complexes initially proposed. In contrast, only one of the two aromatic moieties of fexofenadin[171] can be inserted inside the CD ring.

Recent studies have employed 2D-NOESY[172,173] and/or 2D-ROESY[174–176] data assisted by computational tools to build 3D-models of interaction.

CDs can bind chiral substrates with different binding constants, and this property has been exploited to study the isomers of ibuprofen,[177] lipoic acid[178] or imperan.[179] Akita *et al.*[180] found evidences that the indole ring of D- and L-tryptophan is deeply included inside the CD hollow, whereas the chiral moiety can bind to another CD unit with different strengths. Thus, they observed that the 1 : 1 to 2 : 1 equilibrium is shifted to the 2 : 1 state in the case of the D isomer. The structural properties of some substrates have led several researchers to describe unusual β-CD systems in which the guest is attached to the narrow entrance of the cyclodextrin.[181–183]

NMR measurements have been also employed to evaluate the strength of binding between a given guest and α-, β- or γ-CDs. DOSY methodology has been optimized recently to estimate binding constants from diffusion coefficients[173,184] with higher accuracy in comparison with other common methodologies as UV/Vis methods. Combining DOSY and ROESY experiments, the formation of a non-inclusion complex between riboflavin and 2-hydroxypropil-β-CD[185] was detected, probably by means of weak external CH–π interactions.

NMR titrations have been reported as a reliable analytical method to use CDs for controlling purity of omeprazole or rabaprazole[186] and removing the neurotoxin TETS from water.[187]

The interaction of different guests with structurally modified CDs has been also widely investigated by NMR. Marked modifications in the binding mechanism have been found for wogonin[188] and giberelic acid[189] when complexed by 2-hydroxypropil-β-CD in regard to native β-CD. Ikeda *et al.* showed that the binding properties of deuterated CDs are very similar to pristine CDs, so they can become powerful tools for further NMR studies.[190] Gu *et al.* tested different substituents at the 2 and 6-hydroxyl groups of β-CDs, in complexation with fluoroadamantanes (FA). Moreover, they were able to determine diffusion coefficients of CD–FA complexes and free CDs with great accuracy through the ^{19}F nucleus, and pointed out that, measuring ^{19}F T_1 and T_2 relaxation times, the diffusional behaviour of CDs can be controlled for performing direct experiments in biological media.[191] Other authors have investigated the effect in the complexation capability when the charge of guests or hosts is altered, with the aim of create additional stabilizing electrostatic forces. In this context, Sebestyén *et al.*[192] studied the behaviour of the systems upon titrations between aminoacids and alquilammonium-branched β-CDs at different pHs, whereas the Aachmann group[193] performed studies with native CDs (α-, β-, γ-) and a wide set of structurally modified phenylalanines, tyrosines and tryptophans, as well as small peptides.

Self-inclusion phenomena are also well known when CDs are linked to large and/or flexible substituents. Legros *et al.* have investigated the possible conformations of a CD dimer in which two β-CD units were attached by a symmetric spacer.[194] The NMR data allowed proposing that, besides a symmetric conformation, two more conformations can exist due to the existence of an "inversion mechanism" in which one of the β-CD cavities accommodate the proper linker.

The self-inclusion mechanism occurring in β-CDs conjugated with Gemini surfactant[195] has been described. The interaction of the hydrophobic surfactant structure with the inner part of β-CDs was studied to confirm the self-inclusion phenomenon and to find specific interactions. The interaction of this conjugate with a guest molecule (a curcumin derivative) was also evaluated. Fittingly, although this guest forms a 2 : 1 complex with conventional β-CDs, a 1 : 1 non-inclusion complex is present with the modified β-CD. The β-CD cavity remains occupied by the surfactant.

15 Carbohydrate–nucleic acid interactions

In the last years, NMR studies have been carried out to describe the conformation in aqueous media of a set of oligoamides that can act as DNA minor groove binders.[196] Titrations were performed to check differences in affinities of these ligands for DNA, and the corresponding intramolecular forces (CH–π and hydrogen bonds) were characterized by combining different NMR data (T coefficients, coupling constants, NOE and ROE cross-peaks) with computational modelling. As a result, it was discovered that, despite OH-2 groups from talose and mannose participate in cooperative hydrogen-bonding with intramolecular NH groups, oligoamides with these sugar moieties show slightly lower affinity for polynucleotides mainly composed by adenine–thymine pairs than those with D-Glc or D-Gal. Thus, it was suggested that axial OH-2 groups do not reach base-pair acceptors in the minor groove.

Within the broad and largely studied field of carbohydrate–aromatic interactions, a study has dissected the nature and the relevance of CH–π stacking interactions in formation of sugar–nucleic acid conjugates. Effects of structural modifications on CH–π interactions were investigated by bearing a carbohydrate unit to a single nucleotide.[197] Modifications essentially included the fluorination and methylation of the different hydroxyl groups in the sugar moiety, as well as the use of natural structural variations, as *N*-acetyl or carboxylic groups.

16 Concluding remarks

NMR methods continue providing access to detailed information of structural, conformational and dynamic features of saccharides and glycomimetics, as well as essential information on their molecular recognition processes. It is expected that the continuous advances in NMR-based methodologies will provide scientists with more robust possibilities to address those essential processes in which glycans are involved, with paramount importance in a variety of scientific fields.

Acknowledgements

We thank MINECO (Spain, Projects CTQ2012-32025 and CTQ2015-64597-P-C02-01 and CTQ2015-64597-P-C02-02) and the European Union (Projects DYNANO, GLYCOPHARM, and GLYCOVAX) for financial support.

References

1 D. Solis, N. V. Bovin, A. P. Davis, J. Jimenez-Barbero, A. Romero, R. Roy, K. Smetana, Jr. and H. J. Gabius, *Biochim. Biophys. Acta*, 2015, **1850**, 186.
2 D. Shishmarev and G. Otting, *J. Biomol. NMR*, 2013, **56**, 203.
3 K. Kato and T. Yamaguchi, *Glycoconjugate J.*, 2015, **32**, 505.
4 M. Erdélyi, E. d'Auvergne, A. Navarro-Vázquez, A. Leonov and C. Griesinger, *Chemistry*, 2011, **17**, 9368.
5 S. Yamamoto, T. Yamaguchi, M. Erdelyi, C. Griesinger and K. Kato, *Chemistry*, 2011, **17**, 9280.
6 A. Canales, A. Mallagaray, J. Pérez-Castells, I. Boos, C. Unverzagt, S. André, H.-J. Gabius, F. J. Cañada and J. Jiménez-Barbero, *Angew. Chem., Int. Ed.*, 2013, **52**, 13789.
7 S. Yamamoto, Y. Zhang, T. Yamaguchi, T. Kameda and K. Kato, *Chem. Commun.*, 2012, **48**, 4752.
8 Y. Zhang, S. Yamamoto, T. Yamaguchi and K. Kato, *Molecules*, 2012, **17**, 6658.
9 Y. Zhang, T. Yamaguchi, T. Satoh, M. Yagi-Utsumi, Y. Kamiya, Y. Sakae, Y. Okamoto and K. Kato, *Adv. Exp. Med. Biol.*, 2015, **842**, 217.
10 T. Yamaguchi, Y. Sakae, Y. Zhang, S. Yamamoto, Y. Okamoto and K. Kato, *Angew. Chem., Int. Ed.*, 2014, **53**, 10941.
11 Y. Kamiya, K. Yanagi, T. Kitajima, T. Yamaguchi, Y. Chiba and K. Kato, *Biomolecules*, 2013, **3**, 108.
12 T. Yamaguchi, Y. Kamiya, Y.-M. Choo, S. Yamamoto and K. Kato, *Chem. Lett.*, 2013, **42**, 544.
13 J. Correa, L. F. Pinto, R. Riguera and E. Fernandez-Megia, *Anal. Chem.*, 2015, **87**, 760.
14 M. Zierke, M. Smieško, S. Rabbani, T. Aeschbacher, B. Cutting, F. H. T. Allain, M. Schubert and B. Ernst, *J. Am. Chem. Soc.*, 2013, **135**, 13464.
15 M. D. Battistel, H. F. Azurmendi, M. Frank and D. I. Freedberg, *J. Am. Chem. Soc.*, 2015, **137**, 13444.
16 C. G. Panagos, D. S. Thomson, C. Moss, A. D. Hughes, M. S. Kelly, Y. Liu, W. Chai, R. Venkatasamy, D. Spina, C. P. Page, J. Hogwood, R. J. Woods, B. Mulloy, C. D. Bavington and D. Uhrin, *J. Biol. Chem.*, 2014, **289**, 28284.
17 M. D. Battistel, M. Shangold, L. Trinh, J. Shiloach and D. I. Freedberg, *J. Am. Chem. Soc.*, 2012, **134**, 10717.
18 J. Xia and D. A. Case, *Biopolymers*, 2012, **97**, 276.
19 J. Xia and D. A. Case, *Biopolymers*, 2012, **97**, 289.
20 M. D. Battistel, R. Pendrill, G. Widmalm and D. I. Freedberg, *J. Phys. Chem. B*, 2013, **117**, 4860.
21 D. J. Langeslay, R. P. Young, S. Beni, C. N. Beecher, L. J. Mueller and C. K. Larive, *Glycobiology*, 2012, **22**, 1173.
22 C. N. Beecher, R. P. Young, D. J. Langeslay, L. J. Mueller and C. K. Larive, *J. Phys. Chem. B*, 2014, **118**, 482.
23 C. Fontana, H. Kovacs and G. Widmalm, *J. Biomol. NMR*, 2014, **59**, 95.
24 R. Pendrill, O. W. Sorensen and G. Widmalm, *Magn. Reson. Chem.*, 2014, **52**, 82.
25 K. H. Jonsson, E. Sawen and G. Widmalm, *Org. Biomol. Chem.*, 2012, **10**, 2453.
26 D. Fujinami, J. Nyirenda, S. Matsumoto and D. Kohda, *Carbohydr. Res.*, 2015, **413**, 55.
27 D. S. Patel, R. Pendrill, S. S. Mallajosyula, G. Widmalm and A. D. MacKerell, Jr., *J. Phys. Chem. B*, 2014, **118**, 2851.

28 T. Angles d'Ortoli, N. A. Sjoberg, P. Vasiljeva, J. Lindman, G. Widmalm, M. Bergenstrahle-Wohlert and J. Wohlert, *J. Phys. Chem. B*, 2015, **119**, 9559.
29 M. Frank, P. M. Collins, I. R. Peak, I. D. Grice and J. C. Wilson, *Molecules*, 2015, **20**, 14234.
30 A. Santra, A. Si, R. K. Kar, A. Bhunia and A. K. Misra, *Carbohydr. Res.*, 2014, **391**, 9.
31 T. A. Jackson, V. Robertson and F. I. Auzanneau, *J. Med. Chem.*, 2014, **57**, 817.
32 M. del Carmen Fernandez-Alonso, D. Diaz, M. A. Berbis, F. Marcelo, J. Canada and J. Jimenez-Barbero, *Curr. Protein Pept. Sci.*, 2012, **13**, 816.
33 C. A. Bewley and S. Shahzad-ul-Hussan, *Biopolymers*, 2013, **99**, 796.
34 A. Arda, A. Canales, F. J. Cañada and J. Jimenez-Barbero, Carbohydrate-Protein Interactions: A 3D View by NMR, in *Carbohydrates in Drug Design and Discovery*, The Royal Society of Chemistry, 2015, ch. 1, p. 1.
35 S. André, H. Kaltner, J. Manning, P. Murphy and H.-J. Gabius, *Molecules*, 2015, **20**, 1788.
36 A. Antanasijevic, B. Ramirez and M. Caffrey, *J. Biomol. NMR*, 2014, **60**, 37.
37 L. Unione, S. Galante, D. Diaz, F. J. Canada and J. Jimenez-Barbero, *MedChemComm*, 2014, **5**, 1280.
38 M. Nagae, K. Soga, K. Morita-Matsumoto, S. Hanashima, A. Ikeda, K. Yamamoto and Y. Yamaguchi, *Glycobiology*, 2014, **24**, 368.
39 A. Arda, P. Blasco, D. Varon Silva, V. Schubert, S. Andre, M. Bruix, F. J. Canada, H. J. Gabius, C. Unverzagt and J. Jimenez-Barbero, *J. Am. Chem. Soc.*, 2013, **135**, 2667.
40 B. S. Blaum, J. P. Hannan, A. P. Herbert, D. Kavanagh, D. Uhrin and T. Stehle, *Nat. Chem. Biol.*, 2015, **11**, 77.
41 C. McCullough, M. Wang, L. Rong and M. Caffrey, *PLos One*, 2012, **7**, e33958.
42 S. R. Rauthu, T. C. Shiao, S. André, M. C. Miller, É. Madej, K. H. Mayo, H.-J. Gabius and R. Roy, *ChemBioChem*, 2015, **16**, 126.
43 J. C. Munoz-Garcia, E. Chabrol, R. R. Vives, A. Thomas, J. L. de Paz, J. Rojo, A. Imberty, F. Fieschi, P. M. Nieto and J. Angulo, *J. Am. Chem. Soc.*, 2015, **137**, 4100.
44 T. Onizuka, H. Shimizu, Y. Moriwaki, T. Nakano, S. Kanai, I. Shimada and H. Takahashi, *FEBS J.*, 2012, **279**, 2645.
45 F. Probert, D. A. Mitchell and A. M. Dixon, *FEBS J.*, 2014, **281**, 3739.
46 L. M. Koharudin, K. T. Debiec and A. M. Gronenborn, *Structure*, 2015, **23**, 2143.
47 J. Breitling and M. Aebi, *Cold Spring Harbor Perspect. Biol.*, 2013, **5**, a013359.
48 A. Varki, *Glycobiology*, 1993, **3**, 97.
49 M. A. Wolfert and G.-J. Boons, *Nat. Chem. Biol.*, 2013, **9**, 776.
50 Y. E. Tsvetkov, M. Burg-Roderfeld, G. Loers, A. Arda, E. V. Sukhova, E. A. Khatuntseva, A. A. Grachev, A. O. Chizhov, H. C. Siebert, M. Schachner, J. Jimenez-Barbero and N. E. Nifantiev, *J. Am. Chem. Soc.*, 2012, **134**, 426.
51 B. Schmitz, M. Schachner, Y. Ito, T. Nakano and T. Ogawa, *Glycoconjugate J.*, 1994, **11**, 345.
52 A. Tokuda, T. Ariga, Y. Isogai, S. Komba, M. Kiso, A. Hasegawa, T. Tai and R. K. Yu, *J. Carbohydr. Chem.*, 1998, **17**, 535.
53 P. Lak, S. Makeneni, R. J. Woods and T. L. Lowary, *Chemistry*, 2015, **21**, 1138.
54 R. Marchetti, A. Canales, R. Lanzetta, I. Nilsson, C. Vogel, D. E. Reed, D. P. Aucoin, J. Jimenez-Barbero, A. Molinaro and A. Silipo, *ChemBioChem*, 2013, **14**, 1485.
55 R. Pantophlet and D. R. Burton, *Annu. Rev. Immunol.*, 2006, **24**, 739.

56 M. Pancera, S. Shahzad-Ul-Hussan, N. A. Doria-Rose, J. S. McLellan, R. T. Bailer, K. Dai, S. Loesgen, M. K. Louder, R. P. Staupe, Y. Yang, B. Zhang, R. Parks, J. Eudailey, K. E. Lloyd, J. Blinn, S. M. Alam, B. F. Haynes, M. N. Amin, L. X. Wang, D. R. Burton, W. C. Koff, G. J. Nabel, J. R. Mascola, C. A. Bewley and P. D. Kwong, *Nat. Struct. Mol. Biol.*, 2013, **20**, 804.
57 L. M. Walker, S. K. Phogat, P. Y. Chan-Hui, D. Wagner, P. Phung, J. L. Goss, T. Wrin, M. D. Simek, S. Fling, J. L. Mitcham, J. K. Lehrman, F. H. Priddy, O. A. Olsen, S. M. Frey, P. W. Hammond, G. P. I. Protocol, S. Kaminsky, T. Zamb, M. Moyle, W. C. Koff, P. Poignard and D. R. Burton, *Science*, 2009, **326**, 285.
58 E. Buchinger, D. H. Knudsen, M. A. Behrens, J. S. Pedersen, O. A. Aarstad, A. Tondervik, S. Valla, G. Skjak-Braek, R. Winnmer and F. L. Aachmann, *J. Biol. Chem.*, 2014, **289**, 31382.
59 V. Lombard, H. Golaconda Ramulu, E. Drula, P. M. Coutinho and B. Henrissat, *Nucleic Acids Res.*, 2014, **42**, D490.
60 J. M. Grondin, S. Chitayat, E. Ficko-Blean, S. Houliston, C. H. Arrowsmith, A. B. Boraston and S. P. Smith, *J. Mol. Biol.*, 2014, **426**, 869.
61 J. Marín-Navarro, D. Talens-Perales, A. Oude-Vrielink, F. J. Cañada and J. Polaina, *World J. Microbiol. Biotechnol.*, 2014, **30**, 989.
62 M. Nieto-Domínguez, L. I. de Eugenio, J. Barriuso, A. Prieto, B. Fernández de Toro, Á. Canales-Mayordomo and M. J. Martínez, *Appl. Environ. Microbiol.*, 2015, **81**, 6380.
63 R. D. Bunker, E. M. M. Bulloch, J. M. J. Dickson, K. M. Loomes and E. N. Baker, *J. Biol. Chem.*, 2013, **288**, 1643.
64 M.-C. Pélissier, C. Sebban-Kreuzer, F. Guerlesquin, J. A. Brannigan, Y. Bourne and F. Vincent, *J. Biol. Chem.*, 2014, **289**, 35215.
65 M. B. Poulin, Y. Shi, C. Protsko, S. A. Dalrymple, D. A. R. Sanders, B. M. Pinto and T. L. Lowary, *ChemBioChem*, 2014, **15**, 47.
66 L. Bohlmann, G. D. Tredwell, X. Yu, C.-W. Chang, T. Haselhorst, M. Winger, J. C. Dyason, R. J. Thomson, J. Tiralongo, I. R. Beacham, H. Blanchard and M. von Itzstein, *Nat. Chem. Biol.*, 2015, **11**, 955.
67 S. Singh, P. Peltier-Pain, M. Tonelli and J. S. Thorson, *Org. Lett.*, 2014, **16**, 3220.
68 S. R. Beeren, S. Meier and O. Hindsgaul, *Chemistry*, 2013, **19**, 16314.
69 S. R. Beeren and O. Hindsgaul, *Angew. Chem., Int. Ed.*, 2013, **52**, 11265.
70 S. R. Beeren and S. Meier, *Chem. Commun.*, 2015, **51**, 3073.
71 M. S. Mikkelsen, S. B. Cornali, M. G. Jensen, M. Nilsson, S. R. Beeren and S. Meier, *J. Agric. Food Chem.*, 2014, **62**, 11472.
72 R. D. Cummings, *Mol. Biosyst.*, 2009, **5**, 1087.
73 T. Matsushita, N. Ohyabu, N. Fujitani, K. Naruchi, H. Shimizu, H. Hinou and S.-I. Nishimura, *Biochemistry*, 2013, **52**, 402.
74 J. Hofmann, H. S. Hahm, P. H. Seeberger and K. Pagel, *Nature*, 2015, **526**, 241.
75 M. Schubert, M. J. Walczak, M. Aebi and G. Wider, *Angew. Chem., Int. Ed.*, 2015, **54**, 7096.
76 C. De Castro, I. Speciale, G. Duncan, D. D. Dunigan, I. Agarkova, R. Lanzetta, L. Sturiale, A. Palmigiano, D. Garozzo, A. Molinaro, M. Tonetti and J. L. Van Etten, *Angew. Chem., Int. Ed.*, 2016, **55**, 654.
77 B. Meyer, H. Möller, Conformation of Glycopeptides and Glycoproteins, in *Glycopeptides and Glycoproteins*, ed. V. Wittmann, Springer, Berlin, Heidelberg, 2007, vol. 267, p. 187.
78 M. Wacker, D. Linton, P. G. Hitchen, M. Nita-Lazar, S. M. Haslam, S. J. North, M. Panico, H. R. Morris, A. Dell, B. W. Wren and M. Aebi, *Science*, 2002, **298**, 1790.

79 V. Slynko, M. Schubert, S. Numao, M. Kowarik, M. Aebi and F. H. T. Allain, *J. Am. Chem. Soc.*, 2009, **131**, 1274.
80 A. Reif, S. Siebenhaar, A. Tröster, M. Schmälzlein, C. Lechner, P. Velisetty, K. Gottwald, C. Pöhner, I. Boos, V. Schubert, S. Rose-John and C. Unverzagt, *Angew. Chem., Int. Ed.*, 2014, **53**, 12125.
81 O. Boutureira and G. J. L. Bernardes, *Chem. Rev.*, 2015, **115**, 2174.
82 A. Fernández-Tejada, J. Brailsford, Q. Zhang, J.-H. Shieh, M. S. Moore and S. Danishefsky, Total Synthesis of Glycosylated Proteins, in *Protein Ligation and Total Synthesis I*, ed. L. Liu, Springer International Publishing, 2015, vol. 362, p. 1.
83 F. Liu, B. Vijayakrishnan, A. Faridmoayer, T. A. Taylor, T. B. Parsons, G. J. L. Bernardes, M. Kowarik and B. G. Davis, *J. Am. Chem. Soc.*, 2014, **136**, 566.
84 O. Boutureira, F. D'Hooge, M. Fernandez-Gonzalez, G. J. L. Bernardes, M. Sanchez-Navarro, J. R. Koeppe and B. G. Davis, *Chem. Commun.*, 2010, **46**, 8142.
85 M. Salvadó, B. Amgarten, S. Castillón, G. J. L. Bernardes and O. Boutureira, *Org. Lett.*, 2015, **17**, 2836.
86 F. Y. Avci and D. L. Kasper, *Annu. Rev. Immunol.*, 2010, **28**, 107.
87 M. N. Christiansen, J. Chik, L. Lee, M. Anugraham, J. L. Abrahams and N. H. Packer, *Proteomics*, 2014, **14**, 525.
88 D. H. Dube, J. A. Prescher, C. N. Quang and C. R. Bertozzi, *Proc. Natl. Acad. Sci. U. S. A.*, 2006, **103**, 4819.
89 T. Ju, V. I. Otto and R. D. Cummings, *Angew. Chem., Int. Ed.*, 2011, **50**, 1770.
90 A. Matsumoto-Takasaki, J. Horie, K. Sakai, Y. Furui, R. Sato, H. Kawakami, K. Toma, A. Takayanagi, N. Shimizu and Y. Fujita-Yamaguchi, *BioSci. Trends*, 2009, **3**, 87.
91 N. Yuasa, T. Koyama, G. P. Subedi, Y. Yamaguchi, M. Matsushita and Y. Fujita-Yamaguchi, *J. Biochem.*, 2013, **154**, 521.
92 M. B. Tessier, O. C. Grant, J. Heimburg-Molinaro, D. Smith, S. Jadey, A. M. Gulick, J. Glushka, S. L. Deutscher, K. Rittenhouse-Olson and R. J. Woods, *PLos One*, 2013, **8**, e54874.
93 N. Martínez-Sáez, J. Castro-López, J. Valero-González, D. Madariaga, I. Compañón, V. J. Somovilla, M. Salvadó, J. L. Asensio, J. Jiménez-Barbero, A. Avenoza, J. H. Busto, G. J. L. Bernardes, J. M. Peregrina, R. Hurtado-Guerrero and F. Corzana, *Angew. Chem., Int. Ed.*, 2015, **54**, 9830.
94 D. Madariaga, N. Martínez-Sáez, V. J. Somovilla, L. García-García, M. Á. Berbis, J. Valero-Gónzalez, S. Martín-Santamaría, R. Hurtado-Guerrero, J. L. Asensio, J. Jiménez-Barbero, A. Avenoza, J. H. Busto, F. Corzana and J. M. Peregrina, *Chemistry*, 2014, **20**, 12616.
95 J. Agirre, J. Iglesias-Fernandez, C. Rovira, G. J. Davies, K. S. Wilson and K. D. Cowtan, *Nat. Struct. Mol. Biol.*, 2015, **22**, 833.
96 V. Roldós, F. J. Cañada and J. Jiménez-Barbero, *ChemBioChem*, 2011, **12**, 990.
97 M. Amano, H. Eriksson, J. C. Manning, K. M. Detjen, S. Andre, S. Nishimura, J. Lehtio and H. J. Gabius, *FEBS J.*, 2012, **279**, 4062.
98 H. Coelho, T. Matsushita, G. Artigas, H. Hinou, F. J. Canada, R. Lo-Man, C. Leclerc, E. J. Cabrita, J. Jimenez-Barbero, S. Nishimura, F. Garcia-Martin and F. Marcelo, *J. Am. Chem. Soc.*, 2015, **137**, 12438.
99 J. Angulo, P. M. Enriquez-Navas and P. M. Nieto, *Chemistry*, 2010, **16**, 7803.
100 F. Marcelo, F. Garcia-Martin, T. Matsushita, J. Sardinha, H. Coelho, A. Oude-Vrielink, C. Koller, S. Andre, E. J. Cabrita, H. J. Gabius, S. Nishimura, J. Jimenez-Barbero and F. J. Canada, *Chemistry*, 2014, **20**, 16147.

101 A. B. Yongye, L. Calle, A. Arda, J. Jimenez-Barbero, S. Andre, H. J. Gabius, K. Martinez-Mayorga and M. Cudic, *Biochemistry*, 2012, **51**, 7278.
102 M. C. Rodriguez, S. Yegorova, J. P. Pitteloud, A. E. Chavaroche, S. Andre, A. Arda, D. Minond, J. Jimenez-Barbero, H. J. Gabius and M. Cudic, *Biochemistry*, 2015, **54**, 4462.
103 E. Leclerc, X. Pannecoucke, M. Etheve-Quelquejeu and M. Sollogoub, *Chem. Soc. Rev.*, 2013, **42**, 4270.
104 R. Lahiri, A. A. Ansari and Y. D. Vankar, *Chem. Soc. Rev.*, 2013, **42**, 5102.
105 J. Magnani, Glycomimetic Drugs: A Source of Novel Therapeutics, in *Glycoscience: Biology and Medicine*, ed. T. Endo, P. H. Seeberger, G. W. Hart, C.-H. Wong and N. Taniguchi, Springer, Japan, 2014, p. 1.
106 S. Cecioni, A. Imberty and S. Vidal, *Chem. Rev.*, 2015, **115**, 525.
107 M. del Carmen Fernandez-Alonso, D. Diaz, M. Alvaro Berbis, F. Marcelo, J. Canada and J. Jimenez-Barbero, *Curr. Protein Pept. Sci.*, 2012, **13**, 816.
108 E. M. Sanchez-Fernandez, R. Risquez-Cuadro, C. Ortiz Mellet, J. M. Garcia Fernandez, P. M. Nieto and J. Angulo, *Chemistry*, 2012, **18**, 8527.
109 V. H. Lillelund, H. H. Jensen, X. Liang and M. Bols, *Chem. Rev.*, 2002, **102**, 515.
110 B. Ernst and J. L. Magnani, *Nat. Rev. Drug Discovery*, 2009, **8**, 661.
111 B. Xu, L. Unione, J. Sardinha, S. Wu, M. Ethève-Quelquejeu, A. Pilar Rauter, Y. Blériot, Y. Zhang, S. Martín-Santamaría, D. Díaz, J. Jiménez-Barbero and M. Sollogoub, *Angew. Chem., Int. Ed.*, 2014, **53**, 9597.
112 J. P. Colomer, M. Á. Canales Mayordomo, B. Fernández de Toro, J. Jiménez-Barbero and O. Varela, *Eur. J. Org. Chem.*, 2015, **2015**, 1448.
113 C. H. Liu, C. T. Chien, C. H. Lin and S. T. Hsu, *Biomol. NMR Assignments*, 2015, **9**, 427.
114 H. Ippel, M. C. Miller, M. A. Berbis, D. Suylen, S. Andre, T. M. Hackeng, F. J. Cañada, C. Weber, H. J. Gabius, J. Jimenez-Barbero and K. H. Mayo, *Biomol. NMR Assignments*, 2015, **9**, 59.
115 M. A. Berbis, S. Andre, F. J. Cañada, R. Pipkorn, H. Ippel, K. H. Mayo, D. Kubler, H. J. Gabius and J. Jimenez-Barbero, *Biochem. Biophys. Res. Commun.*, 2014, **443**, 126.
116 K. Saraboji, M. Hakansson, S. Genheden, C. Diehl, J. Qvist, U. Weininger, U. J. Nilsson, H. Leffler, U. Ryde, M. Akke and D. T. Logan, *Biochemistry*, 2012, **51**, 296.
117 E. Ermakova, M. C. Miller, I. V. Nesmelova, L. Lopez-Merino, M. A. Berbis, Y. Nesmelov, Y. V. Tkachev, L. Lagartera, V. A. Daragan, S. Andre, F. J. Canada, J. Jimenez-Barbero, D. Solis, H. J. Gabius and K. H. Mayo, *Glycobiology*, 2013, **23**, 508.
118 P. Vidal, V. Roldos, C. Fernandez-Alonso Mdel, B. Vauzeilles, Y. Bleriot, F. J. Canada, S. Andre, H. J. Gabius, J. Jimenez-Barbero, J. F. Espinosa and S. Martin-Santamaria, *Chemistry*, 2013, **19**, 14581.
119 L. Astorgues-Xerri, M. E. Riveiro, A. Tijeras-Raballand, M. Serova, C. Neuzillet, S. Albert, E. Raymond and S. Faivre, *Cancer Treat. Rev.*, 2014, **40**, 307.
120 L. Calle, V. Roldós, F. J. Cañada, M. L. Uhrig, A. J. Cagnoni, V. E. Manzano, O. Varela and J. Jiménez-Barbero, *Chemistry*, 2013, **19**, 4262.
121 M. J. Lo Fiego, C. Marino and O. Varela, *RSC Adv.*, 2015, **5**, 45631.
122 C. Hamark, J. Landström and G. Widmalm, *Chemistry*, 2014, **20**, 13905.
123 O. Boutureira, G. J. L. Bernardes, M. Fernández-González, D. C. Anthony and B. G. Davis, *Angew. Chem., Int. Ed.*, 2012, **51**, 1432.
124 I. Perez-Victoria, O. Boutureira, T. D. Claridge and B. G. Davis, *Chem. Commun.*, 2015, **51**, 12208.

125 E. P. Gillis, K. J. Eastman, M. D. Hill, D. J. Donnelly and N. A. Meanwell, *J. Med. Chem.*, 2015, **58**, 8315.
126 A. Vulpetti and C. Dalvit, *Drug Discovery Today*, 2012, **17**, 890.
127 A. Vulpetti, U. Hommel, G. Landrum, R. Lewis and C. Dalvit, *J. Am. Chem. Soc.*, 2009, **131**, 12949.
128 E. Matei, S. Andre, A. Glinschert, A. S. Infantino, S. Oscarson, H. J. Gabius and A. M. Gronenborn, *Chemistry*, 2013, **19**, 5364.
129 J. P. Ribeiro, T. Diercks, J. Jimenez-Barbero, S. Andre, H. J. Gabius and F. J. Cañada, *Biomolecules*, 2015, **5**, 3177.
130 I. N'Go, S. Golten, A. Arda, J. Canada, J. Jimenez-Barbero, B. Linclau and S. P. Vincent, *Chemistry*, 2014, **20**, 106.
131 L. P. Calle, B. Echeverria, A. Franconetti, S. Serna, M. C. Fernandez-Alonso, T. Diercks, F. J. Cañada, A. Arda, N. C. Reichardt and J. Jimenez-Barbero, *Chemistry*, 2015, **21**, 11408.
132 A. Canales, A. Mallagaray, M. A. Berbis, A. Navarro-Vazquez, G. Dominguez, F. J. Canada, S. Andre, H. J. Gabius, J. Perez-Castells and J. Jimenez-Barbero, *J. Am. Chem. Soc.*, 2014, **136**, 8011.
133 J. Egger, C. Weckerle, B. Cutting, O. Schwardt, S. Rabbani, K. Lemme and B. Ernst, *J. Am. Chem. Soc.*, 2013, **135**, 9820.
134 S. R. Rauthu, T. C. Shiao, S. Andre, M. C. Miller, E. Madej, K. H. Mayo, H. J. Gabius and R. Roy, *ChemBioChem*, 2015, **16**, 126.
135 A. Ardá, R. Bosco, J. Sastre, F. J. Cañada, S. André, H.-J. Gabius, B. Richichi, J. Jiménez-Barbero and C. Nativi, *Eur. J. Org. Chem.*, 2015, **2015**, 6823.
136 R. Sommer, D. Hauck, A. Varrot, S. Wagner, A. Audfray, A. Prestel, H. M. Möller, A. Imberty and A. Titz, *ChemistryOpen*, 2015, **4**, 756.
137 M. Thepaut, C. Guzzi, I. Sutkeviciute, S. Sattin, R. Ribeiro-Viana, N. Varga, E. Chabrol, J. Rojo, A. Bernardi, J. Angulo, P. M. Nieto and F. Fieschi, *J. Am. Chem. Soc.*, 2013, **135**, 2518.
138 N. Varga, I. Sutkeviciute, C. Guzzi, J. McGeagh, I. Petit-Haertlein, S. Gugliotta, J. Weiser, J. Angulo, F. Fieschi and A. Bernardi, *Chemistry*, 2013, **19**, 4786.
139 K. Pederson, D. A. Mitchell and J. H. Prestegard, *Biochemistry*, 2014, **53**, 5700.
140 M. D. Swanson, D. M. Boudreaux, L. Salmon, J. Chugh, H. C. Winter, J. L. Meagher, S. Andre, P. V. Murphy, S. Oscarson, R. Roy, S. King, M. H. Kaplan, I. J. Goldstein, E. B. Tarbet, B. L. Hurst, D. F. Smee, C. de la Fuente, H. H. Hoffmann, Y. Xue, C. M. Rice, D. Schols, J. V. Garcia, J. A. Stuckey, H. J. Gabius, H. M. Al-Hashimi and D. M. Markovitz, *Cell*, 2015, **163**, 746.
141 J. R. Rosien, W. Seichter and M. Mazik, *Org. Biomol. Chem.*, 2013, **11**, 6569.
142 J. Lippe, W. Seichter and M. Mazik, *Org. Biomol. Chem.*, 2015, **13**, 11622.
143 J. Lippe and M. Mazik, *J. Org. Chem.*, 2015, **80**, 1427.
144 C. Nativi, O. Francesconi, G. Gabrielli, I. De Simone, B. Turchetti, T. Mello, L. Di Cesare Mannelli, C. Ghelardini, P. Buzzini and S. Roelens, *Chemistry*, 2012, **18**, 5064.
145 P. Carrero, A. Ardá, M. Alvarez, E. G. Doyagüez, E. Rivero-Buceta, E. Quesada, A. Prieto, D. Solís, M. J. Camarasa, M. J. Peréz-Pérez, J. Jiménez-Barbero and A. San-Félix, *Eur. J. Org. Chem.*, 2013, **2013**, 65.
146 N. Koch, J.-R. Rosien and M. Mazik, *Tetrahedron*, 2014, **70**, 8758.
147 S. Rieth, M. R. Miner, C. M. Chang, B. Hurlocker and A. B. Braunschweig, *Chem. Sci.*, 2013, **4**, 357.
148 O. Francesconi, M. Gentili and S. Roelens, *J. Org. Chem.*, 2012, **77**, 7548.
149 O. Francesconi, M. Gentili, C. Nativi, A. Arda, F. J. Canada, J. Jimenez-Barbero and S. Roelens, *Chemistry*, 2014, **20**, 6081.

150 O. Francesconi, C. Nativi, G. Gabrielli, M. Gentili, M. Palchetti, B. Bonora and S. Roelens, *Chemistry*, 2013, **19**, 11742.

151 J. Lippe and M. Mazik, *J. Org. Chem.*, 2013, **78**, 9013.

152 G. Joshi and A. P. Davis, *Org. Biomol. Chem.*, 2012, **10**, 5760.

153 B. Sookcharoenpinyo, E. Klein, Y. Ferrand, D. B. Walker, P. R. Brotherhood, C. Ke, M. P. Crump and A. P. Davis, *Angew. Chem., Int. Ed.*, 2012, **51**, 4586.

154 J. D. Howgego, C. P. Butts, M. P. Crump and A. P. Davis, *Chem. Commun.*, 2013, **49**, 3110.

155 C. Ke, H. Destecroix, M. P. Crump and A. P. Davis, *Nat. Chem.*, 2012, **4**, 718.

156 H. Destecroix, C. M. Renney, T. J. Mooibroek, T. S. Carter, P. F. Stewart, M. P. Crump and A. P. Davis, *Angew. Chem., Int. Ed.*, 2015, **54**, 2057.

157 T. J. Mooibroek, J. M. Casas-Solvas, R. L. Harniman, C. M. Renney, T. S. Carter, M. P. Crump and A. P. Davis, *Nat. Chem.*, 2016, **8**, 69.

158 A. Schmitt, B. Chatelet, D. Padula, L. Di Bari, J.-P. Dutasta and A. Martinez, *New J. Chem.*, 2015, **39**, 1749.

159 A. Schmitt, O. Perraud, E. Payet, B. Chatelet, B. Bousquet, M. Valls, D. Padula, L. Di Bari, J. P. Dutasta and A. Martinez, *Org. Biomol. Chem.*, 2014, **12**, 4211.

160 K. Ouchi, S. Saito and M. Shibukawa, *Inorg. Chem.*, 2013, **52**, 6239.

161 K. Ouchi, C. L. Colyer, M. Sebaiy, J. Zhou, T. Maeda, H. Nakazumi, M. Shibukawa and S. Saito, *Anal. Chem.*, 2015, **87**, 1933.

162 Y. Jang, R. Natarajan, Y. H. Ko and K. Kim, *Angew. Chem., Int. Ed.*, 2014, **53**, 1003.

163 M. Rauschenberg, S. Bandaru, M. P. Waller and B. J. Ravoo, *Chemistry*, 2014, **20**, 2770.

164 P. Ding, X. Li, G. Qing, T. Sun and X. Liang, *Chem. Commun.*, 2015, **51**, 16111.

165 A. Banik, P. Gogoi and M. D. Saikia, *J. Inclusion Phenom. Macrocyclic Chem.*, 2011, **72**, 449.

166 H. Ikeda, N. Ikuta, D. Nakata, H. Fukumi and K. Terao, *J. Inclusion Phenom. Macrocyclic Chem.*, 2011, **73**, 443.

167 C. Yuan, Z. Jin and X. Xu, *Carbohydr. Polym.*, 2012, **89**, 492.

168 V. Jahed, A. Zarrabi, A.-K. Bordbar and M. S. Hafezi, *Food Chem.*, 2014, **165**, 241.

169 J. Chao, H. Wang, W. Zhao, M. Zhang and L. Zhang, *Int. J. Biol. Macromol.*, 2012, **50**, 277.

170 C. G. Floare, A. Pirnau and M. Bogdan, *J. Mol. Struct.*, 2013, **1044**, 72.

171 S. M. Ali, S. Khan and G. Crowyn, *Magn. Reson. Chem.*, 2012, **50**, 299.

172 S. Hazra and G. S. Kumar, *RSC Adv.*, 2015, **5**, 1873.

173 R. Ferrazza, B. Rossi and G. Guella, *J. Phys. Chem. B*, 2014, **118**, 7147.

174 A. Lis-Cieplak, J. Sitkowski and W. Kolodziejski, *J. Pharm. Sci.*, 2014, **103**, 274.

175 A. Keniche, M. Slimani, J. Miranda, J. Aizpurua and J. K. Mulengi, *Med. J. Chem.*, 2013, **2**, 620.

176 Y. Izutani, K. Kanaori and M. Oda, *Carbohydr. Res.*, 2014, **392**, 25.

177 M. Pérez-Trujillo, J. C. Lindon, T. Parella, H. C. Keun, J. K. Nicholson and T. J. Athersuch, *Anal. Chem.*, 2012, **84**, 2868.

178 X. Ge, Z. Huang, S. Tian, Y. Huang and C. Zeng, *Carbohydr. Polym.*, 2012, **89**, 208.

179 T. Sohajda, Z. Szakács, L. Szente, B. Noszál and S. Béni, *Electrophoresis*, 2012, **33**, 1458.

180 T. Akita, Y. Matsui and T. Yamamoto, *J. Mol. Struct.*, 2014, **1060**, 138.

181 C. Acuña-Rougier, C. Jullian and C. Olea-Azar, *J. Inclusion Phenom. Macrocyclic Chem.*, 2011, **74**, 157.

182 R. Krishnan, A. M. Rakhi and K. R. Gopidas, *J. Phys. Chem. C*, 2012, **116**, 25004.
183 K. Lomsadze, E. D. Vega, A. Salgado, A. L. Crego, G. K. E. Scriba, M. L. Marina and B. Chankvetadze, *Electrophoresis*, 2012, **33**, 1637.
184 M. Nishizawa, T. Hosoya, T. Hirokawa, K. Shin-ya and S. Kumazawa, *Food Sci. Technol. Res.*, 2014, **20**, 663.
185 M. B. de Jesus, L. F. Fraceto, M. F. Martini, M. Pickholz, C. V. Ferreira and E. de Paula, *J. Pharm. Pharmacol.*, 2012, **64**, 832.
186 J. Redondo, A. Capdevila, I. Latorre and J. Bertrán, *J. Inclusion Phenom. Macrocyclic Chem.*, 2011, **73**, 225.
187 B. P. Mayer, R. L. F. Albo, S. Hok and C. A. Valdez, *Magn. Reson. Chem.*, 2012, **50**, 229.
188 J. Li, J. Chao and M. Zhang, *Spectrochim. Acta, Part A*, 2012, **87**, 25.
189 R. Yang, J.-B. Chen, X.-Y. Dai, R. Huang, C.-F. Xiao, Z.-Y. Gao, B. Yang, L.-J. Yang, S.-J. Yan, Z. Hong-Bin, C. Qing and J. Lin, *Carbohydr. Polym.*, 2012, **89**, 89.
190 A. Ikeda, T. Hida, J.-I. Kikuchi, K. Nobusawa and T. Matsuo, *Org. Lett.*, 2013, **15**, 6194.
191 X. Gu, E. V. Mathias, K. T. H. Nguyen and Y. Ba, *J. Inclusion Phenom. Macrocyclic Chem.*, 2012, **76**, 427.
192 Z. Sebestyén, Á. Buvári-Barcza and J. Rohonczy, *J. Inclusion Phenom. Macrocyclic Chem.*, 2011, **73**, 199.
193 F. L. Aachmann, K. L. Larsen and R. Wimmer, *J. Inclusion Phenom. Macrocyclic Chem.*, 2011, **73**, 349.
194 V. Legros, C. Vanhaverbeke, F. Souard, C. Len and J. Désiré, *Eur. J. Org. Chem.*, 2013, **2013**, 2583.
195 M. Poorghorban, A. H. Karoyo, P. Grochulski, R. E. Verrall, L. D. Wilson and I. Badea, *Mol. Pharmaceutics*, 2015, **12**, 2993.
196 M. T. Blázquez-Sánchez, F. Marcelo, M. C. Fernández-Alonso, A. Poveda, J. Jiménez-Barbero and C. Vicent, *Chemistry*, 2014, **20**, 17640.
197 R. Lucas, P. Peñalver, I. Gómez-Pinto, E. Vengut-Climent, L. Mtashobya, J. Cousin, O. S. Maldonado, V. Perez, V. Reynes, A. Aviñó, R. Eritja, C. González, B. Linclau and J. C. Morales, *J. Org. Chem.*, 2014, **79**, 2419.

Controlled and highly efficient preparation of carbohydrate-based vaccines: squaric acid chemistry is the way to go

Pavol Kováč* and Peng Xu
DOI: 10.1039/9781782626657-00083

Evolution of the method and practical aspects of the conjugation of single amino group-containing carbohydrates to protein carriers is described. Proven protocols for conjugation of low- and high-molecular-weight carbohydrates are also provided. It is concluded from the experimental evidence in authors' laboratory and elsewhere that the squaric acid-based method is the most efficient means for converting proteins to glycoconjugates with predictable carbohydrate-carrier ratios.

1 Introduction

The concept of conjugate vaccines for infectious diseases caused by bacterial pathogens originated from the findings of Avery and Goebel who were first to show[1–4] that the immunogenicity of carbohydrate antigens can be enhanced considerably by their conjugation to proteins. The first glycoconjugate immunogens prepared by these pioneers of coupling of carbohydrates to proteins were based on the simple diazotization reaction. Since those early days of conjugations, there has been a continuous quest for sort of a better mousetrap, namely a simpler and more efficient conjugation method. Until the last quarter of the 20th century, the interest in the conjugation of carbohydrates was mainly academic. Before that, because the knowledge of the structure of bacterial polysaccharides was largely obscure and synthesis of oligosaccharides was much less developed, applying conjugation to making medically useful vaccines was virtually thought of as science fiction. We have made vast progress since those days. While synthesis of large oligosaccharides is still not a routine task, our ability to make them and the knowledge of the structure of bacterial polysaccharide antigens has increased tremendously, thanks largely to advances in chromatography, and sophistication of the instrumental techniques used in structural analysis of carbohydrates. This, together with the ever more recognized importance of neoglycoconjugates,[†] triggered the birth of a new field within the life sciences and appearance of a large body of work related to

NIDDK, LBC, National Institutes of Health, Bethesda, MD 20892-0815, USA.
E-mail: kpn@helix.nih.gov

[†]"Glycoconjugate" is a general term to describe materials where carbohydrates are linked to anything else, including other carbohydrates. Thus, technically, an oligosaccharide, natural or synthetic, is also a glycoconjugate. On the other hand, the term "neoglycoconjugate" is reserved for man-made, *i.e.*, synthetic or semi-synthetic, carbohydrate-containing substances.[5]

conjugation and the use of neoglycoconjugates. A substantial part of that work has been concerned with conjugation methods that have been developed since the days of Avery and Goebel. In this context, it is both surprising and ironic that the squaric acid chemistry-based method for conjugation introduced[6–8] more than two decades ago is not mentioned either in the first[9] or in the recently updated, comprehensive, 1200 pages long 2nd Edition[10] of the otherwise very useful monograph on conjugation techniques, and is completely ignored also in very recent reviews on the synthesis of glycoconjugates.[11,12] Nevertheless, the technique based on squaric acid diesters is now recognized as virtually the most powerful one for the conjugation of amine-containing substances, and as such has been widely used (for example ref. 13–62). The purpose of this short review, which complements the one by Wurm and Klok,[14,‡] is to show how the method invented by Glüsenkamp and Tietze[6,7] evolved into a tool which has a potential to becoming the method of choice for making neoglycoconjugate vaccines from synthetic or bacterial carbohydrate antigens that contain, or can be modified to contain a single amino group. The method is simple to perform, it is reliable, uses inexpensive reagents, and its outcome is predictable and reproducible. In addition, it can be used to prepare well-defined products, or a series thereof in one pot[63,64] with predetermined carbohydrate–protein ratio with unsurpassed efficiency. Moreover, much of the excess of the precious antigen used at the onset of conjugation and left unconjugated, be it a monosaccharide, oligosaccharide or a bacterial polysaccharide, can be recovered and regenerated.[65] No other conjugation method can claim all these attributes.

2 The early days

The principle of the squaric acid-based conjugation method lies in the selective reaction of squaric acid diesters (*e.g.* **1**, Scheme 1) with amines. At pH 7, one of the ester groups reacts to form a monoamide (**2**). At more basic conditions, further amidation takes place to form a diamide (**3**). When the first amine is an amino group-containing carbohydrate and the second amine is anything else, a glycoconjugate is formed. In 1991, Glüsenkamp[6] and Tietze[8] coupled amines to obtain biopolymer conjugate drugs. The conjugation was based on the previous work by Thorpe and co-workers and others' reactions of diethyl squarate with primary or

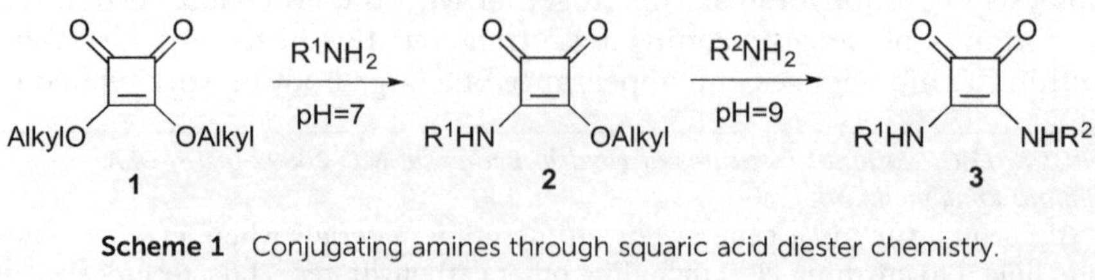

Scheme 1 Conjugating amines through squaric acid diester chemistry.

‡Readers are encouraged to take advantage of this review where the topic is treated from somewhat different perspective.

secondary amines.[66,67] It was shortly followed by Tietze and co-workers' first application of squaric acid chemistry to conjugation of carbohydrates to BSA.[8] These workers showed the potential of the new coupling reagent for making neoglycoconjugates using the ethyl squarate derivative of a *p*-aminophenyl glycoside of a disaccharide, which they attached covalently to BSA (Scheme 2).

Although they lucidly demonstrated that squarate reagents could provide access to a wide spectrum of tools for research in the glycosciences, the importance of squaric acid chemistry for making glycoconjugates was first not duly recognized. The boom of the use of dialkyl squarates for conjugating carbohydrates started about five years later, when Kamath *et al.*[68] virtually rediscovered the method. In their seminal paper, the latter authors highlighted the most salient characteristics of the method, namely that it works under very mild conditions, that the method is very simple to perform, and that it can be used on a very small scale. They showed also that when the starting carbohydrate is ω-alkoxycarbonylalkyl glycoside the amino group required by this method could be readily introduced into the molecule. They used ethylenediamine, but other diamines were later used also.[69,70] Credit to Kamath and co-workers[68] goes also for being the first to show that the average degree of incorporation of the sugar onto protein and the molecular weight distribution can be readily determined using matrix-assisted laser desorption ionization time-of-flight (MALDI-TOF) mass spectrometry because the conjugation in this way follows the single-point attachment model.[71]

The use of MALDI-TOF MS avoids the colorimetric sugar analysis of neoglycoconjugates, whose accuracy with hetero oligosaccharide-protein constructs is arguably questionable. This is particularly true when the carbohydrate part of the conjugate mimics or is a bacterial or other cell-surface polysaccharide, as is frequently the case. These often contain rare sugars that decompose nonspecifically when treated with hot dilute acid[72] or with phenol–sulfuric acid reagent,[73] and a reliable calibration curve is then virtually impossible to construct. Thus, mass spectroscopy should be the method of choice for accurate determination of carbohydrate content in non-cross-linked neoglycoconjugates (Fig. 1), when the molecular masses of the carbohydrate and of the protein carrier are known, and when the molecular mass of the conjugate is within the range of the MS instrumentation at hand.

3 More recent conjugations of synthetic oligosaccharides

Shortly after Kamath *et al.*[68] rediscovered the squaric acid conjugation of carbohydrates, the method was applied by Auzanneau and Pinto[74] who conjugated synthetic oligosaccharides to each of ovalbumin and BSA. They found the method more efficient than the one based on the addition of *N*-acryloylated oligosaccharides to the ε-amino group of lysine in proteins. Other reports of successful use of dialkyl squarates, both ethyl[75] and methyl,[75,76] for conjugation of oligosaccharides to proteins followed

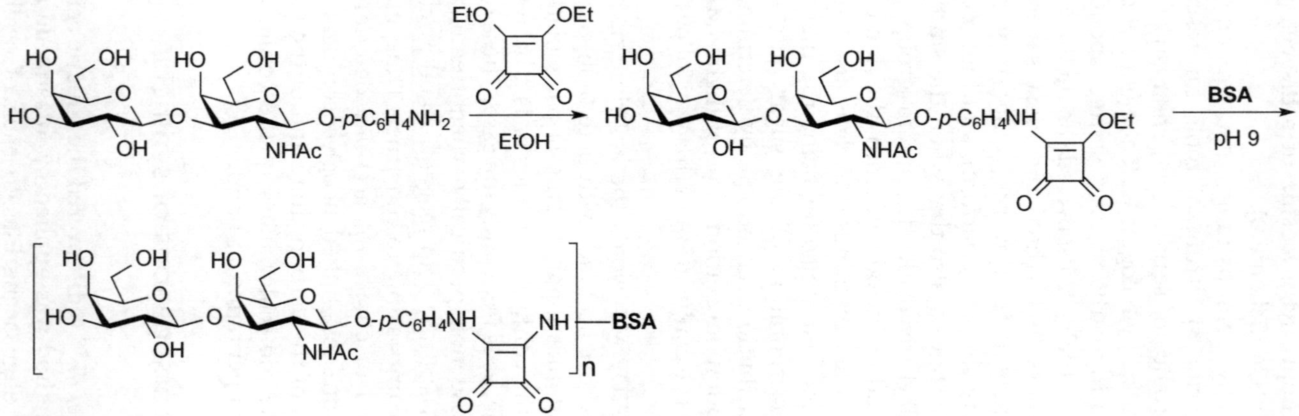

Scheme 2 Conjugation of *p*-aminophenyl glycoside of disaccharide to bovine serum albumin through squaric acid chemistry.

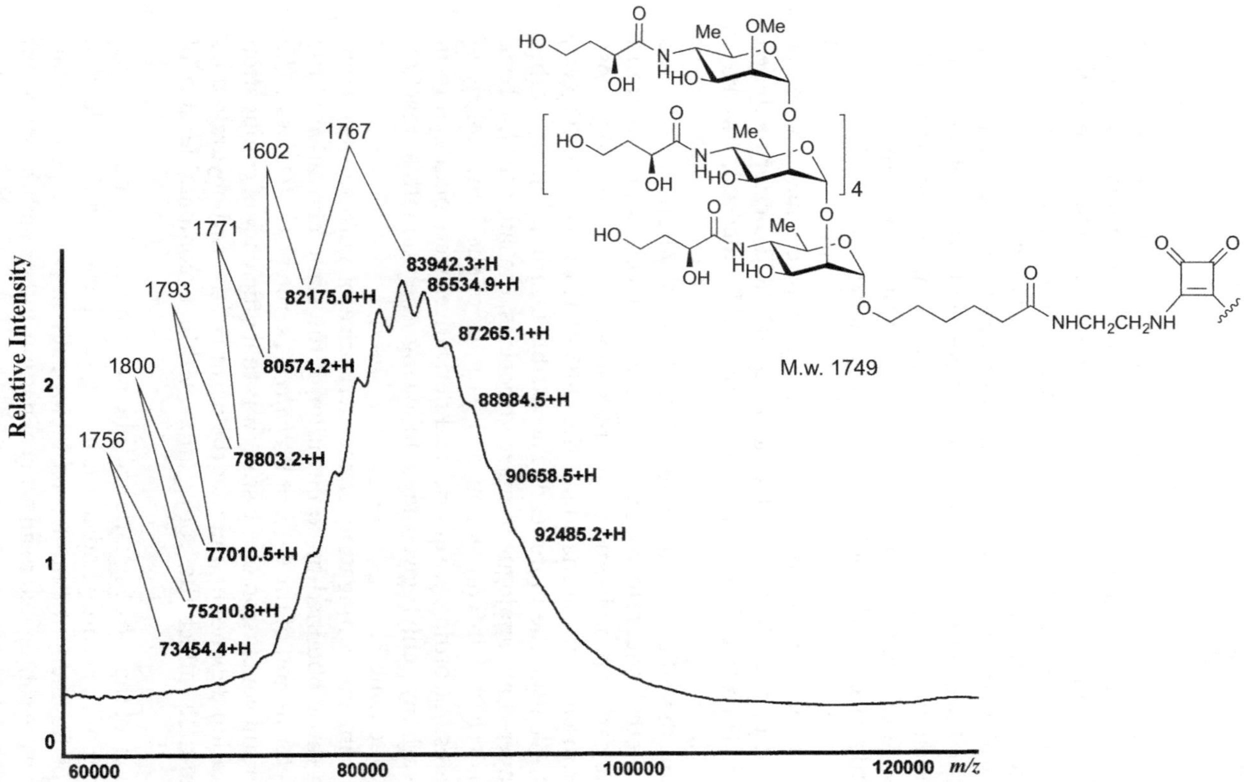

Fig. 1 An example of analysis of a conjugate of the synthetic, upstream, terminal hexasaccharide fragment of the O-specific polysaccharide of *Vibrio cholerae* O1, serotype Ogawa with BSA (M.w. 66 430 DA126) by MALDI-TOF mass spectrometry. The average molecular mass of the conjugate (83942 Da) is given by the peak at about the center of the gravity of the MS peak. The incorporation of the ligand (loading) is calculated by subtracting the molecular mass of the carrier from the average molecular mass of the conjugate and dividing the difference by the molecular mass of the ligand conjugated ($83942-66430=17512$; $17512/1749\approx10$).

soon, including ours.[77] We became involved in the conjugation of synthetic oligosaccharides by squaric acid chemistry within our efforts to develop a conjugate vaccine for cholera from synthetic fragments of the O-PS of *Vibrio cholerae* O1. In the search for a suitable conjugation method for oligosaccharides, we first considered reductive amination and have also developed an easy-to-synthesize spacer (linker),[78] equipped with a latent aldehyde group. However, while our conjugations[79,80] yielded the expected products, we observed that a small amount of the starting aldehyde was reduced during conjugation to the corresponding alcohol due to incomplete reductive chemospecificity of sodium cyanoborohydride. Similar observation has been reported before[81] and also after[82] our work. The purpose of our excursion into conjugation methodology was to make synthetic vaccines from larger, labor-intensive oligosaccharides. We do not deny reductive amination its value for conjugation of polysaccharides but, because of potentially irreversible loss of the very precious synthetic materials through the aforementioned side reaction, we searched for a more suitable conjugation alternative for our work towards vaccines through single-point-attached carbohydrates. An extensive search of the literature revealed that neither the original developers of the methodology nor their early followers have looked at variables that might affect the outcome of squaric acid diester-mediated conjugation. Therefore, in our first closer look at the method,[83] we have studied the effect of reaction time, concentration and the molar excess of hapten upon the efficiency of conjugation of carbohydrates to proteins using the diethyl squarate reagent. We used chicken serum albumin (CSA) as the carrier protein and a linker-equipped derivative of D-glucose and the terminal monosaccharide of the O-PS of *V. cholerae* O1 (Ogawa) as haptens. We used the more readily accessible 6-carbon linker[84] rather than its longer-chain analogue to make glycosides, the latter having been used extensively[85–87] in connection with studies of the immunology of carbohydrates. Accordingly, squaric acid derivatives were obtained from 5-methoxycarbonylpentyl glycosides following the Kamath's reaction sequence[68] (Scheme 3).

Results from the conjugation of the derivative of glucose were compared with those obtained by conjugation of the same derivative of the terminal determinant of the O-PS of *V. cholerae*, serotype O1 (Ogawa). The most important conclusion of the study was that efficiency of conjugation is concentration dependent and that the nature of the carbohydrate does not substantially affect the conjugation within a comparable size of ligands.[83]

A real boom of the use of squaric acid chemistry for conjugation occurred with the turn of the century. Advances in oligosaccharide synthesis lead to syntheses of fragments of numerous biologically important polysaccharides, which were conjugated to an array of natural and recombinant protein carriers applying this methodology.[16,17,20–23,28–31,33–37,45,64,88–100] Because systematic efforts to establish optimum conditions for conjugation had not yet been made, conjugations had been conducted, without any particular rationalization, with large excess of labor-intensive carbohydrate–squarate reagents, in order to maintain a reasonable reaction rate. We developed a

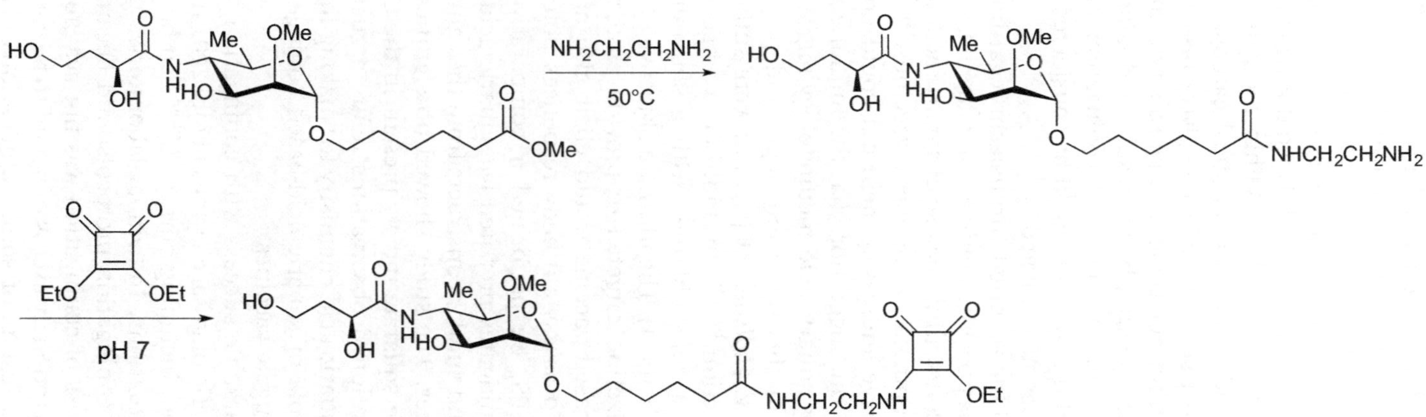

Scheme 3 Conversion of a 5-methoxycarbonylpentyl glycoside of the monosaccharide fragment of the O-PS of *V. cholerae* O1 (Ogawa) to the corresponding ethyl squarate derivative.

protocol allowing recovery of some of the precious synthetic material that was used in excess at the onset of conjugation.[90] Further improvement in that regard was suggested by Bergh and coworkers,[48,70] who used decyl squarate ester derivative. It provided a hydrophobic tag and enabled recovery by means of C18 solid phase extraction. Many examples of successful conjugations using squaric acid diesters, also outside of vaccine development field,[64,99,101] have been published during the first few years of the 21st century.[29–31,33,34,94,97,98,100,102] Within the same period of time, Izumi[13] evaluated the utility of disuccinimidyl carbonate, disuccinimidyl glutarate, disuccinimidyl suberate, ethylene glycol bis-(succinimidylsuccinate), 1,5-difluoro-2,4-dinitrobenzene, diethyl squarate, and thiophosgene as conjugation reagents. Judging by his results and taking into consideration cost of reagents, the efficiency of conjugation and experimental simplicity, squaric acid diethyl ester could be considered the most advantageous among the reagents evaluated.

Notwithstanding all the reported enthusiasm regarding the potency of squaric acid diesters as powerful conjugation reagents, some reservations have been expressed concerning the use of squaric acid diesters as useful reagents in the development of conjugate vaccines.[34,95] We took these issues to heart and have addressed them but could not confirm those concerns. Contrary to the suggestion that the immune response of constructs given rise to by squaric acid methodology might be reduced, our conjugates prepared in this way from oligo- and polysaccharide fragments of the O-PS of *V. cholerae* O1 (Ogawa) were sufficiently immunogenic to render protection.[103,104] In addition, we had synthesized and recently published[104] results obtained with a squarate-labelled linker conjugated to BSA (Scheme 4). The ultimate objective of the study was to determine if immunization targeted to provide protection for cholera with a squaric acid-based vaccine would elicit also anti squaric acid-specific antibodies. To assess antibody responses potentially targeting squarate and not O-PS, we performed a comparative immunization study[104] with glycoconjugates prepared *via* squaric acid chemistry from *V. cholerae* antigens and similar constructs from the same carrier and the squarate-labelled linker. The study showed no significant induction of antibody targeting the squarate moiety present in the O-PS–carrier constructs. The aforementioned observations refute arguably the previous objections to using squaric acid chemistry for making immunogens, and further support the role of squaric acid-based conjugation in development and use of conjugate vaccines.

Despite some remarkable success with conjugation, all conjugations performed with synthetic oligosaccharides before 2001 irrespective of the method of conjugation, including ours,[77,79,80,83] can be characterized as virtual hit or miss situations. Individual laboratories used an array of variously (in)efficient conjugation protocols, and because a method for monitoring the progress of conjugation was not available, the reactions were set up with a discretionary excess of ligand to be conjugated. Table 1 shows conditions and results of some conjugations using squaric acid chemistry. Unmethodical choice of reaction conditions in some of the cases (Table 1) is evident.

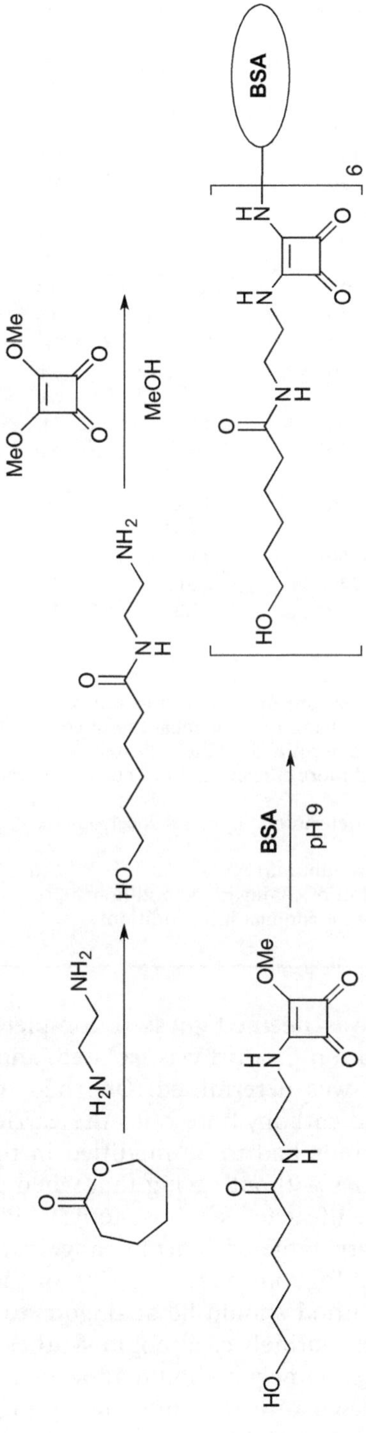

Scheme 4 Synthesis of the linker (spacer)–squarate labeled construct and its conjugation to BSA.

Table 1 Conjugation of oligosaccharides to BSA by squaric acid chemistry.[a]

Ref.	Saccharide	Squarate ester	Onset carbohydrate/ protein [Mol/Mol]	Carbohydrate/ protein reached [Mol/Mol]	Hapten conc. [mM]	Reaction time [h]	Coupling efficiency [%]
8	Di	Et	85	15.3[b]	3.5	3.5	18
74[c]	Hexa	Et	26	16[b]	21	168	64
75	Di	Et	110	12[d]	0.9	18	11
91[e]	Hexa	Et	20	10[d]	NR[i]	72	50
13	Mono	Et	60	25[d]	4.5	120	42
38	Tetra	Et	33[f]	9[d]	3.6	18	27
126	Tri	Et	20	5[d]	0.5	72	25
126	Penta	Et	20	3.7[d]	0.5	72	18.5
126	Hexa	Et	20	0.5[d]	0.5	72	2.5
37	Tri	Et	50	35[b]	30	168	70
20	Penta	Et	30	10[d]	3.6	72	33
20	Hexa	Et	30	9[d]	NR[i]	NR[i]	30
76	Mono	Me	58[f]	23[d]	51	72	40
45	Hexa	Et	30[f]	14[d]	NR[i]	16	46
90[g]	Hexa	Me	80	10.1[d]	20	72	12.5
90[g]	Hexa	Me	80	16.2[d]	20	144	20
32[h]	Di	Me	6	5.3[d]	40	2	88
32[h]	Di	Me	60	45.9[d]	40	72	76.5
32[h]	Di	Me	120	55.6[d]	40	72	46
32[h]	Tri	Et	6	5.3[d]	40	8	88
32[h]	Tetra	Et	6	4.9[d]	40	8	81.6
32[h]	Hexa	Et	7	5.4[d]	40	5	77

[a] Many other published conjugations of carbohydrates to BSA using squaric acid chemistry are not described in sufficient detail to allow meaningful comparison.
[b] Carbohydrate–protein ratio determined by the colorimetric method of Dubois.[73]
[c] Squarate coupling was found more efficient than when N-acryloylated adduct was used as a hapten.
[d] Carbohydrate–protein ratio determined by mass spectrometry.
[e] Conjugation to CRM_{197}.
[f] Conjugation to human serum albumin.
[g] Our work, before optimization of conjugation conditions.
[h] Our work, after optimization of conjugation conditions.
[i] NR, Not reported.

When the reaction was deemed/guessed complete, the mixture was worked up, the conjugation product was isolated, and the carbohydrate–protein ratio (loading) was determined. One had to be satisfied with whatever loading of the carbohydrate onto the carrier was achieved, or the conjugation conditions had to be modified in the hope for a more satisfactory result. It goes without saying that when glycoconjugates are meant to be tools in the life sciences those tools would be more useful if they could be made with targeted, narrow-range carbohydrate loading. That requires a method for routine monitoring of the conjugation reaction. Ideally, such a method should be analogous to thin-layer chromatography, which is used routinely as a tool in syntheses of low molecular mass organic molecules, namely it should allow one to rapidly analyze a minute sample withdrawn directly from the conjugation mixture and provide, within minutes, reliable information about the increasing molecular mass of the conjugate being formed, thereby allowing termination of the process when the desired loading is reached. We found such a tool[90] in the surface-enhanced laser desorption-ionization time-of-flight mass spectrometry (SELDI-TOF MS) in combination with the ProteinChip® System,[105] which was designed to serve a different purpose. It provided us with reliable information about the size of the conjugate being formed. We used it to monitor in real time squarate-mediated conjugations but it can be reasonably assumed that crude reaction mixtures resulting from any conjugation chemistry could be analyzed in this way. Our introduction of this technology into the glycoconjugation field constituted a significant addition to the tools available to conjugation chemists. It brought at the time the art of making glycoconjugates to a new, higher level of sophistication by allowing routine and rapid analysis of the unprocessed conjugation mixture, thereby eliminating the guesswork about the conjugation in progress. Figure 2 shows data obtained during conjugation of a linker equipped hexasaccharide.[90] The current state of technology makes it possible to obtain the same information by engaging MALDI-TOF spectrometry in conjunction with Zip-Tip devices (EMD Millipore Corporation).[106]

The conjugations described above were conducted using a large excess of the precious synthetic antigenic material at the onset of conjugation. Recovery of conjugation reagent from reaction mixtures, which involved different coupling chemistry, had been previously described.[74] It prompted us to attempt recovery of the unconjugated hapten from the squaric ester-based conjugation reaction. The precious synthetic material could be recovered[90] engaging solid-phase extraction technique, thereby increasing conjugation efficiency.

Following the above approach, we prepared three neoglycoconjugates from the hexasaccharide fragment of *V. cholerae* O1, serotype Ogawa, and BSA with hapten/carrier ratio 5, 10, and 15, which were evaluated for their immunological properties. The study showed[103] that all 3 conjugates were immunogenic in mice and showed also protective capacity. The protection was evident with mice immunized with all conjugates, but it was highest in the group of animals that received the conjugate with the lowest level of substitution. Based on the latter results and to allow objective comparisons, the molar carbohydrate–protein ratio ~5 : 1 has become our target loading when the conjugates are to be used for immunization studies.

When neoglycoconjugates are used in areas where the carbohydrate–protein ratio may affect biological properties, such as the development of conjugate vaccines from synthetic antigens, the efficacy *vs.* loading studies require series of conjugates with different, discrete loading. The ability to monitor conjugation reactions in virtually real time allows preparation of a series of such materials from one hapten efficiently in a one-pot reaction. According to the protocol we developed,[63,64] the conjugation is set up with the amount of hapten sufficient to make the whole series of conjugates, and the conjugation reaction is monitored by mass spectrometry (SELDI- or MALDI-TOF). A portion of the reaction mixture can be withdrawn and processed when the lowest desired loading is reached. The rest of the material is allowed to react until the MS analysis

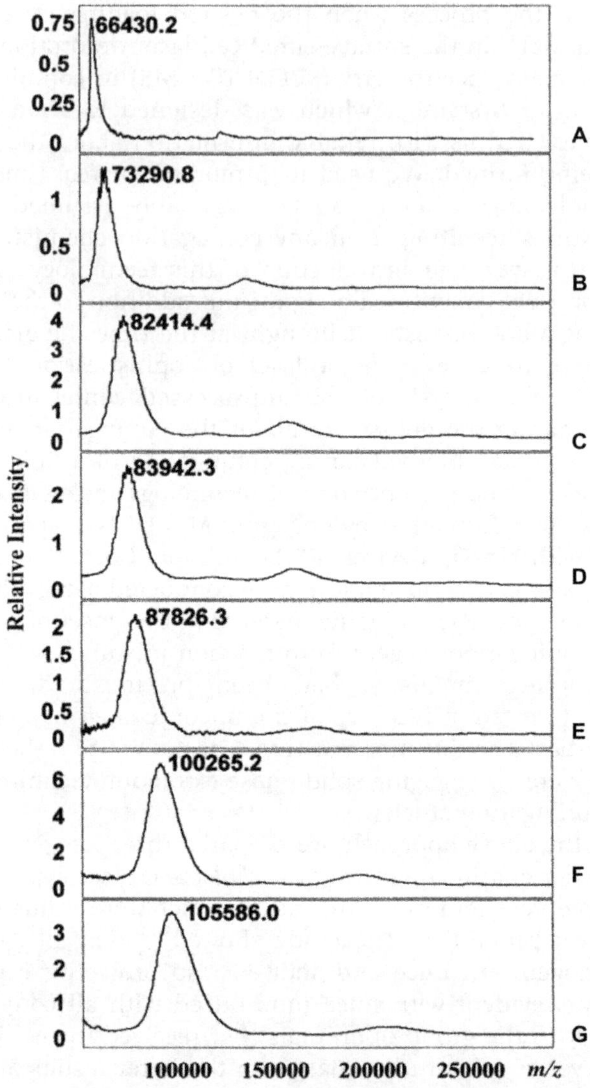

Fig. 2 Progress of conjugation of a hexasaccharide (molecular mass, 1780.79 Da) and BSA (molecular mass 66 430 Da) by monitoring the reaction by SELDI-TOF MS. Spectrum A was taken at the onset of the reaction ($t = 0$); spectra B–G were taken at 1, 3, 7, 9, 27, and 54 h, respectively. (Reproduced from ref. 90 by permission from Elsevier).

shows the next desired molecular mass. In this way, virtually unlimited number of tailor-made conjugates can be prepared in a one-pot manner from one ligand.

Notwithstanding our successful preparation of many conjugates from synthetic oligosaccharides, we have observed some irregularities in the outcome of some conjugations. To those belonged occasional poor reproducibility of reaction rate or faster conjugation of a higher compared to a lower oligosaccharide. Obviously, conjugation was affected by factors we were unaware of. In this context, the observation which prompted us to reexamine[32] the effect of variables involved in conjugation by squaric acid chemistry is worth mentioning. Our previous results[83] suggested that the rate of conjugation was concentration dependent. With the aim to increase the rate of the reaction beyond that observed at hapten concentration 25 mM,[83] we conducted a similar conjugation at 40 mM concentration, using BSA as the carrier. The buffer used was a commercially available 0.05 M buffer. The conjugation was monitored by SELDI-TOF MS, and it showed that the reaction was much slower than expected, and that the loading reached after 25 h was unexpectedly low despite the large excess of hapten used at the onset of the conjugation. Also, hardly any progress was noted within the last 20 h of the 25 h period. Thus, it appeared that the conjugation started normally and then stopped after a few hours of the reaction time. Moreover, unexpected decrease of alkalinity from pH 9 to pH 7.2 was observed. The pH could be adjusted to the desired value (pH 9) by addition of a small amount of pH 9 buffer salts, obtained by freeze-drying the buffer solution. Having done that, continued monitoring by mass spectrometry showed that the conjugation process resumed, and after 12 additional hours almost all hapten was attached to the carrier. This could not suggest anything other than because the conjugation was conducted at high hapten concentration (40 mM) only very small volume of buffer was used, and its capacity was insufficient to keep the pH constant. However, the dramatic decrease of pH during the first 25 h of reaction time (to pH 7.2) could not be explained only by hydrolysis of the squarate derivative because adjustment of alkalinity to pH 9 resulted in resumption and virtual completion of the conjugation process, showing that at the time when the pH was adjusted, a large proportion of unchanged squaric ester was still present. Within search of the possible cause of the pH drop we measured the pH of a solution of the BSA we used, and also of other commercial samples of BSA in the 0.05 M pH 9 buffer at the same concentration of the protein as in the experiment just described. We found that the individual samples of commercially available BSA, one of which was the BSA used in the experiment just described, caused different decrease of pH. Obviously, the acidity of different samples of commercial BSA is a variable that is normally neglected. The possible change of pH when BSA is solubilized is not among specifications made available by suppliers, although it, together with slowly progressing hydrolysis at basic conditions of the squarate monoester, may play a significant role in making glycoconjugates. The easiest and most convenient way to ensure that pH persists during the whole duration of the conjugation conducted at high hapten-concentration was to conduct the conjugation in more concentrated and, therefore, higher capacity 0.5 M buffer. Later conjugations carried out in this way were uneventful and consistently yielded conjugates with the expected loading. This revealed that maintaining pH 9 during the whole duration of conjugation was the most important factor in ensuring that the conjugation is fast, efficient, and reproducible.

Other factors that could affect conjugation by squaric acid chemistry were also examined.[32] Reactivity of dialkyl squarates (methyl, ethyl, butyl and decyl) that have been used for the preparation of neoglycoconjugates have been examined as to their reactivity with terminal amino or

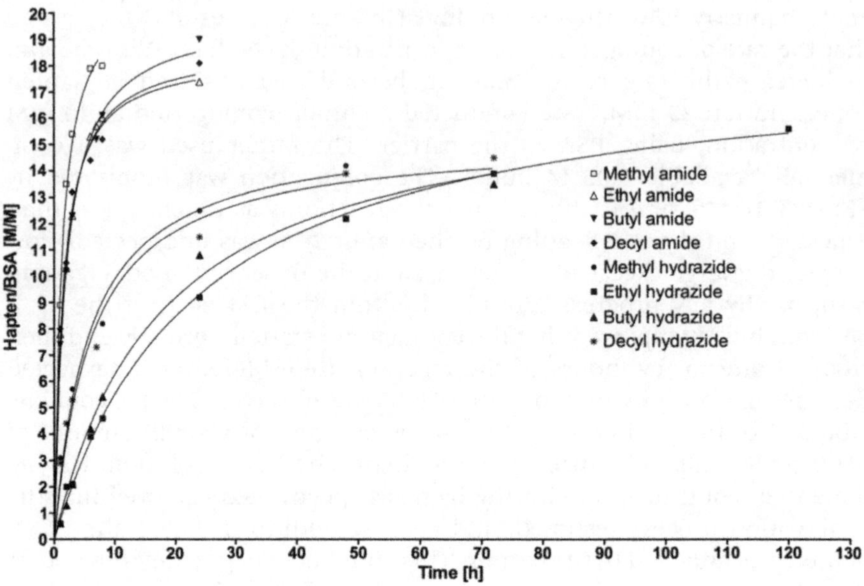

Fig. 3 Conjugation to BSA of methyl, ethyl, butyl and decyl amide and hydrazide squarates in 0.5 M pH 9.00 Buffer at 40 mM concentration of hapten, and initial hapten/BSA = 20/1.32.

hydrazino group in a linker-equipped β-lactose, as well as the potency of the resulting monoamido or monohydrazido alkyl esters to form conjugates with BSA. No definite advantage was found for the use of any of the four dialkyl squarate reagents (methyl-, ethyl-, butyl-, and decyl-) for the conversion of carbohydrate derivatives to conjugation reagents. Based on their reactivity to the model carrier BSA, the eight monoesters (four amido- and four hydrazido esters) could be divided into two groups (Fig. 3): All amido compounds produced conjugates faster and with higher efficiency compared with their hydrazido counterparts.

When the rate of hydrolysis of the alkyl squarate conjugation reagents made from lactose was examined at the conditions of conjugation, the methyl squarate hydrolyzed at pH 9 considerably faster than any other of the squarates tested (Fig. 4, ~25% hydrolyzed after 1 day). Nevertheless, since conjugates from a hexasaccharide with carbohydrate–protein ratio ~5 : 1 could be obtained within just a few hours with close to 80% efficiency, dimethyl squarate – a crystalline, commercially available, easy-to-handle substance, was deemed the preferred reagent. Results obtained during the same study[32] showed also that the size of the oligosaccharide to be conjugated, up to molecular mass ~2000 Da, plays only a minor role in the rate of conjugation. Much more important variable affecting conjugation was found to be the concentration of hapten. Data showed that it was advantageous to work at highest concentration practicable. The validity of the revised conjugation protocol for preparation of neoglycoconjugates from synthetic oligosaccharides with predetermined carbohydrate-carrier ratios was successfully verified with many different oligosaccharides.[35,102,107] The new protocol[32] was also found

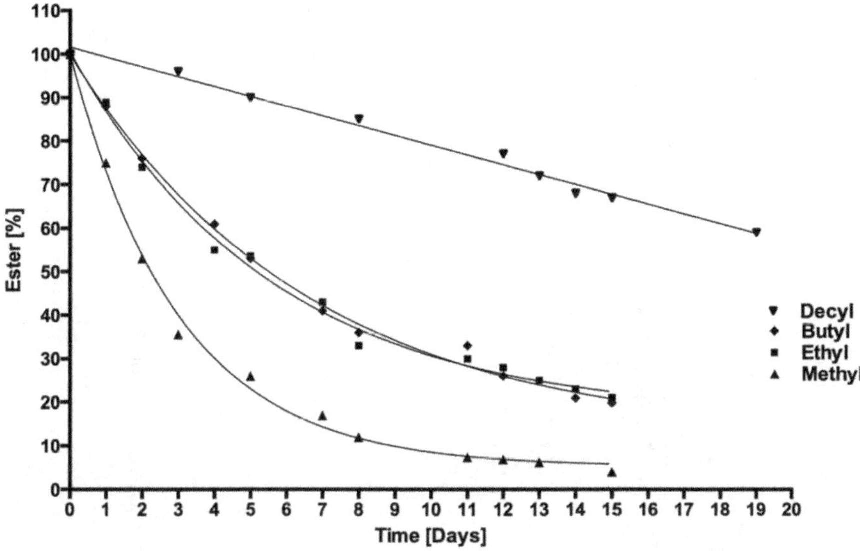

Fig. 4 Hydrolysis of squaric acid monoester conjugation reagents in 0.5 M borate buffer, pH 9.0.

reproducible by others (*e.g.* ref. 36). With oligosaccharide haptens, it was remarkable to find that when targeting loading of 5, only small excess (~20% excess) of the squarate-labeled carbohydrate was required to reach the desired loading, which is comparable to the excess used in syntheses of small organic molecules.

Despite improvements that have been made in the squarate-based methodology of conjugation, the fact that the dialkyl squarate reagents, and to a certain degree also the squarate labeled carbohydrates, are prone to hydrolysis constitutes a minor weak point of the otherwise very attractive method of conjugation. It looked, therefore, very enticing when Wurm *et al.*[26] introduced a novel squaric acid-based reagent, squaric acid di(tri(ethylene glycol) monomethyl ether)ester (**4**, Scheme 5). One of the advantages ascribed to the reagent was higher hydrolytic stability of the squarate amide-esters made from it than of those made from dimethyl squarate **5**.[32] However, aside from the fact that unpurified **4** was used throughout the work,[26] the foregoing claim appears questionable because the kinetics of hydrolysis of the squarates obtained from each of **5** and **4** were not determined at the same conditions (0.05 and 0.5 M, pH 9[32] *vs.* 0.01 M, pD 9.5[26] buffer). Nevertheless, the successful use[26] of the crude preparation of diester **4** for the preparation of conjugation reagents from small ligands and their conjugation to proteins prompted us to examine the utility of **4** in the overall conjugation process. We prepared[108] squarate reagent **4** in the analytically pure state and used it to convert linker-equipped lactose **6**[32] to the corresponding squarate ester amide **7**. We then compared in parallel experiments the hydrolytic stability of **7** and **8**[32] at the optimized conditions for conjugation.[32] We have also used squarates **7** and **8** for conjugation with BSA at the same conditions to obtain conjugate **9**. The data showed that the stabilities toward

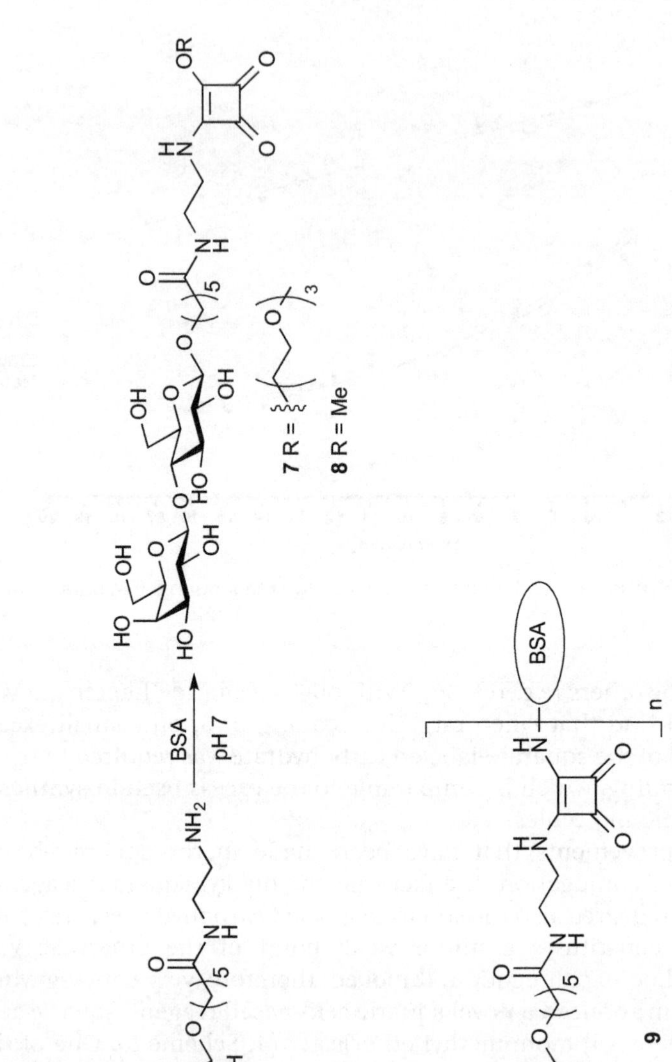

Scheme 5 Conjugation of linker equipped lactose **6** to BSA using two different squarate reagents **7** and **8**.

hydrolysis of the two squarate ester amides and the rate of conjugation were comparable. In other words, we found no advantage to using **4** as squarate labeling reagent. Therefore, because involving impure reagents in protocols used in vaccine preparation would be undesirable and purification of squarate **4** is rather laborious,[108] whereas methyl ester **5** is an inexpensive, commercially available crystalline compound, we still consider dimethyl squarate **5** the preferred reagent for conversion of amines to conjugation reagents.

4 Conjugation of bacterial polysaccharides

The first and the only clinically useful oligosaccharide vaccine made thus far from a fully synthetic carbohydrate antigen[109] provides the proof of principle and evidence that making potent, protective vaccines from synthetic oligosaccharides on a scale large enough to be clinically useful is feasible. However, the time and the overall cost normally involved in syntheses of large fragments of O-specific polysaccharides, which are required[15] for the desired immune response, prohibit largely the use of neoglycoconjugate vaccines on the medically useful scale. Thus, until this situation changes, bacterial pathogens will stay as the primary source of antigens in the industrial production of glycoconjugate vaccines.

Making well-defined glycoconjugates requires single-point attachment of the antigen.[15,71] Despite considerable progress in the conjugation of carbohydrates to proteins, many existing conjugate vaccines do not satisfy that requirement. Synthetic oligosaccharides are normally prepared in spacer-equipped form to be ready for conjugation by single point attachment. The possibilities to obtain bacterial polysaccharides with the same attributes are limited. Consequently, our ability to industrially convert bacterial carbohydrates to uniform, well-characterized conjugate vaccines leaves a lot to be desired. Bundle[15] emphasized the need for well-defined chemistry to form clinically useful conjugate vaccines and reviewed possibilities in that regard. Despite the success in conjugating synthetic oligosaccharides during the last two decades, squaric acid chemistry had not been applied to conjugation of bacterial lipopolysaccharides or fragments prepared therefrom until 2011.[110] Having been encouraged by the ability of squarate diesters to readily form conjugation reagents from synthetic carbohydrates, and the conversion of the latter to well-defined conjugates efficiently in a predictable way,[32] we ventured into conjugating O-specific polysaccharide–core (O-PSc) fragments of some bacterial lipopolysaccharides (LPS) to carrier proteins. Such fragments from *V. cholerae* O1[111] and some other O-PSs are known to contain[112,113] a single free amino group as part of the D-glucosamine residue present in the core. Thus, these materials, where the core or its part may be formally looked upon as the spacer, are amenable to single-point attachment conjugation. The chemical structures of the core of the two strains of *V. cholerae* O1, Inaba and Ogawa, are deemed to be the same,[111] and the number of monosaccharide repeating units in the O-PS varies[114,115] (Fig. 5).

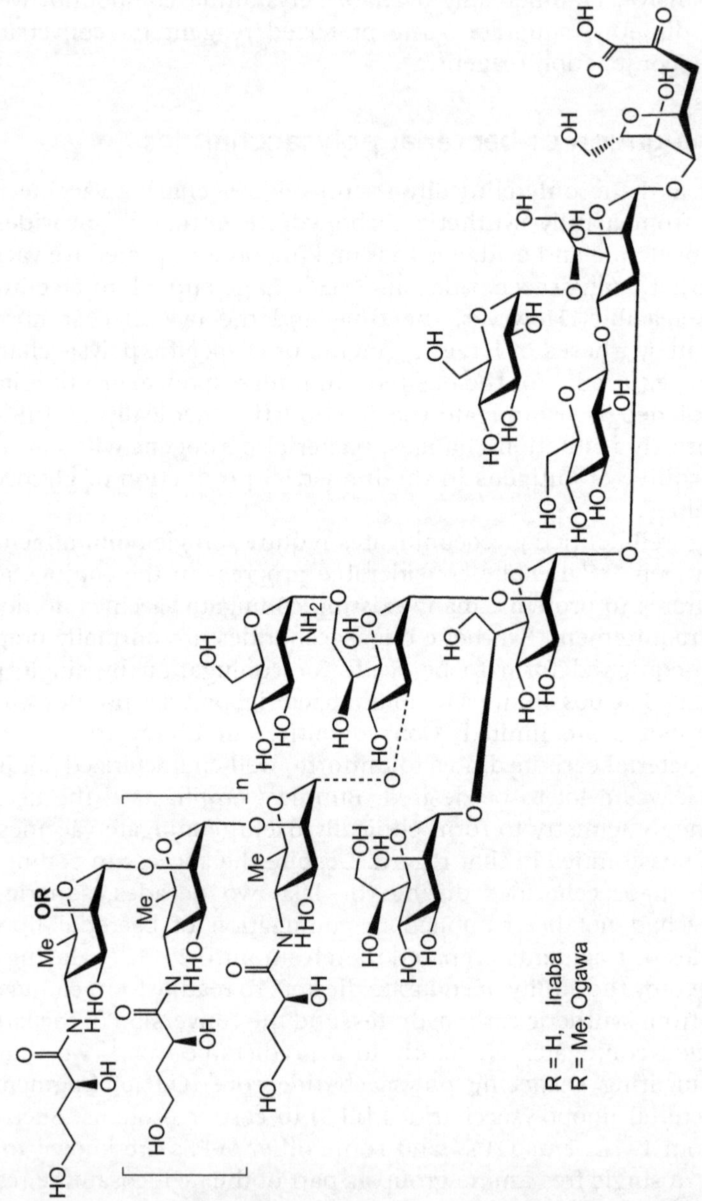

Fig. 5 The structure of the O-PSc of *V. cholerae* O1, serotypes Inaba and Ogawa.

Conjugation according to the single point attachment model[71] of detoxified LPS or of O-PSc of *V. cholera* O1 utilizing the amino group present in the core has been reported.[115,116] However, in each of the works quoted, both the antigen and the carrier protein were derivatized prior to the conjugation. As reasoned in the more recent work,[115] based on results of NMR studies, attaching an additional linker/spacer was thought necessary to compensate for diminished accessibility and reactivity of the free amino group present in the core. When we converted each of the O-PSc of *V. cholerae* O1 (Inaba and Ogawa) to the corresponding methyl squarate derivative directly, *i.e.* without prior derivatization (attachment of linker/spacer) of either component to be conjugated, and then conjugated[110] the squarate obtained to BSA, the conjugation was uneventful, and yielded the expected conjugates. As during similar work with synthetic oligosaccharides, the progress of conjugations could be monitored by SELDI-TOF Mass spectrometry,[90,117] and previous conclusions regarding the effects of some variables, such as concentration and size of antigen, upon the conjugation efficiency and reaction rate were confirmed. During the initial attempts,[110] the conjugation efficiency achieved with O-PSc varied between 23% and 26%, depending on the concentration during the reaction. Immunological studies showed[110] that the neoglycoconjugates thus prepared were recognized by convalescent-phase sera from patients recovering from cholera, and anti-O-PS–core-protein responses correlated with plasma anti-lipopolysaccharide and vibriocidal responses. Overall, these findings opened unprecedented possibilities[41,104,118] for the preparation of conjugate vaccines for infectious diseases from bacterial carbohydrates that contain a single amino group, or from those which can be modified to contain one.

5 The road to the present state of the art

The finding that a bacterial polysaccharide can be directly conjugated to a carrier protein, that is without first attaching a spacer/linker to either component, was a great leap in the conjugate vaccine development. Chemicals used within squaric acid diester-mediated conjugation are inexpensive, and operations involved, including purification of the product vaccine, are simple to perform. Already the first, rudimentary conjugations of O-PS by squaric acid chemistry showed[110] that the conjugation efficiency[§] was as good or better than the conjugation efficiency observed with making similar vaccines through other chemistries.[115] Aside from the cost of labor, the future economy of making vaccines in this way on the industrial scale will depend largely on conjugation efficiency and the cost of the two components conjugated, namely the carrier protein and the bacterial antigen. The advances made in the recombinant technology will determine the cost of the carrier and can be expected

[§]Conjugation efficiency *i.e.* the effective utilization of the precious O-PSc-squarate is calculated as a ratio of the amount O-PSc-squarate attached chemically to the protein to that used at the onset of the conjugation.

to decrease over time. The overall cost of the vaccine will depend on how effectively the precious bacterial antigen might be utilized during the conjugation process.

Critical analysis[65] of the original protocol[110] for conjugation revealed that the chemistry taking place during labeling the amino group-containing carbohydrate polymer (squarate preparation from O-PSc) and conjugation of O-PSc-squarate are more complex and less straightforward than indicated, *i.e.* Scheme 1. Results of accompanying studies revealed[65] that several processes that had not been previously thought of were taking place during the squarate labeling of the O-PSc and conjugation of O-PSc-squarate, and affected negatively the outcome of the conjugation. For example, to compensate for the presumed lower reactivity of the amino group present in the bulkier core, compared to the one in the linker-equipped small molecules (*e.g.* Scheme 3), attempts were made to drive the squarate-forming reaction with bacterial polysaccharides to completion by the use of a large excess of the dimethyl squarate reagent. This, however, resulted in quite unexpected repercussion: while the squarate derivative of the polysaccharide may have initially been formed faster, a large part of the excess of the dimethyl squarate reagent used hydrolyzed. The strong[119] monobasic squaric acid thus formed caused the pH of the reaction medium to drop considerably, the amine became protonated and, therefore, the amine to squarate transformation stopped. We have experimentally determined[65] that when insufficient amount of low concentration buffer (0.05 M, pH 7.0) was used during amine to squarate conversion, as was, inadvertently, the case with our original protocol,[110] the pH of reaction dropped as low as to pH 1.5, and ~50% of the amine remained unconverted to squarate due to amine protonation. Because the unchanged polysaccharide antigen (amine) was not separated from the squarate, unlike in the case of preparation of squarates from oligosaccharides where the product squarate can be purified by chromatography, the polysaccharide amine and the corresponding squarate were isolated together as one product. This made the squarate prepared at those conditions a much less efficient conjugation reagent. Hence, the efficiency of conjugation was ~25%.[110]

Other processes, physical and chemical, which more or less affect the overall conjugation process with antigens of bacterial origin were also identified and dealt with.[65] Eventually, results from studies using model compounds, aided by monitoring of the squarate formation by the fluorescamine assay[120] on the micro scale, led to protocols[65] for preparation of higher quality of the conjugation reagent and, in turn, to increased conjugation efficiency from ~25%[111] to ~50%.[65]

The aforementioned study[65] showed also that the efficiency of conjugation of bacterial polysaccharides by squaric acid chemistry can be increased somewhat by recovering the spent, squarate labeled polysaccharide antigen. Such material is normally used in excess at the onset of conjugation and undergoes hydrolysis to a variable extent during conjugation in the presence of pH 9 buffer. The squarate salt thus formed, can be isolated and treated with MeI in the absence of an additional base, to result in regeneration of the precious conjugation reagent. Immunological properties of vaccines made exclusively from the squarate labeled polysaccharide antigen regenerated in this way were found comparable[65] to those of the immunogen made from the parent conjugation reagent and the same protein, showing that recovery and regeneration of spent squarate-labeled conjugation reagent are feasible. The improved squarate-labeling protocol along with the ability to regenerate the spent squarate labeled antigen further increases the attractiveness of squaric acid chemistry for industrial production of conjugate vaccines from O-PS–core fragments of bacterial lipopolysaccharides.

Successful, direct conjugation of bacterial polysaccharides to obtain immunogens[104,110] makes the preparation of vaccines from other, similar antigens in the same way a logical next step. We have recently applied the same method of conjugation to O-PS–core polysaccharides from *Shigella flexneri* 2a and 3a. These polysaccharides are much larger molecules than the similar materials from *V. cholerae* O1 strains. Nevertheless, they produced conjugates smoothly from their respective squarate derivatives. Contrary to previous approaches,[121–124] as in the case of *V. cholerae* O1 vaccines, derivatization of components to be conjugated was not necessary. A typical mass spectral analysis of one of the constructs we obtained by conjugating O-PS–core polysaccharide from *Shigella flexneri* 2a to BSA is in Fig. 6. Work towards similar constructs from other bacterial pathogens where the antigen contains the required single amino group is in progress.

6 Practical considerations

Preparation of glycoconjugates to generate vaccines involves two fundamentally different types of antigens: synthetic and bacterial. Working with synthetic materials is much easier because they are well-defined, small or medium sized, single molecular species. Contrary to such species, exceptions aside, bacterial antigens are high molecular mass, polymolecular materials. They are seldom fully characterized, and, therefore, their composition (polymolecularity) and physicochemical properties may vary from batch to batch. Problems with polymeric materials start with the task to transform the amino group-containing antigen into the corresponding squarate derivative. As stated above, in the case of small or intermediate molecules the conversion is easy to monitor, and the product is easy to purify. The same tasks are more difficult with bacterial antigens. Without going into details, which will be described elsewhere,[65] suffice it to say that to drive the O-PS–core → O-PS–core–squarate conversion to completion is eminently important, and it can be confidently determined with bacterial polysaccharides only by the use of an ultrasensitive spectrophotometer that can monitor the transformation on a micro scale. The use of the above-mentioned fluorescamine assay on the micro scale is very useful in that regard. One has to bear in mind, however, that that assay does not monitor the formation of product but the consumption of the free amine. As such, it does not distinguish between the formation of the desired squarate from other

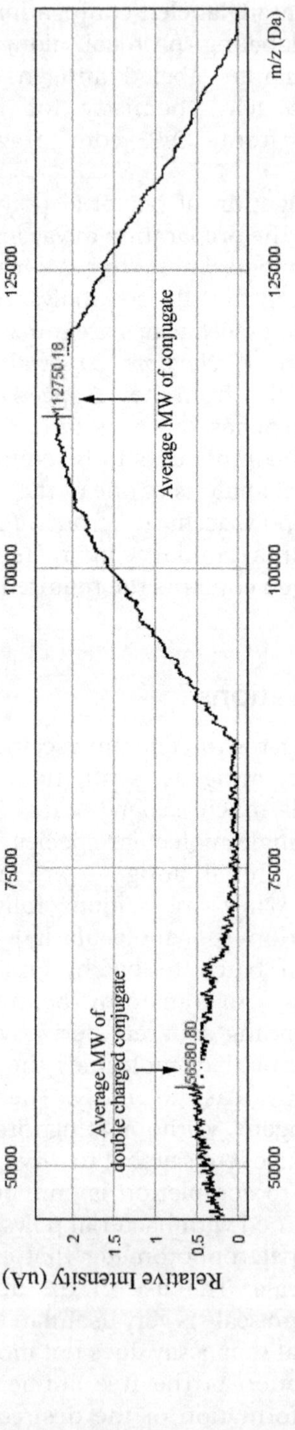

Fig. 6 SELDI-TOF result of a *Shigella flexneri* 2a/BSA conjugation at 92 h of reaction time.

processes that consume the amine, for example protonation of the amine. Because the electrophiles do not react with squaric acid diesters, the fluorescamine assay may produce misleading results when the reaction medium becomes acidic during amine → squarate conversion, when protonation of the amine takes place. Thus, when new synthons are involved, monitoring pH during O-PS–core → O-PS–core–squarate conversions, as well as during conjugations, and readiness to adjust the pH, if necessary, is essential.

As already mentioned, conjugation of amino group-containing substances by squaric acid chemistry is experimentally simple and, normally, predictable. Nevertheless, depending on the nature of the two species to be conjugated, the carrier and the amine-containing carbohydrate, the unexpected can happen, especially when working with a new carrier. Many operations are easy to deal with when working with BSA. It is an inexpensive protein, commercially available in many grades of quality and easy to work with. While initial, exploratory work toward conjugate vaccines often involves BSA, more serious work, whose product is meant to be eventually subjected to regulatory approval, requires working with high-quality carriers, for example recombinant proteins. Physical properties of such materials may not be as favorable as those of BSA, and it is prudent to know them well before planning a conjugation experiment. Carbohydrates are normally well soluble in aqueous media and it is the solubility of the protein at a high concentration of the antigen what might become the limiting factor because too high concentration may lead to technical difficulties (see below). Although chemicals required to effect the conjugation are inexpensive, the two main players, namely the antigen and the recombinant carrier to be conjugated, are usually not. Therefore, it should be possible to do the initial exploration of the conjugation conditions on the smallest practicable scale. Because keeping the pH of the reaction medium at the required level was found[32] to be paramount, to have handy a reliable pH meter and a micro probe allowing pH measurement in a 1–3 μL volume is very useful. Checking the pH both during squarate preparation and conjugation, along with checking the progress of chemical conversions can be useful for troubleshooting, or when one discovers that the conjugation had stopped unexpectedly or that the expected conjugation never started. In one occasion, when we noticed that the latter happened, the culprit was the contaminated freeze-dryer. The ice accumulated on the large trap surface was contaminated with HCl from previous freeze-drying of a free HCl-containing aqueous solution of $R–NH_3^+Cl^-$ salt. Because the freeze dryer was not properly cleaned before the next run, the small amount of the HCl within the freeze-dryer was sufficient to protonate a small sample of a protein that was freeze-dried subsequently. As a result, when such protein was added to a solution of squarate in a small volume of 0.05 M pH 9 buffer, the expected conjugation would not even start.

We also noticed in the past[32] that, for example, some commercially available protein carriers are more acidic than others. Therefore, we recommend measuring pH of the freshly prepared solution made from a new batch of a carrier in the requisite volume of the buffer before addition of the antigen. When the pH had changed, addition of more buffer will not be very helpful, knowing that it is beneficial to conduct conjugation at the highest concentration practicable.[32,83] The easiest way to adjust pH that had run astray, without effecting meaningful concentration change, is to add the respective buffer salt. That can be had by freeze-drying some of each of the pH 7 and pH 9 buffer. Store the salt in a tightly closed container, and have it handy.

Isolation and purification of glycoconjugates is very easy when conjugation mixture does not contain any unchanged carrier because the conjugate is by far the highest molecular mass material present. It is difficult to ensure the absence of the unchanged protein carrier when the targeted loading is low (≤2). Filtration/dialysis through one of the commercially available spin devices, equipped with the requisite filtration membrane, readily removes the low-molecular-weight material, and the conjugate is collected in the retentate. The dialysis should always be done against dilute (0.01 M) ammonium carbonate rather than water, to avoid losses on the plastic membrane owing to hydrophobic interaction.

We found the V-vials (Wheaton Product No. W986273NG) equipped with Spin Vanes to be very convenient vessels for conjugation. The heat transfer from the magnetic stirrer to the bottom of the vial should be prevented by proper insulation, and motion of the content of the vial should be reliable but minimal. Although our experimental evidence shows that the conjugation efficiency increases with the concentration of the ligand, a word of caution is in order. While we have not experienced problems with conjugation of oligosaccharides at high concentration of the ligand (*e.g.* 40 mmol),[32] concentrated solutions of polysaccharides tend to be too viscous and difficult to keep in motion at such high concentration. More importantly, some of the viscous solution tends to dry on the wall of the vessel during the several days long conjugation time and lose contact with the rest of the solution. When this is not corrected[65] the product conjugate may contain starting protein and, in addition, is a much more heterogeneous mixture of conjugates than it would be otherwise.

7 Suggested conjugation protocols

7.1 For synthetic mono- and oligosaccharides

7.1.a Labeling of amino group-containing carbohydrates with the methyl squarate group. A solution of primary amine (1.0 mmol) and 3,4-dimethoxy-3-cyclobutene-1,2-dione (1.5 mmol) in Buffer A (5.0 mL, BuffAR pH 7.0 Reference solution, Mallincrodt, Cat. No. 0031-04) is stirred at room temperature until TLC showed that all the starting amine was consumed (most conversions are complete within a few hours). The mixture is concentrated and the residue is chromatographed to give the methyl squarate labeled amine in good to excellent yield.

7.1.b Conjugation of methyl squarate-labeled carbohydrates to protein carriers (targeted loading, 5.0). Buffer B [0.5 M borate buffer pH 9, made in house (1 L) from boric acid (30.9 g), KCl (26.1 g), and KOH

(8.42 g), and final adjustment to pH 9.0 by addition of solid KOH] is added to a mixture of protein carrier (1.0 eq.) and hapten (6.0 eq., methyl squarate-labeled mono- or oligosaccharide) in the amount to form a 4 mM solution of the hapten. The progress of the reaction is monitored by mass spectrometry (for a typical SELDI-TOF result, see Fig. 7). When average loading ~5.0 is reached,¶ the reaction is worked up by ultrafiltration through 30 kDa cut-off membrane (Millipore, Amicon ultrafiltration devices).

7.2 For bacterial O-PS–core with recovery and reactivation of the spent conjugation reagent

7.2.a Labeling of amino group-containing polysaccharide with the methyl squarate group. A 4 mM solution of bacterial O-PS–core amine (1.0 eq.) and 3,4-dimethoxy-3-cyclobutene-1,2-dione (20 eq.) in Buffer C (Buffer A concentrated to 1/10 of the original volume) is gently stirred at room temperature and the progress of the reaction is monitored by fluorescamine assay.‖ When the conversion is complete, the product is isolated by ultrafiltration, at 4 °C, through 3 kDa cut-off centrifugal devices (Millipore Amicon; Follow manufacturer's guidelines for the use of spin devices. The mixture is ultrafiltered/concentrated to 1/10 of the original volume and filled with ultra-pure water. This is repeated 8 times). After the last run, the retentate is lyophilized, to give the methyl squarate-labeled bacterial O-PS–core as white fluffy solid.

7.2.b Conjugation of methyl squarate-labeled carbohydrates to protein carriers. The reactivity of carrier proteins and antigens largely depends on the accessibility of active reaction sites and sizes of the two species. Because the size and shape of different proteins and antigens varies, the optimum molar ratio of antigen and carrier protein used at the onset of the conjugation has to be determined for each case. Described below is an example of conjugation of *V. cholerae* O1 Inaba O-PS–core squarate (M.w. ~6000 Da) with rTT-Hc, (M.w. 52 002 Da, based on the amino acid sequence[125] using ExPASy Compute pI/Mw tool), targeted for a conjugate having an antigen/protein ratio of ~5.0.

rTT-Hc (1.0 eq.) and methyl squarate labeled antigen (10.0 eq.) is weighed into a V-vial (see above; the size of the vial is selected depending on the anticipated volume, and it should allow for the smallest empty volume practicable over the level of the reaction mixture, to minimize concentration change during the conjugation time due to evaporation). Buffer B is added to make the concentration of the antigen 5 mM, and the mixture is gently shaken occasionally. When a clear solution is formed

¶Approximate reaction times can be deduced from results of conjugations of di-, tetra- and a hexasaccharide to BSA.[32]

‖2 μL of the reaction mixture is withdrawn and diluted with pH 7.2 PBS buffer (KD Medical, RGF-3190) to make the concentration of the antigen 1 mg mL^{-1}. This solution is mixed with a fluorescamine stock solution (3 mg mL^{-1} fluorescamine in DMSO, 1/3 of the volume of the sample solution) and kept at room temp. for 15 min. The resulting solution is used for fluorescamine assay.

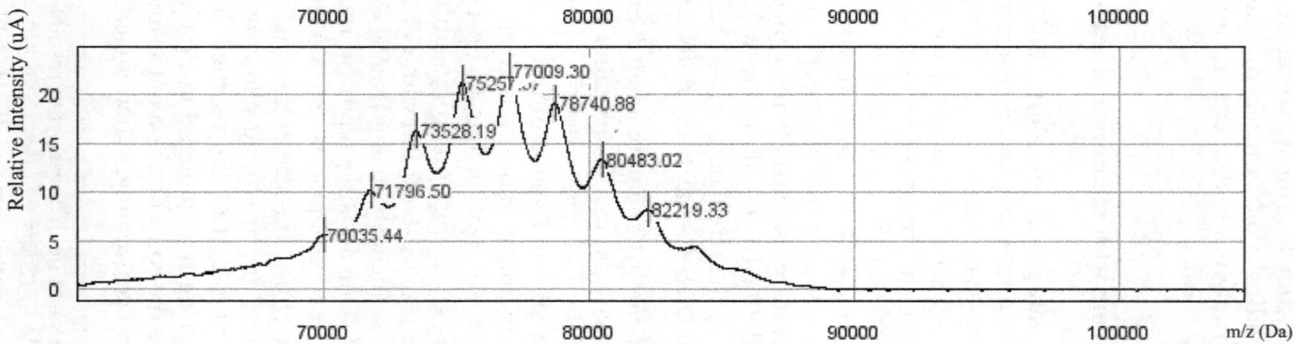

Fig. 7 SELDI-TOF result of conjugation of a hexasaccharide squarate (M.w. 1749 Da) to BSA (M.w. 66 430 Da) at 22 h of reaction time.

(~5 min), the mixture is kept at room temperature with intermittent monitoring of progress of the conjugation by SELDI TOF MS (Fig. 8). After 120 h, the antigen/protein ratio reaches ~4.9. The reaction mixture is diluted with 10 mM aq. $(NH_4)_2CO_3$ and the solution is passed, at 4 °C, through Millipore Amicon Ultra 30 KDa cut-off ultrafiltration devices (Millipore Amicon; Follow manufacturer's guidelines for the use of spin devices.). The material that passes through the membrane is collected separately. The retentate is washed with 10 mM aq. $(NH_4)_2CO_3$ (7×), to ensure that all low-molecular mass material is removed from the retentate (the conjugate). The final retentate is transferred into a glass vial and lyophilized, yielding the conjugate as white fluffy solid.

7.2.c Recovery and reactivation of the spent conjugation reagent. Note: The protocol below was developed for recovery of spent squarate reagent made from the O-PSc fragment of LPS of *V. cholerae* O1, serotype Inaba, and its general applicability to spent squarate reagents made from other polysaccharides is not implied here.

During isolation of the above conjugate by ultrafiltration, the borate buffer is exchanged for the volatile $(NH_4)_2CO_3$ solution. The filtrate, combined with the washings, may contain among other things, a small amount of unchanged squarate reagent (**10**) and a considerable amount of product of its hydrolysis in the form of potassium salt (**11**). Ammonium carbonate present reacts (Scheme 6), very slowly,[65] with **10**, if present, to form the corresponding amide (**12**). Because only **11** can be converted (reactivated) to **10**, the formation of **12** is undesirable because it decreases the efficiency of the reactivation. Therefore, it is desirable to let conjugation run longer than necessary, to allow for the squarate to either attach to the protein to form conjugate **13** or to hydrolyze, to form **11**. Alternatively, the reaction of unchanged **10** with $(NH_4)_2CO_3$ can be minimized by keeping the filtrate combined with the washings refrigerated during work-up of the conjugation mixture or during any time prior to methylation.

The material that had passed through the Millipore Amicon Ultra 30 KDa cut-off membrane, combined with the seven washings, is concentrated at reduced pressure to a smaller volume and transferred into a Millipore Amicon Ultra 3 KDa cut-off ultrafiltration devices and washed (8×), at 4 °C, with ultra-pure water as described above. The filtrate is discarded and the final retentate is transferred, with the aid of water, into a vial for freeze-drying, yielding the spent antigen as white solid.

In a V-vial, the above solid (1.0 eq.), is dissolved in anhydrous DMF to make the concentration of the antigen ~8 mM. A solution of MeI (5.0 eq.) in DMF (1.1 μL) was added from a stock solution prepared from MeI (5.6 volume) and DMF (34.4 volume). The reaction vessel is closed tightly and the mixture is kept at room temperature for 24 h. The mixture is diluted with water (following the manufacturer's guidelines for solvent–membrane compatibility with DMF) and passed, at 4 °C, through a Millipore Amicon Ultra-4 (3 KDa cut-off membrane) device. The retentate is washed/ultrafiltered with pure water (7×) and, after

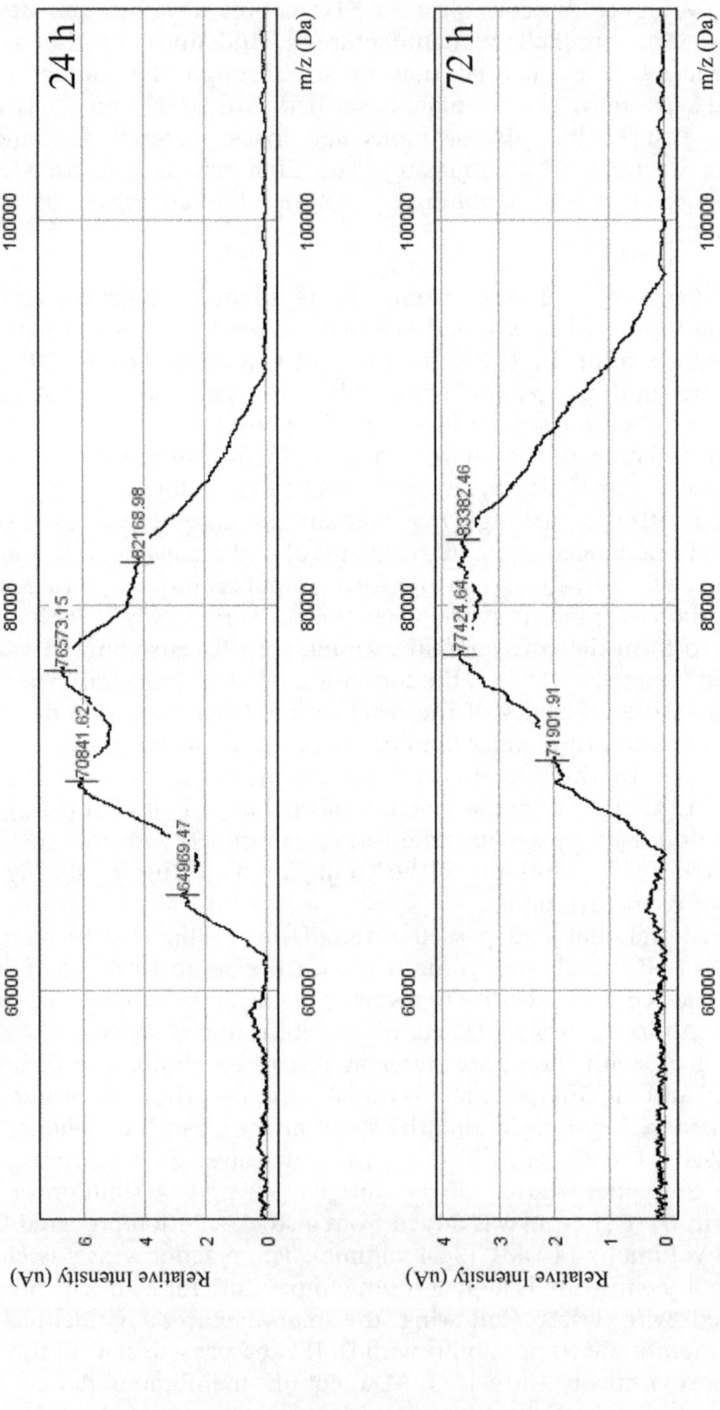

Fig. 8 SELDI-TOF result of a *V. cholerae* O1 Inaba O-PS–core/rTT-Hc conjugation at 24 h and 72 h of reaction time.

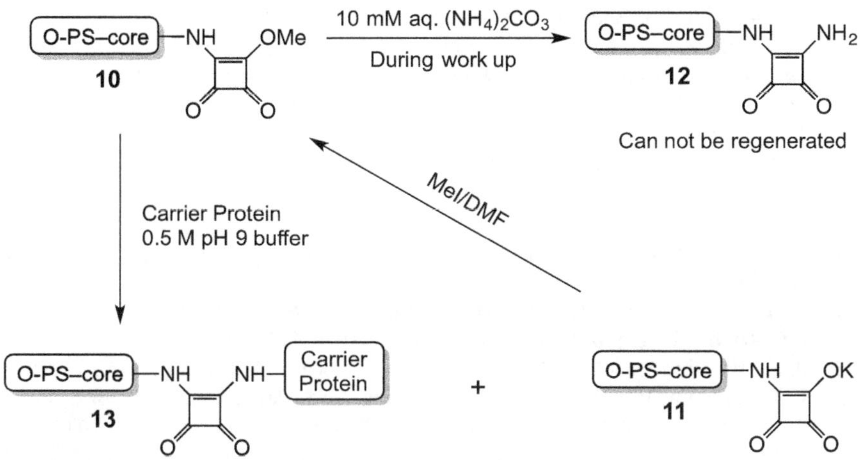

Scheme 6 Conjugation and recovery/regeneration of the spent bacterial O-PS–core squarate.

freeze-drying of the retentate, the regenerated antigen is obtained as white fluffy solid.

Acknowledgements

This work was supported by the Intramural Research Program of the NIH, NIDDK.

References

1 O. T. Avery and W. F. Goebel, *J. Exp. Med.*, 1929, **50**, 533–550.
2 W. F. Goebel and O. T. Avery, *J. Exp. Med.*, 1929, **50**, 521–531.
3 W. F. Goebel, *J. Exp. Med.*, 1939, **69**, 353–363.
4 W. F. Goebel, *J. Exp. Med.*, 1940, **70**, 33–48.
5 Y. C. Lee and R. T. Lee, *Neoglycoconjugates: Preparation and Application*, Academic Press, New York, 1994.
6 K. H. Glüsenkamp, W. Drosdziok, G. Eberle, E. Jähde and M. F. Rajewsky, *Z. Naturforsch., C: J. Biosci.*, 1991, **46**, 498–501.
7 L. F. Tietze, M. Arlt, M. Beller, K. H. Glüsenkamp, E. Jähde and M. F. Rajewsky, *Chem. Ber.*, 1991, **124**, 1215–1221.
8 L. F. Tietze, C. Schröter, S. Gabius, U. Brinck, A. Goerlach-Graw and H. J. Gabius, *Bioconjugate Chem.*, 1991, **2**, 148–153.
9 G. T. Hermanson, *Bioconjugate Techniques*, 1st edn, Academic Press, New York, 1996.
10 G. T. Hermanson, *Bioconjugate Techniques*, 2nd edn, Elsevier Inc., 2008.
11 C. I. C. Crucho, P. Correia-da-Silva, K. T. Petrova and M. T. Barros, *Carbohydr. Res.*, 2015, **402**, 124–132.
12 D. Cancogni, M. Fusari, N. Poggi and L. Lay, in *Polysaccharides*, ed. K. G. Ramawat and J.-M. Mérillon, Springer, 2015.
13 M. Izumi, S. Okumura, H. Yuasa and H. Hashimoto, *J. Carbohydr. Chem.*, 2003, **22**, 317–329.
14 F. R. Wurm and H.-A. Klok, *Chem. Soc. Rev.*, 2013, **42**, 8220–8236.
15 D. R. Bundle, in *Vaccine Design: Innovative Approaches and Novel Strategies*, ed. F. Bagnoli and R. Rappuoli, Caister Academic Press, 2011, pp. 69–107.

16 M. Ohno and S. Eguchi, *Top. Heterocycl. Chem.*, 2006, **6**, 1–37.
17 H. Kunz, S. Dziadek, S. Wittrock and T. Becker, in *Carbohydrate-Based Vaccines*, American Chemical Society, 2008, vol. 989, pp. 293–310.
18 M. Lu, MS, University of Waterloo, Waterloo, Ontario, Canada, 2008.
19 A. A. Karelin, Y. E. Tsvetkov, L. Paulovicova, S. Bystricky, E. Paulovicova and N. E. Nifantiev, *Carbohydr. Res.*, 2009, **344**, 29–35.
20 A. A. Karelin, Y. E. Tsvetkov, L. Paulovicova, S. Bystricky, E. Paulovicova and E. N. Nikolay, *Carbohydr. Res.*, 2010, **345**, 1283–1290.
21 T. K. Lindhorst, K. Bruegge, A. Fuchs and O. Sperling, *Beilstein J. Org. Chem.*, 2010, **6**, 801–809.
22 A. D. Cox, F. St Michael, D. Neelamegan, S. Lacelle, C. Cairns and J. C. Richards, *Glycoconjugate J.*, 2010, **27**, 401–417.
23 P. I. Abronina, N. M. Podvalnyy, T. M. Mel'nikova, A. I. Zinin, K. G. Fedina, V. V. Kachala, V. I. Torgov, L. O. Kononov, E. A. Panfertsev, E. V. Baranova, V. V. Mochalov, V. I. Dyatlova and S. F. Biketov, *Russ. Chem. Bull.*, 2010, **59**, 2333–2337.
24 B. Cao, J. M. White and S. J. Williams, *Beilstein J. Org. Chem.*, 2011, **7**, 369–377.
25 A. C. Araújo, A. Nakhaia, M. Rudab, R. Slättegårdb, P. Gatenholmc and H. Brumer, *Carbohydr. Res.*, 2012, **354**, 116–120.
26 F. Wurm, T. Steinbach and H.-A. Klok, *Chem. Commun.*, 2013, **49**, 7815–7817.
27 N. V. Ganesh, J. M. Sadowska, S. Sarkar, L. Howells, J. McGiven and D. R. Bundle, *J. Am. Chem. Soc.*, 2014, **136**, 16260–16269.
28 O. Cabrera Blanco, M. Cuello Perez, V. G. Sierra Gonzalez, C. R. Soto Rodriguez and M. E. Martinez Pozo, WO 2004043489, 2004.
29 O. Sperling, M. Dubber and T. K. Lindhorst, *Carbohydr. Res.*, 2007, **342**, 696–703.
30 F. Mawas, J. Niggemann, C. Jones, M. J. Corbel, J. P. Kamerling and J. F. G. Vliegenthart, *Infect. Immun.*, 2002, **70**, 5107–5114.
31 F. Sztaricskai, E. Roth, M. Andrei, I. F. Pelyvas and P. Herczegh, *Chem. Lett.*, 2007, **36**, 1012–1013.
32 S. j. Hou, R. Saksena and P. Kováč, *Carbohydr. Res.*, 2008, **343**, 196–210.
33 H. Yan, A. L. Aguilar and Y. Zhao, *Bioorg. Med. Chem. Lett.*, 2007, **17**, 6535–6538.
34 X. Wu, C.-C. Ling and D. R. Bundle, *Org. Lett.*, 2004, **6**, 4407–4410.
35 A. F. G. Bongat, R. Saksena, R. Adamo, Y. S. Z. Fujimoto, D. C. Peterson, K. Fukase, W. F. Vann and P. Kováč, *Glycoconjugate J.*, 2010, **27**, 69–77.
36 N. W. McGill and S. J. Williams, *J. Org. Chem.*, 2009, **74**, 9388–9398.
37 J.-W. Wang, A. Asnani and F.-I. Auzanneau, *Bioorg. Med. Chem.*, 2010, **18**, 7174–7185.
38 O. Blixt and T. Norberg, *Carbohydr. Res.*, 1999, **319**, 80–91.
39 F.-I. Auzanneau, S. Borrelli and B. M. Pinto, *Bioorg. Med. Chem. Lett.*, 2013, **23**, 6038–6042.
40 C. Dingels, F. Wurm, M. Wagner, H.-A. Klok and H. Frey, *Chem. – Eur. J.*, 2012, **18**, 16828–16835.
41 E. T. Ryan, P. Kováč, F. Qadri, P. Xu, S. B. Calderwood, W. F. Vann and D. C. Peterson, WO 2013009826 Al, 2013.
42 S. Niewiadomski, Z. Beebeejaun, H. Denton, T. K. Smith, R. J. Morris and G. K. Wagner, *Org. Biomol. Chem.*, 2010, **8**, 3488–3499.
43 H. Schottenberger, M. Buchmeiser, J. Polin and K. E. Schwarzhans, *Z. Naturforsch., B: Chem. Sci.*, 1993, **48**, 1524–1532.

44 C. R. Chenthamarakshan and A. Ajayaghosh, *Chem. Mater.*, 1998, **10**, 1657–1663.
45 A. Chernyak, S. Oscarson and D. Turek, *Carbohydr. Res.*, 2000, **329**, 309–316.
46 D. M. Corsi, L. Vander Elst, R. N. Muller, H. Van Bekkum and J. A. Peters, *Chem. – Eur. J.*, 2001, **7**, 64–71.
47 Y. He, J. Xu, E. Fasella and L. L. Kiessling, 222nd ACS National Meeting, Chicago, IL, 2001.
48 A. Bergh, S. Bhattacharyya and U. J. Nilsson, *Carbohydr. Res.*, 2002, **337**, 947–949.
49 S. M. C. Johansson, N. Arnberg, M. Elofsson, G. Wadell and J. Kihlberg, *ChemBioChem*, 2005, **6**, 358–364.
50 T. Ishida, T. Shinada and Y. Ohfune, *Synthesis*, 2005, 2723–2729.
51 M. B. Onaran, A. B. Comeau and C. T. Seto, *J. Org. Chem.*, 2005, **70**, 10792–10802.
52 T. Hino, Y. Seida, T. Takahashi and N. Kuramoto, *Polym. Int.*, 2006, **55**, 243–247.
53 M. V. Reddington, *Bioconjugate Chem.*, 2007, **18**, 2178–2190.
54 B. Halton, *Chem. N. Z.*, 2008, **72**(57–59), 62.
55 S. F. Voelker, S. Uemura, M. Limpinsel, M. Mingebach, C. Deibel, V. Dyakonov and C. Lambert, *Macromol. Chem. Phys.*, 2010, **211**, 1098–1108.
56 C. Dingels, F. Wurm, H.-A. Klok and H. Frey, presented in part at 241st ACS National Meeting & Exposition, Anaheim, CA, United States, March 27–31, 2011.
57 C. Luo, Q. Zhou, G. Jiang, L. He, B. Zhang and X. Wang, *New J. Chem.*, 2011, **35**, 1128–1132.
58 F. Wurm, C. Dingels, H. Frey and H.-A. Klok, *Biomacromolecules*, 2012, **13**, 1161–1171.
59 D. V. Yashunskii, Y. E. Tsvetkov and N. E. Nifant'ev, RU Pat. 2476444C1, 2013.
60 K. W. Hill and V. R. Mokler, WO 2013176844A1, 2013.
61 T. Steinbach, F. Wurm and H.-A. Klok, *Polym. Chem.*, 2014, **5**, 4039–4047.
62 B. Palitzsch, S. Hartmann, N. Stergiou, M. Glaffig, E. Schmitt and H. Kunz, *Angew. Chem., Int. Ed.*, 2014, **53**, 14245–14249.
63 R. Saksena, X. Ma and P. Kováč, *Carbohydr. Res.*, 2003, **338**, 2591–2603.
64 X. Ma, R. Saksena, A. Chernyak and P. Kováč, *Org. Biomol. Chem.*, 2003, **1**, 775–784.
65 P. Xu, W. F. Vann, E. T. Ryan and P. Kováč, in preparation, 2016.
66 J. E. Thorpe, *J. Chem. Soc., B*, 1968, 435–436.
67 G. Maahs and P. Hegenberg, *Angew. Chem.*, 1966, **78**, 927–929.
68 V. P. Kamath, P. Diedrich and O. Hindsgaul, *Glycoconjugate J.*, 1996, **13**, 315–319.
69 R. Saksena, J. Zhang and P. Kováč, *Tetrahedron: Asymmetry*, 2005, **16**, 187–197.
70 A. Bergh, B.-G. Magnusson, J. Ohlsson, U. Wellmar and U. J. Nilsson, *Glycoconjugate J.*, 2001, **18**, 615–621.
71 W. E. Dick Jr. and M. Beurret, in *Conjugate Vaccines*, ed. J. M. Cruse and R. E. Lewis Jr., Krager, Basel, 1989, vol. 10, pp. 48–114.
72 L. Kenne, B. Lindberg, P. Unger, B. Gustafsson and T. Holme, *Carbohydr. Res.*, 1982, **100**, 341–349.
73 M. Dubois, K. A. Gilles, J. K. Hamilton, P. A. Rebers and F. Smith, *Anal. Chem.*, 1956, **28**, 350–356.
74 F. I. Auzanneau and M. Pinto, *Bioorg. Med. Chem.*, 1996, **4**, 2003–2010.

75 L. A. Ellis, C. S. Mc Vay, M. A. Probert, Z. Zhang, D. R. Bundle and J. A. Appleton, *Glycobiology*, 1997, **7**, 383–390.
76 V. Pozsgay, E. Dubois and L. Pannell, *J. Org. Chem.*, 1997, **62**, 2832–2846.
77 J. Zhang and P. Kováč, *Carbohydr. Res.*, 1999, **321**, 157–167.
78 J. Zhang and P. Kováč, *Tetrahedron Lett.*, 1998, **39**, 1091–1094.
79 J. Zhang and P. Kováč, *Bioorg. Med. Chem. Lett.*, 1999, **9**, 487–490.
80 J. Zhang, A. Yergey, J. Kowalak and P. Kováč, *Tetrahedron*, 1998, **54**, 11783–11792.
81 R. Roy, E. Katzenellenbogen and H. J. Jennings, *Can. J. Biochem. Cell Biol.*, 1984, **62**, 270–275.
82 M. Dubber and T. K. Lindhorst, *Synthesis*, 2001, 327–330.
83 J. Zhang, A. Yergey, J. Kowalak and P. Kováč, *Carbohydr. Res.*, 1998, **313**, 15–20.
84 G. Solladie and C. Ziani-Cherif, *J. Org. Chem.*, 1993, **58**, 2181–2185.
85 D. R. Bundle and S. Josephson, *Carbohydr. Res.*, 1980, **80**, 75–85.
86 R. U. Lemieux, D. R. Bundle and D. A. Baker, *J. Am. Chem. Soc.*, 1975, **97**, 4076–4083.
87 R. U. Lemieux, D. A. Baker and D. R. Bundle, *Can. J. Biochem.*, 1977, **55**, 507–512.
88 H. J. Vermeer, J. P. Kamerling and J. F. G. Vliegenhart, *Tetrahedron: Asymmetry*, 2000, **11**, 539–547.
89 H. J. Vermeer, K. M. Halkes, J. A. van Kuik, J. P. Kamerling and J. F. G. Vliegenthart, *Perkin 1*, 2000, 2249–2263.
90 A. Chernyak, A. Karavanov, Y. Ogawa and P. Kováč, *Carbohydr. Res.*, 2001, **330**, 479–486.
91 M. Nitz and D. R. Bundle, *J. Org. Chem.*, 2001, **66**, 8411–8423.
92 D. J. Lefeber, PhD, University of Utrecht, 2001.
93 D. J. Lefeber, J. P. Kamerling and J. F. G. Vliegenthart, *Chem. – Eur. J.*, 2001, **7**, 4411–4421.
94 A. A. Karelin, Y. E. Tsvetkov, G. Kogan, S. Bystricky and N. E. Nifantiev, *Russ. J. Bioorg. Chem.*, 2007, **33**, 110–121.
95 J. R. Rich, W. W. Wakarchuk and D. R. Bundle, *Chem. – Eur. J.*, 2006, **12**, 845–858.
96 B. Sharma, H. Keula, H. Hockera, T. Loontjensb and R. van Benthem, *Polymer*, 2005, **46**, 1775–1783.
97 R. B. Hossany, M. A. Johnson, A. A. Eniade and B. M. Pinto, *Bioorg. Med. Chem.*, 2004, **12**, 3743–3754.
98 A. Tevyashova, F. Sztaricskai, G. Batta, P. Herczegh and A. Jeney, *Bioorg. Med. Chem. Lett.*, 2004, **14**, 4783–4789.
99 J. Y. Wang, A. H. C. Chang, H.-K. Guttormsen, A. L. Rosas and D. L. Kasper, *Vaccine*, 2003, **21**, 1112–1117.
100 N. Nagahori, R. T. Lee, S. I. Nishimura, D. Pagé, R. Roy and Y. C. Lee, *ChemBioChem*, 2002, **3**, 836–844.
101 K. Sato, K. Seio and M. Sekine, *J. Am. Chem. Soc.*, 2002, **124**, 12715–12724.
102 R. Saksena, R. Adamo and P. Kováč, *Bioorg. Med. Chem.*, 2007, **15**, 4283–4310.
103 A. Chernyak, S. Kondo, T. K. Wade, M. D. Meeks, P. M. Alzari, J. M. Fournier, R. K. Taylor, P. Kováč and W. F. Wade, *J. Infect. Dis.*, 2002, **185**, 950–962.
104 M. A. Sayeed, M. K. Bufano, P. Xu, G. Eckhoff, R. C. Charles, M. M. Alam, M. Kelly, T. Sultana, M. R. Rashu, A. Berger, G. Gonzalez-Escobedo, A. Mandlik, T. R. Bhuiyan, D. T. Leung, R. C. LaRocque, J. B. Harris, S. B. Calderwood, Qadri, W. F. Vann, P. Kováč and E. T. Ryan, *PLoS Neglected Trop. Dis.*, 2015, **9**, e3881.

105 J. G. L. Wright, L. H. Cazares, S. M. Leung, S. Nasim, B. L. Adam, T. T. Yip, P. F. Schellhammer, L. Gong and A. Vlahou, *Prostate Cancer and Prostate Diseases*, 1999, **2**, 264–276.
106 Y. Zhang, A. R. Sinaiko and G. L. Nelsestuen, in *Methods in Molecular Biology*, ed. D. Josic and D. C. Hixson, 2012, vol. 909, pp. 141–150.
107 A. R. Parameswar, I. H. Park, R. Saksena, P. Kováč, M. H. Nahm and A. V. Demchenko, *ChemBioChem*, 2009, **10**, 2893–2899.
108 P. Xu, M. Trinh and P. Kováč, in preparation, 2016.
109 V. Verez-Bencomo, V. Fernández-Santana, E. Hardy, M. E. Toledo, M. C. Rodrıguez, L. Heynngnezz, A. Rodriguez, A. Baly, L. Herrera, M. Izquierdo, A. Villar, Y. Valdés, K. Cosme, M. L. Deler, M. Montane, E. Garcia, A. Ramos, A. Aguilar, E. Medina, G. Torano, I. Sosa, I. Hernandez, R. Martınez, A. Muzachio, A. Carmenates, L. Costa, F. Cardoso, C. Campa, M. Diaz and R. Roy, *Science*, 2004, **305**, 522–525.
110 P. Xu, M. M. Alam, A. Kalsy, R. C. Charles, S. B. Calderwood, F. Qadri, E. T. Ryan and P. Kováč, *Bioconjugate Chem.*, 2011, **21**, 2179–2185.
111 A. Cox and M. B. Perry, *Carbohydr. Res.*, 1996, **290**, 59–65.
112 O. Holst, in *Bacterial Lipopolysaccharides*, ed. Y. A. Knirel and M. A. Valvano, Springer, Wien, 1999, pp. 21–39.
113 O. Holst and A. Molinaro, in *Microbial Glycobiology*, Elsevier, 2009, pp. 29–55.
114 P. A. Manning, U. H. Stroeher and R. Morona, in *Vibrio cholerae and Cholera: Molecular to Global Perspectives*, ed. I. K. Wachsmuth, P. A. Blake and O. Olsvik, American Society for Microbiology, Washington, D.C., 1994, pp. 77–94.
115 C. Grandjean, A. Boutonnier, B. Dassy, J.-M. Fournier and A. L. Mulard, *Glycocnjugate J.*, 2009, **26**, 41–55.
116 R. K. Gupta, S. C. Szu, R. A. Finkelstein and J. B. Robbins, *Infect. Immun.*, 1992, **60**, 3201–3208.
117 R. Saksena, A. Chernyak, A. Karavanov and P. Kováč, in *Methods in Enzymology*, ed. Y. C. Lee and R. Lee, Academic Press, 2003, vol. 362, pp. 125–139.
118 M. M. Alam, M. K. Bufano, P. Xu, A. Kalsy, Y. Yu, Y. W. Freeman, T. Sultana, M. R. Rashu, I. Desai, G. Eckhoff, D. T. Leung, R. C. Charles, R. C. LaRocque, J. B. Harris, J. D. Clements, S. B. Calderwood, F. Qadri, W. F. Vann, P. Kováč and E. T. Ryan, *PLOS Neglected Trop. Dis.*, 2014, **8**, e2683.
119 S. Cohen and S. G. Cohen, *J. Am. Chem. Soc.*, 1966, **88**, 1533–1536.
120 S. Udenfriend, S. Stein, P. Bohlen, W. Dairman, W. Leimgruber and M. Weigele, *Science*, 1972, **178**, 881–882.
121 J. Kubler-Kielb, E. Vinogradov, C. Mocca, V. Pozsgay, B. Coxon, J. B. Robbins and R. Schneerson, *Carbohydr. Res.*, 2010, **345**, 1600–1608.
122 D. N. Taylor, A. C. Trofa, J. Sadoff, C. Chu, D. Bryla, J. Shiloach, D. Cohen, S. Ashkenazi, Y. Lerman, W. Egan, R. Schneerson and J. B. Robbins, *Infect. Immun.*, 1993, **61**, 3678–3687.
123 A. Phalipon, M. Tanguy, C. Grandjean, C. Guerreiro, F. Bélot, D. Cohen, P. J. Sansonetti and A. L. Mulard, *J. Immunol.*, 2009, **182**, 2241–2247.
124 J. H. Passwell, S. Ashkenzi, Y. Banet-Levi, R. Ramon-Saraf, N. Farzama, L. Lerner-Geva, H. Even-Nir, B. Yerushalmi, C. Chu, J. Shiloach, J. B. Robbins and R. Schneerson, *Vaccine*, 2010, **28**, 2231–2235.
125 P. C. McCarthy, R. Saksena, D. C. Peterson, C.-H. Lee, Y. An, J. F. Cipollo and W. F. Vann, *Glycoconjugate J.*, 2013, **30**, 857–870.
126 D. Safari, H. A. T. Dekker, J. A. F. Joosten, D. Michalik, A. C. de Souza, R. Adamo, M. Lahmann, A. Sundgren, S. Oscarson, J. P. Kamerling and H. Snippe, *Infect. Immun.*, 2008, **76**, 4615–4623.

Recent advances in Kdo-glycoside formation

Paul Kosma

DOI: 10.1039/9781782626657-00116

This chapter reports selected relevant contributions to Kdo glycoside chemistry with an emphasis on progress in stereoselectivity, yield and suppression of the elimination side reaction. It mainly covers examples from the past decade.

1 Introduction

1.1 Occurrence of Kdo

Since its first discovery and structural characterization, Kdo (3-deoxy-D-*manno*-oct-2-ulosonic acid) has been found as a major carbohydrate in bacterial cell wall polysaccharides.[1–3] The abbreviation "Kdo" has been derived from the original term denoting this saccharide as "keto-deoxy-octosonic acid". While Kdo is absent in mammalian organisms, it has been detected in plants as a constituent of rhamnogalacturonan II and in green algae.[4,5] Within Gram-negative bacteria, Kdo occurs in lipopolysaccharides (LPS), capsular polysaccharides (CPS) and exopolysaccharides (EPS). The abundance of Kdo in bacteria is reflected in a recent statistical analysis of a bacterial carbohydrate structure data base.[6] For a detailed overview on the structures of Kdo-containing bacterial oligo- and polysaccharides the reader is referred to in-depth reviews.[7]

This chapter summarizes and illustrates recent and most significant developments achieved in the field of Kdo-glycoside formation relevant to these bacterial poly- and oligosaccharides.

1.2 Structural features of Kdo glycosides in bacteria

Kdo is a member of the family of 3-deoxy-ald-2-ulosonic acids and shares structural features and reactivity properties with related sugars such as *N*-acetyl-neuraminic acid, Kdn as well as with legionaminic and pseudaminic acids. In contrast to the latter saccharides, Kdo has been found in both anomeric pyranoside configurations, but also in a small number of examples in furanoside form. Specifically, α-pyranosidic Kdo constitutes the conserved linkage sugar of the inner core region of bacterial lipopolysaccharides, connecting the core domain to the Lipid A part.[8] In addition, α-Kdo has also been detected as a component in the outer core region,[9] whereas β-Kdo and α-Kdo were found as constituents of O-antigenic polysaccharides in a few bacterial strains.[10] While α-Kdo is prevalent in LPS, in repeating units of capsular polysaccharides and some exopolysaccharides, β-anomeric Kdo has more frequently been reported than α-anomeric Kdo.[11,12] Kdo may occur within chains and as side-chain

University of Natural Resources and Life Sciences-Vienna, Muthgasse 18, A-1190 Vienna, Austria. E-mail: paul.kosma@boku.ac.at

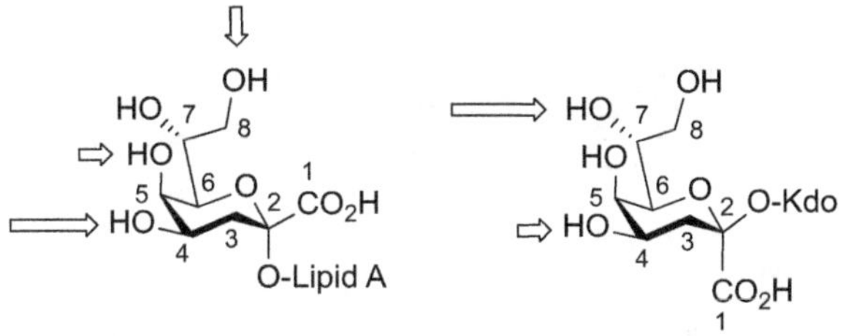

Fig. 1 Main linkage sites found for oligomeric α-Kdo in LPS and β-Kdo units in CPS.

unit, respectively, and harbours four glycosyl acceptor positions, all of which have been found to be glycosylated in bacterial oligo/polysaccharides. In addition, the Kdo unit of the core region may further be substituted by non-carbohydrate appendices such as phosphate and 2-aminoethylphosphate residues, which add additional challenges to the synthesis of the native Kdo ligands. Notably, oligomeric structures of Kdo units have also been detected, involving inter-residue linkages of both the exocyclic diol system as well as the pyranose hydroxyl groups (Fig. 1). The conserved nature of the α-Kdo*p*-(2→4)-α-Kdo*p* disaccharide[8] as constituent of the LPS inner core of Gram-negative bacteria as well as its linkage to the distal position of the glucosamine disaccharide of the Lipid A fragment has rendered this region as the main target for the majority of synthetic approaches in the past. Substitution at position 8 of α-Kdo has been found in a family-specific Kdo antigen in Chlamydiaceae[13] but also in an *Acinetobacter lwoffii* strain.[14] Kdo linkages to position 5 have been identified in other *Acinetobacter* strains.[15] Thus far, the size of α-linked Kdo-oligomers in LPS does not exceed the tetrasaccharide level. Much less attention has been devoted to the assembly of complex β-Kdo glycosides. In capsular polysaccharides, a β-(2→7)-linked Kdo polysaccharide of low-molecular weight (4–4.5 kDa) has been described for a *Sinorhizobium* strain[16] and recently, Kdo-oligomers up to the size of a nonasaccharide comprising putative β-(2→4) and/or β-(2→7)-linkages have been found, which are directly connected to the polysialic acid in *Escherichia coli* K1 and *Neisseria meningitidis* group B capsular polysaccharides, respectively.[17]

2 Chemistry of Kdo glycoside formation

2.1 General aspects

The present review is focused on major developments achieved in the past decade including the synthesis of 4,5-dibranched Kdo units, thereby updating previous reports in the field.[18] Glycoside chemistry of Kdo is challenging beyond the generally important parameters to be considered in glycosylation reactions such as appropriate choice of leaving and protecting groups, solvent, promoter, temperature and additives. Specifically, the absence of a stereodirecting group at carbon 3 substantially reduces the anomeric selectivity. The electron withdrawing effect of the carboxylic group not only deactivates the anomeric center but also leads to facile formation of the α,β-unsaturated ester, which reduces the overall

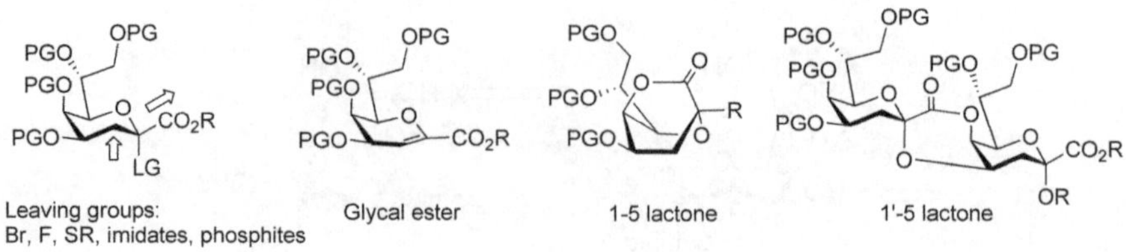

Fig. 2 Common Kdo donors and byproducts.

yield and requires additional efforts in separating the 2,3-dehydro compound from glycoside products. In addition, the ester group exerts steric strain in the formation of the ketosidic linkage and is also prone to intraresidue and interresidue lactonization. The latter lactone type may form in acidic conditions or even during chromatography on silica gel (Fig. 2).[19] While the lactones may easily be reopened under alkaline conditions, they are cumbersome byproducts during deprotection of alkali-labile Kdo–Lipid A conjugates.

The sensitivity of the ketosidic bond toward acidic conditions also limits the application of acid-cleavable protecting groups. Thus, preparation of suitable Kdo donors and glycosylation of Kdo glycosyl acceptor derivatives is not a trivial task in order to solve these issues related to yield, anomeric selectivity, purification and byproduct formation, as has also been briefly summarized in a recent review.[20] While glycosylation reactions of primary alcohols with Kdo donors are usually high-yielding, byproduct formation is increasingly observed with sterically hindered and less reactive acceptor molecules.

2.2 Kdo bromide donors

Previously, Kdo bromides have extensively been used in the 1980s and 1990s for the assembly of α-Kdo units of the LPS inner core. The peracetylated Kdo bromide methyl ester **2** is easily accessible in three steps from ammonium Kdo **1** and has been used to generate α-Kdo*p*-(2→4)-α-Kdo*p* and α-Kdo*p*-(2→8)-α-Kdo*p* units, mostly in modest yields and with low anomeric selectivity, as outlined in previous reviews.[18] The use of Kdo bromide donors is illustrated in a more recent example of regioselective glycosylations of Kdo diol and triol acceptor derivatives, which reduce the effort for protecting group manipulations but result in lower product yields and require additional purification steps. Thus, exploiting the more reactive hydroxy groups at position 8 and 4, respectively, the branched α-(2→8)- and α-(2→4)-linked Kdo-trisaccharide **9** was formed under Helferich conditions.[21] The low yield of **5** in the first coupling of **2** to diol **3** was due to concomitant formation of the glycal ester **4**, the β-(2→8)-disaccharide as well as trisaccharide products (Scheme 1). Subsequent 7-*O*-acetylation and isopropylidene hydrolysis gave a 25% yield of diol acceptor **7**, which was reacted with bromide **2** to produce trisaccharide **8**. Following global deprotection of **8**, the allyl glycoside was converted into the 2-aminoethyl-3-thiopropyl spacer derivative and coupled to BSA to give **10**. Immunization of mice with the

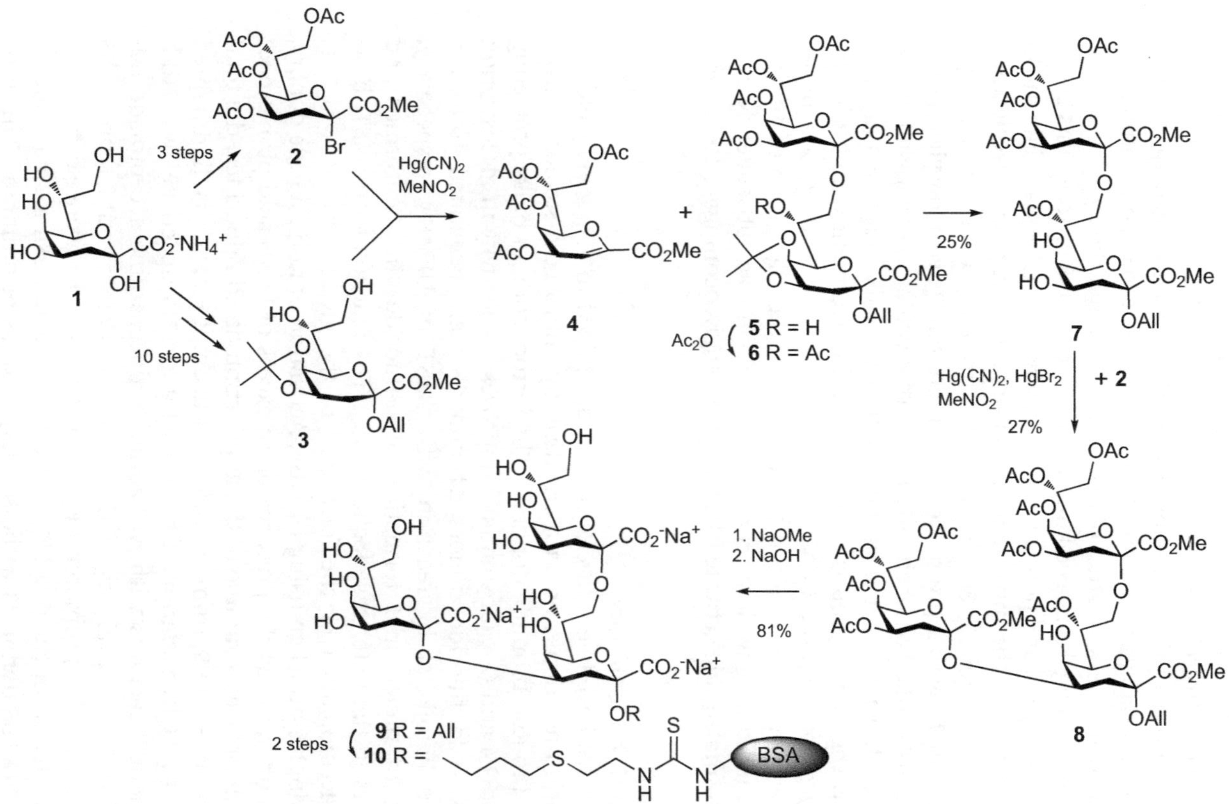

Scheme 1 Synthesis of the *Chl. psittaci* specific branched Kdo trisaccharide **9**.[21] Adapted from P. Kosma, A. Hofinger, S. Müller-Loennies and H. Brade, Synthesis of a neoglycoconjugate containing a *Chlamydophila psittaci*-specific branched Kdo trisaccharide epitope, *Carbohydr. Res.*, **345**(5), 704–708, Copyright (2010), with permission from Elsevier.

synthetic neoglycoconjugate **10** afforded monoclonal antibodies discriminating *Chlamydia trachomatis* from *Chlamydophila psittaci* elementary bodies, thus being of interest in human and veterinarian diagnostics.[22]

Bromide **2** was also used in the synthesis of a branched trisaccharide containing 4-amino-4-deoxy-β-L-arabinose (Ara4N) which corresponds to LPS fragments from *Burkholderia* and *Proteus*.[23] Coupling of the *N*-phenyltrifluoroacetimidate donor **13** to the primary alcohol **12** was promoted by a catalytic amount of TMSO-triflate at low temperature, and produced an anomeric mixture of disaccharide **14** and the 7-*O*-glycosylated regioisomer (arising from acetyl migration to position 8 of acceptor **12**) in a combined 74% yield and poor anomeric selectivity (axial/equatorial ratio of 1.6 : 1) (Scheme 2). The mixture was resolved after Zemplén deacetylation to give the 4,5,7-triol **15** (36% overall yield from **12**). Reaction of the glycosyl acceptor **15** with **2** promoted by mercury salts gave a low yield (19%) of the α-(2→4)-linked trisaccharide **16**, 5% of the α-(2→7)-regioisomer and glycal ester **4**. To retain the allylic function for subsequent conjugation chemistry, the benzyl groups were removed by treatment with $TiCl_4$, followed by global deprotection and reduction of the azido group under Staudinger conditions to afford **18**. Reaction of the allyl group of **18** with 1,3-propanedithiol gave a spacer glycoside, which was subsequently conjugated to maleimide-activated BSA to give neoglycoconjugate **19**.

2.3 Kdo fluoride donors

Kdo fluorides have been developed and optimized to give good yields in glycosylation reactions and with significantly improved α-stereoselectivity compared to Kdo bromide donors.[24] The introduction of bulky protecting groups at *O*-4 and *O*-5 substantially improves α-selectivity in glycosylation reactions. Kdo fluoride donors **21** and **23** were generated from the respective hemiketal by reaction with DAST of DFMBA, respectively, resulting in preferential formation of α-fluorides in the former and β-anomers in the latter case (Scheme 3).[25] The reaction with DAST, however, is accompanied by the formation of the glycal ester **24**, which is largely suppressed when using DFMBA. Activation of Kdo fluoride donors, however, often needs the presence of an excess of Lewis-acid promoters.

As an impressive example of the application of Kdo fluorides, the total synthesis of R*e* LPS from *E. coli* by Kusumoto *et al.* is highlighted (Scheme 4).[26] Glycosidation of donors **21** and **23**, respectively, was studied in model reactions with *N*,*O*-acylated glucosamine acceptor derivatives using 2–4 equivalents of the respective donor and 1.1 up to 2.5 equivalents of the promoter $BF_3 \cdot Et_2O$. Best results were seen in the glycodesilylation of the 6-*O*-TES-derivative **26**, when using 2 equivalents of the 4,5-*O*-TBS donor **23** without a large excess of promoter. This way, the protected α-Kdo-(2→6)-GlcN disaccharide **27** was obtained in 81% yield with good anomeric selectivity (α/β ratio 93 : 7). Application of similar reaction conditions to the coupling with tetra-acylated glucosamine disaccharide acceptor **28** gave the α-(2→6)-linked trisaccharide **29** in 89% yield. The regioselective introduction of the Kdo residue at

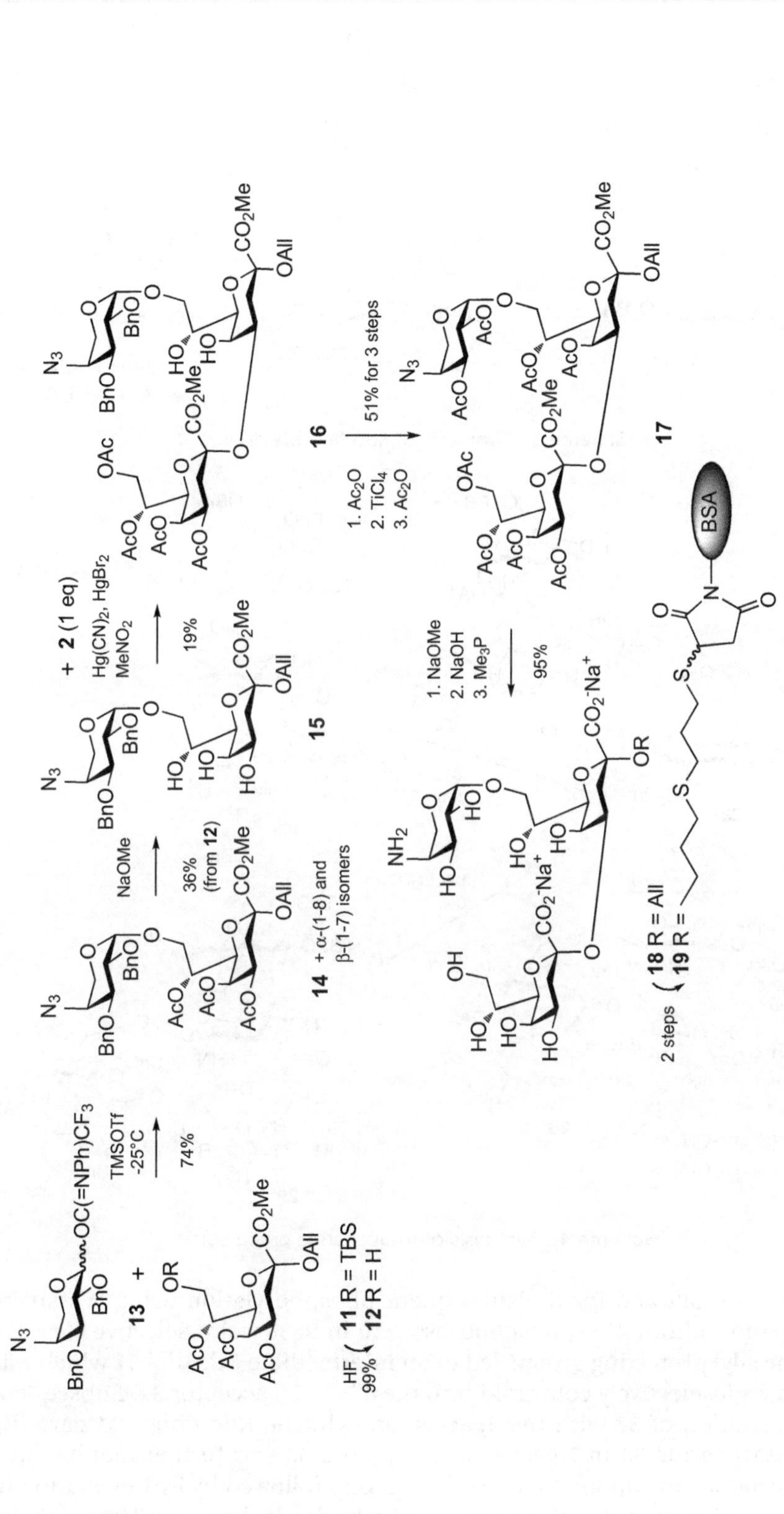

Scheme 2 Synthesis of neoglycoconjugate **19** related to *Proteus* LPS.[23]

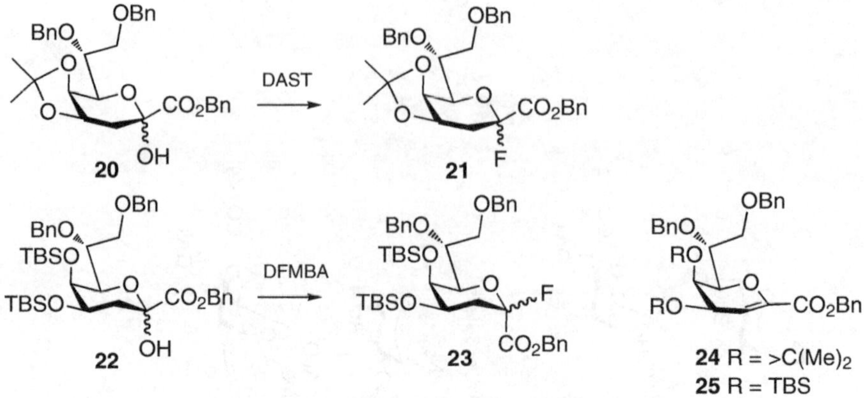

Scheme 3 Synthesis of Kdo fluoride donors.

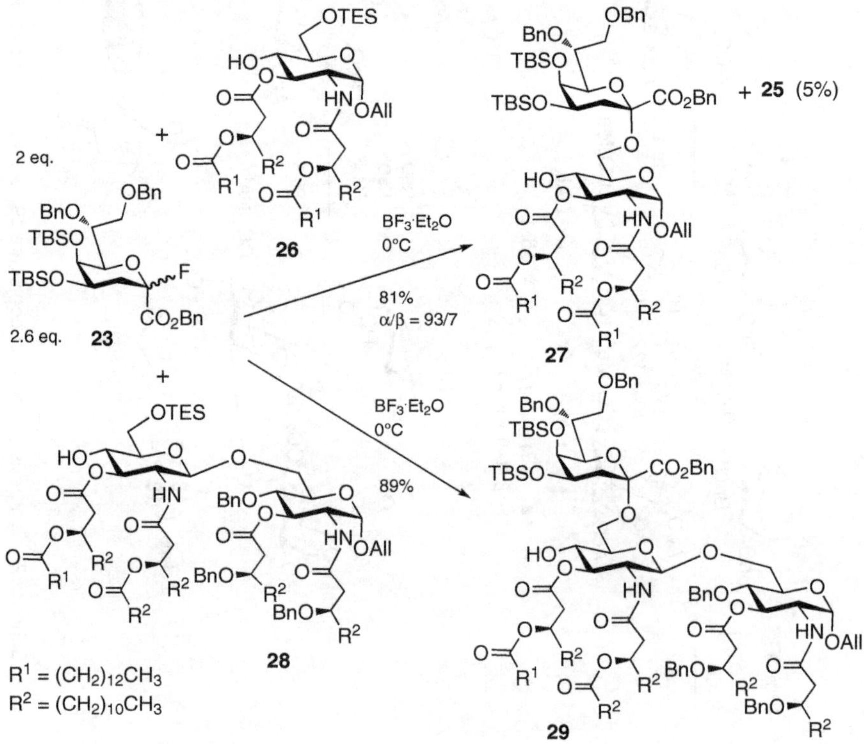

Scheme 4 Synthesis of trisaccharide precursor **29**.

position 6 allowed for the subsequent phosphorylation using Watanabe reagent to afford 4′-*O*-phosphotriester **30** in 96% yield. Selective cleavage of the silyl protecting groups led to an intermediate Kdo diol **31** which was then regioselectively converted into the 4″-*O*-TES acceptor **32**. Subsequent glycosylation of **32** with the 4,5-*O*-isopropylidene Kdo donor **21** gave the tetrasaccharide **33** in 75% yield. Compound **33** was further manipulated to introduce the anomeric phosphate group followed by hydrogenation to produce the target *E. coli* Re lipopolysaccharide (Scheme 5). The synthetic

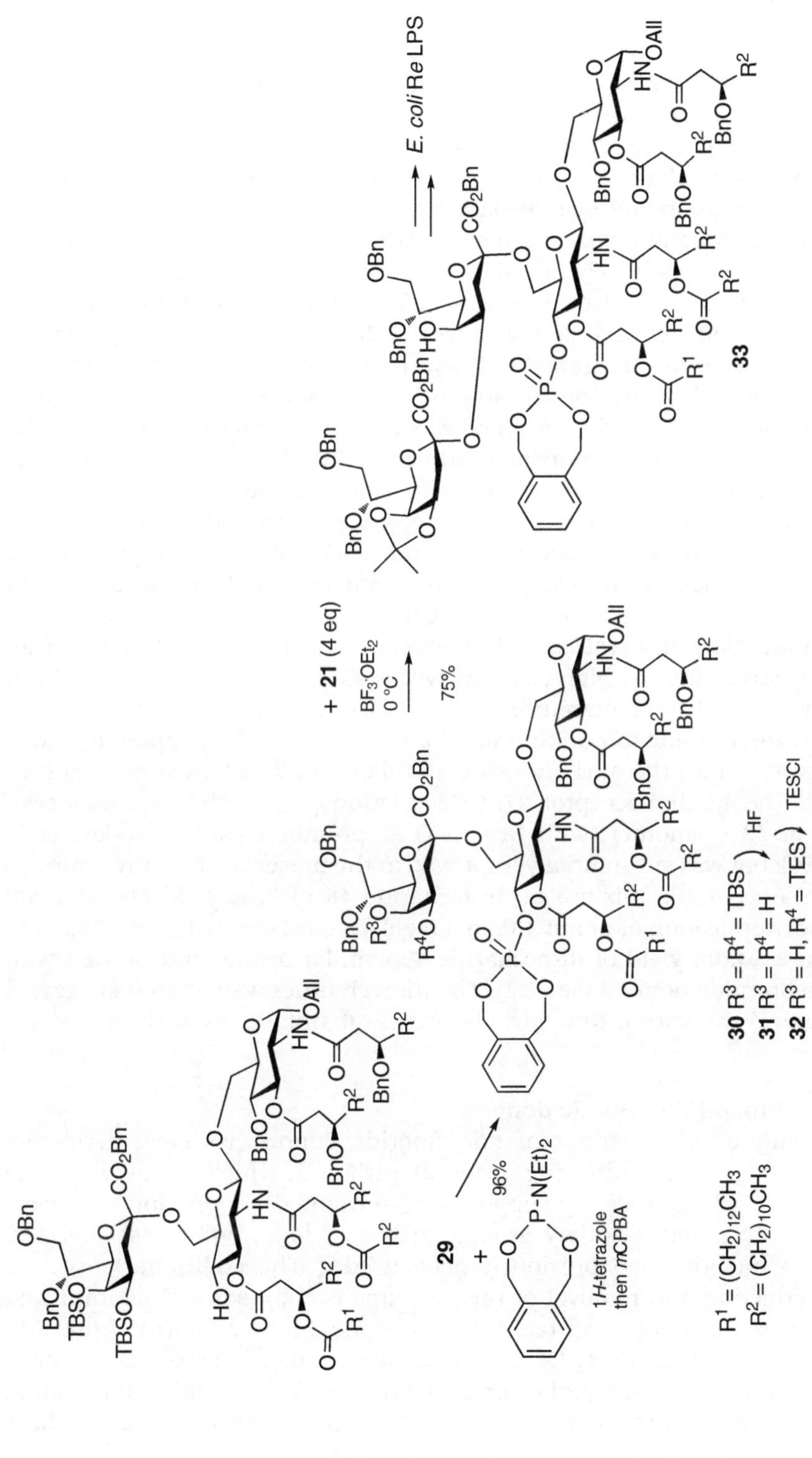

Scheme 5 Synthesis of *E. coli* Re LPS.[26]

compound – equipped with a defined and homogeneous fatty acid acylation pattern – showed biological activities close to the native material.

In a similar approach, the synthesis of a part structure of 4′-O-dephosphorylated *Helicobacter pylori* LPS had been accomplished by Fukase *et al.* (Scheme 6).[25] Coupling of donor **23** to the disaccharide acceptor **34** was performed in the presence of 5 Å molecular sieves which captures the released fluoride ions. Performing the reaction with 2.6 equivalents of donor **23** in the presence of 10 equivalents of promoter gave the allyl trisaccharide **35** in a high yield. De-*O*-allylation of **35** followed by phosphorylation and hydrogenation gave the target trisaccharide **36.** The deprotected compound had antagonistic effects toward the activation of TLR4 and competitively inhibited the binding of *E. coli* LPS.

Fluoride **21** was also used as an efficient α-selective Kdo donor in a regioselective glycosylation step by the Boons group.[27] Glycosylation of the diol acceptor **37** gave the trisaccharide **38** in good yield and high anomeric selectivity. Separation of the anomers was achieved after phosphorylation to give **39** which was eventually converted into the fully acylated and bisphosphorylated Kdo–lipid A derivative **40** corresponding to a fragment of *Neisseria meningitidis* LPS (Scheme 7).

The related 7,8-di-*O*-benzoyl-protected Kdo fluoride donor **43** was elaborated by Ichiyanagi *et al.* from 1-*O*-acetyl-2,3-*O*-isopropylidene mannofuranose **41** in 6 steps *via* the hemiketal **42.** Reaction of **42** with DAST afforded a separable anomeric mixture of fluoride **43** accompanied by formation of glycal ester **44** (Scheme 8).[28] It is to be noted that an alternative efficient high-yielding synthesis of the Kdo hemiketal **45** in only 3 steps from D-mannose has been disclosed recently, which offers a more direct route to Kdo fluoride donor derivatives.[29] Compound **42** was converted into the allyl glycoside **46** followed by TFA treatment to generate the 4,5-diol acceptor **47.** Glycosylation of **47** with 2 equivalents of donor **43** (α-anomer) with $BF_3 \cdot Et_2O$ as promoter gave a modest yield (30%), but was substantially improved in the presence of triethylamine to give a separable mixture of disaccharide **48** in 72% yield and 5 : 1 α/β ratio. Application of donor **43α** in a regioselective glycosylation of 7,8-diol **3** gave a poor yield of disaccharide **49,** similar to the reaction of **3** with Kdo bromide donor **2** (see **2.2**). The disaccharides were then deprotected to afford the known Kdo allyl glycosides **50** and **51,** respectively.[30,31]

2.4 3-Iodo-Kdo fluoride donor

Recently another variant of Kdo fluoride donors has been developed exploiting a stereodirecting group at carbon 3, which act as α-specific donors and significantly suppress the unwanted elimination side reaction. Temporary auxiliary groups such as 3-iodo, 3-phenylselenyl and 3-thio moieties had previously been used.[32] The additional effort for introduction and removal of these groups is compensated by improved donor performance and facile access of the 3-iodo-2-fluoride from Kdo glycal ester derivatives. Of note, only stoichiometric amounts of donor are required. Thus, glycal ester **4** – conveniently accessible from ammonium Kdo in three steps[33] – was subjected to alkoxyiodination which

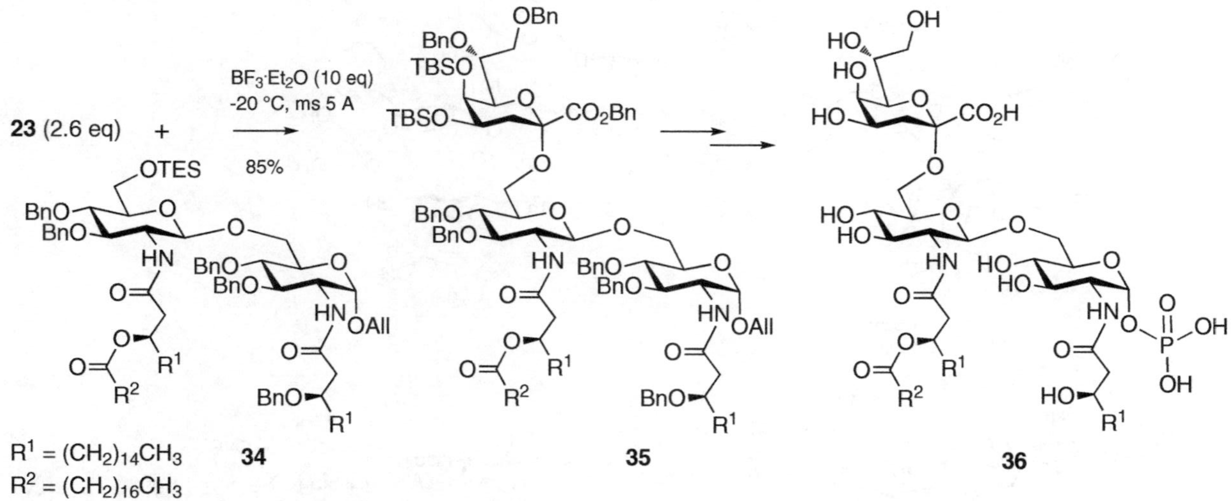

Scheme 6 Synthesis of *Helicobacter pylori* LPS fragment **36**.[25]

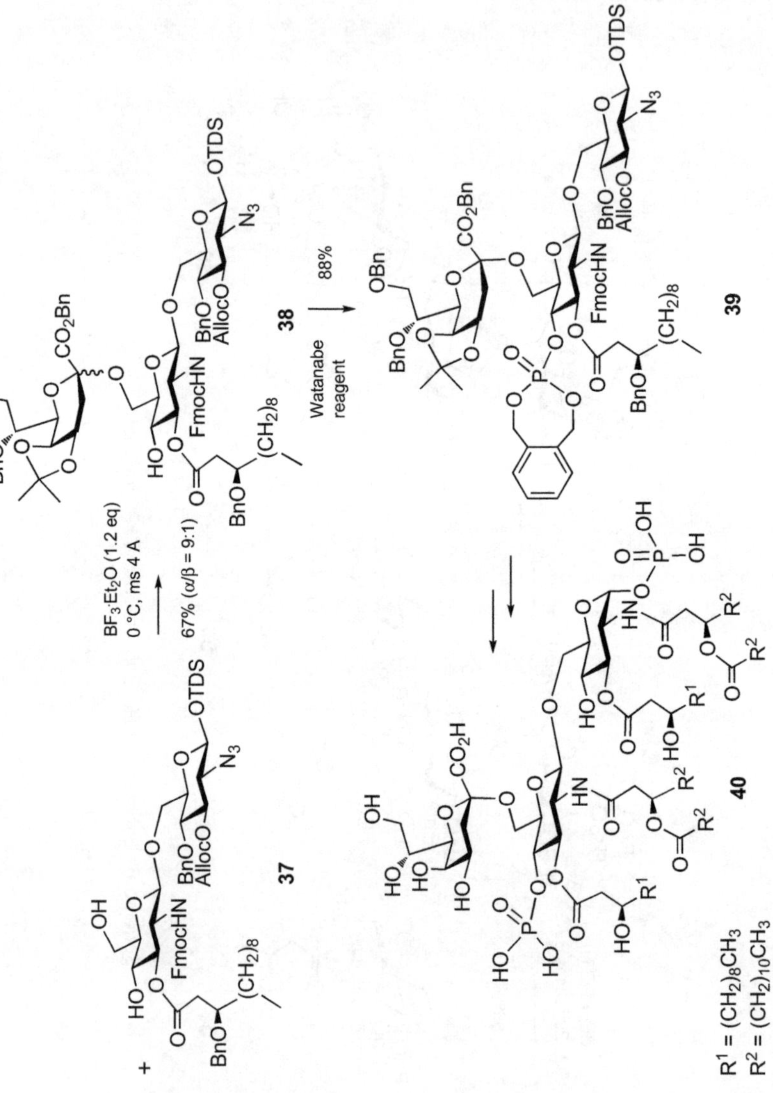

Scheme 7 Synthesis of *Neisseria meningitidis* LPS fragment **40**.[27]

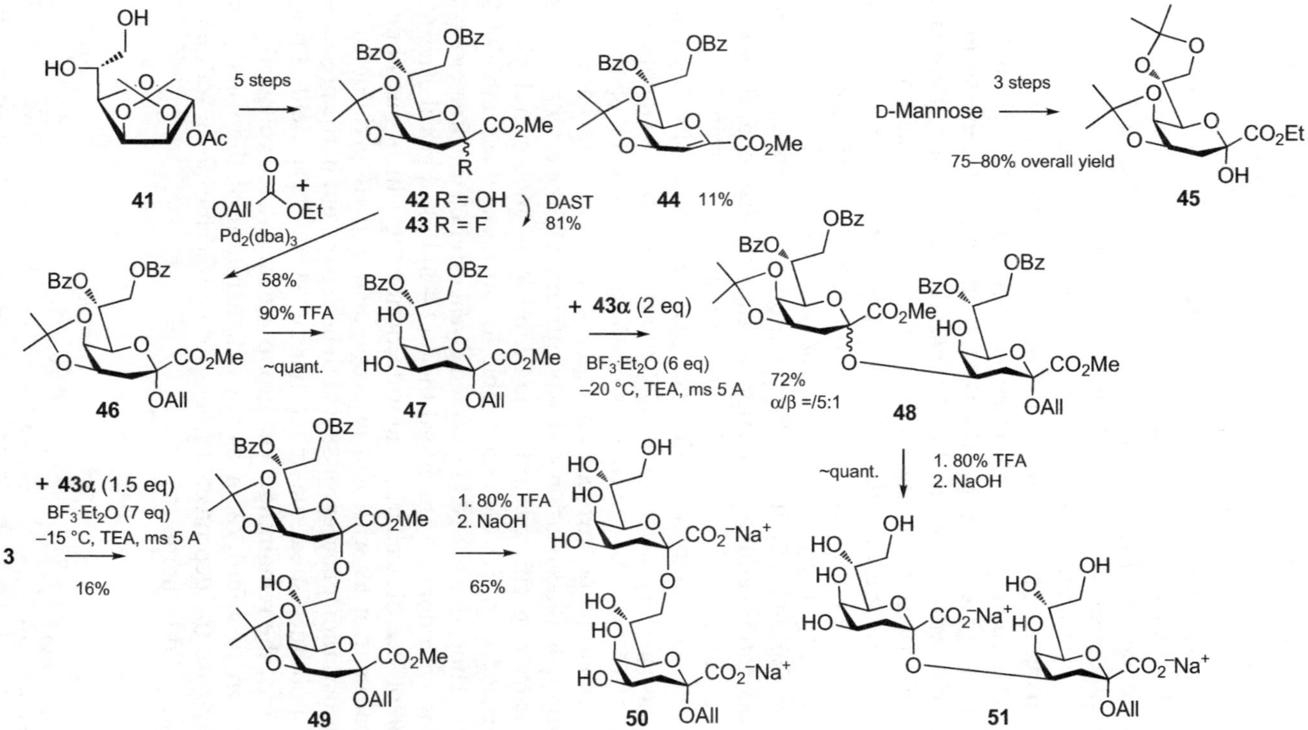

Scheme 8 Synthesis of α-(2→4)- and α-(2→8)-linked Kdo disaccharides.[28]

afforded the 2,3-*trans* diaxial iodoacetate **52** in 78% yield.[34] By using HF-pyridine, the anomeric acetate was then converted – without formation of elimination product – into the 3-iodo Kdo fluoride **53** as a single anomer and as bench-stable glycosyl donor in excellent yield (Scheme 9).[35]

Donor **53** was first evaluated in a test reaction with 2-propanol in the presence of $BF_3 \cdot Et_2O$ as promoter, which gave the α-glycoside **54** in 83% yield with a low amount of elimination product **4** (5%). Removal of the 3-iodo-group was accomplished under hydrogen atom transfer conditions to furnish the Kdo glycoside **55** (Scheme 10). The method was further extended to include the azido spacer derivative **56** and the dehalogenated spacer derivative **57** in similar yields.

Donor **53** was then employed for the assembly of the α-(2→4)- and α-(2→8)-linkages present in enterobacterial and chlamydial LPS (Scheme 11). Coupling of the 7,8-*O*-carbonyl derivative **58** with equimolar amounts of **53** gave the α-(2→4)-linked 3-iodo disaccharide **59** in 78% yield, which was followed by 5-*O*-acetylation (→**60**). Subsequent dehalogenation with lauroyl peroxide in refluxing cyclohexane and global deprotection gave known[36] disaccharide **61** in 80% yield (for 3 steps).

As shown above for the Kdo fluorides **21** and **23**, donor **53** could also be favorably used for a glycodesilylation of the 8-*O*-TES Kdo derivative **62** which gave the α-(2→8)-linked disaccharide **63** in 62% yield followed by dehalogenation and deprotection to furnish disaccharide **64**. The favorable properties of donor **53** were further demonstrated in a synthesis of the 4,8-dibranched Kdo-trisaccharide in a remarkable double regioselective glycosylation of the 4,5,7,8-tetraol acceptor **65** (Scheme 12). Coupling of **65** with 2.4 equivalents of **53** in the presence of $BF_3 \cdot Et_2O$ gave trisaccharide **66**, which was then converted *via* *O*-acetylation, deiodination, de-*O*-acetylation and saponification of the ester groups into the *Chl. psittaci* specific trisaccharide ligand **67** in good overall yield.

The methodology was also applicable for an iterated assembly of Kdo oligomers. Thus, the 8-*O*-TES protected glycal ester **68** was first elongated by reaction with donor **53** to afford the α-(2→8)-linked glycal ester **69** as precursor of the disaccharide fluoride donor **70**, which was obtained upon treatment of **69** with NIS in acetic acid and HF-pyridine in 57% yield. Glycosylation of acceptor **58** had to be carried out in the presence of TMSOTf and furnished the trisaccharide **71** in 73% yield. Prior to deprotection, the remaining 5-OH group was acetylated followed by dehalogenation to afford **72** in 85% yield. Removal of the ester groups eventually gave the deprotected linear Kdo trisaccharide as sodium salt **73** in good yield (Scheme 13).

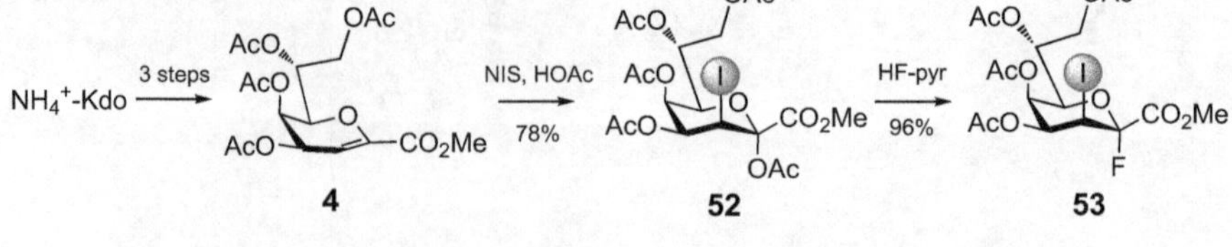

Scheme 9 Synthesis of 3-iodo-Kdo fluoride donor **53**.[35]

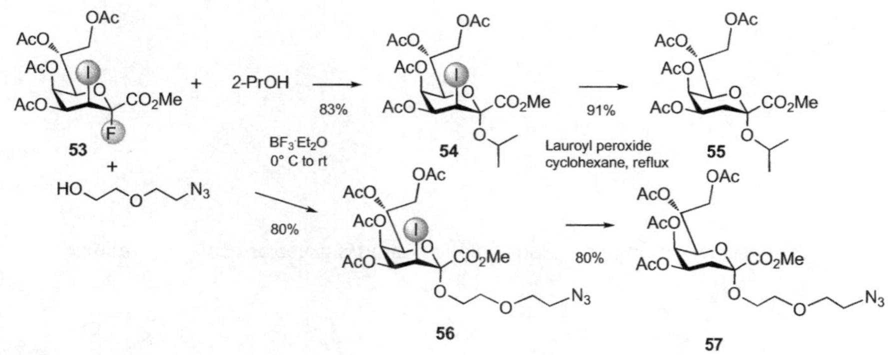

Scheme 10 Model glycosylation reaction with donor **53**.[35]

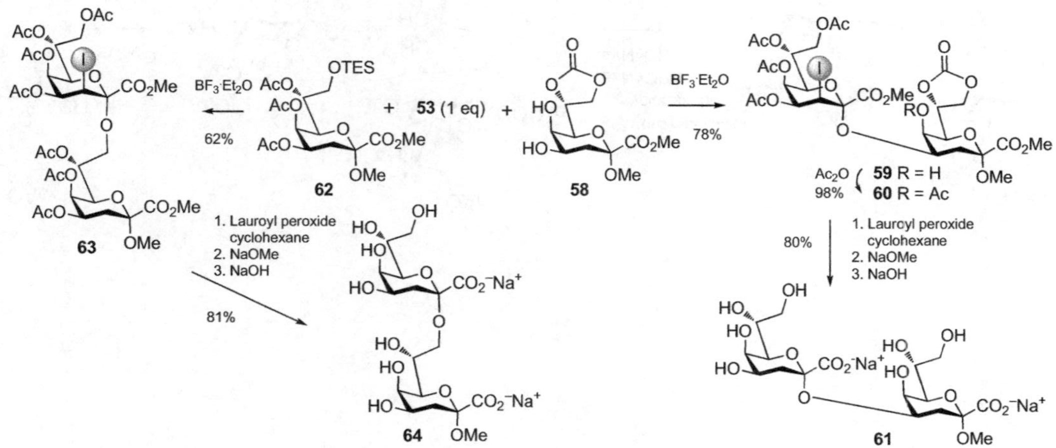

Scheme 11 Synthesis of α-(2→4)- and α-Kdo*p*-(2→8)-linked disaccharides.[35]

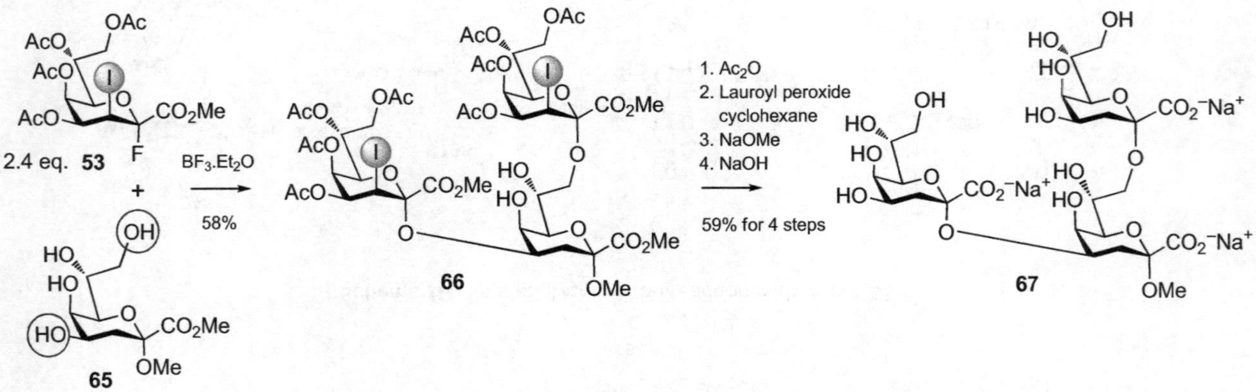

Scheme 12 Regioselective synthesis of 4,8-dibranched Kdo trisaccharide **67**.[35]

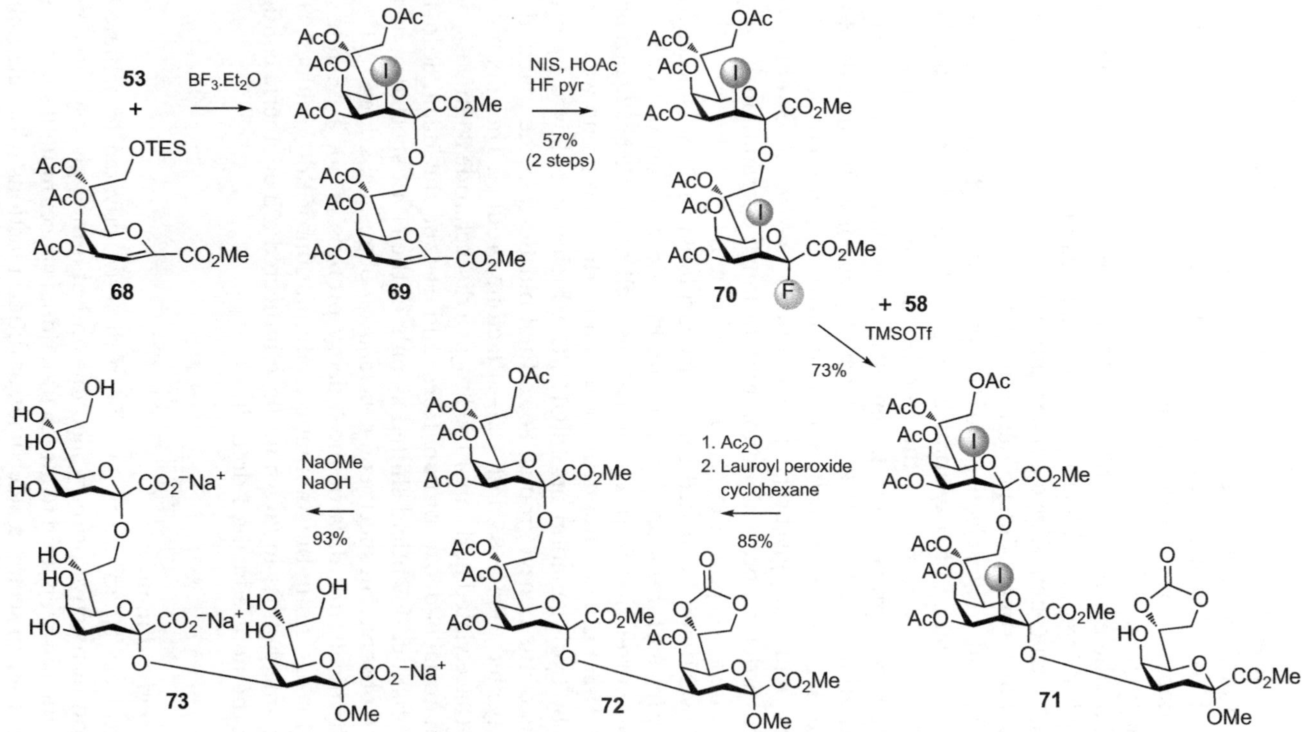

Scheme 13 Synthesis of the linear *Chlamydia* specific Kdo trisaccharide **73**.[35]

While the electron-deficient double bond in glycal ester **68** was compatible with the glycosylation conditions, electron-rich alkenes, however, may also undergo an iodonium ion transfer reaction, which excludes allylic protecting groups in this system.[37] Donor **53** was also used for connection of Kdo to position 6 of the glucosamine derivative **74** to give disaccharide **75** in 89% yield (Scheme 14). Deiodination as described and hydrogenolysis of the benzyl group (→**76**) was followed by phosphorylation using the amidite protocol to furnish **77** in 58% yield (3 steps). Global deprotection of **77** gave disaccharide **78** in near theoretical yield.

Based on the performance of donor **53**, the challenging coupling to the sterically hindered, axial 5-OH group of Kdo was next undertaken.[38] The armed 5,7,8-tri-*O*-benzyl glycosyl acceptor **80** was derived from **58** to first give the 4,5-*O*-isopropylidene derivative **79** in 3 steps, followed by TFA treatment and regioselective benzyl protection of *O*-4 to give **80** (Scheme 15). Since glycosylation attempts of **80** with the "disarmed" donor **53**, however, failed, the per-*O*-benzylated Kdo donor **83** was elaborated from glycal ester **4**. Zemplén deacetylation and subsequent perbenzylation gave **81** in 77% yield, which was converted into the 3-iodo 2-acetate **82**. Treatment with HF-pyridine then afforded donor **83** in excellent yield. The glycosylation conditions had to be carefully optimized in order to prevent donor and product degradation at higher temperatures. Keeping the temperature in a range between −40 and −10 °C eventually furnished the disaccharide **84** in 71% isolated yield with only minor formation of glycal ester **81**. The structure was confirmed by HMBC correlation from H-5 to C-2′ as well as by a Noesy correlation between H-5 and H-6′. Reductive removal of the iodo-group was accomplished by hydrogenation on $Pd(OH)_2/C$, which was followed by hydrogenolysis of the benzyl groups. For a final purification the material was re-*O*-acetylated to give **85** and a 1′→4 lactone product. The mixture was fully de-*O*-acetylated and saponified to give disaccharide **86** in 90% yield. The equatorial 3-deoxy proton H3*eq* and H-4 of the distal Kdo unit showed distinct downfield-shifted signals in the NMR spectrum, as had also been observed in spectra of *Acinetobacter* oligosaccharides.[15a] The ^{1}H NMR chemical shift data of the 3-deoxy protons is not a reliable parameter for assigning the anomeric configuration of Kdo residues, since their chemical shifts in **86** would have indicated a β-anomeric configuration of the non-reducing Kdo unit.

2.5 Kdo glycal donors

Kdo glycal ester derivatives have been directly activated to achieve glycosylation of reactive primary alcohols and of armed glycosyl acceptors, and has mostly been used to attach functional spacer moieties onto a Kdo unit. In order to secure α-selectivity, addition reactions with selenonium and iodonium species in the presence of strong acids were performed, as first developed by Achiwa.[32a] Seeberger *et al.* generated the 3-phenylselenyl derivative **87** by reaction of glycal ester **4** with 5-*N*-benzyl-*N*-benzoxycarbonylamino-1-pentanol and phenylselenyl chloride in the

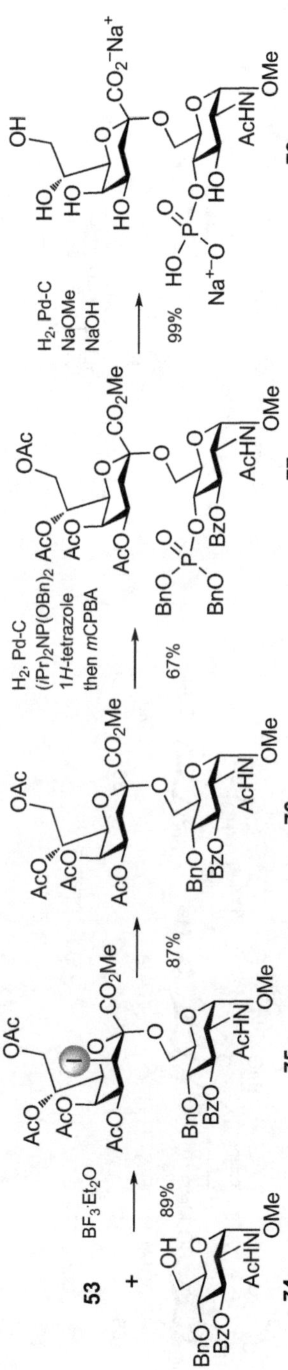

Scheme 14 Synthesis of disaccharide 4-*O*-phosphate **78**.[37]

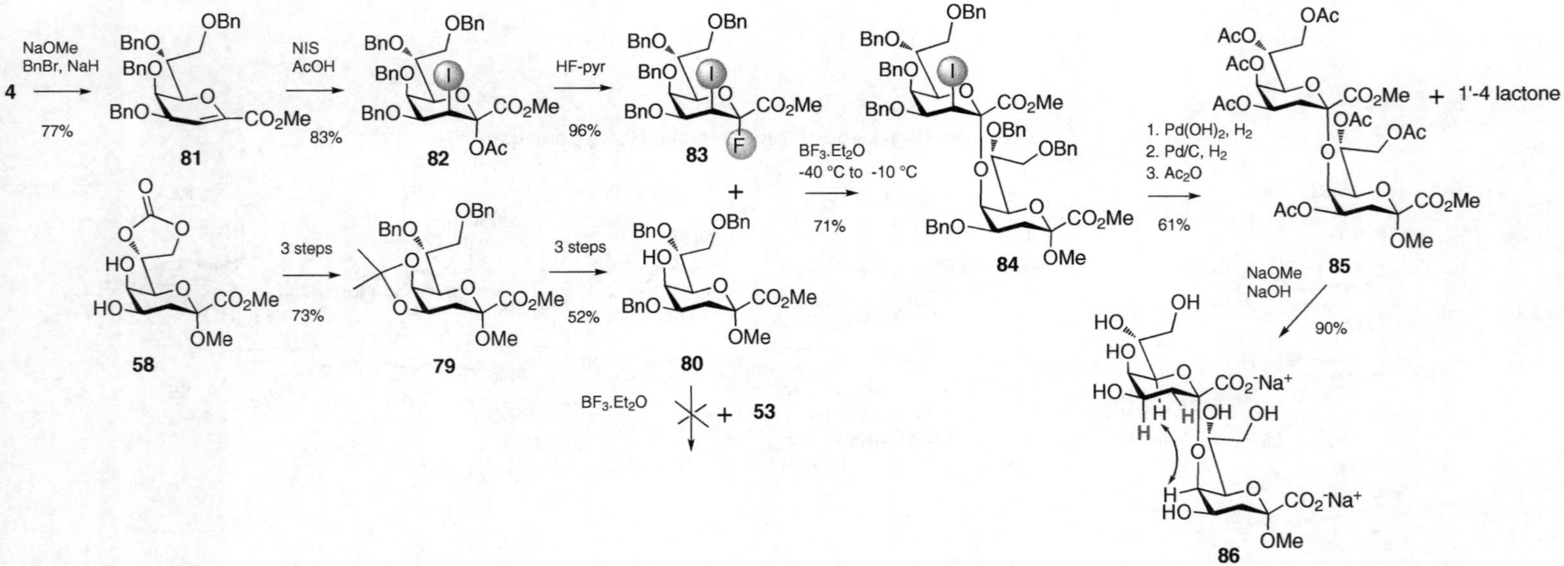

Scheme 15 Synthesis of α-(2→5)-linked Kdo disaccharide **86**.[38]

presence of Ag-triflate and TMSOTf. Compound **87** was then reduced under radical conditions to give the Kdo spacer glycoside **88** (Scheme 16). This building block was further equipped with protecting groups at the 4,7 and 8 positions to serve as glycosyl acceptor **89** for subsequent attachment of heptosyl units. This way, LPS fragments (albeit without the lateral Kdo unit) of *Yersinia pestis*, *Haemophilus influenzae*, *Proteus* and *Neisseria meningitidis*, respectively, were efficiently synthesized and converted into neoglycoconjugates for immunization studies.[39] The 2-(4-trifluoroacetamidophenyl)ethyl spacer derivative **91** was obtained in excellent 91% yield by Oscarson *via* a similar reaction of **4** with *in situ*-generated phenylselenyl triflate.[32d] The related 3-iodo spacer glycoside **92** was obtained in 53% yield by reaction of **4** with NIS and an excess of triflic acid, which was needed for activation of the electron-deficient double bond. The 3-iodo group was then subjected to dehalogenation to generate α-spacer glycoside **91** in 85% yield.[40]

Iodoalkoxylation of glycal esters promoted by stoichiometric amounts of triflic acid was further developed by Takahashi *et al.* in an approach that allows for an iterative assembly of α-(2→8)-linked Kdo oligomers.[41] The glycosylation was carried out effectively by an iodoalkoxylation reaction of a Kdo glycal ester with suitable electron-rich glycosyl acceptor derivatives. This way, the benzylated glycal ethyl ester **93** was coupled to "armed" *gluco*-, *galacto*- and Kdo glycosyl acceptors in the presence of TfOH and NIS to give good yields of 3-iodo-Kdo glycosides **94–96** with very high α-selectivity (Scheme 17).

The synthesis of oligomeric α-(2→8)-linked Kdo units was then achieved through the coupling of an open-chain, reactive Kdo acceptor – protected at the anomeric center by a sulfoxide and a thioalkyl moiety – to an *in situ* formed iodonium species (Scheme 18). Glycosyl acceptor **103** was prepared in a multistep synthesis from 2,3-di-*O*-benzyl-4,6-*O*-benzylidene mannitol **97**, which was converted into the primary alcohol **98** in 60% yield followed by iodination to give **99** in 82% yield. The 1-deoxy-1-iodo-mannitol **99** was then elongated by reaction with the 1,3-dithiane-2-carboxylate reagent **100** followed by benzylidene cleavage to produce the diol **102** in 71% yield. Oxidation of **102** gave the diastereomeric sulfoxide **103** in 95% yield serving as glycosyl acceptor for the ensuing reaction with glycal ester donor derivatives. Reaction of **103** with glycal ester **93**, promoted by NIS in the presence of stoichiometric amounts of triflic acid, afforded pseudo-disaccharide **104** in 89% yield. Cyclization of **104** with elimination of the thio-protecting groups gave the disaccharide glycal ester **105** in 77% yield. Repetition of the sequence led to the pseudo-trisaccharide **106**, which was subsequently converted into trisaccharide glycal ester **107**. Similar to the synthesis of **94**, the Kdo trisaccharide building block was successfully connected to *O*-6 of protected glucoside **108** (*cf.* Scheme 17) to furnish tetrasaccharide **109** in 55% yield with reduced stereoselectivity. Global deprotection was achieved *via* hydrogenolysis/dehalogenation with $Pd(OH)_2$, reacetylation, Zemplén transesterification and alkaline ester hydrolysis to give tetrasaccharide **110**.

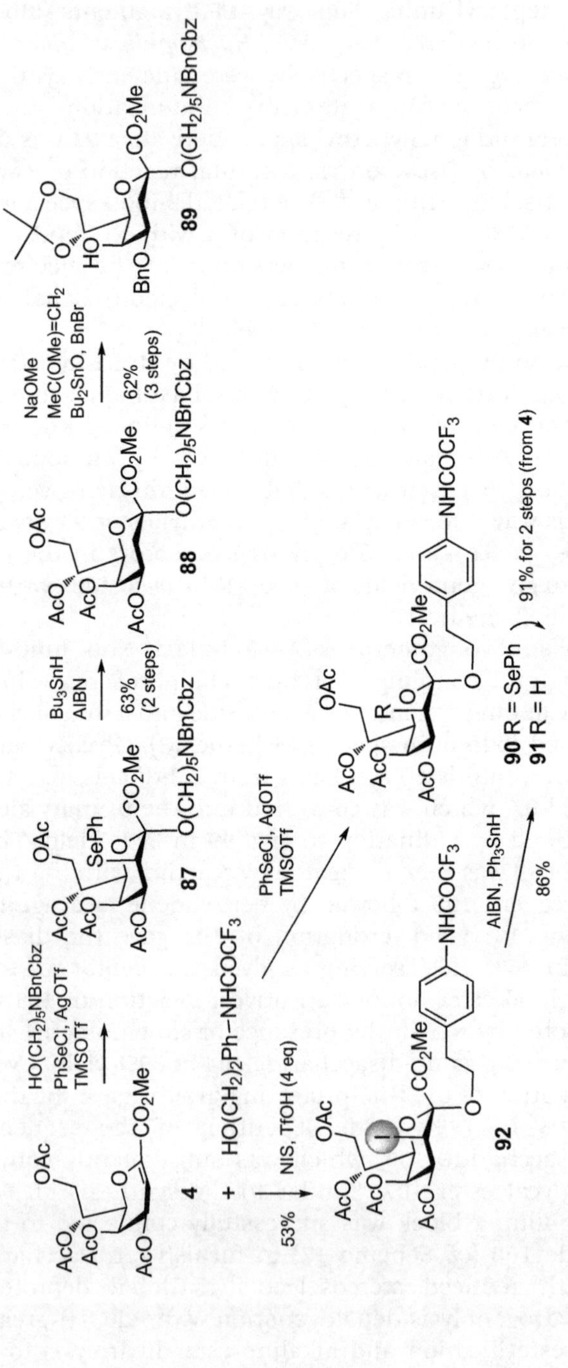

Scheme 16 Synthesis of Kdo spacer glycosides from glycal ester **4**.

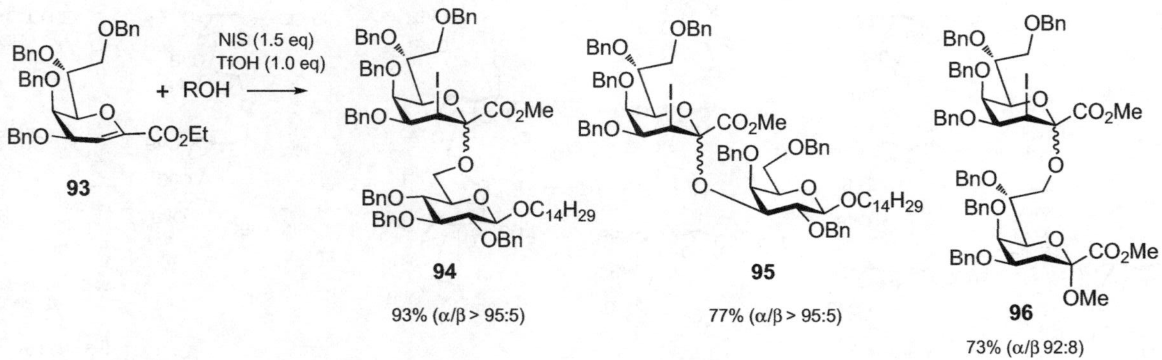

Scheme 17 Iodoalkoxylation of glycal ester **93**.[41]

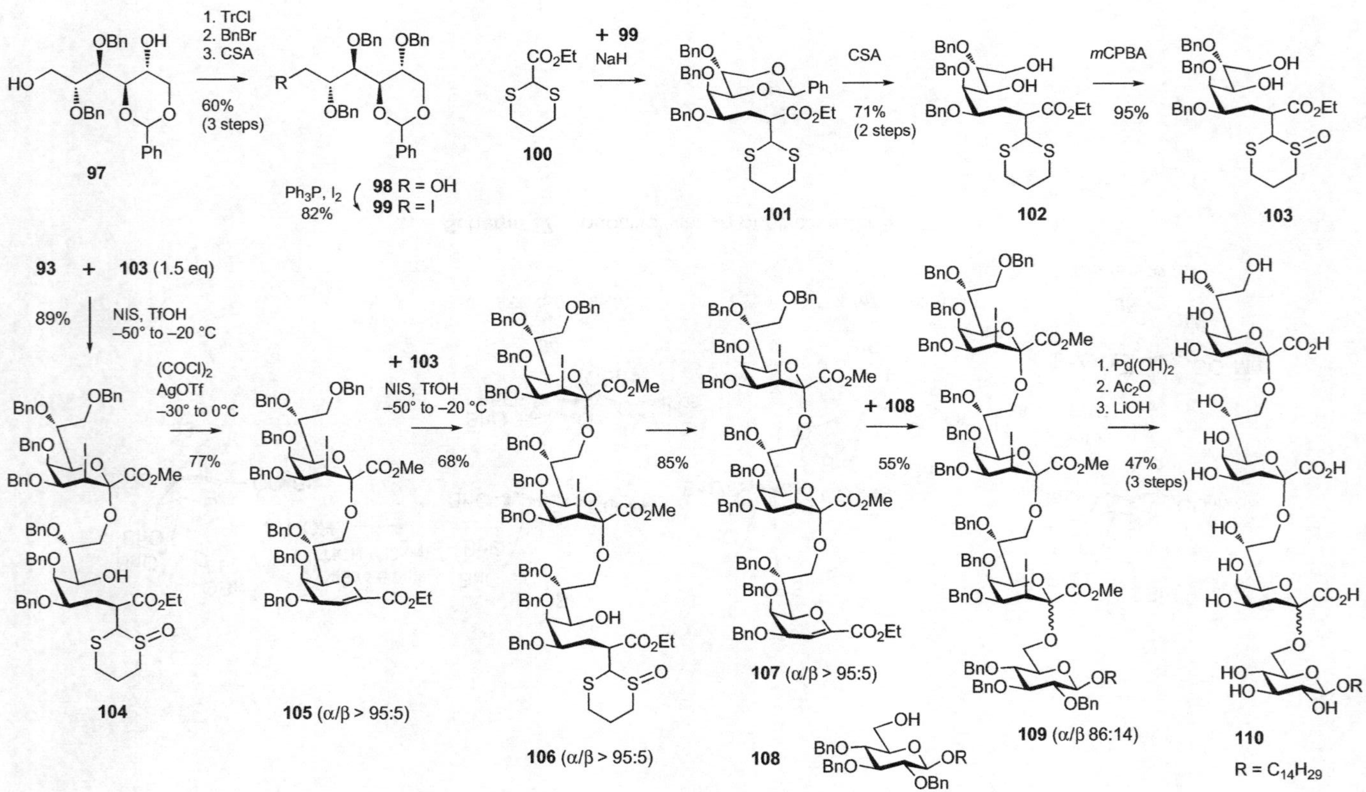

Scheme 18 Synthesis of oligomeric α-(2→8)-linked Kdo derivatives.[41]

2.6 Kdo thioglycoside donors

Compared to the frequently used Kdo bromide and fluoride donors for the synthesis of LPS oligosaccharides, Kdo thioglycoside donors are under-represented in the literature. Extending a first report,[42] wherein a Kdo thioglycoside gave β-linked products when treated with NIS/TfOH, Oscarson *et al.* later undertook a detailed assessment of various Kdo thioglycoside donors and glycosylation conditions with respect to yields, selectivity and byproduct formation.[40] Kdo thioglycosides can be prepared by treatment of the anomeric acetate **111** with Lewis acids such as $ZnCl_2$, which gave a high yield of the anomeric ethyl 2-thio-ketosides with preferential formation of the β-product (Scheme 19). Notably, treatment of the mixture in nitromethane resulted in an inverted α/β ratio. Subsequent Zemplén transesterification gave **113** followed by *O*-benzoylation or reaction with dimethoxypropene to afford donor variants **114** and **115**, respectively. In addition, the *N*-xanthate **116** was derived from Kdo bromide **2**. Reaction of the Kdo chloride **117** (or bromide **2**) with thiophenol in the presence of base selectively gave the acetylated phenyl 2-thio-ketoside **118**.[43] Next, donors **2** and **116** were tested in model glycosylation reactions with 2-(4-trifluoroacetamidophenyl)ethanol in CH_2Cl_2 at −15 °C and in Et_2O at 0 °C using DMTST as promoter. Yields of 81–84% for **91, 119** and **120,** were obtained in the former solvent with preferred formation of β-products (α/β ratios 1 : 3 and 1 : 2.6), whereas the yields were significantly lower in ether with slightly improved α/β ratios but more pronounced glycal formation (**4, 121, 122**). Next, donors **114** and **115** were activated with IBr/AgOTf in 3 : 2 MeCN–CH_2Cl_2 solvent mixture as well as in neat ether at −70 °C and ambient temperature, respectively. Yields for glycosylation at higher temperature in MeCN–CH_2Cl_2 were consistently lower and substantial elimination was observed. Reaction of donor **114** at −70 °C selectively gave product **119** as β-isomer in 81% yield. The isopropylidene-protected donor **115** gave the α-product **120** as the major isomer at room temperature (α/β 2 : 1).

The α-(2→8)-linked Kdo disaccharide thioglycoside donor **125** was elaborated by Baasov *et al. via* reaction of fluoride **123** with phenyl 2-thio-ketoside **124** to give disaccharide **125** in 65% yield, which was then attached to a tri-glucosamine acceptor **126** by NIS-promoted coupling in 45% yield (Scheme 20). Compound **127** was obtained as 1 : 1 anomeric mixture and converted to a deprotected pentasaccharide with GTPase activity.[44]

Based on the α-directing effect of a 4,6-*O*-dibutylsilylene-group in *galacto*-configured carbohydrates,[45] the Yang group recently developed a highly α-selective Kdo thioglycoside donor equipped with a 5,7-*O*-DTBS group (Scheme 21).[46] Despite a lengthy synthesis of the donors and restrictions in further application to 5-*O*-substituted inner core fragments (which would require additional protecting group diversification between *O*-5 and *O*-7, respectively), the donors were of sufficient reactivity to glycosylate all positions of Kdo glycosyl acceptors. Starting from ethyl 2-thio β-ketoside **113**, the primary alcohol position was protected by dimethoxytritylation (→**128**) followed by attachment of the 5,7-*O*-DTBS group to afford **129** in high yield. Subsequent 4-*O*-benzoylation to give

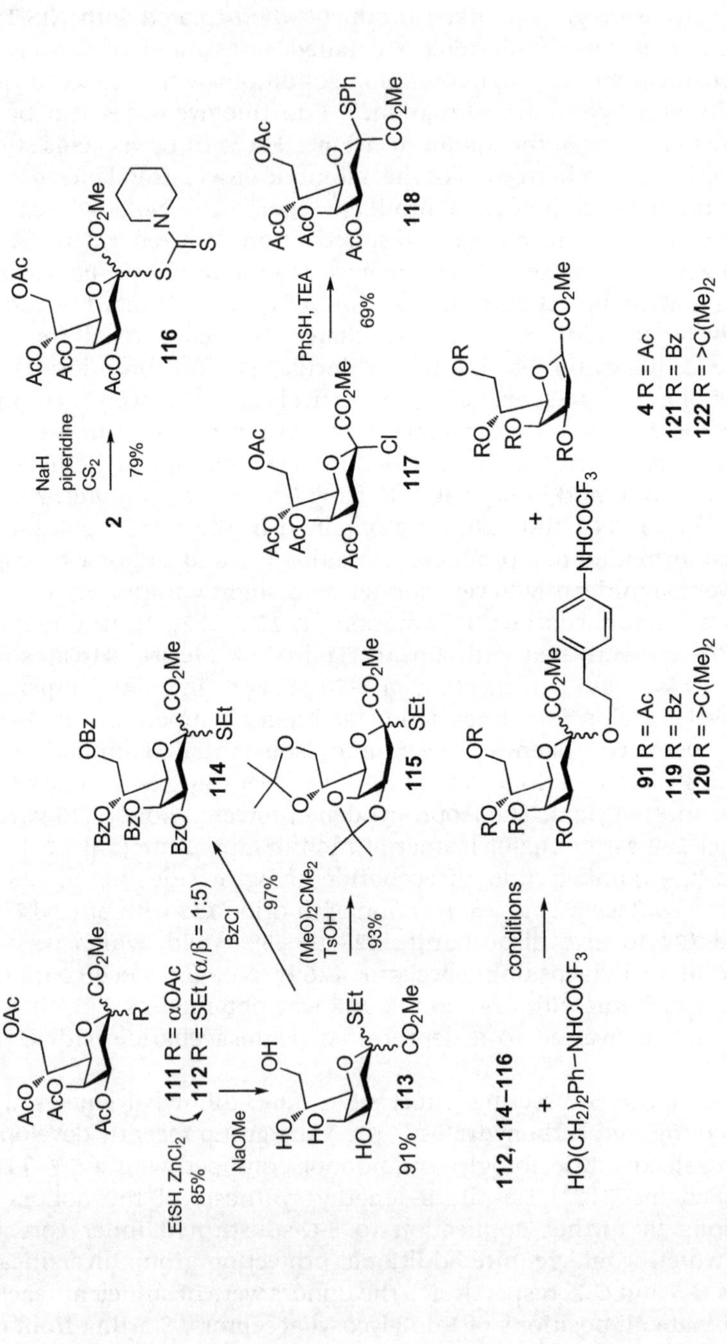

Scheme 19 Synthesis of Kdo thioglycosides and model glycosylation reactions.

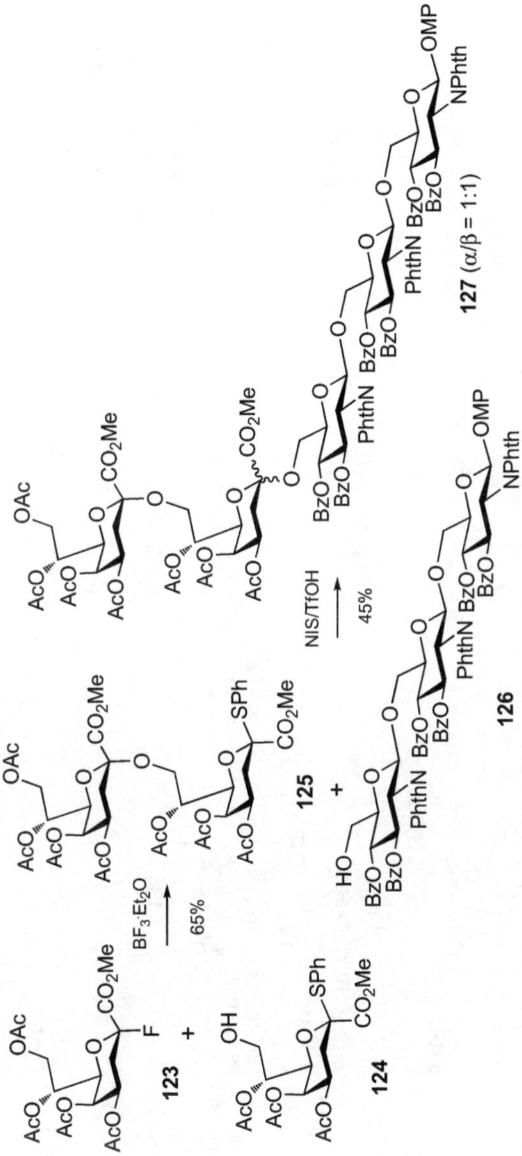

Scheme 20 Synthesis of pentasaccharide 127.[44]

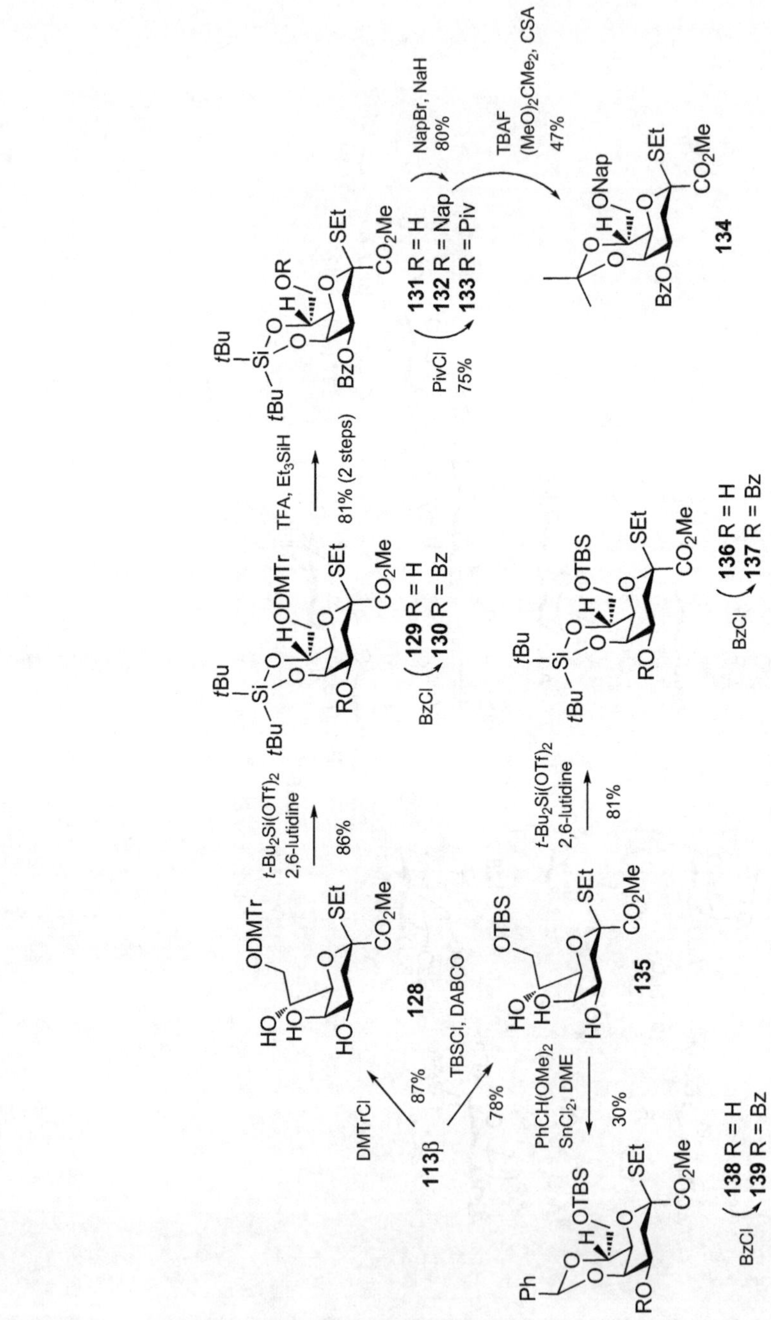

Scheme 21 Synthesis of 5,7-O-DTBS protected Kdo thioglycoside donors.[46]

130 was continued by cleavage of the DMTr group to produce **131** to allow for subsequent introduction of 8-*O*-2-methylnaphthyl (**132**) and 8-*O*-pivaloyl groups (**133**), respectively. For comparison, the cyclic 5,7-*O*-isopropylidene donor **134** was synthesized *via* DTBS cleavage and reaction of the intermediate diol with dimethoxypropane under acid catalysis in 47% yield. A second series of donors was also prepared containing an 8-*O*-TBS group (**113→135→136→137**) in similar yields. Replacement of the 5,7-*O*-TBS tether by a benzylidene acetal *via* reaction of **135** with benzaldehyde dimethylacetal and stannous chloride in refluxing DME could be generated in low yield only to give **138.** The donors were then tested in glycosylation reactions using an excess of donor (2 equivalents) and 4 equivalents of NIS, catalytic amounts of triflic acid (0.1 equivalents) in CH_2Cl_2 at −70 ° to −50 °C. Glycosides were formed in high yields (based on the respective acceptor) and excellent α-selectivity, however, substantial amounts of elimination products (40–60% based on the donor) were also produced.[†] As example, the synthesis of protected Kdo di- and trisaccharides is illustrated (Scheme 22). As acceptor derivatives, suitably protected α- and β-allyl glycosides were prepared using the 5,7-*O*-DTBS thioglycoside donors and subjected to glycosylation with donor **132.** Since excess of donor had been used, yields were based on the respective glycosyl acceptor. This way, the α-(2→8)- and α-(2→7)-linked disaccharides **142** and **144** were isolated in good yields and with excellent stereoselectivity. The 5,7-*O*-benzylidene acceptor **145** was more reactive than its 5,7-*O*-DTBS counterpart **147**, since coupling of **147** with donor **132** with NIS/TfOH as promoter failed, and the authors had to rely on activation with TBPA in MeCN at ice-bath temperature to achieve formation of **148** in 73% yield. The latter promoter was also more effective in the glycosylation of the unreactive 5-OH group of the 4-*O*-benzyl β-allyl ketoside **149** to give the α-(2→5)-connected disaccharide **150** in 71% yield as a single anomer, whereas the former condition gave a modest yield of glycoside and substantial elimination product (67% of **140**). Donor **132** was also successfully applied to the synthesis of the protected inner core fragment from *Proteus mirabilis* LPS. First, donor **132** was subjected to α-selective anomeric allylation with TBPA as promoter followed by DDQ-oxidation to furnish the primary alcohol **151** in 74% yield for 2 steps (Scheme 23).

Reaction of **151** with known[23] 4-amino-4-deoxy-L-arabinopyranosyl NPTFA donor **13** in the presence of TMSOTf gave a **1:1.5** anomeric mixture of the disaccharide **152** in excellent yield. Separation of the isomers and removal of the 4-*O*-benzoyl substituent gave β-L-Ara4N glycosyl acceptor **153.** Again, TBPA as activator for donor **132** had to be used to get the protected trisaccharide **154** in 67% yield as single α-anomer, which is a significant improvement as compared to the reaction of the 4,5,7-triol acceptor **15** with Kdo bromide **2** (see Scheme 2). The straightforward access to 4,8-glycosylated Kdo units was eventually also

[†]The high propensity of 5,7-*O*-DTBS Kdo β-thioglycosides for elimination has also been noted in the laboratory of the author (A. Derler, Master thesis, University of Natural Resources and Life Sciences Vienna, 2013).

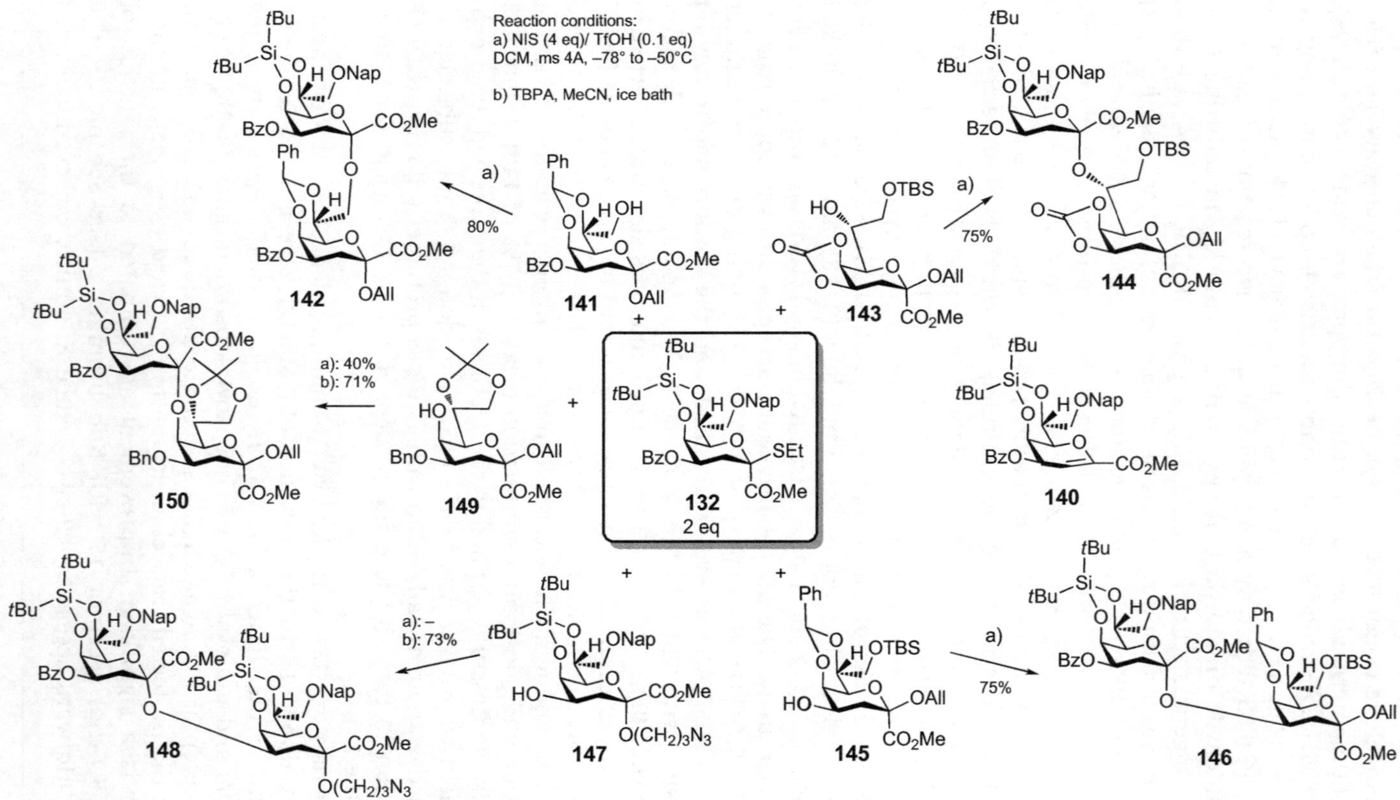

Scheme 22 Synthesis of protected α-(2→4)-, α-(2→5)-, α-(2→7)- and α-(2→8)-linked Kdo disaccharides.[46]

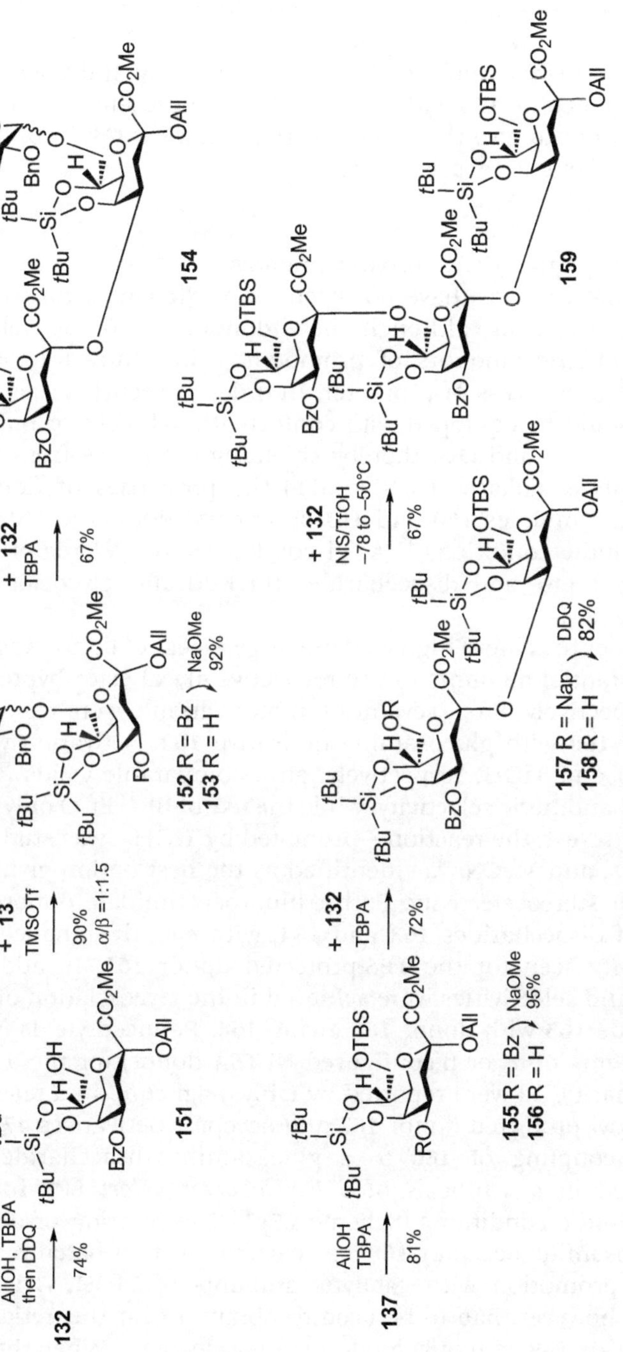

Scheme 23 Synthesis of *Proteus* and *Chlamydia* related LPS building blocks.[46]

shown for the *Chlamydia* specific group antigen. Along similar lines as for **154**, the 8-*O*-TBS protected allyl glycoside **155** was prepared, de-*O*-benzoylated at *O*-4 to give **156.** Similar to glycosylation of compound **153,** TBPA was used as promoter to introduce the α-(2→4)-linkage to give disaccharide **157** in 72% yield. The terminal Kdo unit was then assembled *via* oxidative removal of the 8′-*O*-Nap group and glycosylation of the resulting alcohol **158** with donor **132** in the presence of NIS/TfOH at low temperature to give the protected trisaccharide **159** in 67% yield, again with excellent stereoselectivity.

2.7 Kdo *N*-phenyltrifluoroacetimidates

Trihaloacetimidates have not been used extensively for the preparation of Kdo glycosides, although these donors can be activated with substoichiometric amounts of promoters – in contrast to Kdo fluorides requiring an excess of promoter. In order to secure α-selectivity, the donors should be equipped with conformational locks or bulky protecting groups at *O*-4 and *O*-5, thereby shielding the β-face from the incoming nucleophile. Fukase *et al.* studied the properties of Kdo *N*-phenyltrifluoroacetimidates **160** and **161** as glycosyl donors in batch and under microfluidic conditions.[47] As glycosyl acceptors *N*-Troc protected glucosamine mono- and disaccharide and Kdo allyl glycosides were tested (Scheme 24).

In order to achieve high yields, a large excess of donor was used leading to substantial amounts of the respective glycal ester byproducts **24** and **25**, respectively. First, several promoter variants were tested in the reaction of **160** with glucosamine derivative **162.** Activation with TMSOTf, TBSOTf and TfOH, respectively, gave comparable yields (61–75%) with modest anomeric selectivity, while the use of $BF_3 \cdot Et_2O$ only gave traces of product. Next, the reaction – promoted by TfOH – was studied in various solvents, and MeCN was identified as the best option giving high yields and fair stereoselectivity. Both trifluoroacetimidate donors gave a high yield of disaccharides **163** and **164,** with slightly enhanced α-anomeric selectivity seen for the TBS-protected donor **161.** In addition, similar yields and selectivities were achieved in the glycosylation of the Kdo allyl glycoside **165** with donor **160** giving **166.** Reduced yields (when using 2 equivalents of α- or β-configured NPTFA donor) for the α-(2→4)-linked disaccharide **48** were reported by Ichiyanagi *et al.* in a reaction of 7,8-*O*-dibenzoyl protected donor **167** and acceptor derivatives **47.**[48]

The coupling of **160** to a glucosamine disaccharide was further extended in a synthesis of a *Helicobacter pylori* LPS fragment using microfluidic conditions (Scheme 25).[49] The reaction of donor **160** with diglucosamine acceptor **168** was carried out in a batch reaction at 0 °C under promotion with catalytic amounts of TfOH. A large excess of donor, however, had to be used to obtain a near theoretical yield of trisaccharide **169** with only modest stereoselectivity. When the reaction was executed in a microflow reactor in cyclopentyl methylether, the anomeric selectivity was substantially improved. Furthermore, only 1.5 equivalents of donor **160** were needed to still achieve a good product yield. The

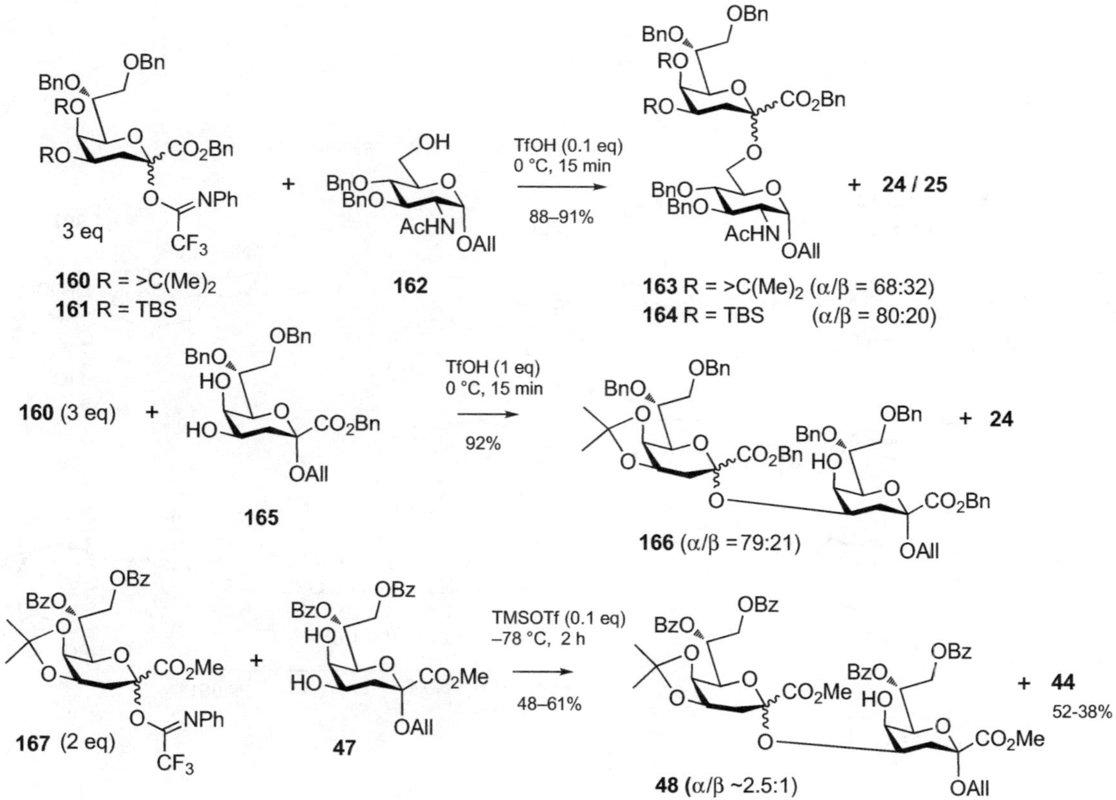

Scheme 24 Glycosylation reactions of *N*-phenyltrifluoroacetimidate donors.[47]

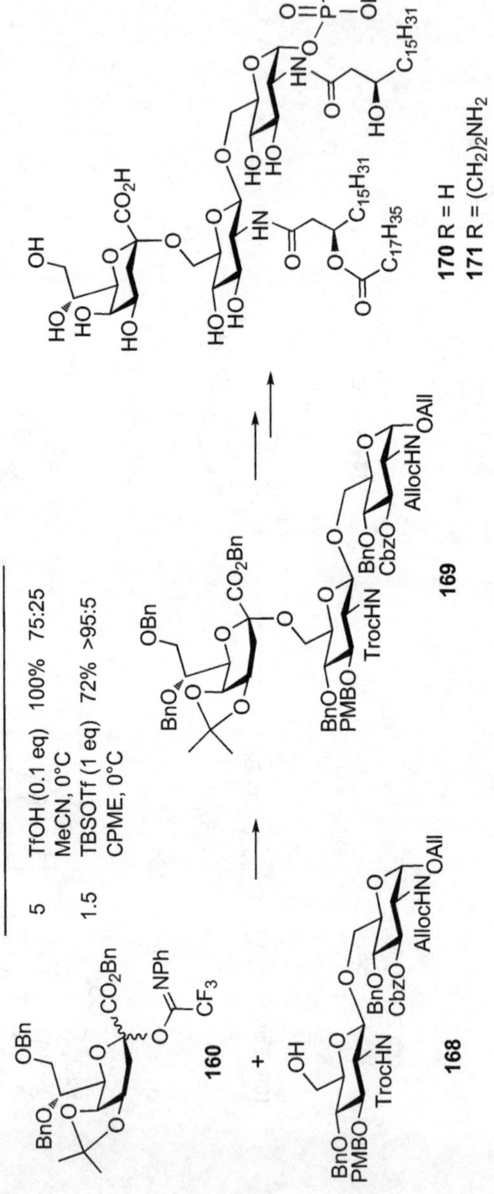

Scheme 25 Synthesis of *Helicobater pylori* LPS structures.[49]

authors state that the superior outcome of the microflow approach was due to improved mixing of reagents and efficient heat transfer thus limiting the impact of the competing β-hydrogen elimination reaction. The protected intermediate **169** was then converted in a multistep sequence to *H. pylori* LPS fragments **170** and **171** containing anomeric phosphate and 2-aminoethyl-phosphate substituents.

2.8 Kdo phosphite donors

Thus, far, Kdo phosphite glycosyl donors have found only very limited interest. The dibenzylphosphite Kdo donor **172** had been prepared by Ichiyanagi *et al.* by reaction of the hemiketal **42** with dibenzyl *N,N*-diisopropylphosphoramidite and 1*H*-tetrazole in 56% yield. Subsequent glycosylation reactions with diol **47**, to produce disaccharide **48**, were catalyzed either by TMSOTf or equimolar $BF_3 \cdot Et_2O$ but gave poor yields and anomeric selectivities (Scheme 26).

2.9 Synthesis of 4,5-dibranched Kdo units

The majority of published syntheses of Kdo-containing LPS fragments do not comprise the lateral α-(2→4)-connected Kdo residue in the presence of a 5-*O*-glycosyl substituent, although this is a conserved structural motif in many Gram-negative bacteria. The first successful synthesis had been achieved in pioneering studies by Paulsen *et al.*[50] The strategy relied on first preparing the 5-*O*-glycosylated Kdo unit followed by a second glycosylation of the 4-OH group using a Kdo bromide donor. According to Paulsen, the inverted sequence, *i.e.* coupling of a heptosyl donor to position 5 of a Kdo disaccharide was not possible. Recently, however, this approach could successfully be worked out by Ichiyanagi *et al.*, who synthesized three branched trisaccharides containing α-Man*p*, α-Hep*p* and α-Gal*p*N_3 residues attached to *O*-5 of the Kdo disaccharide (Scheme 27).[48] The glycosidic linkages were assembled using an excess of the respective trichloroacetimidate donors **173–175**, which were activated with catalytic amounts of TMSOTf to give the α-(1→5)-linked trisaccharides **176–178** in excellent to fair yields. Reaction of the less reactive heptosyl donor **173** required a higher amount of promoter and a higher temperature in order to avoid formation of the heptosyl orthoester byproduct.

Very recently sterically congested oligosaccharide donors have also been successfully employed by the same group (Scheme 28).[51] Thus, based on the results as summarized in Scheme 28, di- and trisaccharide trichloroacetimidate donors **179–182** were employed for glycosylation of acceptor **48** in the presence of catalytic amounts of TMSOTf. In contrast to the peracetylated *manno*-configured donors **173** and **174**, the acetyl-protected lactosyl donor **179** was unreactive. The corresponding armed per-*O*-benzylated donor **180**, however, was of sufficient reactivity to give the α-(1→5)-linked tetrasaccharide **183** in low yield. In addition, the lactosyl-(1→3)-heptosyl donor **185** could also be attached to the Kdo disaccharide acceptor **48** and afforded pentasaccharide **184** in 26% yield. Glycosylation of **48** with the heptobiosyl donor **182** furnished the central

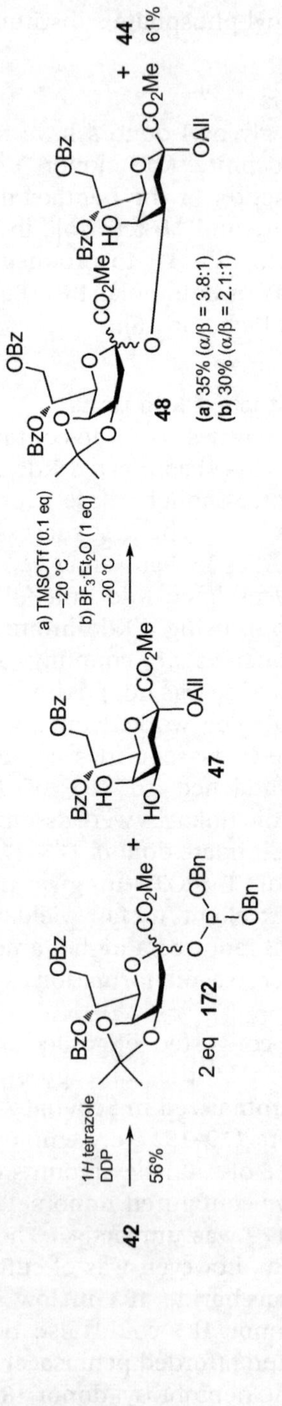

Scheme 26 Glycosylation reaction of Kdo phosphite **172**.

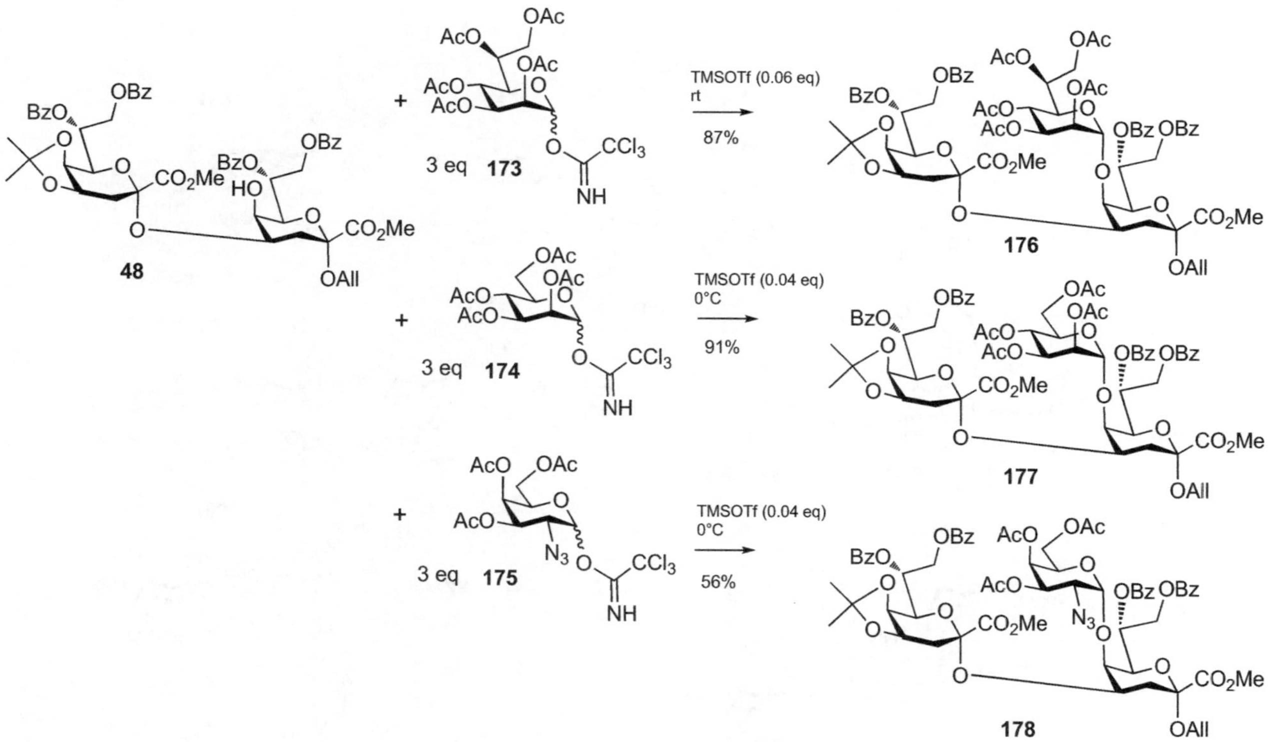

Scheme 27 Synthesis of 4,5-dibranched Kdo trisaccharides.[48]

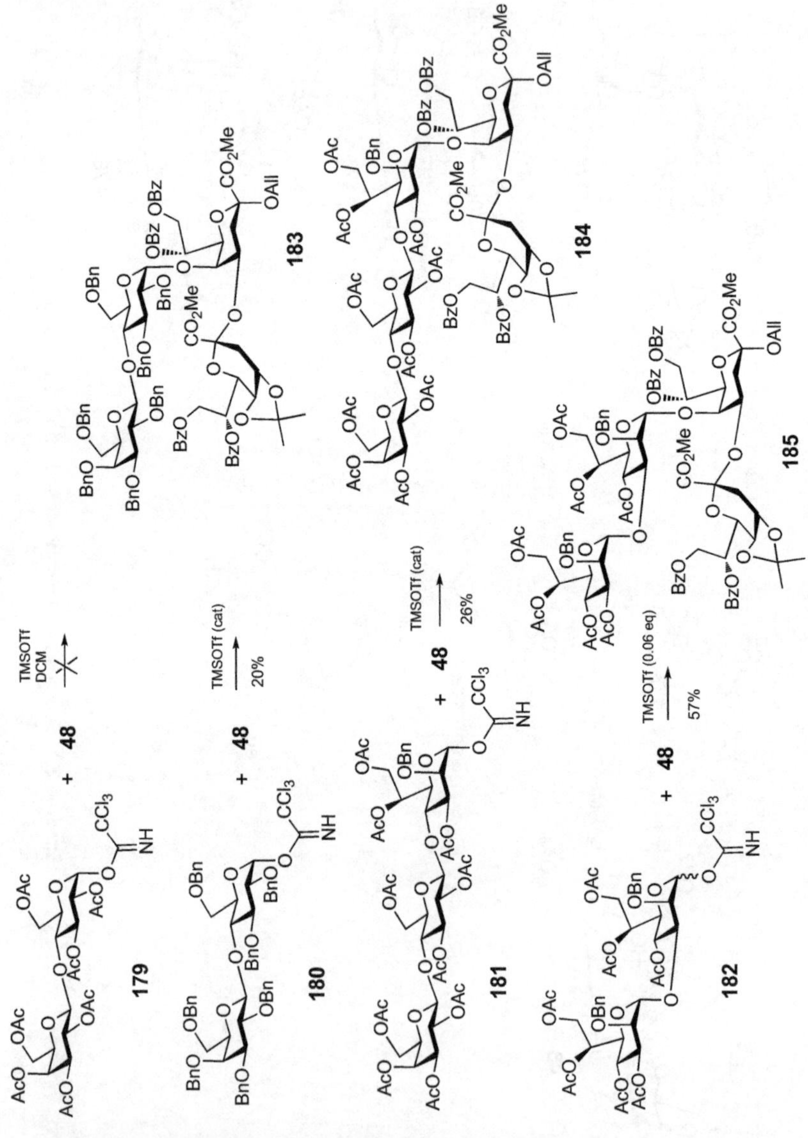

Scheme 28 Synthesis of 4,5-dibranched Kdo oligosaccharides.[51]

LPS core unit **185** as a single isomer in excellent yield (57%). All compounds were then globally deprotected to give the target compounds as propyl glycosides (Scheme 29).[51]

In a similar approach, attachment of a 5-*O*-linked α-glucosyl moiety has recently been performed.[52] Coupling of the 3-iodo-Kdo fluoride donors **53** or **83**, respectively, to known[53] disaccharide acceptors **186** and **187**, however, could not achieved. By contrast, reaction of the Kdo disaccharide acceptor **189** with the 4,6-*O*-benzylidene glucopyranosyl trifluoroacetimidate donor **188** promoted by TfOH gave a separable 4 : 1 anomeric mixture of the trisaccharide **190** in 91% yield. Reductive opening of the benzylidene acetal provided **191** and allowed for subsequent chain elongation. Reaction of **190** with glucosyl NPTFA donors **192** and **193**, respectively, furnished tetra- and pentasaccharide derivatives **194** and **195**. The deprotected oligosaccharides **196–198** correspond to LPS fragments of *Acinetobacter haemolyticus*.

Very recently, the challenging synthesis and the immunochemical evaluation of neoglycoconjugates containing heptobiosyl Kdo disaccharide ligands has been accomplished.[54] Similar to previous strategies (see 2.3.), the key Kdo disaccharide was formed using the bis-isopropylidene protected Kdo fluoride donor **199**. First, fluoride **199** was converted into the diol spacer glycoside **200** followed by attachment of the lateral Kdo unit which afforded a modest yield of the disaccharide **202** and glycal ester **201** among other byproducts. The peracetylated heptobiosyl bromide donor **203** could then be successfully coupled to *O*-5 of Kdo disaccharide acceptor **202** followed by global deprotection to afford tetrasaccharide **205** (Scheme 30).[54]

2.10 Synthesis of Kdo β-glycosides

The synthesis of β-configured Kdo glycosides of primary alcohols has previously been achieved from peracetylated Kdo-bromide **2** under heterogeneous conditions using silver carbonate or silver-silicate promoters.[32e,55] Reactions of Kdo thioglycosides may also be steered toward formation of β-glycosides, *e.g.* when using DMTST[32d] or NIS/AgOTf as promoters.[56] The preparation of complex glycosides corresponding to fragments of capsular polysaccharides is by far more demanding. Recently, the synthesis of a β-(2→7) linked Kdo trimer in a protected form has been published.[57] First, the isopropylidene protected glycal ester **122** was produced in 6 steps from bis(isopropylidene)-mannofuranose **206** in 30% overall yield (Scheme 31). Selective removal of the side-chain acetonide was followed by regioselective 8-*O*-benzylation *via* the stannylidene intermediate to give the 7-OH acceptor **207**. In addition, the benzyl ester **208** was prepared. The β-glycosides **209–213** were synthesized *via* NIS/TMSOTf promoted reaction of the glycal ester with the respective alcohols producing diastereomeric mixtures of the D-*glycero*-β-D-*galacto*-derivatives and their 3-epimeric D-*glycero*-α-D-*talo*-isomers. Changing the solvent from CH_2Cl_2 to a mixture of CH_2Cl_2–MeCN and increasing the amount of NIS to two equivalents substantially improved the yield and the diastereomeric ratio in favor of the 1,2-*trans*-diequatorial products.

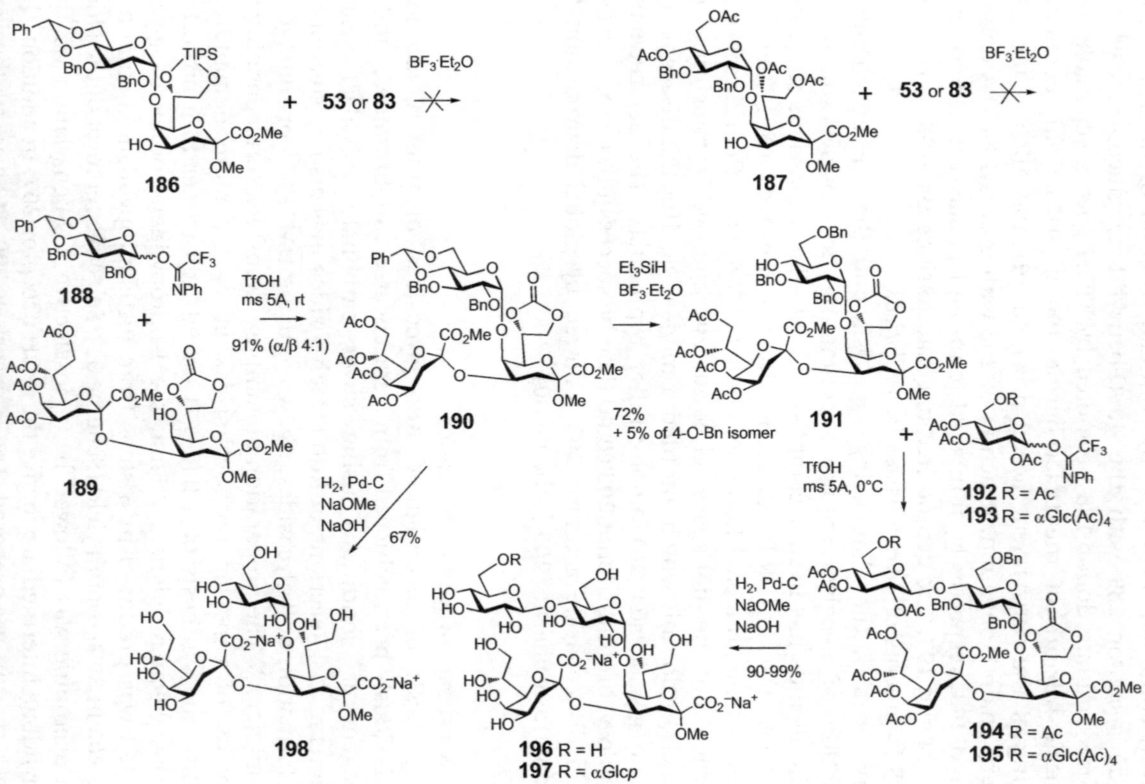

Scheme 29 Synthesis of LPS fragments from *Acinetobacter haemolyticus* LPS.[52,53]

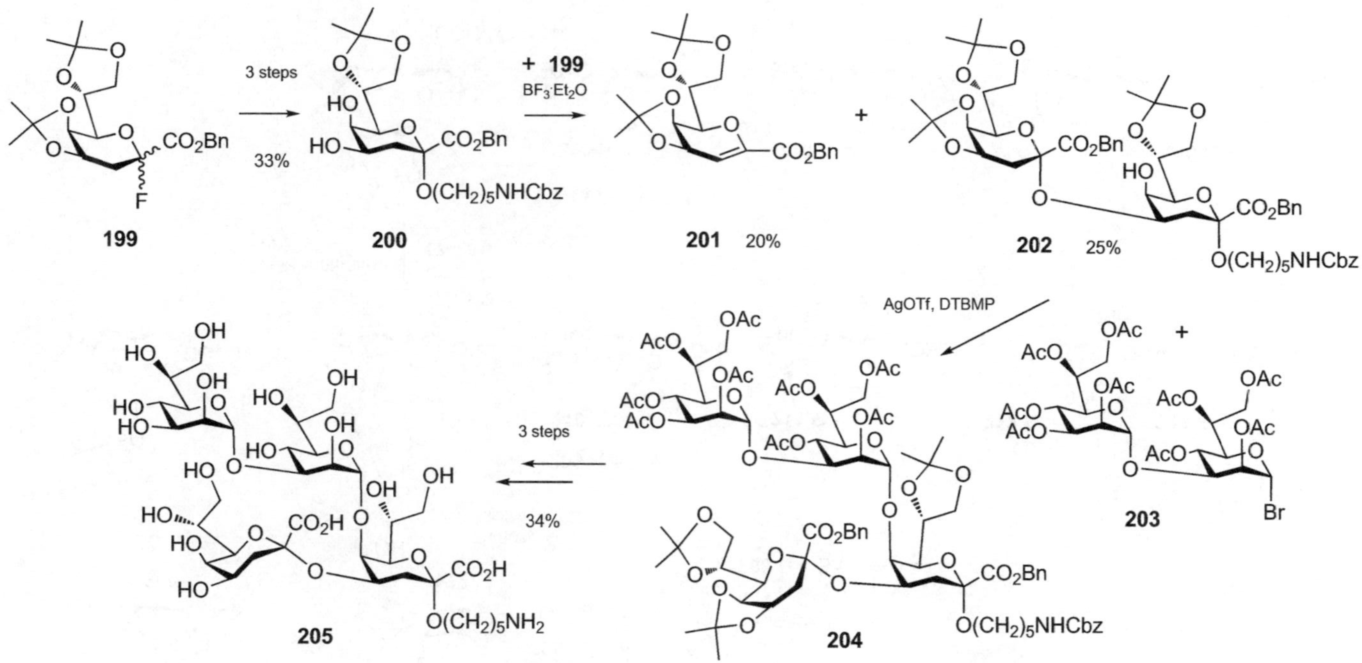

Scheme 30 Synthesis of the inner core LPS fragment **205**.

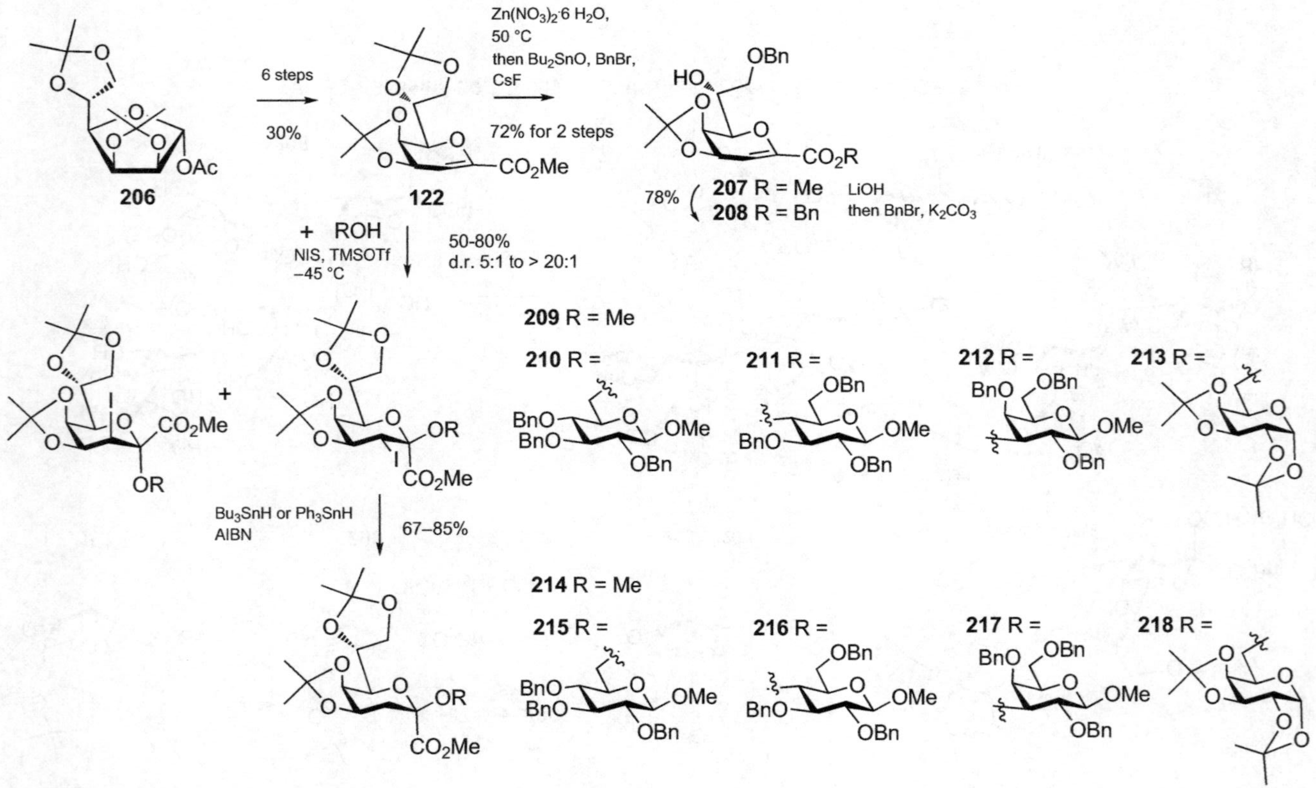

Scheme 31 Synthesis of disaccharides containing β-Kdo units.[57] Adapted with permission from T. K. Pradhan, C. C. Lin and K.-K. T. Mong, *Org. Lett.*, 2014, **16**, 1474–1477. Copyright (2014) American Chemical Society.

The assignment of the anomeric configuration of the products is not trivial and was based on the values of the heteronuclear coupling constants $J_{C\text{-}1,\,H\text{-}3ax}$ (>5 Hz). Eventually the 3-iodo substituents were removed under radical conditions to give the Kdo glycosides **215–218.** Compound **217** corresponds to a protected precursor of CPS fragments from *Rhizobium fredii* and *E. coli.*[58] For the assembly of the β-(2→7) linked Kdo oligomers (Scheme 32), the benzyl glycal ester **207** was subjected to the same protocol to give the 3-iodo derivative **219** in fair yield and selectivity. Subsequent coupling with donor **122** gave the disaccharide **220** in 60% yield. The side chain isopropylidene group was then manipulated to generate the disaccharide acceptor **221**, which was elongated to give the bis-iodo trisaccharide **222**, which was eventually converted into the Kdo-trimer **223** as a protected fragment of Kdo oligomers occurring in *Sinorhizobium meliloti*,[16] *E. coli* and *N. meningitidis* capsular polysaccharides.[17]

An indirect approach to generate β-glycosides of Kdo has been based on oxidation of β-anomeric *C*-aryl glycosides which were derived from a bis-*O*-isopropylidene protected glycal.[58] Glycal **228** was synthesized from mannofuranose **224** by a Wittig reaction with ylide **225** followed by ring-closure of the resulting *E*/*Z* aryl-alkenes **226** with NIS to give the epimeric iodo-products **227** in 66% yield. Subsequent base-catalyzed elimination afforded the glycal donor **228** in good yields (Scheme 33).

Addition reactions of various alcohols to the *C*-*p*-methoxyphenyl glycal **228** in the presence of NIS at low temperature proceeded in excellent yields and anomeric selectivities. An example is shown for the methyl derivative **229.** The preferred formation of the β-adducts was rationalized on the basis of the crystal structure of glycal **228**, which revealed a close proximity of one methyl group to the olefinic carbon, which would thus block the β-side and direct the incoming iodonium species to the bottom site. The subsequent formation of an *epi*-iodonium intermediate (or involving related transition state ionic species) would then be followed by the glycosylation step with the incoming nucleophile from the top position to produce the β-glycoside. Eventually the iodo group of **229** was removed under radical conditions to give **230**, which was oxidized to give the protected Kdo methyl β-ketopyranoside **231** in good yield.

3 Conclusions and outlook

Considerable progress has been achieved in the past few years with dedicated protocols to generate both anomeric ketosides of Kdo in improved yields and excellent stereoselectivities. For many approaches the formation of elimination products and the number of synthetic steps to produce suitable Kdo donors is still unsatisfactory. Further developments, however, will need in-depth theoretical and experimental mechanistic studies, with a focus on the formation of intermediate covalently bound or ion-paired Kdo triflates or related reactive species.

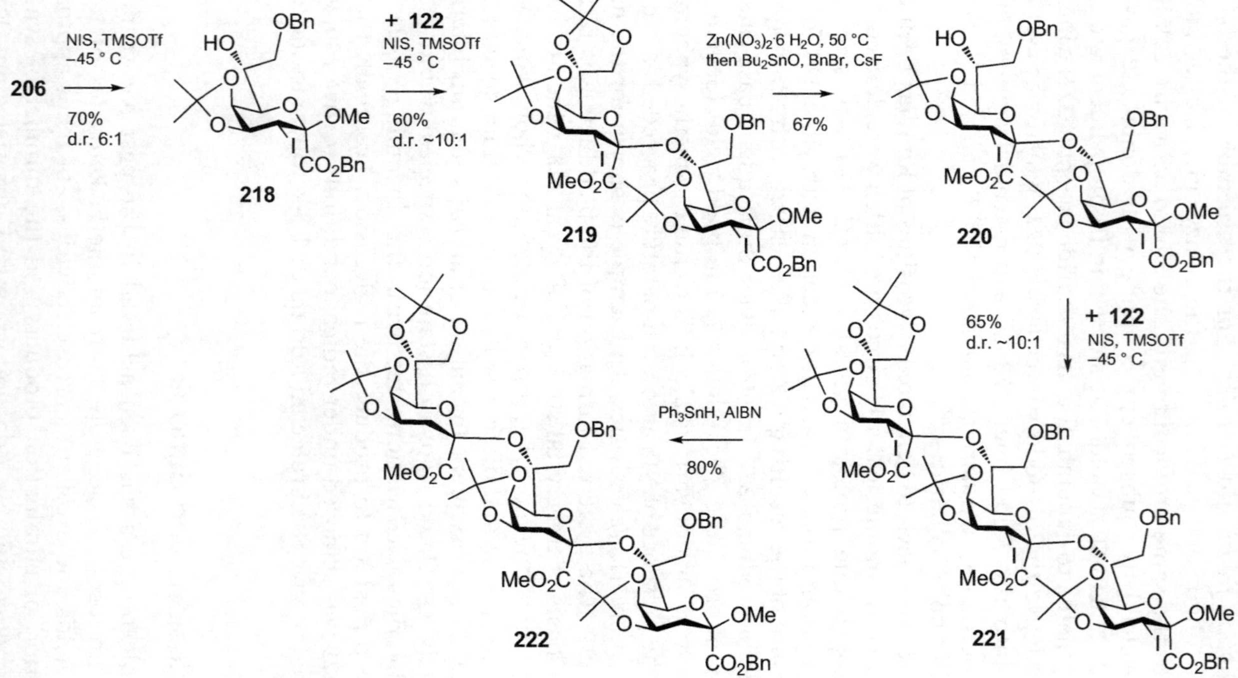

Scheme 32 Synthesis of β-(2→7)-linked Kdo di- and trisaccharides.[57] Adapted with permission from T. K. Pradhan, C. C. Lin and K.-K. T. Mong, *Org. Lett.*, 2014, **16**, 1474–1477. Copyright (2014) American Chemical Society.

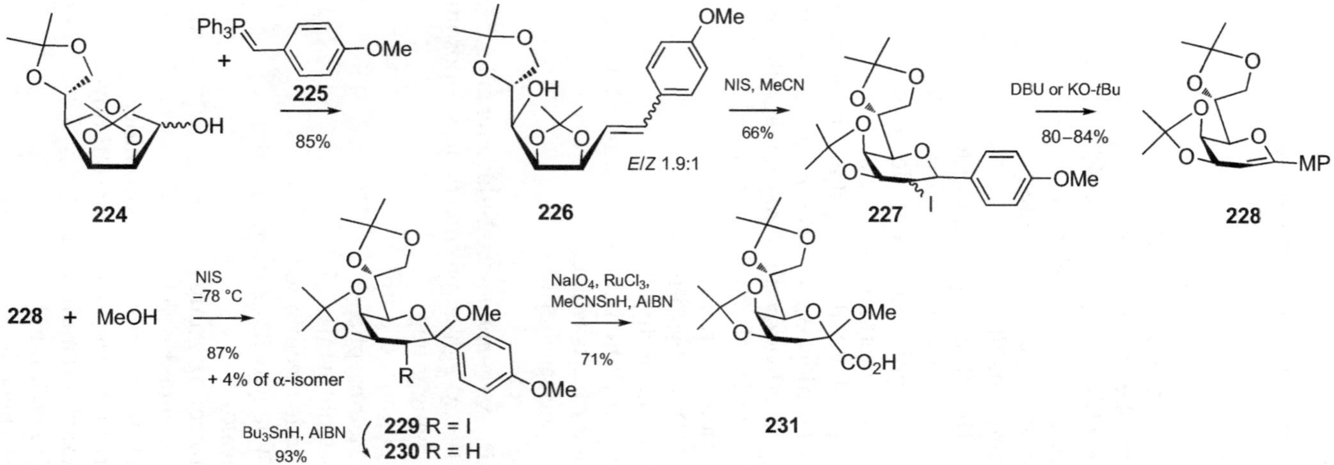

Scheme 33 Synthesis of β-Kdo glycosides from a 1-*C*-aryl glycal derivative.[59] Adapted with permission from Y. Qian, J. Feng, M. Parvez and C.-C. Ling, *J. Org. Chem.*, 2012, **77**, 96–107. Copyright (2012) American Chemical Society.

Abbreviations

AIBN	2,2′-Azobisisobutyronitrile
All	Allyl
Alloc	Allyloxycarbonyl
Ara4N	4-Amino-4-deoxy-L-arabinose
Bn	Benzyl
Bz	Benzoyl
BSA	Bovine serum albumin
*t*BuOOH	*tert*-Butylhydroperoxide
Cbz	Carbobenzoxycarbonyl
*m*CPBA	*meta*-Chloroperbenzoic acid
CPME	Cyclopentyl methylether
CPS	Capsular polysaccharide
CSA	Camphorsulfonic acid
DABCO	1,4-Diazabicyclo[2.2.2]octane
DAST	*N,N*-Diethylaminosulfur trifluoride
DBU	1,8-Diazabicyclo[5.4.0]undec-7-ene
DCM	Dichloromethane
DDP	Dibenzyl *N,N*-diisopropylphosphoramidite
DDQ	2,3-Dichloro-5,6-dicyano-*p*-benzoquinone
DFMBA	*N,N*-Diethyl-α,α-difluoro-(*m*-methylbenzyl)amine
DMAP	4-*N,N*-Dimethylaminopyridine
DME	Dimethoxyethane
DMF	*N,N*-Dimethylformamide
DMTr	Dimethoxytrityl
DMTST	Dimethyl(methylthio)sulfonium triflate
DTBS	Di-*tert*butylsilylene
EPS	Exopolysaccharide
Im	Imidazole
Kdo	3-Deoxy-D-*manno*-oct-2-ulosonic acid
Kdn	3-Deoxy-D-*glycero*-D-*galacto*-non-2-ulosonic acid
LG	Leaving group
LPS	Lipopolysaccharide
MP	4-Methoxyphenyl
Nap	2-Naphthylmethyl
NIS	*N*-iodosuccinimide
NPTFA	*N*-Phenyltrifluoroacetimidate
Piv	Pivaloyl
PG	Protecting group
PMB	4-Methoxybenzyl
TBAB	Tetrabutylammonium bromide
TBPA	Tris(4-bromophenyl)ammoniumyl hexachloroantimonate
TBS	*tert*Butyldimethylsilyl
TEA	Triethylamine
TES	Triethylsilyl
Tf	Trifluoromethanesulfonyl
TFA	Trifluoroacetic acid

THF	Tetrahydrofuran
TIPS	Tetraisopropyl-disiloxane-1,3-diyl
TMS	Trimethylsilyl
Tr	Triphenylmethyl
Troc	Trichloroethoxycarbonyl

Acknowledgements

Financial support of work performed in the author's laboratory by the Austrian Science Fund FWF (P 22909, P24921 and P 26919) is gratefully acknowledged.

References

1 D. H. Levin and E. Recker, *J. Biol. Chem.*, 1959, **234**, 2532.

2 E. C. Heath and M. A. Ghalambor, *Biochem. Biophys. Res. Commun.*, 1963, **10**, 340.

3 M. J. Osborn, *Proc. Natl. Acad. Sci. U. S. A.*, 1963, **50**, 499.

4 (a) W. S. York, A. G. Darvill, M. McNeil and P. Albersheim, *Carbohydr. Res.*, 1985, **138**, 109; (b) F. Delmas, M. Séveno, J. G. B. Northey, M. Hernould, P. Lerouge, P. McCourt and C. Chevalier, *J. Exp. Bot.*, 2007, **59**, 2639.

5 B. Becker, J. P. M. Lommerse, M. Melkonian, J. P. Kamerling and J. F. G. Vliegenthart, *Carbohydr. Res.*, 1995, **267**, 313.

6 S. Herget, P. V. Toukach, R. Ranzinger, W. E. Hull, Y. A. Knirel and C.-W. von der Lieth, *BMC Struct. Biol.*, 2008, **8**, 35.

7 (a) F. M. Unger, *Adv. Carbohydr. Chem. Biochem.*, 1981, **38**, 323; (b) C. R. H. Raetz and C. Whitfield, *Annu. Rev. Biochem.*, 2002, **71**, 635; (c) O. Holst, in *Bacterial Lipopolysaccharides*, ed. Y. A. Knirel and M. A. Valvano, Springer, Wien, New York, 2011, pp. 21–39; (d) O. Holst and A. Molinaro, in *Microbial Glycobiology*, ed. A. P. Moran, O. Holst, P. J. Brennan and M. von Itzstein, Elsevier, Amsterdam, 2009, pp. 29–55; (e) X. Wang, P. J. Quinn and A. Yan, *Biol. Rev.*, 2015, **90**, 408.

8 (a) H. Brade, C. Galanos and O. Lüderitz, *Eur. J. Biochem.*, 1983, **131**, 201; (b) O. Holst, *FEMS Microbiol. Lett.*, 2007, **271**, 3; (c) H. Paulsen, M. Stiem and F. M. Unger, *Tetrahedron Lett.*, 1986, **27**, 1135.

9 (a) E. Vinogradov and M. B. Perry, *Carbohydr. Res.*, 2001, **335**, 291; (b) E. Vinogradov, E. Fridrich, L. L. MacLean, M. B. Perry, B. O. Petersen, J. Ø. Duus and C. Whitfield, *J. Biol. Chem.*, 2002, **277**, 25050; (c) E. L. Zdorovenko, E. Vinogradov, G. M. Zdorovenko, B. Lindner, O. V. Bystrova, A. S. Shashkov, R. Klaus, U. Zähringer and Y. A. Knirel, *Eur. J. Biochem.*, 2004, **271**, 4968; (d) L. S. Forsberg and R. W. Carlson, *J. Biol. Chem.*, 1998, **273**, 2747; (e) G. Pieretti, S. Carillo, B. Lindner, R. Lanzetta, M. Parrilli, N. Jimenez, M. Regué, J. M. Tomás and M. M. Corsaro, *Carbohydr. Res.*, 2010, **345**, 2523.

10 (a) N. A. Kocharova, O. G. Ovchinnikova, A. Torzewska, A. S. Shashkov, Y. A. Knirel and A. Rozalski, *Carbohydr. Res.*, 2007, **342**, 665; (b) L. L. MacLean, E. Vinogradov, F. Pagotto, J. M. Farber and M. B. Perry, *Biochem. Cell Biol.*, 2009, **87**, 927.

11 For examples of β-Kdo in CPS and EPS see: (a) B. Jann and K. Jann, Bacterial capsules, in *Current Topics in Microbiology and Immunobiology*, ed. P. D. K. Jann and D. B. Jann, Springer, Berlin, Heidelberg, 1990, pp. 19–42; (b) P. Cescutti, G. Impallomeni, D. Garozzo, L. Sturiale, Y. Herasimenka,

C. Lagatolla and R. Rizzo, *Carbohydr. Res.*, 2003, **338**, 2687; (c) M. Nimtz, V. Wray, T. Domke, B. Brenneke, S. Häussler and I. Steinmetz, *Eur. J. Biochem.*, 1997, **250**, 608; (d) H. Masoud, M. Ho, T. Schollaardt and M. B. Perry, *J. Bacteriol.*, 1997, **179**, 5663; (e) N. Fraysse, F. Couderc and V. Poinsot, *Eur. J. Biochem.*, 2003, **270**, 1365; (f) R. P. Gorshkova, E. L. Nazarenko, A. A. Zubkov, A. S. Shashkov, Y. A. Knirel, N. Paramonov, S. V. Meshkov and E. P. Ivanova, *Carbohydr. Res.*, 1997, **299**, 69; (g) L. S. Forsberg and B. L. Reuhs, *J. Bacteriol.*, 1997, **179**, 5366; (h) L. M. Beynon, J. C. Richards and M. B. Perry, *Carbohydr. Res.*, 1991, **220**, 185.

12 For examples of α-Kdo in CPS see: (a) H. J. Jennings, R. Roy and R. E. Williams, *Carbohydr. Res.*, 1984, **129**, 243; (b) F. Lelchat, S. Cérantola, C. Brandily, S. Colliec-Jouault, A.-C. Baudoux, T. Ojima and C. Boisset, *Carbohydr. Polym.*, 2015, **124**, 347.

13 (a) H. Brade, L. Brade and F. E. Nano, *Proc. Natl. Acad. Sci U. S. A.*, 1987, **84**, 2508; (b) Y. Fu, M. Baumann, P. Kosma, L. Brade and H. Brade, *Infect. Immun.*, 1992, **60**, 1314.

14 A. Hanuszkiewicz, G. Hübner, E. Vinogradov, B. Lindner, L. Brade, H. Brade, J. Debarry, H. Heine and O. Holst, *Chem. – Eur. J.*, 2008, **14**, 10251.

15 (a) E. V. Vinogradov, B. O. Petersen, J. E. Thomas-Oates, J. Ø. Duus, H. Brade and O. Holst, *J. Biol. Chem.*, 1998, **273**, 28122; (b) E. V. Vinogradov, J. Ø. Duus, H. Brade and O. Holst, *Eur. J. Biochem.*, 2002, **269**, 422; (c) E. Fregolino, G. Fugazza, E. Galano, V. Gargiulo, P. Landini, R. Lanzetta, B. Lindner, L. Pagani, M. Parrilli, O. Holst and C. DeCastro, *Eur. J. Org. Chem.*, 2010, 1345; (d) S. Leone, A. Molinaro, E. Pessione, R. Mazzoli, C. Giunta, L. Sturiale, D. Garozzo, R. Lanzetta and M. Parrilli, *Carbohydr. Res.*, 2006, **314**, 582.

16 N. Fraysse, B. Lindner, Z. Kaczynski, L. Sharypova, O. Holst, K. Niehaus and V. Poinsot, *Glycobiology*, 2005, **15**, 101.

17 L. M. Willis and C. Whitfield, *Proc. Natl. Acad. Sci. U. S. A.*, 2013, **110**, 20753.

18 (a) S. Oscarson and J. Hansson, *Curr. Org. Chem.*, 2000, **4**, 535; (b) S. Oscarson, *Carbohydr. Chem.*, 2012, **45**, 40; (c) P. Kosma in *Microbial Glycobiology*, ed. A. P. Moran, O. Holst, P. J. Brennan and M. von Itzstein, Elsevier, Amsterdam, 2009, pp. 429–454; (d) L. Cipolla, L. Gabrielli, D. Bini, L. Russo and N. Shiakh, *Nat. Prod. Rep.*, 2010, **27**, 161; (e) P. Kosma and A. Zamyatina in *Bacterial Lipopolysaccharides*, ed. Y. A. Knirel and M. A. Valvano, Springer, Wien, New York, 2011, pp. 131–161; T. K. Pradhan and T. K. K. T. Mong, *Isr. J. Chem.*, 2015, **55**, 285.

19 G. Sixta, K. Wimmer, A. Hofinger, H. Brade and P. Kosma, *Carbohydr. Res.*, 2009, **344**, 1660.

20 H. M. Christensen, S. Oscarson and H. H. Jensen, *Carbohydr. Res.*, 2015, **408**, 51.

21 P. Kosma, A. Hofinger, S. Müller-Loennies and H. Brade, *Carbohydr. Res.*, 2010, **345**, 704.

22 S. Gerstenbruch, C. L. Brooks, P. Kosma, L. Brade, C. R. MacKenzie, S. V. Evans, H. Brade and S. Müller-Loennies, *Glycobiology*, 2010, **20**, 461.

23 M. Blaukopf, B. Müller, A. Hofinger and P. Kosma, *Eur. J. Org. Chem.*, 2012, 119.

24 (a) M. Imoto, N. Kusunose, S. Kusumoto and T. Shiba, *Tetrahedron Lett.*, 1988, **29**, 2227; (b) M. Imoto, N. Kusunose, Y. Matsuura, S. Kusumoto and T. Shiba, *Tetrahedron Lett.*, 1987, **28**, 6277; (c) T. J. Boltje, W. Zhong, J. Park, M. A. Wolfert, W. Chen and G.-J. Boons, *J. Am. Chem. Soc.*, 2012, **134**, 14255; (d) W. Rosenbrook Jr., D. A. Riley and P. A. Lartey, *Tetrahedron Lett.*, 1985, **26**, 3.

25 Y. Fujimoto, M. Iwata, N. Imakita, A. Shimoyama, Y. Suda, S. Kusumoto and K. Fukase, *Tetrahedron Lett.*, 2007, **48**, 6577.
26 H. Yoshizaki, N. Fukuda, K. Sato, M. Oikawa, K. Fukase, Y. Suda and S. Kusumoto, *Angew. Chem., Int. Ed.*, 2001, **40**, 1475.
27 Y. Zhang, J. Gaekwad, M. A. Wolfert and G.-J. Boons, *Chem. – Eur. J.*, 2008, **14**, 558.
28 T. Ichiyanagi, M. Fukunaga, R. Tagashira, S. Hayashi, M. Nanjo and R. Yamasaki, *Tetrahedron*, 2011, **67**, 5964.
29 Y. Feng, J. Dong, F. Xu, A. Liu, L. Wang, Q. Zhang and Y. Chai, *Org. Lett.*, 2015, **17**, 2388.
30 P. Kosma, J. Gass, G. Schulz, R. Christian and F. M. Unger, *Carbohydr. Res.*, 1987, **167**, 39.
31 P. Kosma, G. Schulz and H. Brade, *Carbohydr. Res.*, 1988, **183**, 183.
32 (a) K. Ikeda, S. Akamatsu and K. Achiwa, *Chem. Pharm. Bull.*, 1990, **38**, 279; (b) K. Ikeda, S. Akamatsu and K. Achiwa, *Carbohydr. Res.*, 1989, **189**, C1; (c) Y. Yang, C. E. Martin and P. H. Seeberger, *Chem. Sci.*, 2011, **3**, 896; (d) K. Ekelöf and S. Oscarson, *Carbohydr. Res.*, 1995, **278**, 289; (e) P. A. M. van der Klein, G. J. P. G. H. Boons, G. H. Veeneman, G. A. van der Marel and J. H. van Boom, *Synlett*, 1990, 311.
33 A. Claesson and K. Luthman, *Acta Chem. Scand., Ser. B*, 1982, **36**, 719.
34 P. Kosma, H. Sekljic and G. Balint, *J. Carbohydr. Chem.*, 1996, **15**, 701.
35 B. Pokorny and P. Kosma, *Chem. – Eur. J.*, 2015, **21**, 305.
36 F. M. Unger, D. Stix and G. Schulz, *Carbohydr. Res.*, 1980, **80**, 191.
37 B. Pokorny and P. Kosma, *ChemistryOpen*, 2015, **4**, 722.
38 B. Pokorny and P. Kosma, *Org. Lett.*, 2015, **17**, 110.
39 (a) Y. Yang, S. Oishi, C. E. Martin and P. H. Seeberger, *J. Am. Chem. Soc.*, 2013, **135**, 6262; (b) A. Reinhardt, Y. Yang, H. Claus, C. L. Pereira, A. D. Cox, U. Vogel, C. Anish and P. H. Seeberger, *Chem. Biol.*, 2015, **22**, 38.
40 K. Mannerstedt, K. Ekelöf and S. Oscarson, *Carbohydr. Res.*, 2007, **342**, 631.
41 H. Tanaka, D. Takahashi and T. Takahashi, *Angew. Chem., Int. Ed.*, 2006, **45**, 770.
42 G. J. P. H. Boons, F. L. Delft, P. A. M. van der Klein, G. A. van der Marel and J. H. van Boom, *Tetrahedron*, 1992, **48**, 885.
43 T. Wåglund and A. Claesson, *Acta Chem. Scand.*, 1992, **46**, 73.
44 D. Solomon, M. Fridman, J. Zhang and T. Baasov, *Org. Lett.*, 2001, **3**, 4311.
45 (a) D. Kumagai, M. Miyazaki and S. I. Nishimura, *Tetrahedron Lett.*, 2001, **42**, 1953; (b) A. Imamura, A. Kimura, H. Ando, H. Ishida and M. Kiso, *Chem. – Eur. J.*, 2006, **12**, 8862.
46 J.-S. Huang, W. Huang, X. Meng, X. Wang, P.-C. Gao and J.-S. Yang, *Angew. Chem., Int. Ed.*, 2015, **54**, 10894.
47 A. Shimoyama, Y. Fujimoto and K. Fukase, *Synlett*, 2011, 2359.
48 R. Yi, A. Ogaki, M. Fukunaga, H. Nakajima and T. Ichiyanagi, *Tetrahedron*, 2014, **70**, 3675.
49 A. Shimoyama, A. Saeki, N. Tanimura, H. Tsutsui, K. Miyake, Y. Suda, Y. Fujimoto and K. Fukase, *Chem. – Eur. J.*, 2011, **17**, 14464.
50 (a) H. Paulsen and M. Brenken, *Liebigs Ann. Chem.*, 1991, 1113; (b) H. Paulsen and A. C. Heitmann, *Liebigs Ann. Chem.*, 1989, 655.
51 R. Yi, H. Narimoto, M. Nozoe and T. Ichiyanagi, *Biosci., Biotechnol., Biochem.*, 2015, **79**, 1931.
52 B. Pokorny and P. Kosma, *Carbohydr. Res.*, 2016, **422**, 5.
53 B. Pokorny and P. Kosma, *Carbohydr. Res.*, 2014, **391**, 66.
54 L. Kong, B. Vijayakrishnan, M. Kowarik, J. Park, A. N. Zakharova, L. Neiwert, A. Faridmoayer and B. J. Davis, *Nat. Chem.*, 2016, **8**, 242.

55 (a) H. Paulsen and C. Krogmann, *Carbohydr. Res.*, 1990, **205**, 31; (b) P. Waldstätten, R. Christian, G. Schulz, F. M. Unger, P. Kosma, C. Kratky and H. Paulsen, *ACS Symp. Ser.*, 1983, **231**, 6.
56 A. Laroussarie, B. Barycza, H. Andriamboavonjy, M. T. Kenfack, Y. Blériot and C. Gauthier, *J. Org. Chem.*, 2015, **80**, 10386.
57 T. K. Pradhan, C. C. Lin and K.-K. T. Mong, *Org. Lett.*, 2014, **16**, 1474.
58 B. L. Reush, R. W. Carlson and J. S. Kim, *J. Bacteriol.*, 1993, **175**, 3570.
59 Y. Qian, J. Feng, M. Parvez and C.-C. Ling, *J. Org. Chem.*, 2012, 77, 96.

Chemical and enzymatic approaches to the synthesis of cyclic oligosaccharides

Krishnagopal Maiti, Gopal Kumar Samanta, Gour Chand Daskhan and N. Jayaraman*

DOI: 10.1039/9781782626657-00165

Cyclic oligosaccharides, such as naturally-occurring cyclomaltodextrins or cyclodextrins, cycloisomaltooligosaccharides or cyclodextrans, cycloinulooligosaccharides, have attracted a repertoire of interest cutting across various fields, as a result of their immense industrial importance arising from their molecular and supramolecular properties. As opposed to the production of cyclic oligosaccharides biosynthetically in an exquisite manner, their chemical synthesis remains a challenge. Whereas macrocyclization and cyclooligomerization have provided necessary tools to approach their synthesis, modifications in their monosaccharide moieties add further to protract the synthesis overall. Yet significant progress has been made in chemical and enzymatic synthesis during the last three decades, particularly on the synthesis of tailor-made or made-to-order cyclic oligosaccharides. Manifold values of cyclic oligosaccharide macrocycles are imminent, resulting from their reduced hydrolytic rates to hydrophobic cavities, backbone modifications assume even greater importance. Such backbone modified cyclic oligosaccharides have been demonstrated to alter functions in several instances. Focusing on synthesis of cyclic oligosaccharides alone, there is a continuous evolution of methods and strategies that facilitate their synthesis. Both chemical and enzymatic routes are fore-runners in the advancements towards fully synthetic cyclic oligosaccharides. This review compiles logical advancements in cyclic oligosaccharide synthesis, particularly over the last decade, with emphasis on modifications of glycosidic bonds or individual sugar moieties.

1 Introduction

Among the varieties of naturally-occurring macrocyclic molecules, the particular class of polyhydroxylated cyclic oligosaccharides occupies a prime interest and importance. As macrocycles, the cyclic oligosaccharides offer the most valuable host properties, which can be taken advantage elegantly to form stable host–guest supramolecular complexes, in addition to the facilities to covalently modify the macrocycles with the availability of multiple free hydroxy groups. Covalent modification of the exocyclic hydroxy groups thus leads to modify the host properties of cyclic oligosaccharides. Among many naturally-occurring cyclic oligosaccharide macrocyles, α-, β- and γ-cyclodextrins (CDs), with 6, 7 and 8 α-D-(1→4) glycosidic linkages, respectively, are the most common. The CDs are studied most intensively with respect to their molecular modifications of available hydroxy groups, leading to altered molecular properties. Simultaneously, CDs are exploited very extensively in many chemical and interfacial areas as molecular containers to uptake functionally important ions and molecules, so as to benefit from such host–guest complexes. A number of review articles and compilations appear periodically, which update the progress and development of studies focused on chemical

Department of Organic Chemistry, Indian Institute of Science, Bangalore 560 012, India. E-mail: jayaraman@orgchem.iisc.ernet.in

modification of exocyclic substituents and attendant changes in host-guest properties.[1] As opposed to phenomenal progress in the utilitarian values of readily available, naturally-occurring cyclic oligosaccharides, strategies to chemically or enzymatically synthesize cyclic oligosaccharides in a 'made-to-order' fashion lags behind severely. The requirement (i) to install protecting groups judiciously on hydroxy groups; (ii) to implement reactions with preferred anomeric selectivity in the case of chemical synthesis and (iii) to identify appropriate enzymes that permit covalent attachment of sugar moieties, leading to the macrocycle formation in the case of enzymatic synthesis are few impediments to the success of the synthesis of tailor-made cyclic oligosaccharides.[2] Developments towards achieving facile synthesis of cyclic oligosaccharides with desired constitutions are continuing themes, given the fact that the resulting tailor-made cyclic oligosaccharide derivatives could, in principle, have molecular and supramolecular properties outweighing the challenges in their synthesis alone.[3] The present compilation targets on progress that expands the frontiers of chemical and enzymatic synthesis of cyclic oligosaccharides. The article is focused specifically on the backbone or skeletal modifications leading to the formation of such backbone modified cyclic oligosaccharides. The backbone modifications refer to changes either in the glycosidic bond or in the nature of individual sugar moieties constituting the cyclic molecule, as opposed to modifications on the hydroxy groups of a pre-formed cyclic oligosaccharide.[4] In the development of synthesis and methods, it is important that synthesis of native CDs are achieved as a proof-of-concept. Thus, this article is based broadly on two major types of synthesis, (i) synthesis leading to retain the glycosidic oxygen linkage, with or without modification of individual sugar moiety and (ii) synthesis leading to replace the glycosidic bond with other heteroatom linkages.

Macrocyclization reactions could occur in two pathways, with each one elegant in its own merit: (i) macrocyclization through pre-formed linear oligomer having defined number of monomer moieties, as shown in Fig. 1. A step-wise elongation of monomer moieties to a linear oligomer, followed by macrocyclization of the linear oligomer would lead to the macrocycle with pre-determined number of monomer sugar moieties in the macrocycle. The merit of the method lies in the product formation, predominantly that of one macrocyclic oligomer. In the second method, chosen saccharide monomer is allowed to undergo oligomerization, leading to the formation of oligomers that are then allowed to proceed cyclization, in one-pot, so as to result in multiple number of cyclic oligomers, as shown in Fig. 2.

As established in the case of maltoheptaose oligosaccharide helical structure,[5] the growing linear oligomers assume pre-defined conformational preferences. Such conformational preferences might favorably be disposed towards cyclization, such that several macrocycles could form, in principle, and in multiples of the initiator monomer. Both the methods have attracted the interest in the overall development of the synthesis of cyclic oligosaccharides. Thus, both the glycosidic bond oxygen retained cyclic oligosaccharides and the glycosidic oxygen replaced cyclic oligosaccharides syntheses involve both the above synthetic methods interdependently.

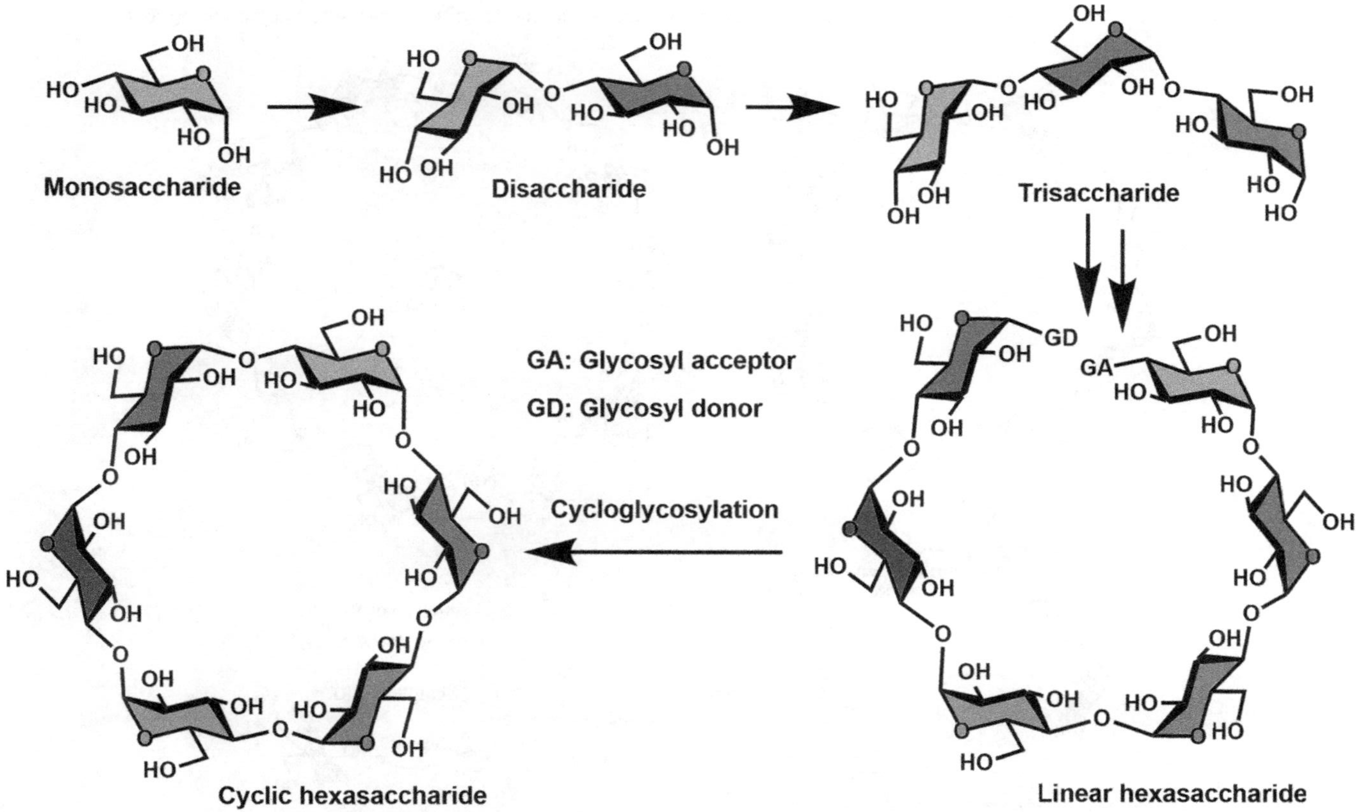

Fig. 1 Schematic representation of cyclic oligosaccharides synthesis through pre-formed linear oligomer.

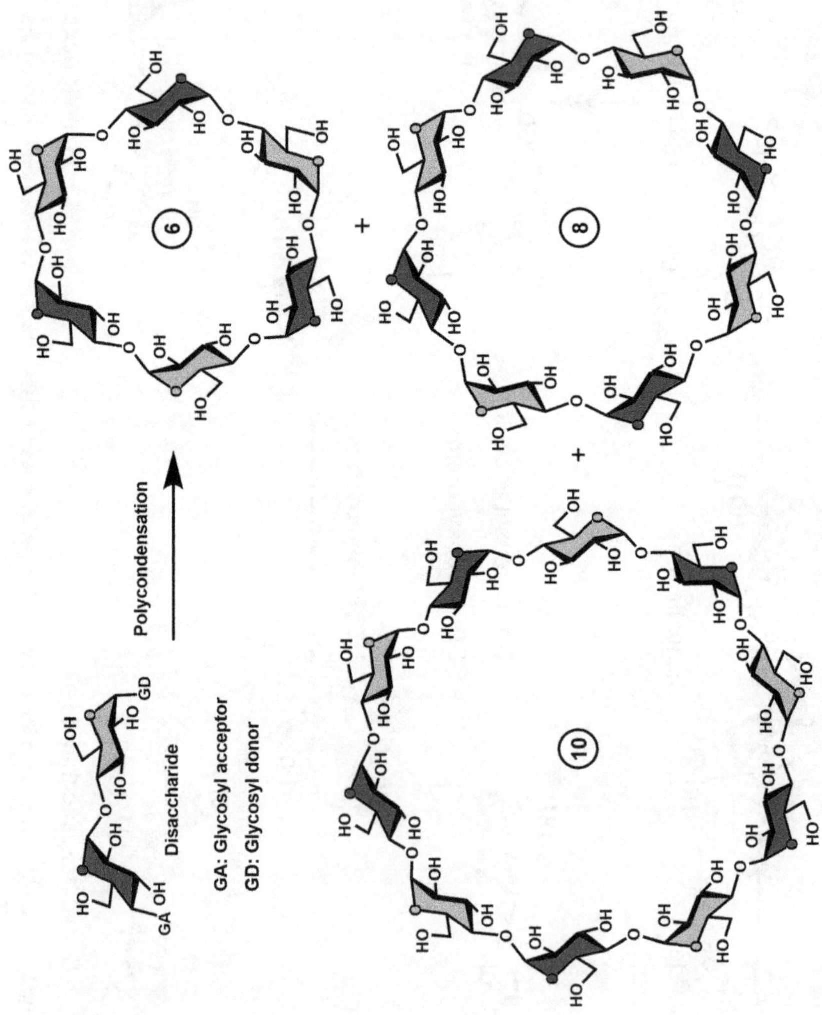

Fig. 2 Schematic representation of cyclic oligosaccharides synthesis through one-pot polycondensation method.

2 Cyclic oligosaccharides retained with glycosidic oxygen

Synthesis of first instances of native CDs occurred several decades later than developments pertaining to exploiting the molecular and supramolecular properties of CDs. The first synthesis of a CD was reported by Takahashi and Ogawa in mid-1980's.[6] The total synthesis of a CD was approached through a macrocyclization of a pre-formed linear oligomer **1**, as shown in Fig. 3. Synthesis of the cyclic octamer **2**, corresponding to the γ-CD, warranted preparation of a linear octamer **1**, with suitably installed glycosyl donor and acceptor sites. The linear octaoside **1** was prepared in several steps and the final macrocyclization was performed in the presence of AgOTf, $SnCl_2$ and molecular sieves. The desired cyclomaltooctaose **2** was obtained in 8.4% yield, with remaining yields of the reaction corresponded to hydrolyzed lactol, 1,2-unsaturated and 1,6-anhydro- linear sugar derivatives. The linear maltohexaose **3**, with glycosyl fluoride donor moiety at the reducing end and acceptor alcohol at *C*-4 of the non-reducing end, led to the formation of cyclic hexasaccharide, namely, α-CD in the hydroxy group protected form **4**, in 19% yield (Fig. 4). The glycosylation reaction also led to the formation of α-(1→6)-linked cyclic oligosaccharides **5**, in 47% yield.[6]

The above early synthesis also included modification of sugar hydroxy groups with varied moieties in the linear hexasaccharide, which upon macrocyclization, would lead to the cyclic oligosaccharide product, incorporated with the modified sugar moiety. Thus, as depicted in Fig. 5, cyclic hexasaccharide with an azido-functionality (**7**) at *C*-6 of one of the sugar moieties could also be prepared, through modification incorporated in the linear oligosaccharide precursor **6**. The multi-step synthesis encountered many side products, compromising the yield of the required linear and cyclic products. Irrespective of this truncation in synthesis, formation of cyclic products **2**, **4**, **5** and **7** is considered as a formidable success to chemically synthesize cyclic oligosaccharides for the first time.[7]

The success of CD synthesis led to employ the methodology for the preparation of constitutionally varied cyclic oligosaccharides by Ogawa and coworkers.[8] The *manno*-configuration analogue of α-CD was prepared by macrocyclization of linear mannosyl hexasaccharide having a donor and acceptor moieties at the reducing and non-reducing sugars, respectively, within the linear hexasaccharide **8** (Fig. 6). The cyclization in this instance occurred impressive, in 64% yield of the cyclic mannohexaose **9**, as opposed to the same reaction occurring in 21% yield in the case of the corresponding glucoside hexamer.[8]

Preparation of linear oligomer, followed by macrocyclization was demonstrated in the case of cyclic pentaose. A cyclomaltopentaose synthesis was achieved by Nakagawa and coworkers through preparation of a linear pentasaccharide **10** (Fig. 7).[9] The linear pentasaccharide was obtained by first utilizing judiciously equipped maltose derivative **I** (Fig. 8) as an acceptor in a glycosylation with a suitably protected

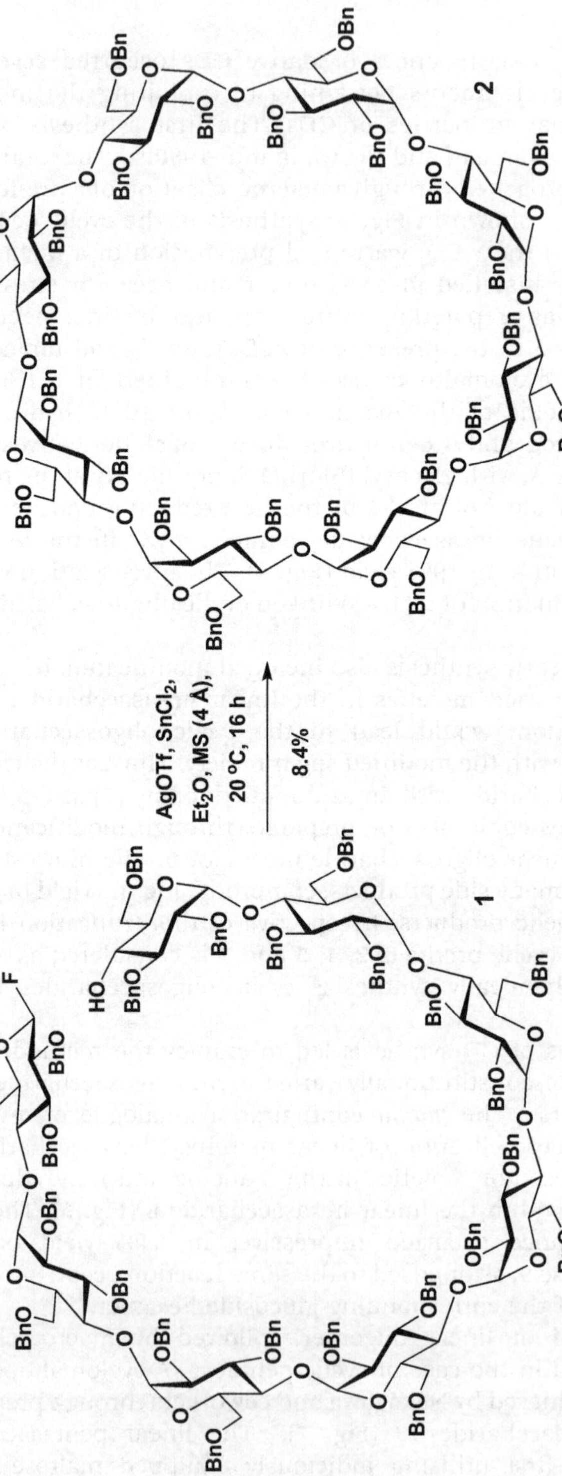

Fig. 3 Synthesis of cyclic octamer from linear octaoside.[6]

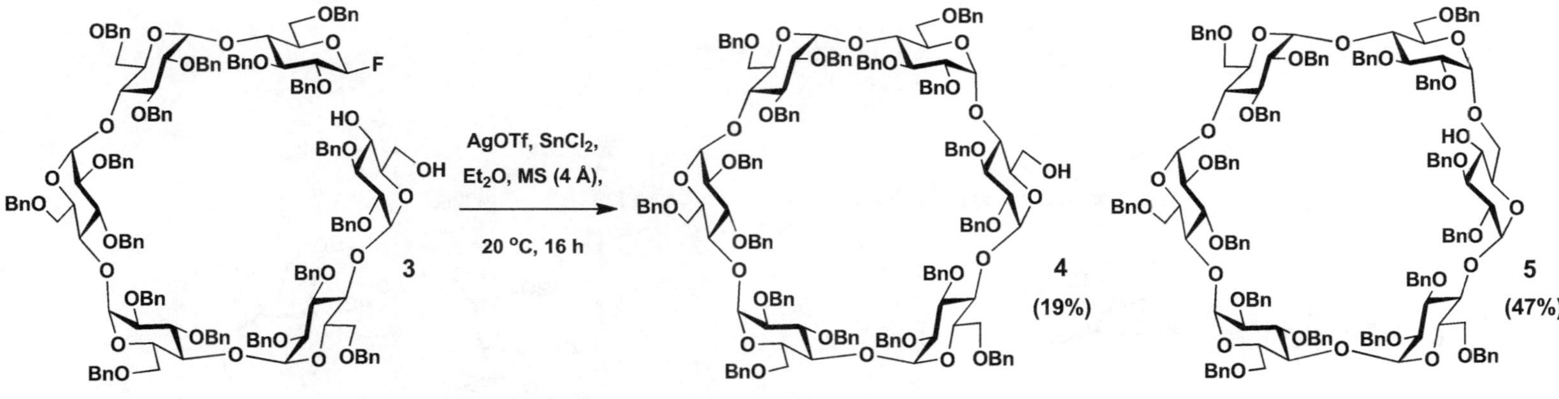

Fig. 4 Synthesis of cyclic hexasaccharides from linear maltohexaose.[6]

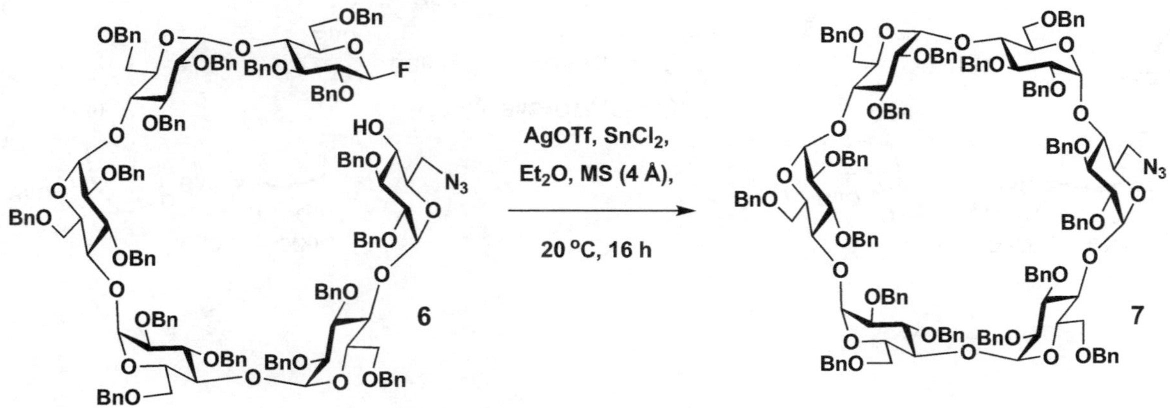

Fig. 5 Synthesis of cyclic hexasaccharide with an azido-functionality.[7]

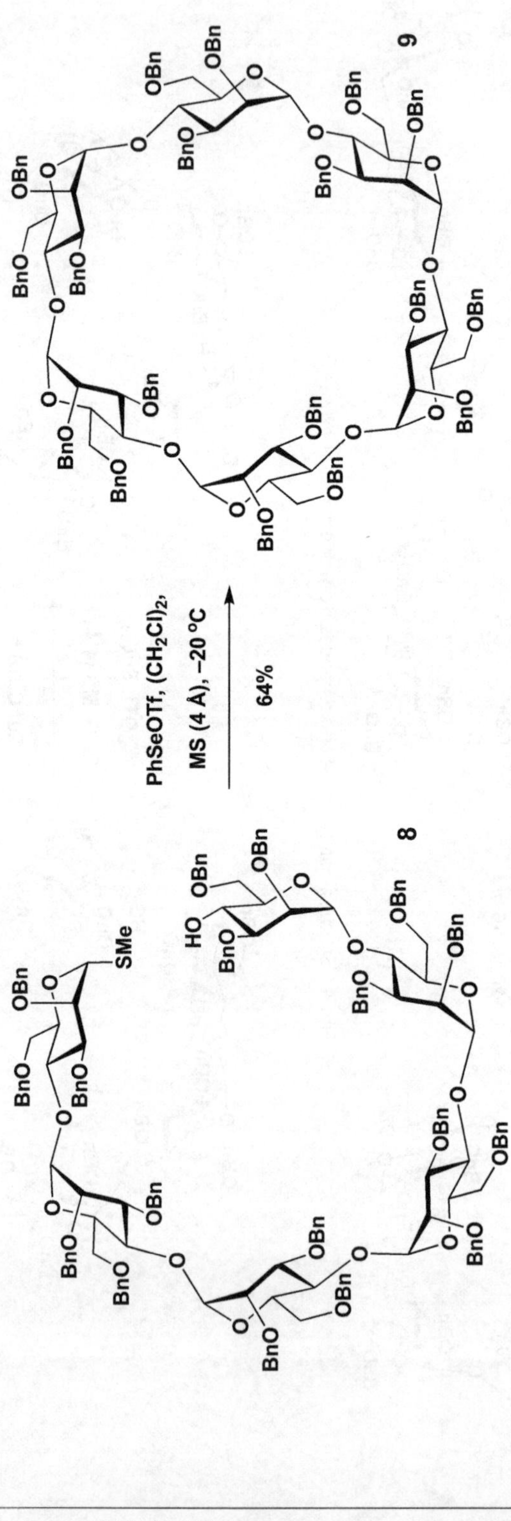

Fig. 6 Synthesis of cyclic mannohexaose from linear hexasaccharide.[8]

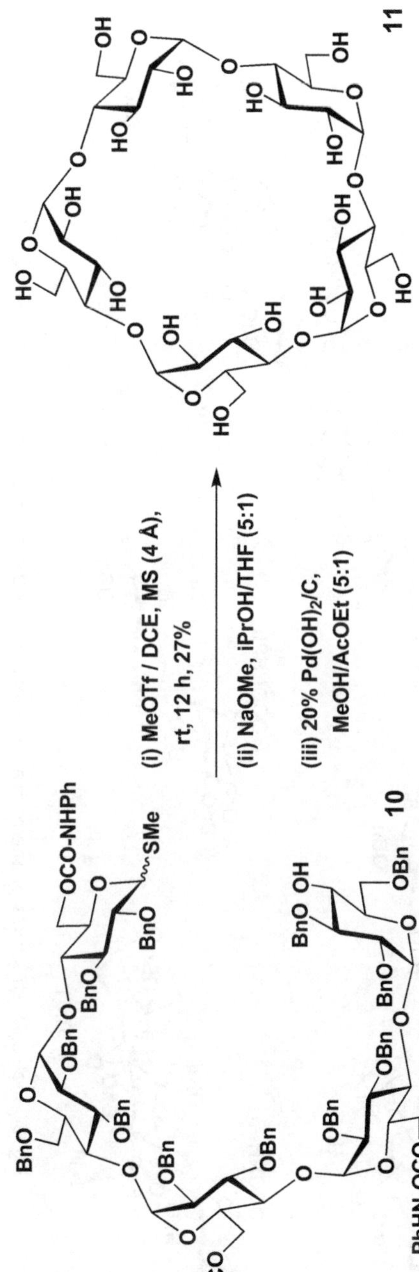

Fig. 7 Synthesis of cyclomaltopentaose 11.[9]

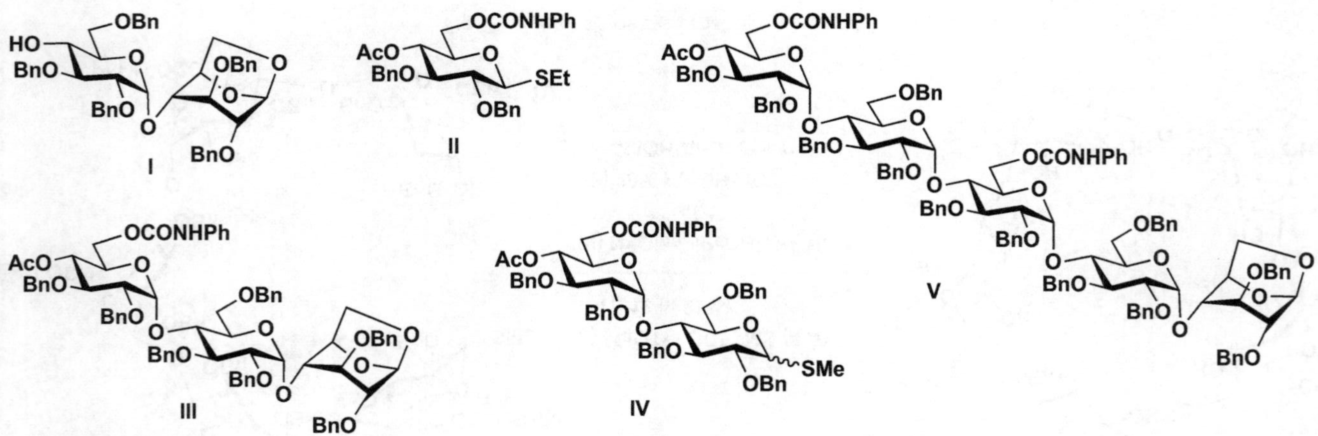

Fig. 8 Molecular structures of intermediate compounds for the synthesis of cyclomaltopentaose **11**.[9]

ethylthioglycopyranoside **II** to afford a trisaccharide, which was then converted to a glycosyl acceptor to react with a disaccharide glycosyl donor **IV**, leading to the formation of linear pentasaccharide derivative **V** (Fig. 8). Judicious manipulation of **V** led to installation of reactive thioglycoside at the reducting end and acceptor alcohol at the non-reducing end (**10**). Macrocyclization of **10**, followed by removal of protecting groups, led to the isolation of cyclomaltopentaose **11**, in a moderate yield of 27% for the cyclization step (Fig. 7).

Earlier to the synthesis of (1-4)-linked cyclic oligosaccharides, Vignon and coworkers demonstrated the first synthesis of cyclooligomer with the preparation of β-(1→6)-linked cyclic tetraose, namely, cyclogentiotetraose **13**, using linear tetrasaccharide monomer **12** (Fig. 9). This report represents a first example of the strategy of cyclization of linear monomer to the cyclic oligomer.[10]

Yet another early example differing in glycosidic linkage is the synthesis of (1→3)-β-linked cycloglucohexaose **15**, prepared from linear hexasaccharide **14** (Fig. 10). Step-wise addition of individual sugar moieties to a growing linear oligomer afforded the desired hexasaccharide **14**. Treatment of glycosyl bromide **14** with AgOTf afforded cyclohexaose **15**, in 30% yield.[11]

That the linear oligomer approach to cyclooligomer formation turns out to be valuable is illustrated in the preparation of cyclo-L-rhamnohexaose. It was demonstrated that a linear oligomer **16** constituted with the desired monosaccharide moieties could undergo cyclization to afford the cyclo-L-rhamnohexaose **17**, in 17% (Fig. 11), representing the first example of a cyclic oligosaccharide with a constitutionally different monosaccharide building block, namely, the L-series sugar.[12]

A conceptually new approach to an efficient glycosylation reaction that could be utilized to perform cyclization of a linear oligomer was demonstrated by Fukase and coworkers.[13] A 'molecular clamp' method for intramolecular glycosylation of a glycosyl donor at the anomeric carbon and an acceptor hydroxy group was developed in this instance, as shown in Fig. 12. The individual donor and acceptor moieties are linked together by a linker, either a phthaloyl moiety or a bis(*tert*-butyl)silyl moiety. The resulting clamped donor and acceptor containing derivative **VI** undergoes a stereoselective glycosylation to **VII**, in favor of either α-anomeric linkage with phthaloyl linker, or β-anomeric linkage in the case of bis(*tert*-butyl)silyl linker. Utilizing this observation, preparation of linear hexa- (**18**) and nona- (**19**) saccharides, and the corresponding cyclic hexamer **20** and nonamer **21** were prepared elegantly in this study (Fig. 13).

Linear oligomer synthesis and subsequent cyclization were utilized in many instances in order to prepare small and medium ring cyclic oligosaccharides. An elegant example is the preparation of β-(1→6)-thio-linked cycloglucopyranoside dimer (**22**), trimer (**23**) and tetramer (**24**) by Fan and Hindsgaul, as shown in Fig. 14.[14] Each of these cyclic sugars were secured in an excellent yield of more than 90%, from the corresponding linear bifunctional di- (**VIII**), tri- (**IX**) and tetrasaccharides (**X**), respectively. The synthesis illustrates clearly the

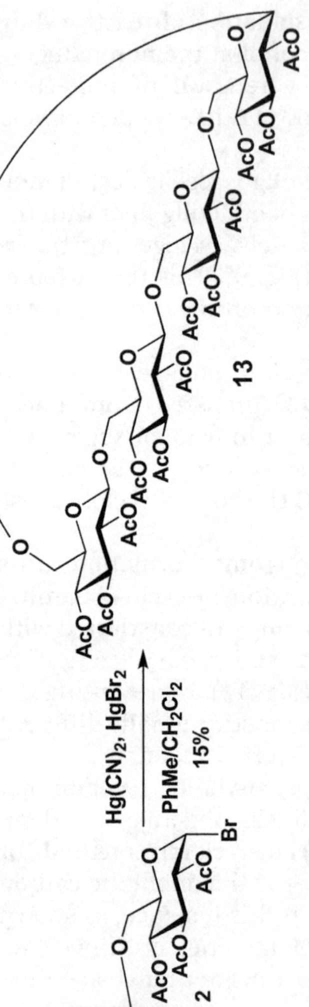

Fig. 9 Synthesis of cyclogentiotetraose 13 from a linear tetrasaccharide monomer 12.[10]

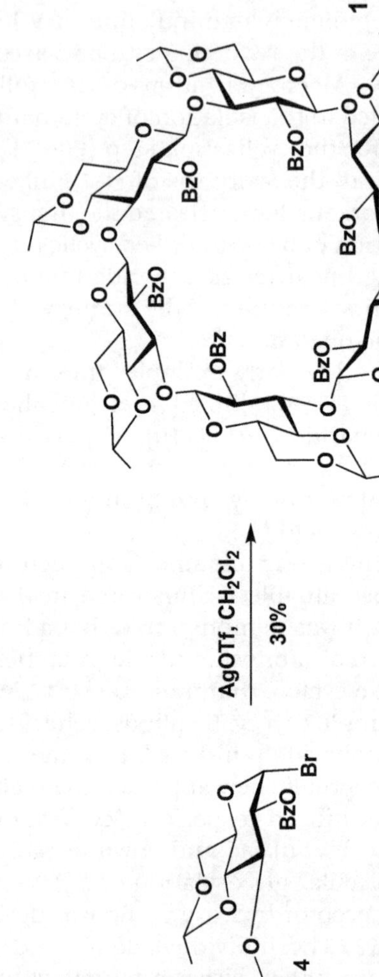

Fig. 10 Synthesis of cycloglucohexaose 15 from linear hexasaccharide 14.[11]

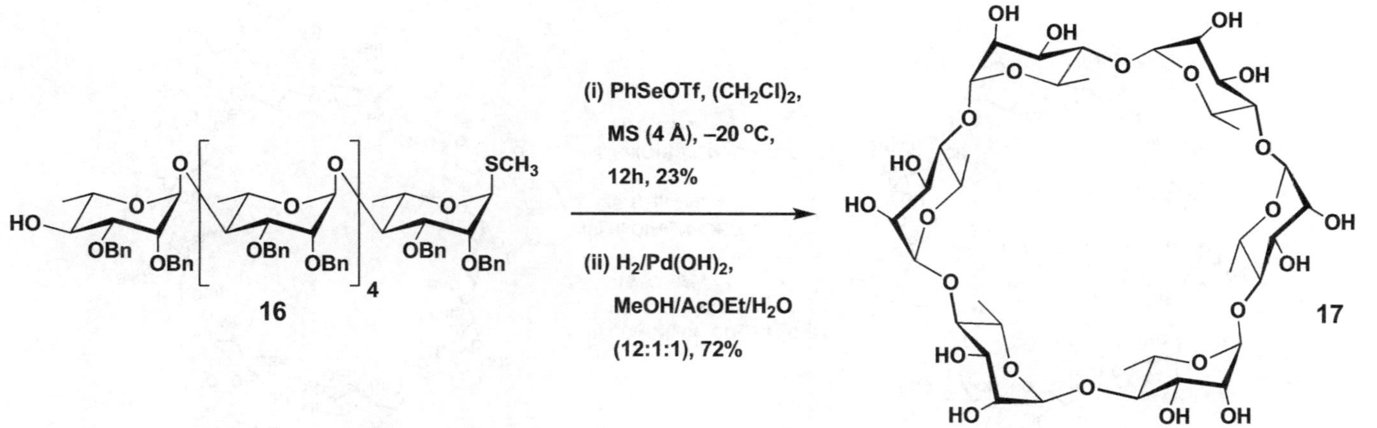

Fig. 11 Synthesis of cyclo-L-rhamnohexaose 17.[12]

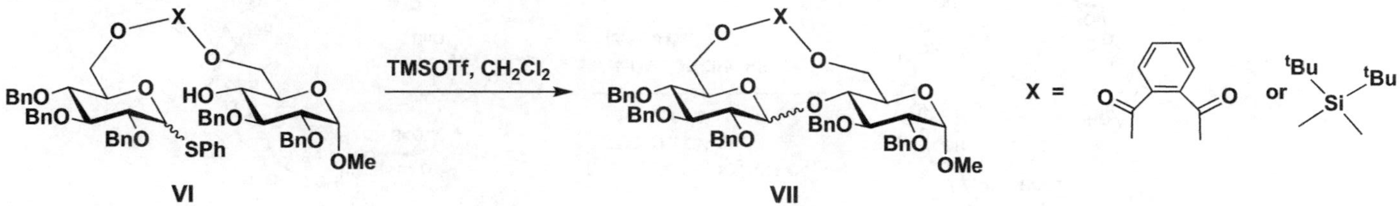

Fig. 12 Stereoselective glycosylation by "molecular clamp" method.[13]

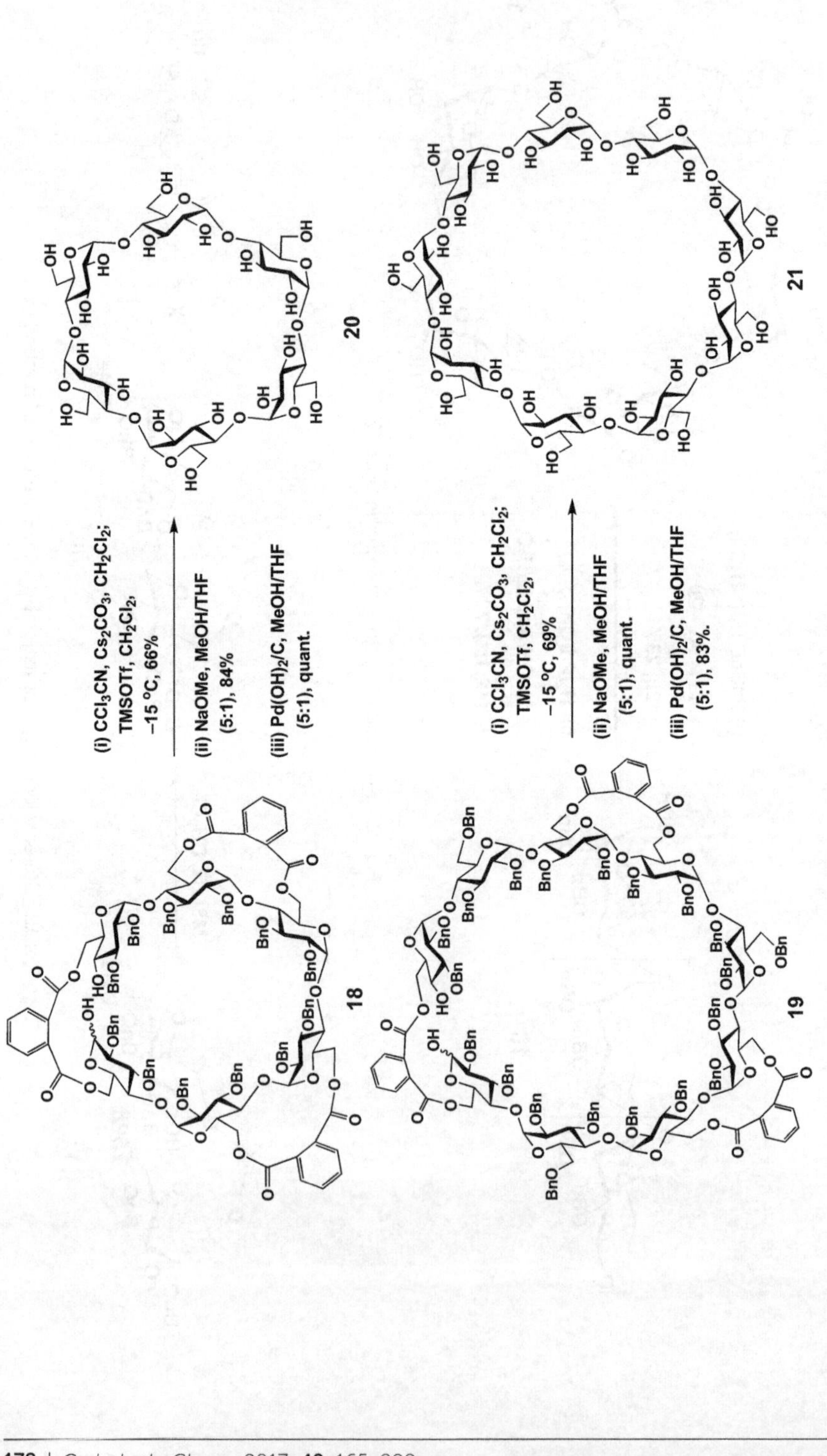

Fig. 13 Synthesis of cyclic hexamer **20** and nonamer **21** by using the "molecular clamp" method.[13]

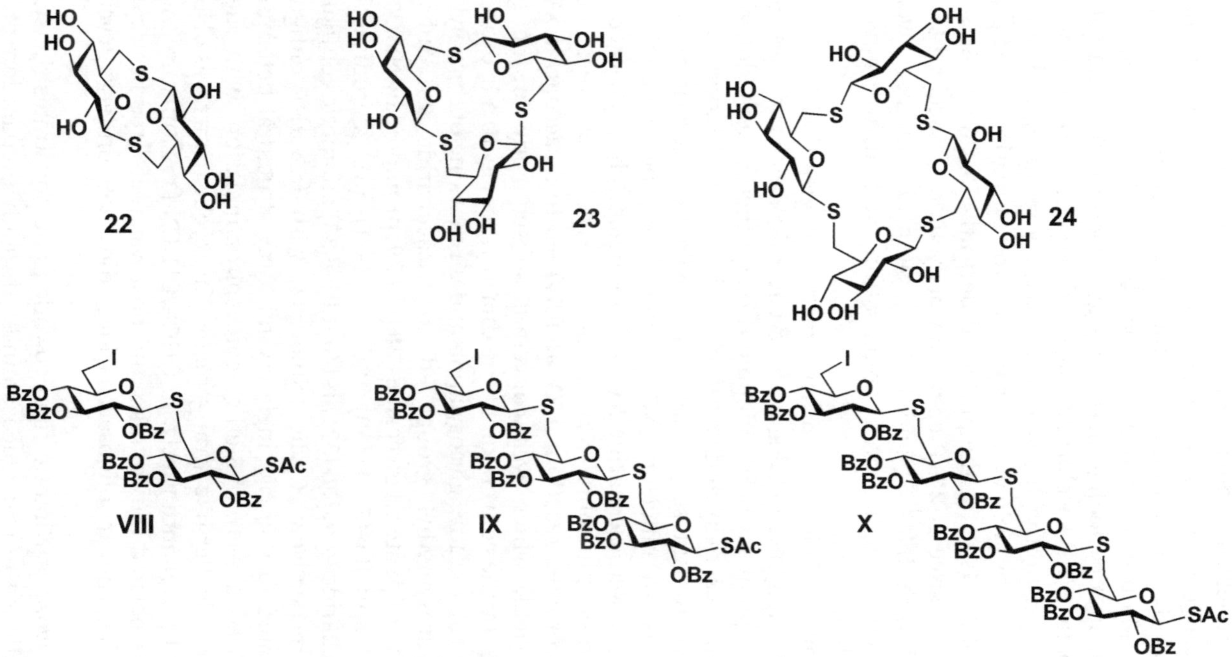

Fig. 14 Molecular structures of β-(1→6)-thio-linked cycloglucopyranosides and their corresponding linear monomers.[14]

advantage of linear oligomer preparation of desired number of monomer moieties, followed by cyclization leading to the desired cyclic oligomer.

Cyclic oligomers composed of galactofuranoside monomer moieties were synthesized by Kim and coworkers, taking impetus that galactofuranoside moiety is a constituent part of arabinogalactan component of mycobacterial cell-wall polysaccharides.[15] Figure 15 shows the cyclic β-(1→6)-linked galactofuranoside tetramer **25**, hexamer **26** and octamer **27**. Using in-house developed 2′-carboxybenzyl (CB) glycoside as a glycosyl donor in a cyclooligomerization approach, linear tetra (**XI**), hexa- (**XII**) and octa- (**XIII**) saccharides afforded respective cyclic product. The cyclooligomerization yields were progressively reducing as the ring size became larger. The yield was thus 91% for the formation of the protected derivative of **25**, 79% for the protected derivative of **26** and 66% for the protected derivative of **27**. Preparation of cyclic oligosaccharides **25–27**, constituted with β-(1→5) and β-(1→6) glycosidic linkages were also demonstrated through a cyclocondensation method, using two varied disaccharide monomers.[15]

Development of elegant synthetic methodologies finds an interest to implement such methodologies to cyclic oligosaccharide preparation. The case in point is the development of synthetic methodologies that overcome low reactivity and stereoselectivity of glycosylations of ketoses, as opposed to aldose sugars. Thiazolylketose derivative **XIV** was developed as an efficient glycosylation donor by Dondoni and coworkers.[16] α-D-Ketosaccharide **XV**, as an efficient glycosyl acceptor, and thiazolylketosyl phosphite **XVI** as the glycosyl donor, were prepared. $BF_3.Et_2O$-promoted reaction of **XV** and **XVI** led to disaccharide **XVII** in a moderate yield, which was converted subsequently to the glycosyl acceptor **28**, by the conversion of the thiazolyl moiety to a hydroxymethyl group (Fig. 16). Iterative glycosylation and un-masking of thiazolyl moiety to the hydroxymethyl group led to trisaccharide **29**. Subsequent glycosylation of pentenyl glycosides **28** and **29**, promoted by NIS/TMSOTf, afforded α- and β-D-(2→1)-(**30**) and α-D-(2→1)- (**31**) spirodisaccharide, spirotrisaccharide (α,α,α)-D-(**32**) and (α,α,β)-D-(**33**) anomeric configurations (Fig. 17), respectively. Cyclic trimer **32**, with (α,α,α)-D-configuration, was elucidated further by single crystal X-ray structural determination, which showed a 9-crown-3-ether core spiroannelation, with the sugar moiety in a propeller-like arrangement. This cyclic trimer was also found to uptake Mg^{2+} quantitatively, in $CHCl_3/CH_3CN$ (1 : 1) mixture. Addition of water dissociated the complex and host was recovered back quantitatively. Higher linear oligosaccharides did not undergo cyclization reaction.[16]

A recent work of Nifantiev and coworkers demonstrates the synthesis of cyclic oligo-(1→6)-β-D-glucosamines, through preparation of corresponding linear oligomers. Thus, di- (**34**) and trisaccharides (**35**) underwent cycloglycosylation, promoted by NIS/TfOH reagents, to afford the cyclic di- (**36**) and tri- (**37**) saccharides, with β-anomeric linkages, respectively (Fig. 18).

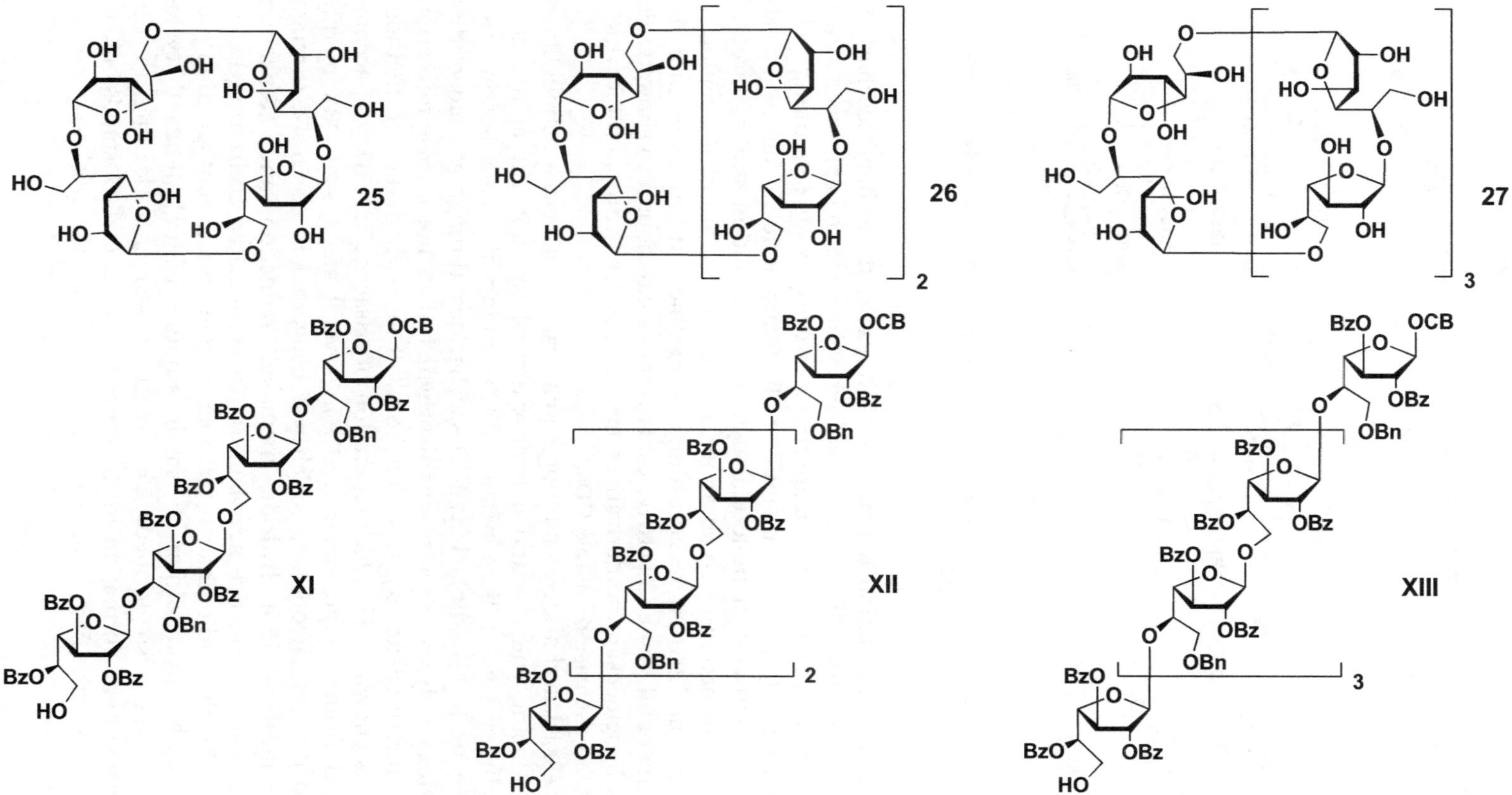

Fig. 15 Molecular structures of cyclic β-(1→6)-linked galactofuranoside and the corresponding linear precursors.[15]

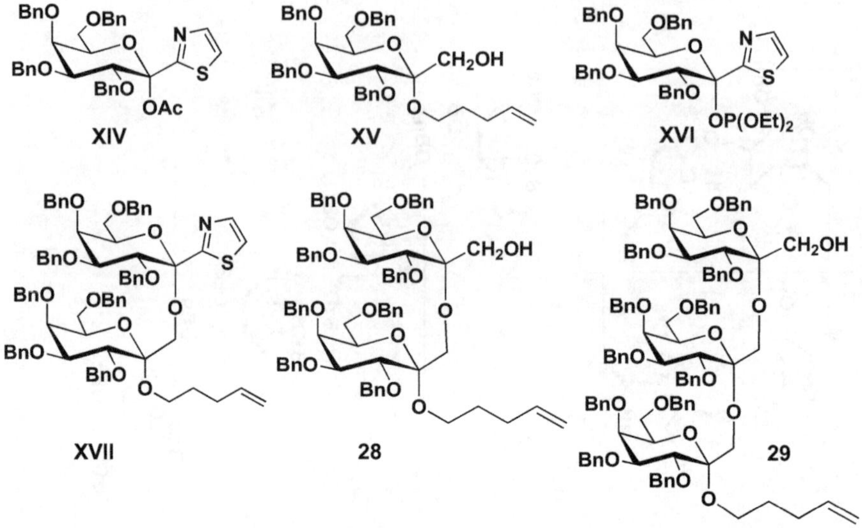

Fig. 16 Molecular structures of monosaccharide building blocks **XIV–XVI** and pentenyl di- and trisaccharide glycosides **28** and **29**.[16]

Similar glycosylation reactions with linear tetra- to heptasaccharides led to the isolation of the corresponding cyclic tetra- to heptasaccharides, that were then deprotected to afford free-hydroxy group containing cyclic oligosaccharides, predominantly with β-anomeric linkages (**38–41**) (Fig. 19). Subsequent molecular dynamics simulation studies revealed rigid conformation to cyclic dimer, a moderately rigid conformation to cyclic tri- and tetramer and a range of flexible conformations to cyclic pentamer and higher macrocycles. Further calculations also showed that cyclooligoglucosamine derivatives lacked distinct hydrophobic cavities, unlike in the case of native CDs.[17]

Linear trisaccharide containing acyclic carboxylic acid functionality at the reducing end *C*-1 and a hydroxymethyl functionality at the non-reducing end was utilized by Liu and Li in an ingenious manner, in order to prepare β-(1→6)-linked cyclic trisaccharides, through a protocol that describes a glycosylation mechanism which does not involve formation of an intermediate oxycarbenium ion.[18] The newly developed methodology is shown in Fig. 20. Preparation of trisaccharide **XIX** was accomplished from the precursor **XVIII**, in several steps. Acid **XIX** readily afforded macrolactone **XX**, which was converted in acetal **42**, through diisobutylaluminium hydride mediated reduction and acetylation (Rychnorsky's reductive acetylation). TfOH-mediated cyclization led to regain the glycosidic linkage, as in **43**, in β-anomeric configuration predominantly. In addition, a small amount of (α,b,β)-linked cyclotrisaccharide also formed. The reaction sequence leading to the glycosidic bond formation in the crucial cyclic acetal **42** stands-out, as this report is the first example of such an approach to synthesize a cyclic oligosaccharide.[18]

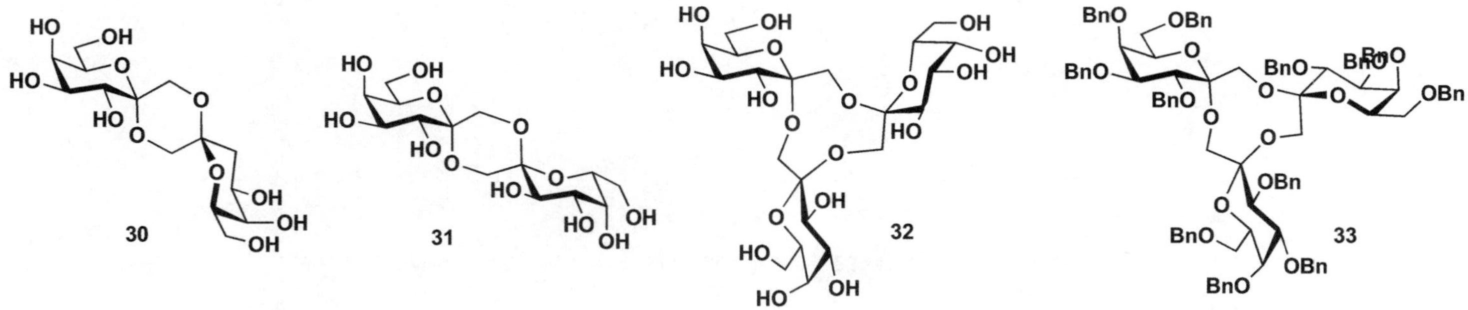

30 31 32 33

Fig. 17 Molecular structures of spirodisaccharides 30, 31 and spirotrisaccharides 32, 33.[16]

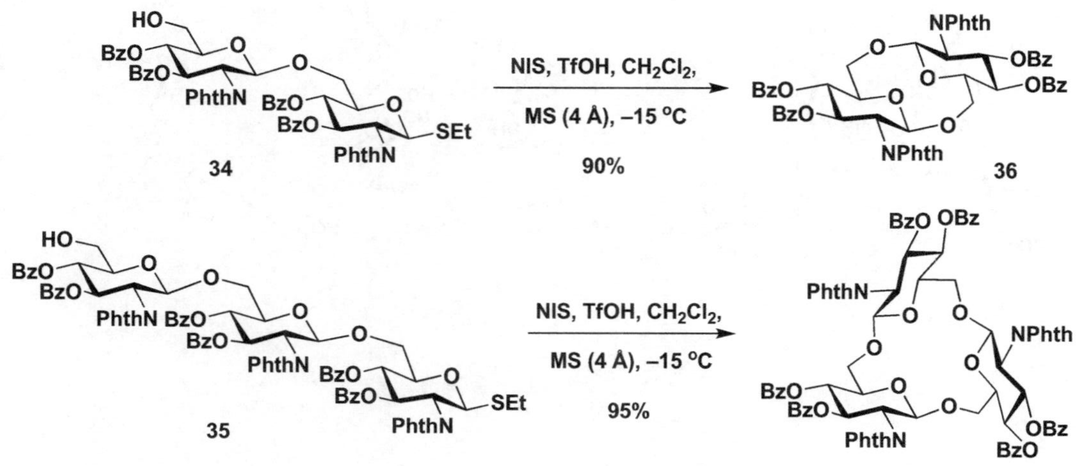

34 36 35 37

Fig. 18 Synthesis of cyclic di- and trisaccharides 36 and 37, with β-anomeric linkages.[17]

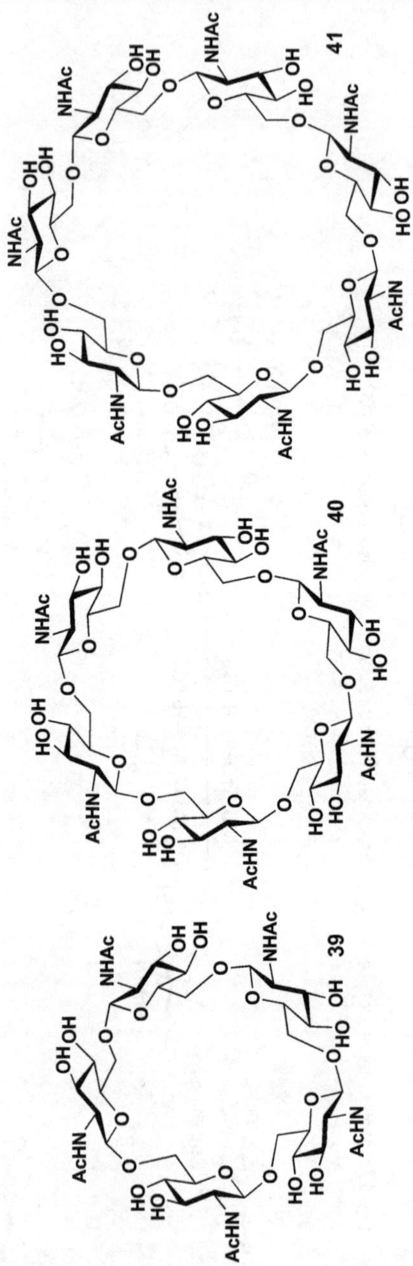

38 39 40 41

Fig. 19 Synthesis of cyclic tetra- to heptasaccharides 38–41, with β-anomeric linkages.[17]

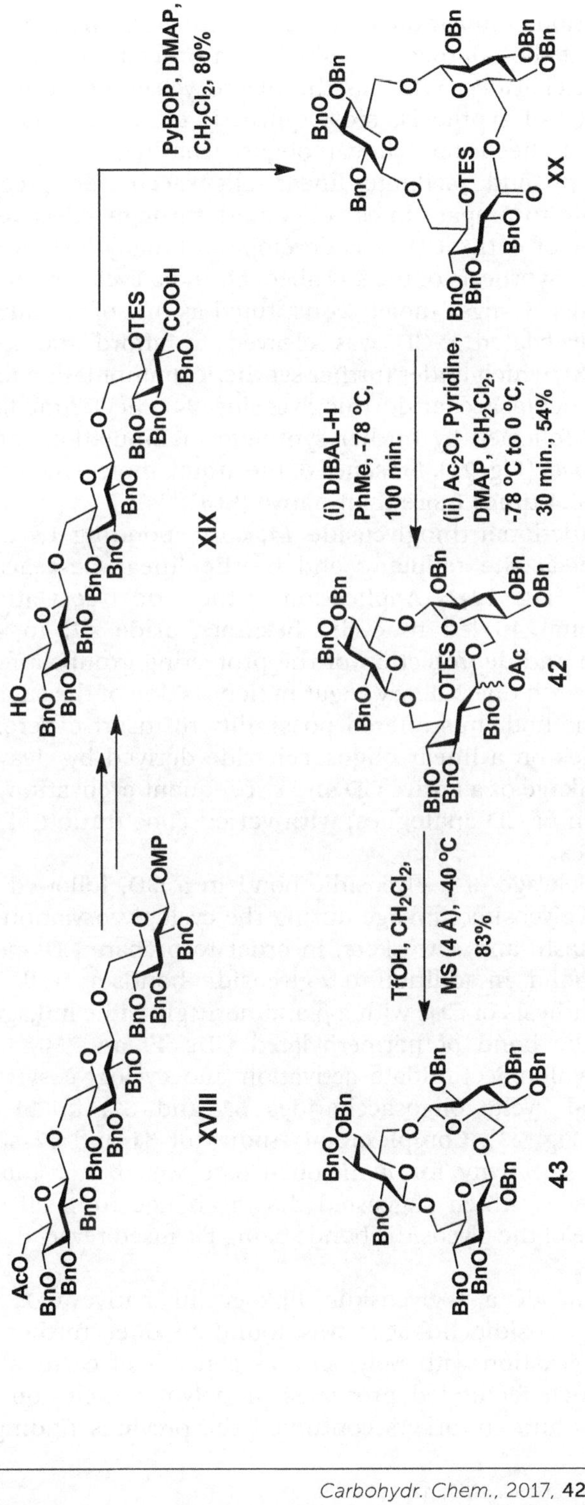

Fig. 20 Synthesis of β-(1→6)-linked cyclic trisaccharide 43.[18]

3 Linear oligosaccharides formed through glycosidic bond cleavage of native CDs and subsequent cyclization to cyclic oligosaccharides

Macrocyclization of linear oligomers as described in the previous section illustrates that the approach is valuable in order to secure the desired cyclic oligosaccharides. Whereas the macrocyclization occurs as one of the final steps of synthesis, most synthetic efforts are devoted to the preparation of the desired linear oligosaccharide. In order to reduce synthetic steps and facilitate linear oligosaccharide precursors, an alternate route to prepare linear oligomers through selective glycosidic bond cleavage of native CDs was developed in many instances. An early example is the synthesis of the so-called 'chimera cyclodextrins', wherein a 2-amino-2-deoxy sugar moiety constituted as one of the sugar units of β-CD.[19] Peracetylated α-CD was cleaved to afford the linear hexasaccharide **XXI**, which under further synthetic manipulation led to secure bifunctional heptasaccharide thioglycoside **44**. MeOTf-mediated cycloglycosylation followed by further synthetic manipulations afforded the chimera β-CD **45** (Fig. 21). Presence of the amine moiety increased the aq. solubility by 4.4 times more than native β-CD.[19]

Using bifunctional thioglycoside **44**, corresponding 1,2-unsaturated sugar moiety at the reducing end of the linear hexasaccharide **46** was prepared (Fig. 22).[20] Application of the iodoglycosylation method on bifunctional **46** led to cyclic hexasaccharide **47**, in 48% yield. Deiodination and deprotection of the protecting groups afforded α-CD analogue **48**, with one 2-deoxy sugar moiety as one of the sugar moieties in **48**. The method thus offered possibility to insert different types of sugar moieties on a linear oligosaccharide derived by cleavage of the glycosidic linkage of a native CD and subsequent cyclization, leading to the formation of CD analogues, with varied constitution of one of the sugar moieties.

Selective cleavage of a glycosidic bond in a CD, followed by modification of the glycosidic linkage during the cycloglycosylation was investigated by Akashi and coworkers, in order to prepare CDs possessing a β-glycosidic bond, in addition to α-glycosidic bonds as well.[21] Figure 23 shows the synthesis of CDs with a β-anomeric glyosidic linkage. Cleavage of a glycosidic bond of permethylated CDs **49** and **50**, followed by anomeric trichloroacetimidate activation and cycloglycosylation led to newly formed cyclic oligosaccharides **51** and **52**, in 14 and 18%, respectively (Fig. 23). Complexation studies of **51** and **52** showed that the binding selectivity for *m*-nitrobenzoate was better than that for *p*-nitrobenzoates, which suggested that a change from all α-anomeric linkage to one of the glycosidic bonds being β caused reversal in the guest selectivity.[21]

Introduction of a β-glycosidic linkage in native CDs with only α-anomeric glycosidic linkages was found to offer further advantage for the complexation with polymers, as a result of cavity shape being elliptical, which facilitated processes of polymer inclusion inside the cavity. Akashi and coworkers continued the previous finding that CDs

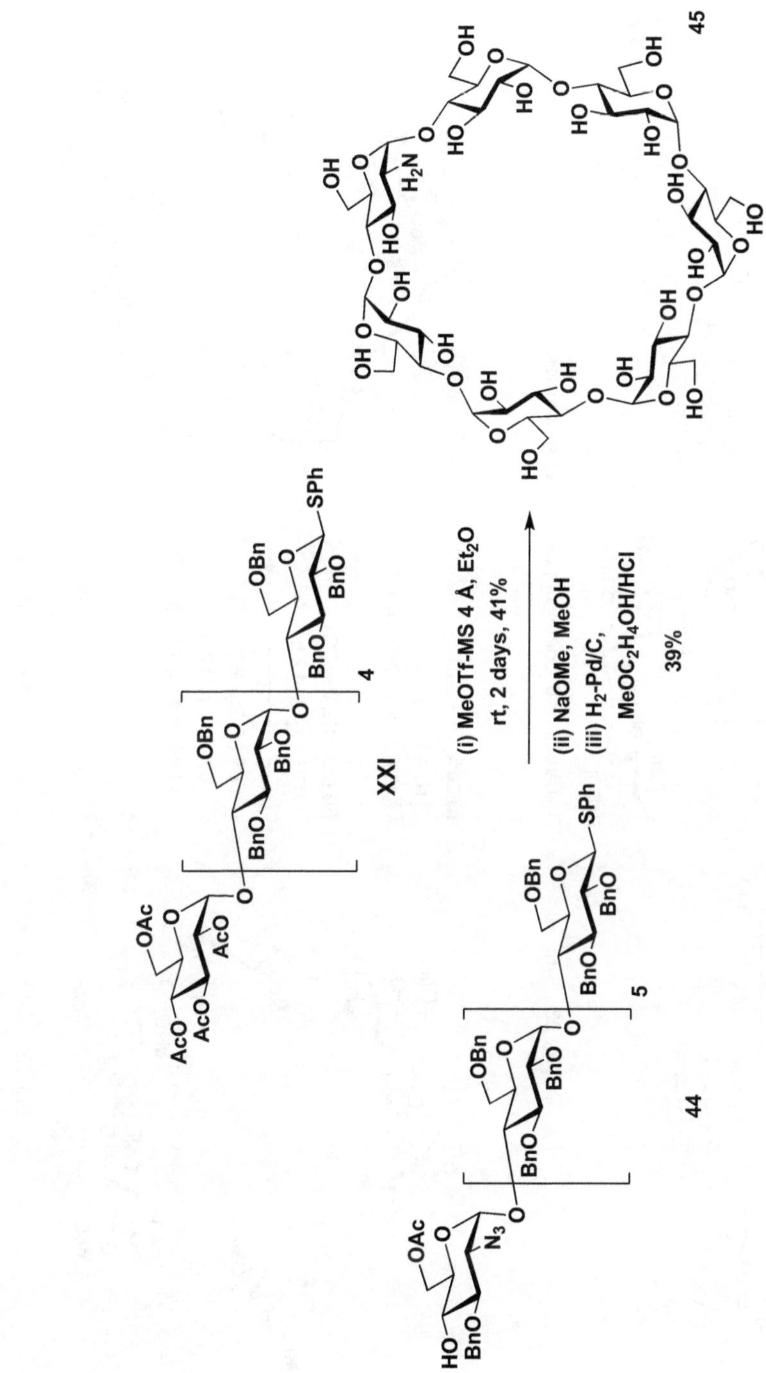

Fig. 21 Synthesis of chimera β-CD 45.[19]

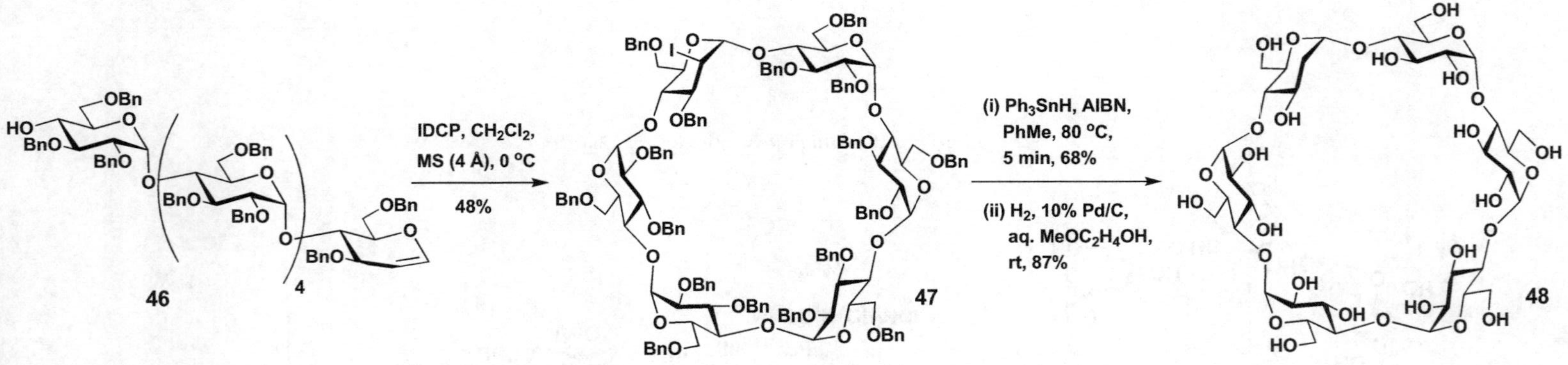

Fig. 22 Synthesis of α-CD analogue 48, containing one deoxy sugar moiety.[20]

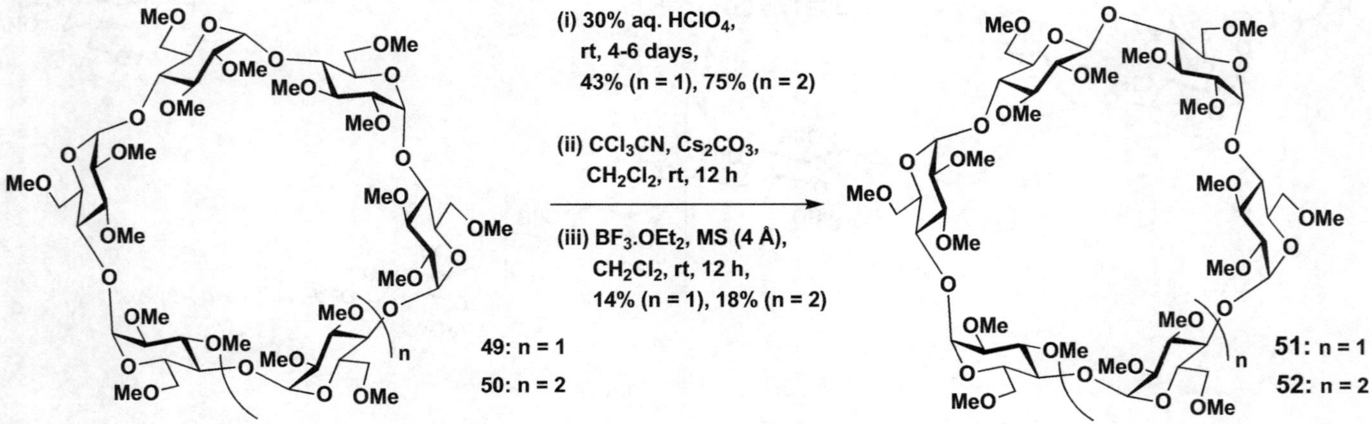

Fig. 23 Synthesis of CD analouges 51 and 52, with one β-anomeric glyosidic linkage.[21]

with a β-anomeric linkage are suitable to polymer encapsulation.[22] Thus, CD analogue **51** was investigated regarding its ability to form a complex with polymers and was found to undergo complex formation in the case of poly(acrylic acid) and poly(propylene glycol), and such a complexation did not occur in the case of permethylated α-CD, as evidenced through X-ray diffraction studies, ^{1}H NMR and FT-IR spectroscopies. The study thus illustrated altered inclusion complex formation by skeletal modified CDs.

Replacement of the glycosidic oxygen with a biologically benign disulfide motif in a β-CD was explored by Davis and coworkers.[23] The impetus in this instance was to identify whether a subtle change in the cavity size and symmetry of β-CD would benefit stochastic sensing of analytes when incorporated in protein pores. Synthesis of a disulfide-linked β-CD was accomplished from linear maltoheptasaccharide **53**, which upon a series of modifications led to two di(acetylsulfanyl) derivatives, **54** and **55**. Aereal oxidation of thiols led to the cyclic products with a disulfide linkage, a concomitant *O*-deacetylation afforded disulfide bond-containing cyclic oligosaccharides **56** and **57** (Fig. 24). As anticipated, altered conformations of the macrocycles in the region of the disulfide bond were observed, whereas regions about the remaining sugar residues in the macrocycle remained as that of β-CD, as deduced by NMR studies and by molecular dynamics simulation studies. After identifying localized variations in the conformations of macrocyles **56** and **57** as a result of the disulfide linkage, their non-covalent binding to α-hemolysin protein pores was investigated by single-channel current (*I*)-voltage (*V*) measurements. Significant current blockades were observed when **56** and **57** were bound to the pores, by a bimolecular interaction mechanism. Subsequent studies to identify guest binding abilities showed a selective binding of the protein pore containing **56** and **57** to sodium deoxycholate, as opposed to no binding of modified CDs inside the protein pore with 2-aminoadamantane. The disulfide-containing CDs **56** and **57** thus exhibited an asymmetric cavity size and bound sodium deoxycholate stronger. This study illustrated a benefit of altered cavity size and symmetry as implemented through backbone modification of the glycosidic bond in a cyclic oligosaccharide with a disulfide bond.[23]

4 Polycondensation of designed oligosaccharide monomers

A successful strategy to prepare designed cyclic oligosaccharides is the polycondensation, leading to cyclooligomers that contain number of monomeric units in multiples of the number of such units present in the initiator monomer. The polycondensation protocol was popularized through early studies reported by Kochetkov and coworkers[24,25] and later by Stoddart and coworkers.[26–29] As shown in Fig. 2, a linear monomer of di- and trisaccharide would undergo cyclooligomerization intramolecularly to produce cyclooligomers with number of monomer units in multiples of initiator di- and trisaccharides. The first successful synthesis using

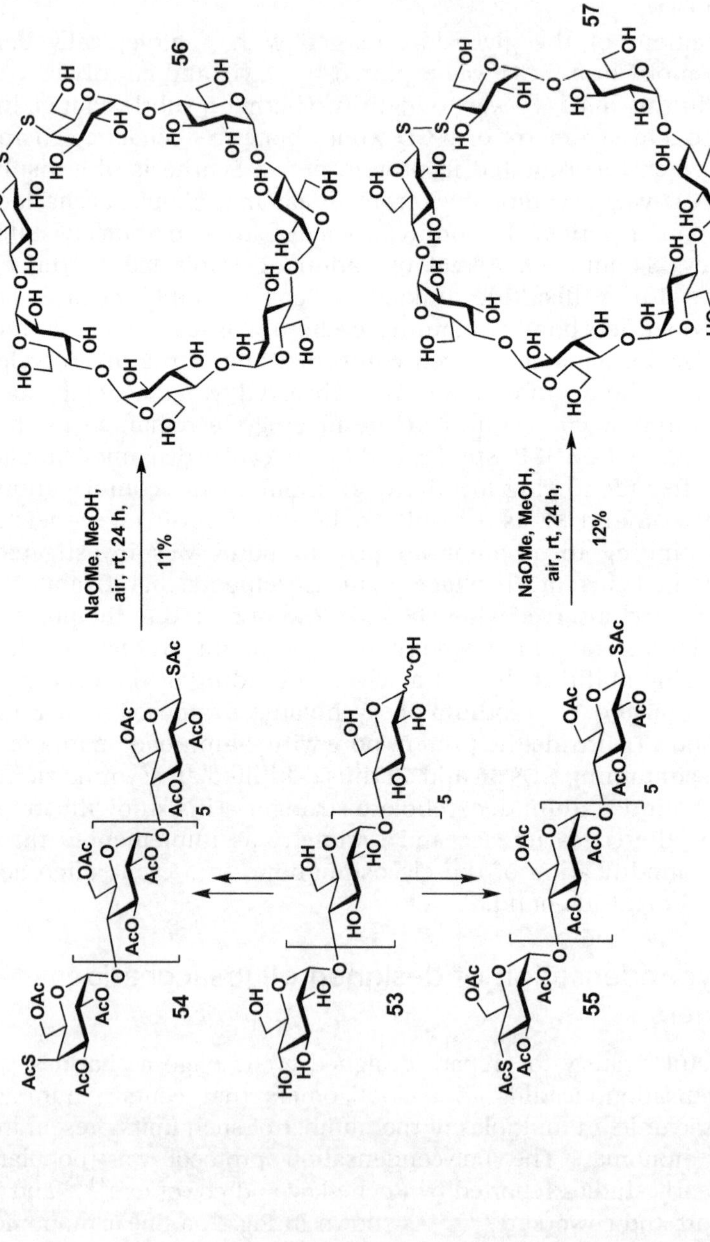

Fig. 24 Synthesis of β-CD analogues constituted by one disulfide bond in the backbone.[23]

this approach to prepare a cyclic oligosaccharide is concerned with the synthesis of cyclo[(1→6)-β-D-galactofurano]oligosaccharides **58–61** (Fig. 25).

Judicious installation of glycosyl donors and acceptors in the derivatized forms permitted the cyclooligomerization through polycondensation. The donor at the reducing end and acceptor at the non-reducing end were equipped innovatively as cyanoethylidene and tritylated moieties, respectively. Generation of an active glycosyl acceptor through *O*-detritylation mediated by a triflate reagent, followed by attack of donor moiety at *C*-1 of cyanoethylidene acceptor moiety promoted a glycosylation reaction, that result in the removal of a nitrile functionality and formation of *O*-acetyl moiety at *C*-2. The scheme allowed synthesis of cyclic dimer to octamer (Fig. 25), from a disaccharide monomer. Polycondensation promoted by AgOTf, followed by deprotection afforded **58–61**, that corresponded to cyclic dimer, tetramer, hexamer and octamer, respectively. Formation of these defined number of cyclic oligomers characterizes the polycondensation method. Sufficient mobility differences among various cyclooligomers exist that permit isolation of each of the oligomer through chromatography. Interestingly, monosaccharide monomer did not undergo polycondensation, possibly as a result of intrinsic strain. This work laid the foundation for further progress in cyclic oligosaccharide synthesis of varied constitutions.[25]

The polycondensation method was popularized further through elegant preparation of cyclic oligosaccharides with designed monomer building blocks by Stoddart and coworkers.[26–28] Preparation of cyclic oligosaccharides with alternate D- and L-configurations illustrates the developments in the polycondensation methodology. The synthetic method stands out, for example, with the formation of cyclo[4)-α-L-Rha*p*-(1→4)-α-D-Man*p*-(1→]$_n$, corresponding to cyclic hexamer ($n=3$), octamer ($n=4$), decamer ($n=5$) and dodecamer ($n=6$), in one-pot through cyclization of corresponding disaccharide monomer **62**, as shown in Fig. 26. Polycondensation, followed by deprotection led to the isolation of **63–66**. Advantage of this method is that the monomer could be subjected to required synthetic manipulation at the di- and trisaccharide monomer levels that reflects eventually in the newly cyclized products. Thus, a disaccharide monomer, with D-rhamnose and L-rhamnose configurations, underwent polycondensation to afford cyclic oligosaccharides that are in multiples of the disaccharide monomer. Similarly, monomer **67**, with appropriately positioned cyanoethylidene and trityl moieties, when subjected to trityl perchlorate-mediated polycondensation and subsequent deprotection of protecting groups, afforded cyclic hexa- (**68**), octa- (**69**), deca- (**70**), dodeca- (**71**) and tetradeca- (**72**) oligomers (Fig. 27).

In a similar manner, alternating D- and L-mannopyranose moieties, the macrocycle **74** was prepared from disaccharide monomer **73** (Fig. 28), although in low yield (9%), as the only polycondensation product.[29] It is interesting to observe that only the α-anomeric glycosidic linkage is formed by the cyclooligomerization reactions described above. The structural features of few of the above new forms of

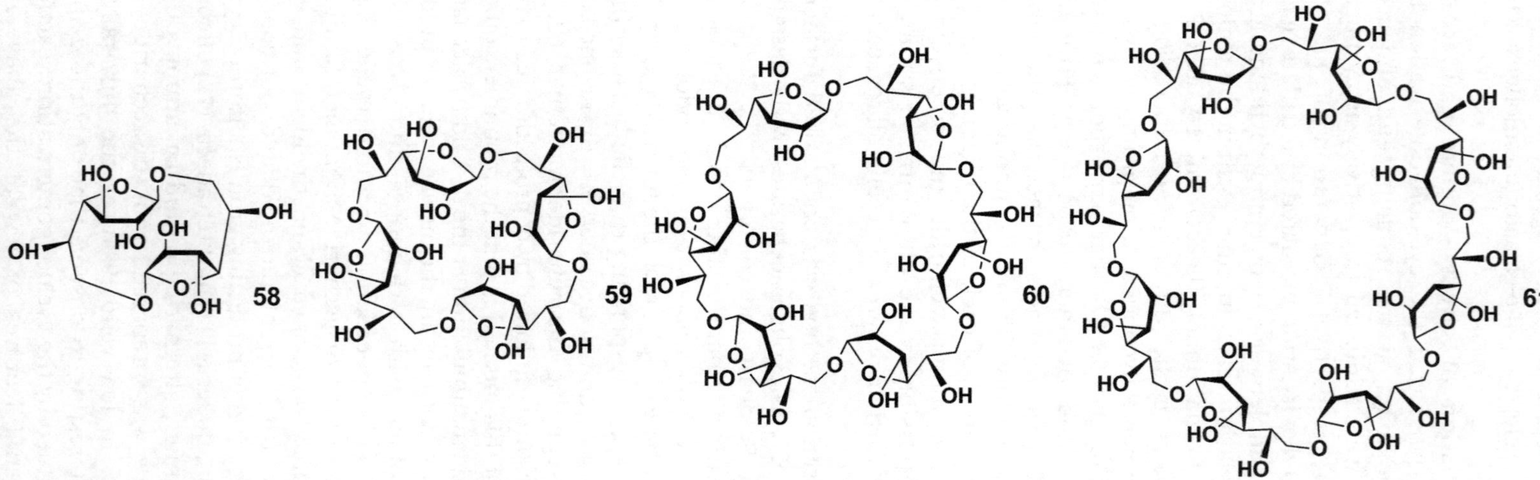

Fig. 25 Molecular structures of cyclo[(1→6)-β-D-galactofurano]oligosaccharides 58–61.[24]

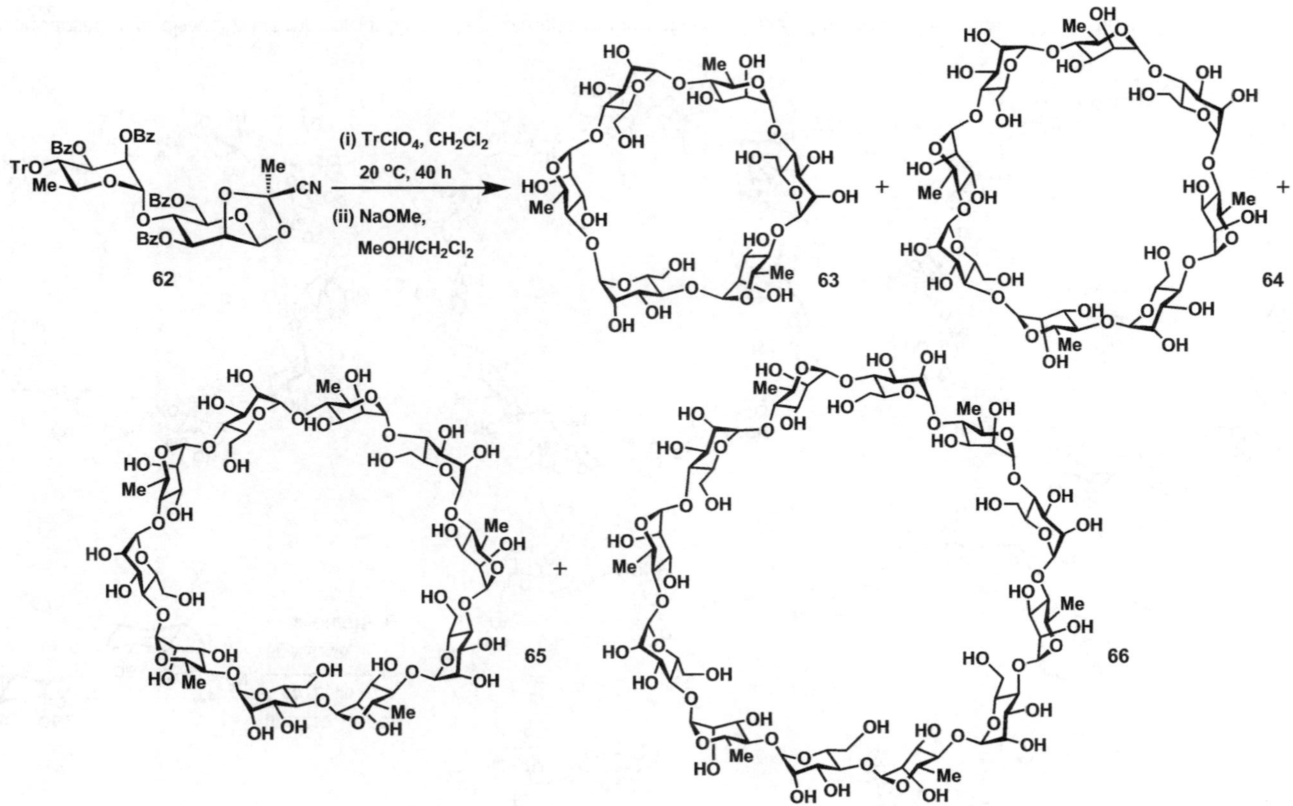

Fig. 26 Synthesis of cyclic oligomers **63–66** from a disaccharide monomer **62**.[26]

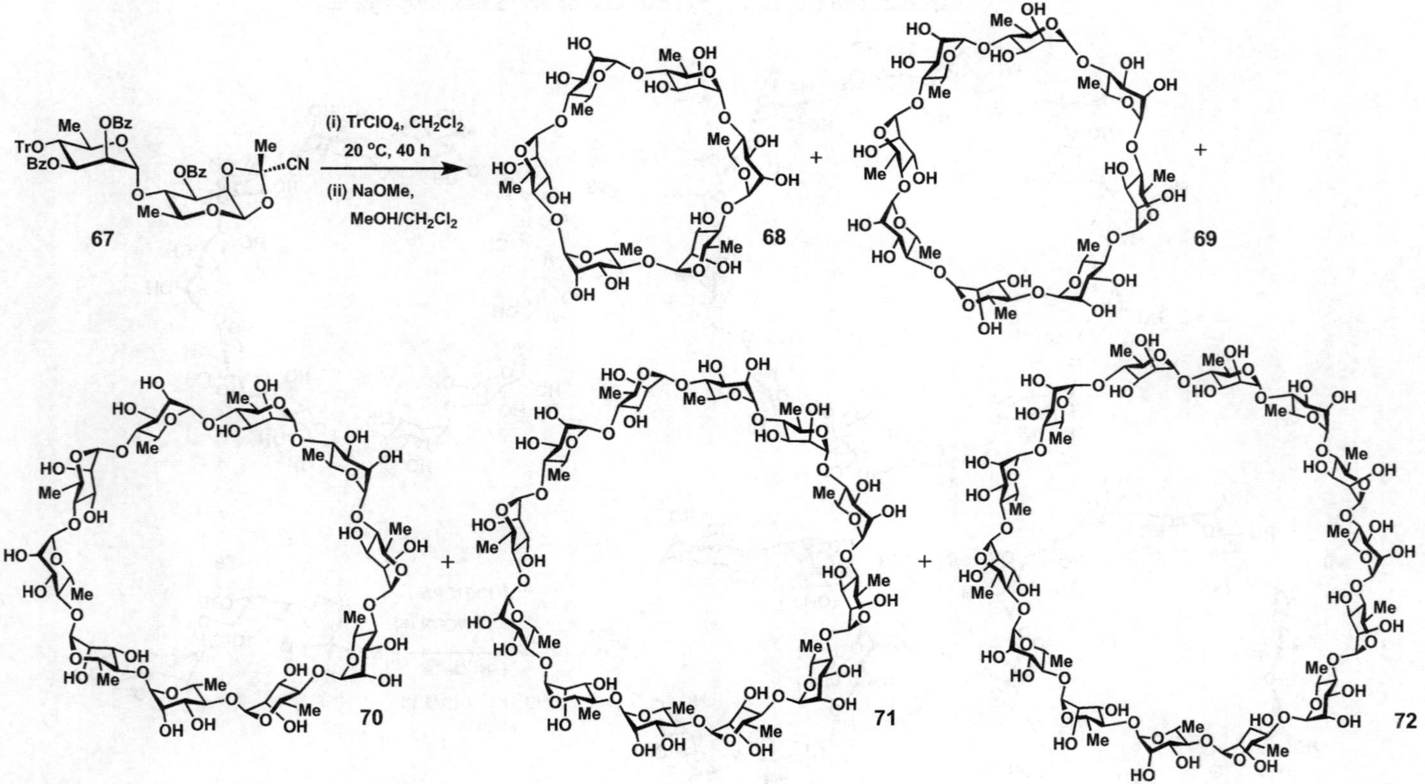

Fig. 27 Polycondensation of disaccharide monomer **67**, leading to the formation of cyclic hexa- (**68**), octa- (**69**), deca- (**70**), dodeca- (**71**) and tetradecasaccharide (**72**) oligomers.[27,28]

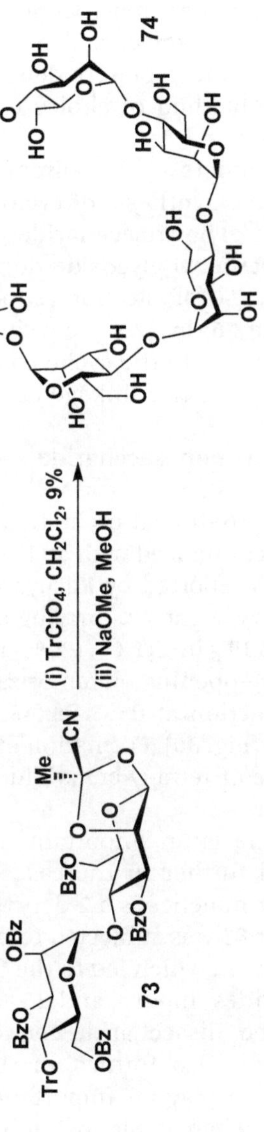

Fig. 28 Synthesis of cyclic hexamer with alternating D- and L-mannopyranose units.[29]

cyclodextrin analogues were determined by solution phase, single crystal X-ray structural determinations and molecular dynamics analysis. Solid state structures show that the secondary hydroxy groups occupy both the rims, as a result of the alternating D- and L-configurations of the sugar moieties, cylindrical shapes and differing internal cavity sizes, ranging from 8 to 13 Å, depending on the CD analogue. Whereas larger macrocycles **64**, **69** and **70** form infinite stacks, leading to a nanotube formation, relatively smaller macrocycle **74** showed a parquet-like arrangement of molecules in the crystal lattice. These and more examples illustrate the advantage of the polycondensation method, which stands out to be an alternate method of choice for the chemical synthesis of cyclic oligosaccharides.

Trisaccharide monomers for polycondensation were reported earlier in the case of the synthesis of cyclomaltotri- and -hexaoside by Houdier and Vottéro.[30] The trisaccharide α-(1→6)-linked bifunctional monomer **75** with *O*-pentenyl glycoside donor at the reducing end and free hydroxy group at *C*-6 of the non-reducing end afforded α-(1→6)-linked cyclic hexaoside **76**, in 12% yield, in addition to intramolecular glycosylation of **75**, leading to the cyclic trisaccharide **77**, with α- (47%) and **78** with β- (24%) anomeric linkage at the newly formed glycosidic bond (Fig. 29).

In a rare example, a monosaccharide monomer was used in oligocondensation, leading to a cyclic tetrasaccharide through dimerization of the monomer. An α-(1→6)-linked cyclic trisaccharide **80**, with a 2-deoxy sugar, and further functionalized with α-(1→4)-linked glycopyranoside at each sugar moiety, was reported by Kasuya and Hatanaka.[31] The monomer 1,6-anhydro-2-deoxy sugar **79**, carrying an α-D-glycopyranosyl moiety at *C*-4, was treated with PF_5 in CH_2Cl_2 (20%) initiator, under high vacuum condition. Cationic ring-opening oligomerization was initiated under the conditions and the reaction at 0 °C for 48 h or more led to the cyclic oligomer, in 89% yield (Fig. 30). Formation of this product was thought to be due to the presence of tetra-*O*-benzylglucopyranosyl moiety at *C*-4 of the anhydrosugar moiety.

That the disaccharide is an important monomer for cyclooligomerization was illustrated further in the case of maltose disaccharide, in which one of the sugar moieties is a 2-deoxyglycopyranose moiety.[32] The disaccharide monomer **81** was subjected to polycondensation under NIS/TfOH promoted condition, which led to the formation of cyclic tetra- (**82**) and hexa- (**83**) saccharides, in 40% and 25% yield, respectively (Fig. 31). In addition, hydrolyzed disaccharide monomer, disaccharide **84** and linear tetramer **85** were also formed in the reaction. The protecting groups were observed to play an important role in the cyclooligomerization reaction. Thus, whereas methylation at available hydroxy groups of the monomer **81** led to successful polycondensation, a change to the benzyl groups, as in **86**, afforded linear tetrasaccharide **87**, in addition to the hydrolyzed product **88** (Fig. 32). Only trace amount of cyclic oligosaccharide **89** was observed.

A recent study shows the utility of thioglycoside as a useful activated glycosylation donor for the polycondensation. A backbone modified

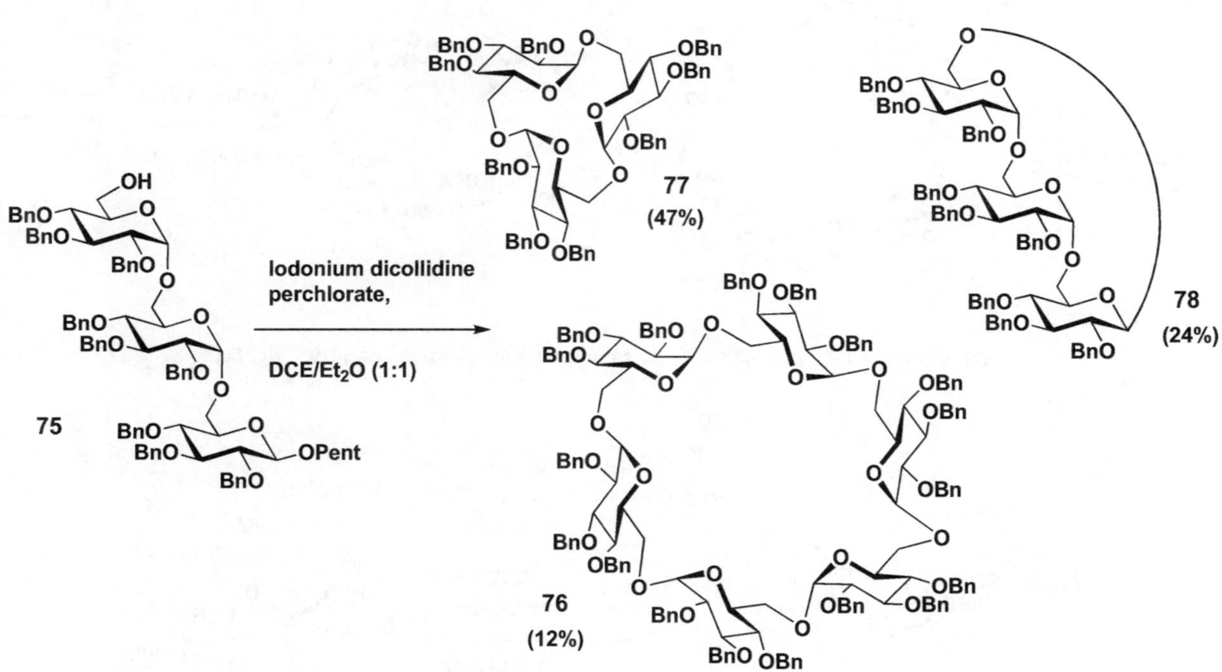

Fig. 29 Synthesis of cyclomaltotri- (77, 78) and –hexaoside (76) from a trisaccharide monomer 75.[30]

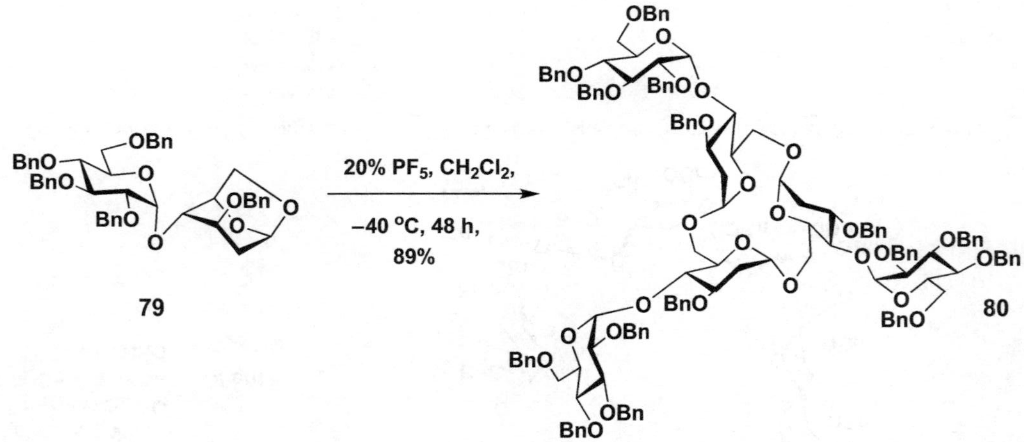

Fig. 30 Synthesis of α-(1→6)-linked cyclic trisaccharide with 2-deoxy sugar **80**.[31]

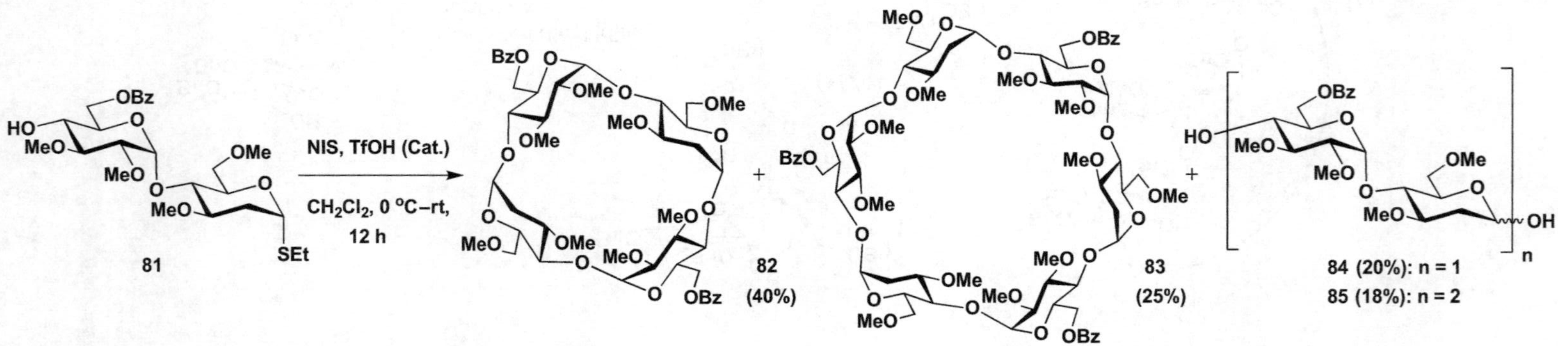

Fig. 31 Synthesis of cyclic tetra- (**82**) and hexasaccharide (**83**), by cyclocondensation of disaccharide monomer **81**.[32]

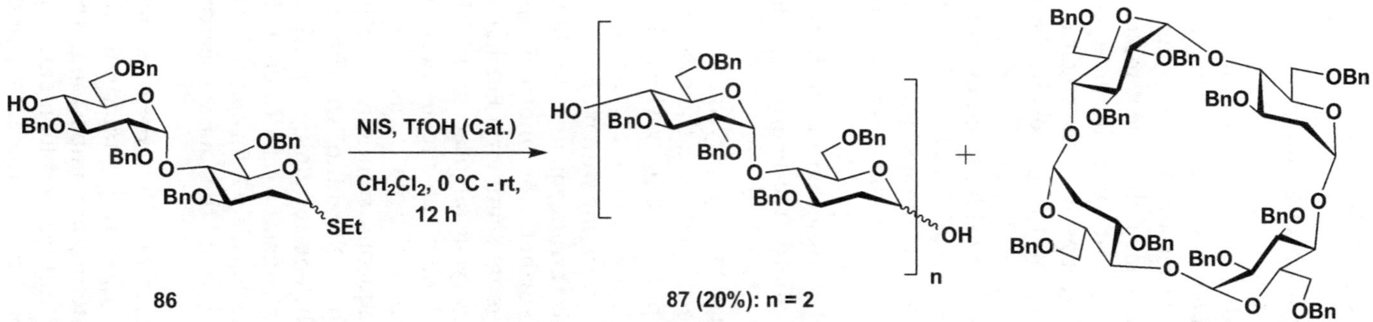

86

87 (20%): n = 2
88 (75%): n = 1

89 (trace)

Fig. 32 Polycondensation of disaccharide monomer **86** leading to the formation of linear tetrasaccharide **87** and trace amount of cyclic tetrasaccharide **89**.[32]

linear disaccharide **90**, which possesses an additional methylene moiety between glycosidic oxygen and *C*-4 of the sugar moiety, was prepared and subjected to polycondensation reaction, mediated by NIS/AgOTf or TMSOTf. The reaction led to cyclic tetramer **91**, in 64% yield, in addition to cyclic disaccharide **92**, in 11% yield (Fig. 33). Cyclic tetramer **91** was deprotected to secure free hydroxy group containing cyclooligomer **93**. The backbone modified cyclic tetramer **93** was observed to be amphiphilic in nature, wherein the tetramer was found to be soluble in both organic solvents and aqueous solutions. This amphiphilic property allowed encapsulation of water insoluble pyrene dye in an aqueous solution and organic solvent insoluble guest molecule L-tyrosine in an organic solution containing host. Amphiphilic cyclic oligosaccharides retained with free hydroxyl groups at all the sugar moieties thus became feasible through expansion of the glycosidic bond with additional methylene moiety.[33]

In a similar strategy, the glycosidic bond expanded trisaccharide **94** was prepared, in a stepwise glycosylation of the corresponding monosaccharide. Trisaccharide **94**, having acceptor site at the non-reducing end and thiocresyl donor site at the reducing end, was allowed to undergo glycosylation promoted by NIS/TfOH reagent system. The glycosylation afforded a cyclic trisaccharide derivative, which was subjected to deprotection to afford free hydroxy group containing cyclic trisaccharide **95**, in 52% overall yield (Fig. 34).[34] The structure of cyclic trisaccharide, as determined by single crystal X-ray diffraction, showed that the molecule was cone-shaped, as in the case of α-CD. However, the cone-shape was more sharp than α-CD, by which the upper rim-to-lower rim ratio was more for **95**, than that for α-CD. The molecular dimensions of **95** are given in Fig. 35, along with molecular arrangement in the crystal lattice. The molecular arrangement in the lattice was found to be of the brick-wall type. Further, the molecule possessed a rather uncommon perfect *P*3 trigonal symmetry. Further evaluation of guest binding abilities of this new cyclic oligosaccharide showed that the host **95** bound with 1-aminoadamantane much more effectively than the binding of the guest with native β-CD, as evaluated by isothermal titration calorimetry, in aqueous solutions. The host–guest complex formation corresponded to a 1 : 1 ratio. On the other hand, host **95** complexed with hexamethylene tetramine in a host-to-guest ratio of 2 : 1, and also the binding constant was lower than that for β-CD. In addition to the cyclic trisaccharide **95**, formation of cyclic hexamer and nonamer were also observed in the mass spectrum of the reaction mixture.

The above studies illustrated the merit of polycondensation or cyclooligomerization as a powerful strategy to the construction of cyclic oligosaccharides differing in composition. The strategy provides elegant opportunities for further development, primarily due to the facility to derive monomers, that are either mono- or di- or trisaccharides, modified with desired functionalities at the pyranoside or furanoside monosaccharides.

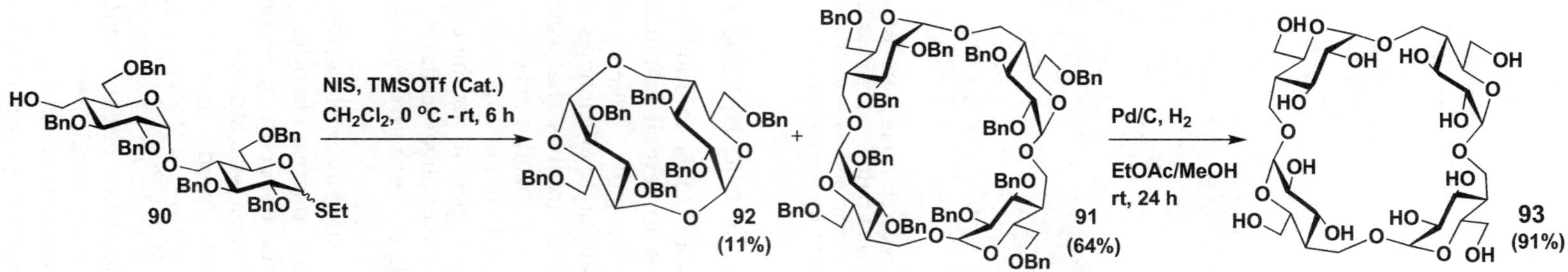

Fig. 33 Synthesis of glycosidic bond expanded cyclic di- (92) and tetrasaccharide (93) by polycondensation of disaccharide 90.[33]

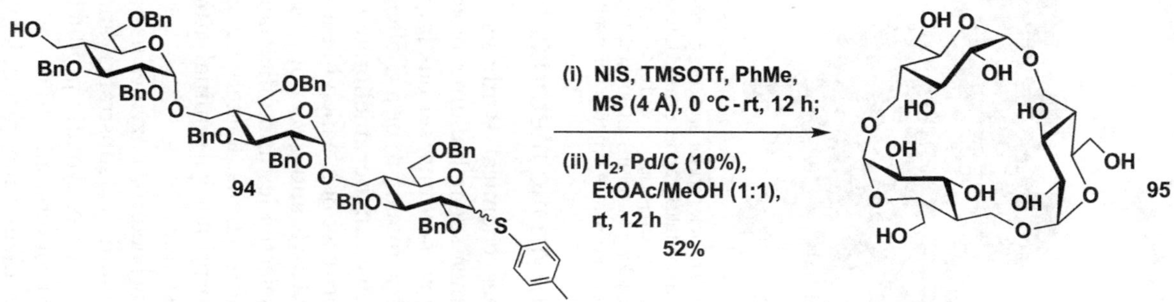

Fig. 34 Synthesis of glycosidic bond expanded cyclic trisaccharide 95 from a linear trisaccharide monomer 94.[34]

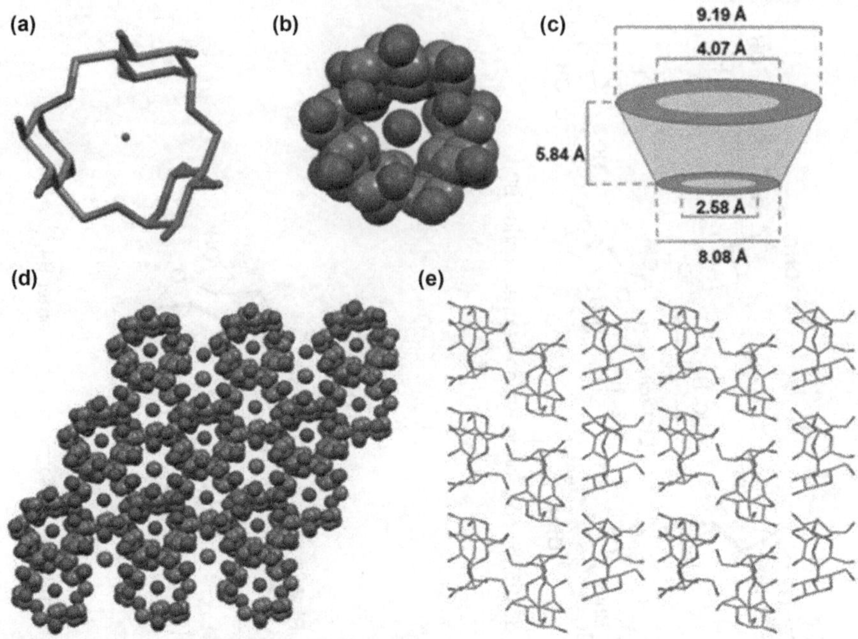

Fig. 35 (a) Stick model and (b) space-fill model of the crystal structure of **95**, as viewed along the crystallographic *c*-axis; (c) molecular dimensions in a cartoon representation; (d) packing diagram included with solvents, namely, methanol (gray) and water (red); (e) wire-frame packing diagram of crystal lattice, as viewed along the crystallographic *b*-axis. Hydrogen atoms are omitted for clarity.[34]

5 Enzymatic methods to prepare cyclic oligosaccharides

Whereas chemical synthesis provides necessary tools to modify the monomer sugar building blocks, as described in several instances in the above sections, serious impediments relating to the efficiency of the reactions remain to be a challenge. A partial solution to efficiencies of the reactions can be ameliorated with the use of enzymes that catalyze specific reactions, which in the case of cyclooligomerization reactions are the glycosyl- and glycanotransferases. The reactions are catalyzed with high selectivities and efficiencies, with enzymes specific for the chosen type of reactions. Bacterial enzymes mediate cyclic oligosaccharide formation primarily, seminal article of French is authoritative for several decades now.[35]

Many bacterial enzymes use polysaccharides, such as amylose, dextran and inulin as substrates and produce cyclic oligosaccharides. The best known enzymatic production is CDs, a recent report by Ferrarotti and coworkers, illustrates a continuous process with advancements in the membrane technology.[36] The enzymatic production of cyclic oligosaccharides produce α-(1→4)-, as in CDs,[35] α-(1→6)-, as in cycloisomaltooligosaccharides (cyclodextrans)[37] and β-(2→1)- linkages as in cycloinulooligosaccharides,[38] using the corresponding glycano- and glycosyltransferases. Interest in such cyclic oligosaccharides arises from their manifold functions that are relevant to many industrial purposes. Alternating (1→6)- and (1→3)-linked α-D-glucose polysaccharide as a branched structure in dextrans,[39] called alternan, was found to be hydrolyzed by a bacterial enzyme alternase to produce a cyclic tetrasaccharide, namely, cyclo[-6)-α-D-Glc*p*-(1→3)-α-D-Glc*p*-(1→6)-α-D-Glc*p*-(1→3)-α-D-Glc*p*-(1→] **96** (Fig. 36), in large proportion, along with yet another branched derivative **97**.[40] Formation of cyclic isomaltooligosaccharides, such as **97** (Fig. 36), by *Bacillus* and *Paenibacillus* bacterial strains was reported recently by Funane and coworkers.[41]

Identification of enzymes and sugars produced thereof occupies prominence in the development of enzymatic synthesis of cyclic oligosaccharides. Nishimoto and coworkers identified that α-1,4-glucan would function as a substrate to 1,6-α-glucosyltransferase and α-(1→3)-isomaltosyltransferase double enzymatic system, leading to the formation of cyclic tetrasaccharide, cyclo[-6)-α-D-glucopyranosyl-(1→3)-α-D-glucopyranosyl-(1-6)-α-D-gluco-pyranosyl-(1→3)-α-D-gluycopyranosyl-(1→] **96** (Fig. 36a).[42] The observation herein was that the α-1,4-glycosidic linkage in α-1,4-glucan changes to α-1,3- and α-1-6-glycosidic linkages through the action of glucanotransferase, even when this substrate was known to produce CDs in previous occasions, with α-(1→4)-glycosidic linkage by cyclomaltodextrin glucanotransferase.[35]

Two enzyme systems could produce cyclic oligosaccharides markedly different from that produced by a single enzyme system as observed in the case of the formation of cyclic tetrasaccharide cyclo[-6)-α-D-Glc*p*-(1→3)-α-D-Glc*p*-(1→6)-α-D-Glc*p*-(1→3)-α-D-Glc*p*-(1→] **96**, having an alternate α-(1→3) and α-(1→6) glycosidic linkages, reported by Kitaoka and coworkers (Fig. 36a).[43] This particular cyclic tetrasaccharide is known to be produced from alternan polysaccharide by the action of cycloalternan-forming enzyme alternase.[44] The cycloalternan-forming enzyme, in conjunction with a disproportionating enzyme, the α-(1→6)-D-glucopyranosyl transferase, synergistically use starch and maltodextrins and produce cycloalternan.[43]

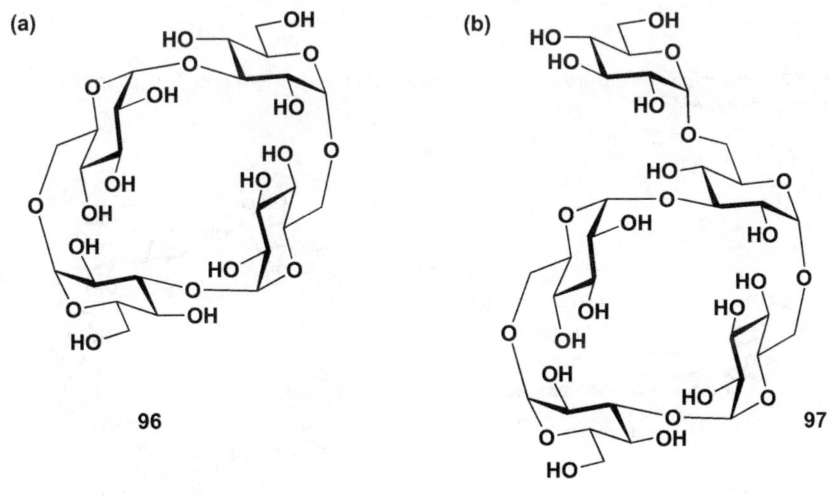

Fig. 36 Molecular structures of (a) cyclo[-6)-α-D-Glc*p*-(1→3)-α-D-Glc*p*-(1→6)-α-D-Glc*p*-(1→3)-α-D-Glc*p*-(1→] **96** and (b) cyclic isomaltooligosaccharide **97**.[40–43]

Cyclic oligosaccharides produced from starch by enzymes assume great interest, as the substrate starch is available in plenty form natural resources. Investigation of Watanabe and coworkers demonstrated that a bacterial strain AM7 produced two cyclic oligosaccharides, cyclo[-6)-α-D-Glc*p*-(1→4)-α-D-Glc*p*-(1→4)-α-D-Glc*P*-(1→4)-α-D-Glc*p*-(1→4)-α-D-Glc*p*-(1→] **98** and a higher homologue **99** (Fig. 37), having one α-(1→6) glycosidic linkage in its structure, differing distinctly from that of CDs.[45]

A series of early work by Driguez and coworkers advanced the enzymatic synthesis of derivatized CDs, with the aid of cyclodextrin glycosyltransferases (CGTase). A first regiospecific synthesis of 6^a, 6^c, 6^c-tri-*O*-methylated cyclodextrin analogue **101** was reported through a polycondensation method.[46] The desired monomethylmaltosyl fluoride **100** was subjected to CGTase catalyzed polycondensation, which led to the formation of **101**, in 43% yield (Fig. 38). In addition, linear oligomer was also formed in the reaction. This enzymatic synthesis thus opened-up a facile route to prepare cyclic oligosaccharides, starting from a disaccharide monomer carrying the desired modification on the sugar.[46]

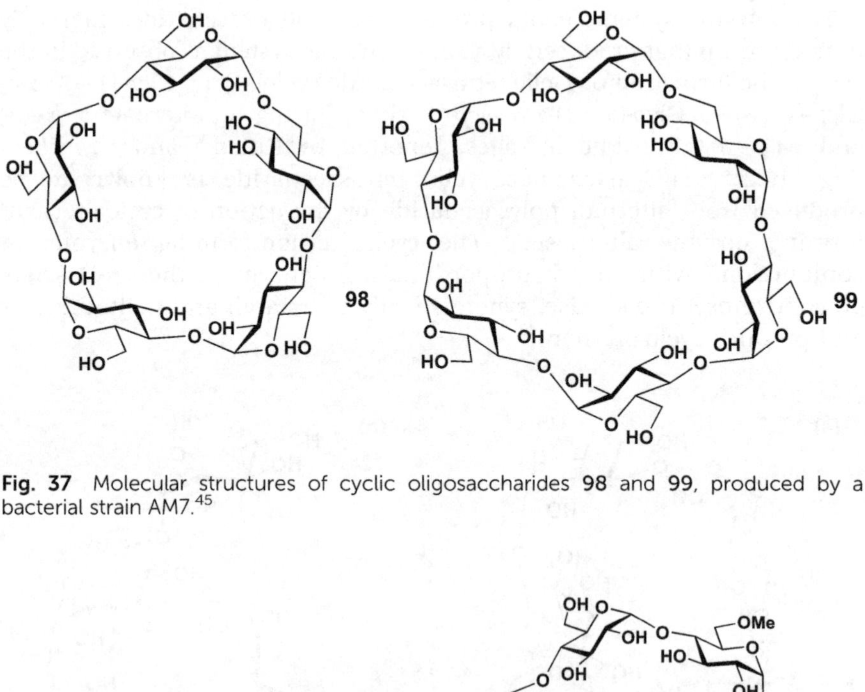

98

99

Fig. 37 Molecular structures of cyclic oligosaccharides **98** and **99**, produced by a bacterial strain AM7.[45]

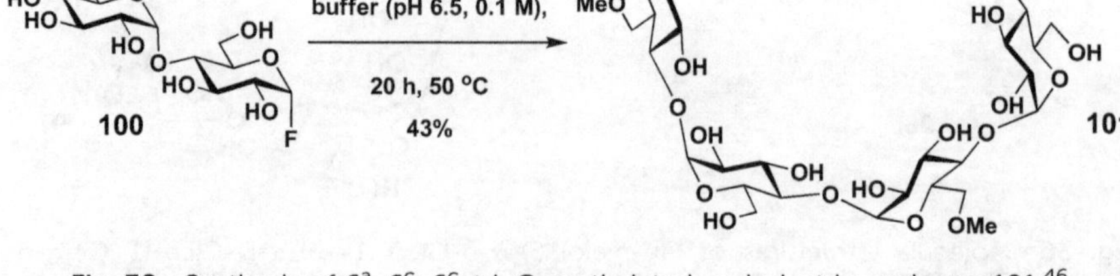

100

101

Fig. 38 Synthesis of 6^a, 6^c, 6^c-tri-*O*-methylated cyclodextrin analogue 101.[46]

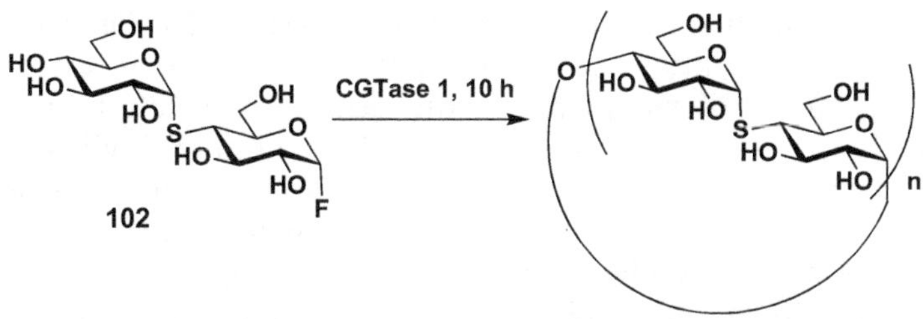

Fig. 39 Synthesis of cyclothiomaltins **103–105**.[49]

The above enzymatic polycondensation reaction on **100** was reported later on to form the β- and γ-CD analogues. 6′-*O*-Acetylmalotosyl fluoride was also found to undergo the enzymatic reaction to afford 6^{a}, 6^{c}, 6^{e}-tri-*O*-acetylcyclohexaamylose, although in 3% yield.[47]

The enzymatic reaction of maltosyl fluoride leading to the formation of native α-, β- and γ-CDs was demonstrated much earlier by Hehre, Mizokami and Kitahata, thus laying the foundation to the enzymatic synthesis of CDs through the polycondenzation method.[48]

An efficient synthesis of cyclothiomaltotetra-, penta- and hexaoside was also developed by Driguez and coworkers. Thus, the reaction of 4-thio-maltosyl fluoride **102**, catalyzed by CGTase 1 from *B. circulans* strain, afforded cyclothiomaltins **103–105** (Fig. 39).[49]

The most preference of the substrate in enzymatic reactions was also encountered in the above types of enzymatic polycondensation reaction. Thus, when interglycosidic linkage in the maltosyl fluoride **100** or the thio analogue **102** was changed to *C*-maltosyl fluoride, the reaction did not lead to cyclooligomerization, rather only a linear oligomerization occurred, leading to multiple linear *C*-maltooligosaccharides.[50]

6 Conclusion

The above compilation describes developments in cyclic oligosaccharide synthesis by varied strategies, either by chemical or enzymatic means. Cyclic oligosaccharides, whether in the native constitution or in a tailor-made modification, generate manifold interest, not only due to the challenges in their synthesis, but also to their interesting value-addition possibilities to varied disciplines. For example, recent investigations establish that native CDs exhibit properties above that of well-known host properties, wherein the CDs are found to intervene lysosomal storage, neurodegeneration and brain tumor diseases and thus exhibit immense potential for therapeutic applications.[51] Cyclic oligosaccharides, that are non-reducing sugars by nature, act as stabilizers to acidolytic cleavage in food industries, CDs are prime example herein.[52] Cyclic oligosaccharides with constitutional modifications might exhibit functions over and above those known in the case of native cyclic oligosaccharides. In this respect, enzymatic synthesis has the most potential to partake chemical synthesis, as increasing number of reports uncover the value of bacterial enzymes to synthesize modified and branched cyclic oligosaccharides. Availability of inexpensive and bench-top sugars and polysaccharides as substrates provides impetus for the enzymatic production. On the other hand, chemical tools potentiate synthesis of finely-tuned cyclic oligosaccharides in a variety of ways and modifications, only left to the imagination of the practitioner. For example, substitution of few sugar moieties with varied types of non-sugar moieties is a thriving area of advancement, a recent review article discusses such modifications greatly.[53] Synthesis of large cyclic oligosaccharides having ten or more sugar moieties remains to be a challenging task, even when higher flexibility of the macrocycle could bring with it advantages not normally achievable with smaller ring macrocyles.[54] Cyclic oligosaccharide synthesis may be traced back during last over three decades. However, the synthesis is slow but steady, clearly awaiting to reach newer heights. Sustained advancements point to the promise that cyclic oligosaccharides will remain to be a topical area of intense investigations in carbohydrate chemistry, aimed at both fundamental and value-added utilitarian developments in the near future.

Abbreviations

AgOTf	Silver trifluoromethanesulfonate
AIBN	Azobisisobutyronitrile
CD	Cyclodextrin
CGTase	Cyclodextrin glycosyltransferases
DCE	1,2-Dichloroethane
DMAP	4-Dimethylaminopyridine
IDCP	Iodonium dicollidine perchlorate
MeOTf	Methyl trifluoromethanesulfonate
MP	4-Methoxy phenyl
MS	Molecular sieves
NIS	*N*-Iodosuccinimide
Pent	Pentenyl
PhSeOTf	Phenylselenyl trifluoromethanesulfonate
Ph_3SnH	Triphenyltin hydride
Phth	Phthalimido
PyBOP	Benzotriazol-1-yl-oxytripyrrolidinophosphonium hexafluorophosphate
$SnCl_2$	Stannous chloride
TES	Triethylsilyl
TfOH	Trifluoromethanesulfonic acid
THF	Tetrahydrofuran
TMSOTf	Trimethylsilyl trifluoromethanesulfonate
$TrClO_4$	Triphenylmethyl perchlorate.

Acknowledgements

We thank Department of Science and Technology and Council of Scientific and Industrial Research (CSIR), New Delhi, for a financial support. CSIR is acknowledged for a research fellowship to KM, GKS and GCD.

References

1 (a) J. Szejtli, *Chem. Rev.*, 1998, **98**, 1743–1754; (b) J. Szejtli, *Pure Appl. Chem.*, 2004, **76**, 1825–1845; (c) G. Crini, *Chem. Rev.*, 2014, **114**, 10940–10975.

2 (a) D. Armspach, G. Gattuso, R. Koniger and J. F. Stoddart, in *Bioorganic Chemistry: Carbohydrates*, ed. S. M. Hecht, Oxford University Press, New York, 1999, pp. 458–488; (b) G. Gattuso, S. A. Nepogodiev and J. F. Stoddart, *Chem. Rev.*, 1998, **98**, 1919–1958.

3 S. Immel, K. Fujita, H. J. Lindner, Y. Nogami and F. W. Lichtenthaler, *Chem. – Eur. J.*, 2000, **4**, 2327–2333.

4 (a) G. Wenz, B.-H. Han and A. Muller, *Chem. Rev.*, 2006, **106**, 782–817; (b) G. Wenz, *Beilstein J. Org. Chem.*, 2015, **11**, 271–272; (c) J. Wang, Z. Qiu, Y. Wang, L. Li, X. Guo, D.-T. Pham, S. F. Lincoln and R. K. Prud'homme, *Beilstein J. Org. Chem.*, 2016, **12**, 50–72.

5 E. Goldsmith, S. Sprang and R. Fletterick, *J. Mol. Biol.*, 1982, **156**, 411–427.

6 Y. Takahashi and T. Ogawa, *Carbohydr. Res.*, 1987, **169**, 127–149.

7 Y. Takahashi and T. Ogawa, *Carbohydr. Res.*, 1987, **164**, 277–296.

8 (a) M. Mori, Y. Ito and T. Ogawa, *Tetrahedron Lett.*, 1989, **30**, 1273–1276; (b) M. Mori, Y. Ito and T. Ogawa, *Tetrahedron Lett.*, 1990, **31**, 3029–3030; (c) M. Mori, Y. Ito, J. Uzawa and T. Ogawa, *Tetrahedron Lett.*, 1990, **31**, 3191–3194.

9 T. Nakagawa, K. Ueno, M. Kashiwa and J. Watanabe, *Tetrahedron Lett.*, 1994, **35**, 1921–1924.

10 G. Excoffier, M. Paillet and M. Vignon, *Carbohydr. Res.*, 1985, **135**, C10–C11.

11 P. M. Collins and M. H. Ali, *Tetrahedron Lett.*, 1990, **31**, 4517–4520.

12 (a) M. Nishizawa, H. Imagawa, Y. Kan and H. Yamada, *Tetrahedron Lett.*, 1991, **32**, 5551–5554; (b) M. Nishizawa, H. Imagawa, E. Morikuni, S. Hatakeyama and H. Yamada, *Chem. Pharm. Bull.*, 1994, **42**, 1365–1366.

13 M. Wakao, K. Fukase and S. Kusumoto, *J. Org. Chem.*, 2002, **67**, 8182–8190.

14 L. Fan and O. Hindsgaul, *Org. Lett.*, 2002, **4**, 4503–4506.

15 K. S. Kim, B.-Y. Lee, S. H. Yoon, H. J. Jeon, J. Y. Baek and K.-S. Jeong, *Org. Lett.*, 2008, **10**, 2373–2376.

16 (a) A. Dondoni and M.-C. Scherrmann, *J. Org. Chem.*, 1994, **59**, 6404–6412; (b) A. Dondoni and A. Marra, *Chem. Commun.*, 1999, 2133–2145; (c) A. Dondoni, A. Marra, M.-C. Scherrmann and V. Bertolasi, *Chem. – Eur. J.*, 2001, **7**, 1371–1382.

17 (a) M. L. Gening, D. V. Titov, A. A. Grachev, A. G. Gerbst, O. N. Yudina, A. S. Shashkov, A. O. Chizhov, Y. E. Tsvetkov and N. E. Nifantiev, *Eur. J. Org. Chem.*, 2010, 2465–2475; (b) M. L. Gening, Y. E. Tsvetkov, D. V. Titov, A. G. Gerbst, O. N. Yudina, A. A. Grachev, A. S. Shashkov, S. Vidal, A. Imberty, T. Saha, D. Kand, P. Talukdar, G. B. Pier and N. E. Nifantiev, *Pure Appl. Chem.*, 2013, **85**, 1879–1891.

18 H. Liu and X. Li, *Tetrahedron Lett.*, 2014, **55**, 5525–5528.

19 N. Sakairi, L.-X. Wang and H. Kuzuhara, *J. Chem. Soc., Chem. Commun.*, 1991, 289–290.

20 (a) N. Sakairi and H. Kuzuhara, *J. Chem. Soc., Chem. Commun.*, 1992, 510–512; (b) N. Sakairi, L.-X. Wang and H. Kuzuhara, *J. Chem. Soc., Perkin Trans. 1*, 1995, 437–443.

21 A. Kikuzawa, T. Kida, Y. Nakatsuji and M. Akashi, *J. Org. Chem.*, 2005, **70**, 1253–1261.

22 A. Kikuzawa, T. Kida and M. Akashi, *Macromolecules*, 2008, **41**, 3393–3395.

23 W.-W. Li, T. D. W. Claridge, Q. Li, M. R. Wormald, B. G. Davis and H. Bayley, *J. Am. Chem. Soc.*, 2011, **133**, 1987–2001.
24 (a) N. K. Kochetkov, S. A. Nepogodiev and L. V. Backinowsky, *Tetrahedron*, 1990, **46**, 139–150; (b) L. V. Backinowsky, S. A. Nepogodiev and N. K. Kochetkov, *Bioorgan. Khim.*, 1985, **14**, 1122–1124.
25 N. K. Kochetkov, *Tetrahedron*, 1987, **43**, 2389–2436.
26 P. R. Ashton, C. L. Brown, S. Menzer, S. A. Nepogodiev, J. F. Stoddart and D. J. Williams, *Chem. - Eur. J.*, 1996, **2**, 580–591.
27 S. A. Nepogodiev, G. Gattuso and J. F. Stoddart, in *Proceedings of the Eighth International Symposium on Cyclodextrins*, ed. J. Szejtli and L. Szente, Kluwer Academic, Dordrecht, 1996, pp. 89–94.
28 P. R. Ashton, S. J. Cantrill, G. Gattuso, S. Menzer, S. A. Nepogodiev, A. N. Shipway, J. F. Stoddart and D. J. Williams, *Chem. - Eur. J.*, 1997, **3**, 1299–1314.
29 G. Gattuso, S. Menzer, S. A. Nepogodiev, J. F. Stoddart and D. J. Williams, *Angew. Chem., Int. Ed. Engl.*, 1997, **36**, 1451–1454.
30 S. Houdier and P. J. A. Vottéro, *Angew. Chem. Int., Ed. Engl.*, 1994, **33**, 354–356.
31 M. C. Kasuya and K. Hatanaka, *Tetrahedron Lett.*, 1998, **39**, 9719–9722.
32 S. Paul, S. Raghothama and N. Jayaraman, *Carbohydr. Res.*, 2009, **344**, 177–186.
33 G. C. Daskhan and N. Jayaraman, *Chem. Commun.*, 2014, **50**, 8554–8557.
34 K. Maiti and N. Jayaraman, *J. Org. Chem.*, 2016, **81**, 4616–4622.
35 D. French, *Adv. Carbohydr. Chem.*, 1957, **12**, 189–260.
36 J. A. R. Gastón, H. Costa and S. A. Ferrarotti, *Biotechnol. Prog.*, 2015, **31**, 695–699.
37 T. Oguma, T. Horiuchi and M. Kobayashi, *Biosci., Biotechnol., Biochem.*, 1993, **57**, 1225–1227.
38 T. Kuramoto, Y. Tamura and K. Mizutani, *Carbohydr. Res.*, 1989, **192**, 83–90.
39 A. Misaki, M. Torii, T. Sawai and I. J. Goldstein, *Carbohydr. Res.*, 1980, **84**, 273–285.
40 G. L. Côté and P. Biely, *Eur. J. Biochem.*, 1994, **226**, 641–648.
41 K. Funane, K. Terasawa, Y. Mizuno, H. Ono, S. Gibu, T. Tokashiki, Y. Kawabata, Y.-M. Kim, A. Kimura and A. Kobayashi, *Biosci., Biotechnol., Biochem.*, 2008, **72**, 3277–3280.
42 T. Nishimoto, H. Aga, K. Mukai, T. Hashimoto, H. Watanabe, M. Kubota, S. Fukuda, M. Kurimoto and Y. Tsujisaka, *Biosci., Biotechnol., Biochem.*, 2002, **66**, 1806–1818.
43 Y.-K. Kim, M. Kitaoka, K. Hayashi, C.-H. Kim and G. L. Côté, *Carbohydr. Res.*, 2003, **338**, 2213–2220.
44 P. Biely, G. L. Côté and A. Burgess-Cassler, *Eur. J. Biochem.*, 1994, **226**, 633–639.
45 H. Watanabe, T. Nishimoto, T. Sonoda, M. Kubota, H. Chaen and S. Fukuda, *Carbohydr. Res.*, 2006, **341**, 957–963.
46 S. Cottaz and H. Driguez, *J. Chem. Soc., Chem. Commun.*, 1989, 1088–1089.
47 S. Cottaz, C. Apparu and H. Driguez, *J. Chem. Soc. Perkin Trans 1*, 1991, 2235–2241.
48 E. J. Hehre, K. Mizokami and S. Kitahata, *Denpun Kagaku*, 1983, **30**, 70, (*Chem. Abstr.*, 1983, **99**, 84236m).
49 L. Bornaghi, J.-P. Utille, D. Penninga, A. K. Schmidt, L. Dijkhuizen, G. E. Schulz and H. Driguez, *Chem. Commun.*, 1996, 2541–2542.
50 L. Bornaghi, J.-P. Utille, E. D. Rekaï, J.-M. Mallet, P. Sinaÿ and H. Driguez, *Carbohydr. Res.*, 1997, **305**, 561–568.

51 M. Vecsernyés, F. Fenyvesi, I. Bácskay, M. A. Deli, L. Szente and E. Fenyvesi, *Arch. Med. Res.*, 2014, **45**, 711–729.
52 G. Astray, C. Gonzalez-Barreiro, J. C. Mejuto, R. Rial-Otero and J. Simal-Gándara, *Food Colloids*, 2009, **23**, 1631–1640.
53 J. Xie and N. Bogliotti, *Chem. Rev.*, 2014, **114**, 7678–7739.
54 T. Endo, *Trends Glycosci. Glycotechnol.*, 2011, **23**, 79–92.

Ferrier rearrangement: an update on recent developments

Ana M. Gómez,* Silvia Miranda and J. Cristobal López*
DOI: 10.1039/9781782626657-00210

The transformation of glycals (1,2-unsaturated cyclic carbohydrate derivatives) into 2,3-unsaturated glycosyl derivatives, currently termed Ferrier rearrangement, is a well-established synthetic procedure with ample use in the fields of carbohydrate and organic chemistry. This article highlights the developments in the Ferrier rearrangement published in the literature since the last review, in 2013, to early 2016.

1 Introduction

The transformation of pyranose glycals (1,5-anhydrohex-1-enitols, *e.g.* **1**) into allylic glycosides, *e.g.* **3**,[1] a formal allylic glycosylation, has come to be known as Ferrier rearrangement (FR)[2] or Ferrier I reaction (Scheme 1).[3] This topic, which continues to be the most investigated chemical transformation of glycals, has been the subject of several reviews over the years,[4] the latest having appeared in 2013.[5,6] The present account summarizes developments in the Ferrier rearrangement that have emerged in the literature over the last two years.

In general, the Ferrier reaction can be considered as an allylic rearrangement, which implies displacement of the C-3 substituent of a glycal, followed by attack of a given (*O*-, *N*-, *C*- or *S*-) nucleophile at the anomeric (C-1) position of an intermediate allylic oxycarbenium ion, **2**. The latter reacts with nucleophilic species to provide mixtures of adducts (**3**) in which the pseudoaxial glycosides predominate.

Early studies on the FR were carried out with simple Lewis acids such as $BF_3 \cdot OEt_2$, as promoters.[7] Subsequently, a large number of catalysts including a variety of Lewis acids, Brønsted acids, and oxidants have been reported. More recently, transition-metal activation of suitably designed glycals has also been reported.

2 Pd-catalyzed Ferrier rearrangement

The difficulty in the formation and the poor reactivity of Pd-π-allyl intermediates in electron-rich glycal systems have been recognized as major challenges for the palladium-catalyzed Ferrier rearrangement.[6]

In a series of recent contributions, Liu and coworkers have evaluated the potential of the palladium-catalyzed decarboxylative reaction in the synthesis of allylic glycosyl derivatives. In this context, they described a palladium-catalyzed stereo- and regioselective *C*-glycosylation process based on an intramolecular decarboxylative coupling.[8] Initial studies with glycal derived β-ketoesters, *e.g.* **4**, allowed the authors to identify

Instituto de Química Orgánica General, (IQOG-CSIC), Juan de la Cierva 3, Madrid 28006, Spain. E-mail: ana.gomez@csic.es; jc.lopez@csic.es

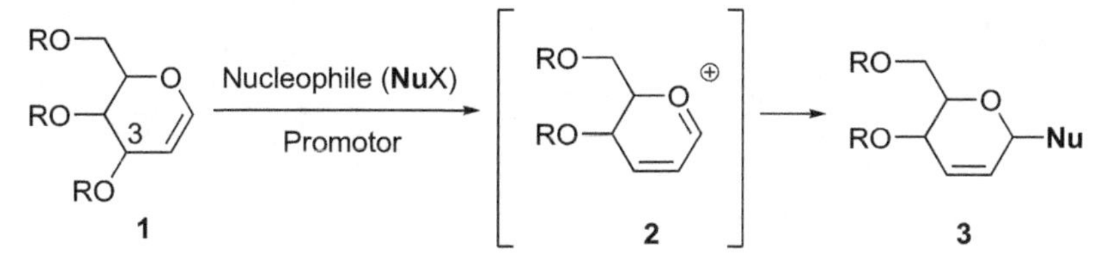

Scheme 1 Ferrier rearrangement of glycals (**1**) leading to unsaturated glycosides, **3**.

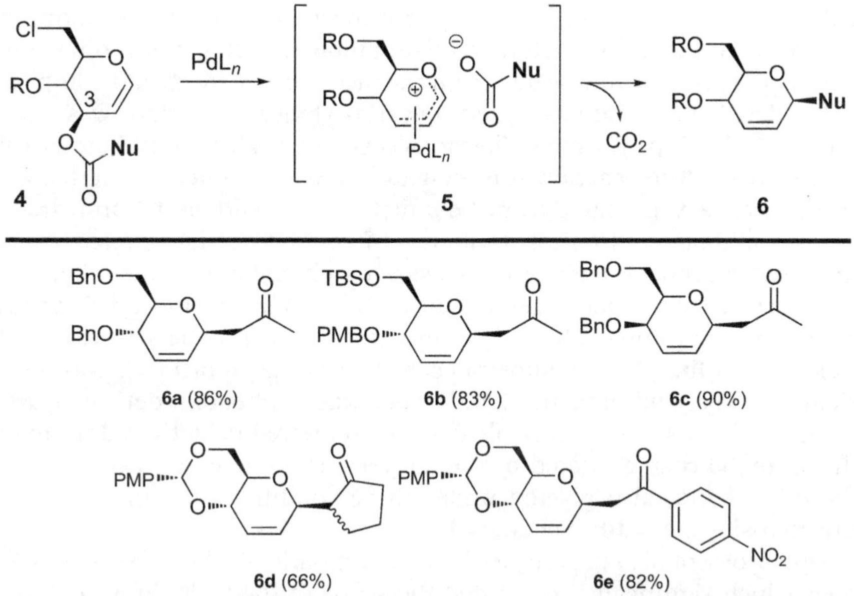

Scheme 2 Liu and coworkers stereoselective palladium-catalyzed decarboxylative *C*-glycosylation leading to β-*C*-glycosyl derivatives **6**.

$Pd(OAc)_2$ and 1,1′-bis(diisopropylphosphino)ferrocene (D*i*PPF) (toluene, 60 °C, 2 h) as the optimized reagents and conditions for the completely β-stereoselective, preparation of *C*-glycosyl derivatives type **6** (Scheme 2). This reaction was applied to the stereoselective preparation of a range of β-*C*-glycosyl derivatives **6a–e**, which could be obtained in moderate to good yields (Scheme 2).[8]

In order to prove that a palladium-catalyzed decarboxylative allylation mechanism, rather than an intramolecular Carroll rearrangement,[9] was taking place, the authors designed a crossover experiment involving the decarboxylative coupling reaction between two different glycal derived β-ketoesters. The reaction took place leading to four (rather than two) different *C*-glycosyl derivatives, thus showing that complete scrambling of the anomeric substituents had taken place. This result indicated that the palladium-promoted ionization leading to **5** (Scheme 2) takes place at the beginning of the reaction and an intramolecular rearrangement pathway was not involved.[8]

The method, however, when applied to C-3 epimeric β-ketoesters, *i.e.* **7**, did not lead exclusively to α-*C*-glycosyl derivatives, and instead mixtures

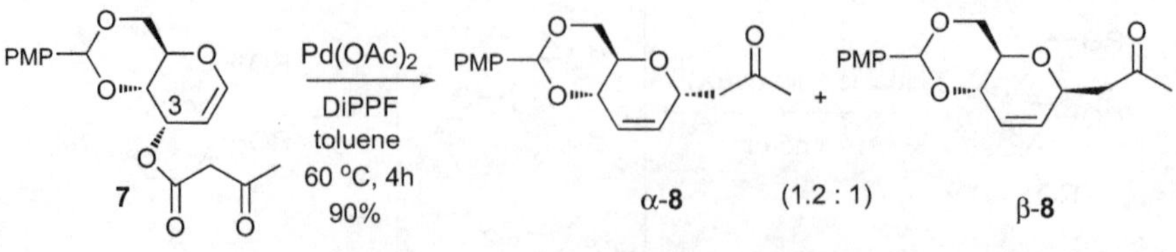

Scheme 3 Decarboxylative *C*-glycosylation of compound 7.

of α/β glycosyl derivatives were obtained (Scheme 3).[8] The authors ascribed this behavior to an outer-sphere mechanism in the reaction, in which the attack of the ketone enolate anion resulting from ionization might approach the allyl group from the face opposite to the Pd-complex.

A related palladium-catalyzed decarboxylative allylation was subsequently developed, which allowed access to β-*O*-glycosides from glycal carbonates.[10] The reaction was first explored with phenolic carbonate substrates, *i.e.* **9**, prepared from 4,6-*p*-methoxybenzylidene-1,5-anhydro-2-deoxy-D-*arabino*-hex-1-enitol (not recommended trivial name: 4,6-*p*-methoxybenzylidene D-glucal)[†] and various phenolic (thiophenolic) acyl chlorides under the optimized conditions shown in Scheme 4. Substrates with alkyl substituted phenolic groups furnished products in excellent yields, *i.e.* **10b**, whereas substrates with bromide moieties gave lower yieds, *i.e.* **10c, 10d**, and the 2-iodo substituted phenolic derivative **10e** could not be obtained, probably due to a preferred oxidative addition of the aromatic ring, as stated by the authors. Thiophenolic substrates, on the other hand, also reacted under these conditions leading to allylic thioglycosides, *e.g.* **10f** (Scheme 4).[10]

This protocol also proved useful with aliphatic alcohols, besides phenols, which significantly expanded the scope of the palladium-catalyzed decarboxylation by allowing the preparation of disaccharides. The optimized reaction conditions for the new transformation involved a change in the ligand from D*i*PPF to 1,1′-bis(di-*tert*-butylphosphino)ferrocene (D*t*BPF), and an increase in the reaction temperature (Scheme 5). The method proved to be compatible with a variety of protecting groups.[10]

Liu and coworkers, have been able to obviate the preparation of the complex starting carbonate substrates employed in these glycosylations by using a glycal-derived ethyl carbonate, *i.e.* **13**, in the presence of a base and an external nucleophile (NuH, Scheme 6).[11] The mechanism, as proposed by the authors, starts with the traditional palladium-catalyzed decarboxylative allylation reaction to generate the palladium intermediate **14**, which in the presence of an external nucleophile evolves (by a proton transfer between the ethoxide anion and the NuH) to yield Pd-intermediate **15**. The latter then proceeds to β-glycoside **6** (Scheme 6).

[†]The term "glucal" is a not recommended trivial name for the cyclic enol ether with a double bond between carbon atoms 1 and 2 of the ring, synthesized starting from glucose derivatives, because it does not exhibit a *gluco* configuration. The generic term glycal can be applied because it does not imply any specific configuration.

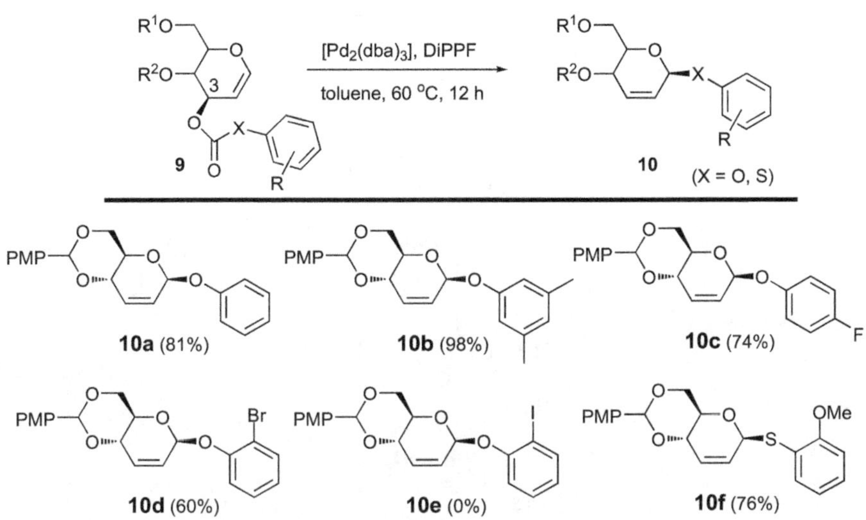

Scheme 4 Palladium-catalyzed *O*-glycosylation of glycal (thio)carbonates **9** derived from (thio)phenols.

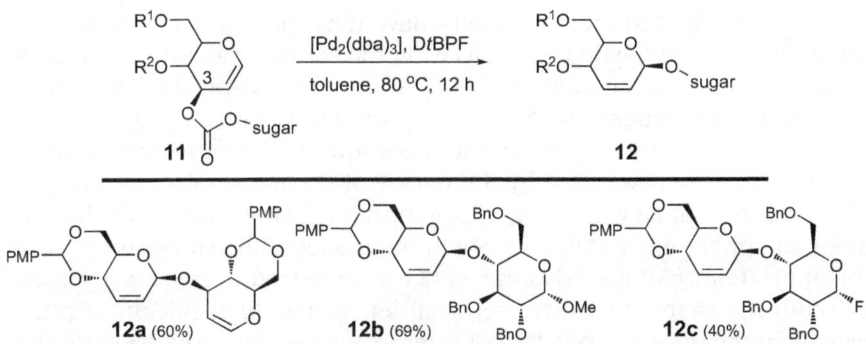

Scheme 5 Palladium-catalyzed *O*-glycosylation of glycal carbonates **11** leading to disaccharides **12**.

This transformation proved to be sensitive to the temperature and the base employed (chemical properties and loading). Thus, when less than 0.5 equiv. of base was used, the α-isomer was obtained as the major product, while exclusive β-selectivity was observed with more than 0.5 equiv. of base. The optimized reaction conditions were found to be Cs_2CO_3 (2.0 equiv.) as the base, $Pd(dba)_3$ (10 mol%) as the catalyst, and D*t*BPF (20 mol-%) as the ligand, in toluene at 60 °C.[11]

Accordingly, the reaction could be applied to phenols and aliphatic alcohols as external nucleophiles, allowing access to a variety of *O*-glycosides. A remarkable example of the usefulness of this approach is the synthesis of trisaccharide **21** (Scheme 7). Thus, iterative palladium-catalyzed intermolecular *O*-glycosylation of glycal **17** with glycal carbonates **16** and **20**, the latter arising by protecting group manipulation of the initially obtained disaccharide **18**, paved the way for the stereoselective formation of **21** (Scheme 7).[11]

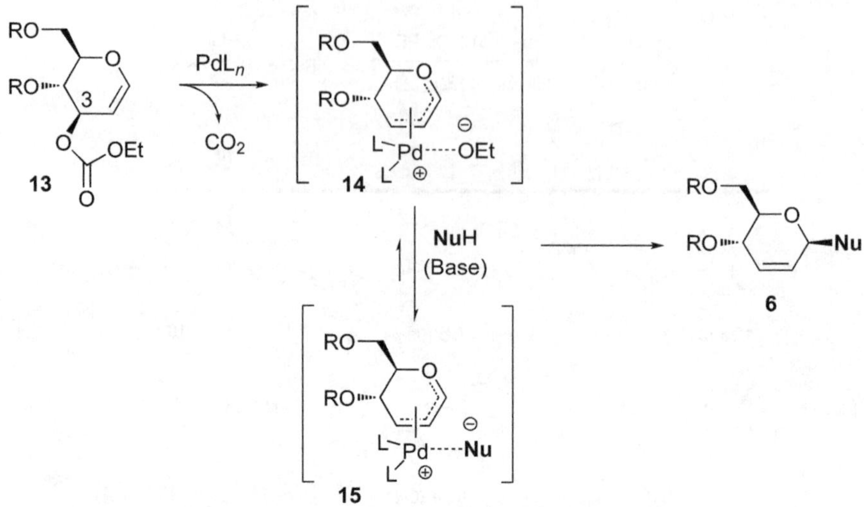

Scheme 6 Liu and coworkers' proposed mechanism for the intermolecular palladium-catalyzed *O*-glycosylation of glycal carbonates **13** leading to *C*-glycosyl derivatives **6**.

More recently, Liu and coworkers have developed an elegant strategy for stereoselective allylic *O*-glycosylation based on a palladium-catalyzed (inner-sphere or outer-sphere) glycosylation with 3-*O*-picoloyl glycal type donors, *e.g.* **22** (Scheme 8).[12] In their protocol, the picoloyl group functioned both as a leaving group and a coordinating group. The (α- or β-) selectivities were controlled by the nature of the nucleophiles employed. Thus, reaction of **22** with softer nucleophiles (phenols) proceeds through an outer-sphere mechanism in which the nucleophilic substitution from the opposite face of the Pd-complex (**23**) is preferred to give a α-glycoside **24** (path A, Scheme 8). Conversely, harder nucleophiles (alcohols, phenoxides) in the presence of a base, prefer to coordinate to the palladium(II) center to generate an intermediate **26** from which a β-glycoside **27** is formed by intramolecular nucleophilic addition (path B, Scheme 8).[12]

The optimized reaction conditions with soft nucleophiles (*e.g.*, phenols) to give α-glycosides **24**, were found to be $Pd(PPh_3)_4$ (10 mol%) as catalyst, and 1,4-bis(diphenylphosphino)butane (dppb) (20 mol%) as ligand in CH_2Cl_2 at 60 °C for 48 h (Scheme 9). The presence of Et_3N under, otherwise, identical reaction conditions led to improved β-selectivities in the reaction of harder nucleophiles leading to β-glycosides **27** (Scheme 9).[12]

The methodology when applied to the C-3 epimer of **22** took place according to the mechanistic aspects proposed by Liu and coworkers. Therefore, reaction of *C3-epi* **22** (Scheme 10) with phenols and alcohols provided β-glycosides, and α-glycosides, respectively, in a stereoselective manner.[12]

3-*O*-Picoloyl glycal derivative **22** has also been successfully used as the glycosyl donor on a palladium-catalyzed allylation reaction with various N-heterocyclic nucleophiles leading to β-*N*-glycosyl derivatives type **28**, (Scheme 11).[13]

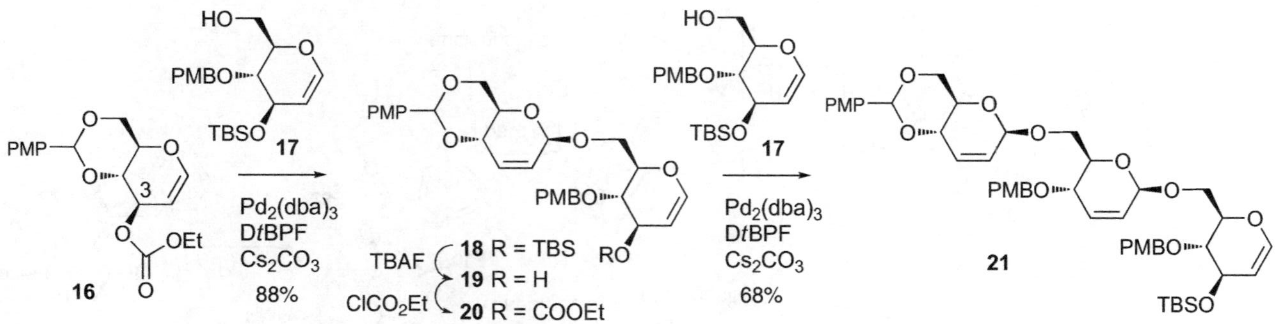

Scheme 7 Liu and coworkers' iterative palladium-catalyzed glycosylation approach to trisaccharide **21**.

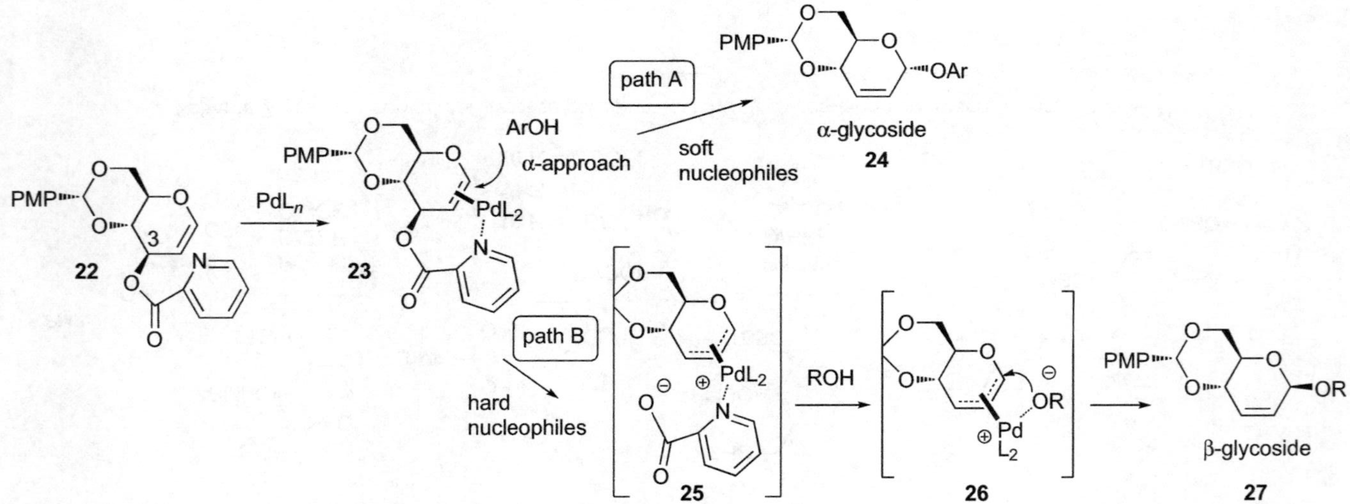

Scheme 8 Liu and coworkers' stereoselective α- or β-glycosylation with C-3-*O*-picoloyl glycals and soft or hard *O*-nucleophiles, respectively.

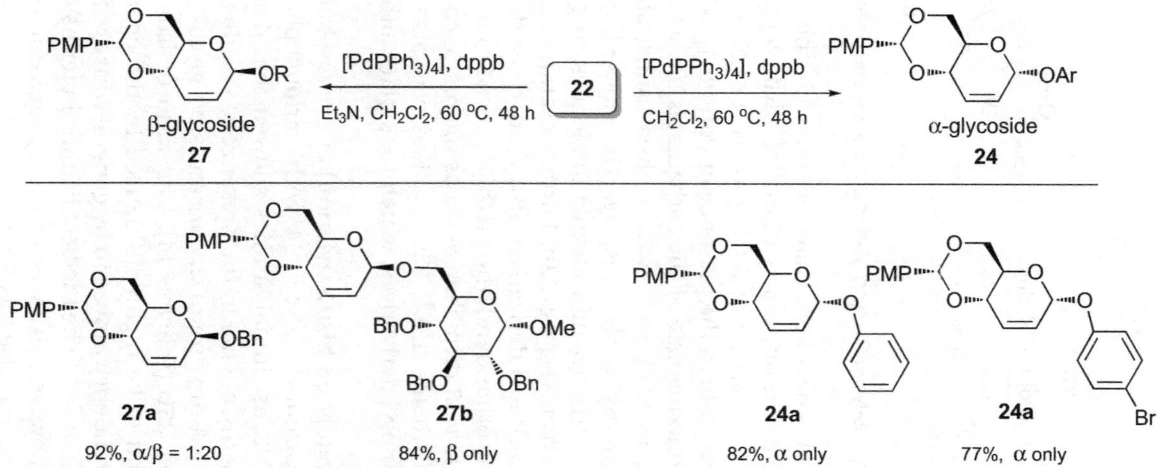

Scheme 9 Palladium-catalyzed *O*-glycosylation of alcohols and phenols with **22**, leading to β- or α-glycosides, **27** and **24**, respectively.

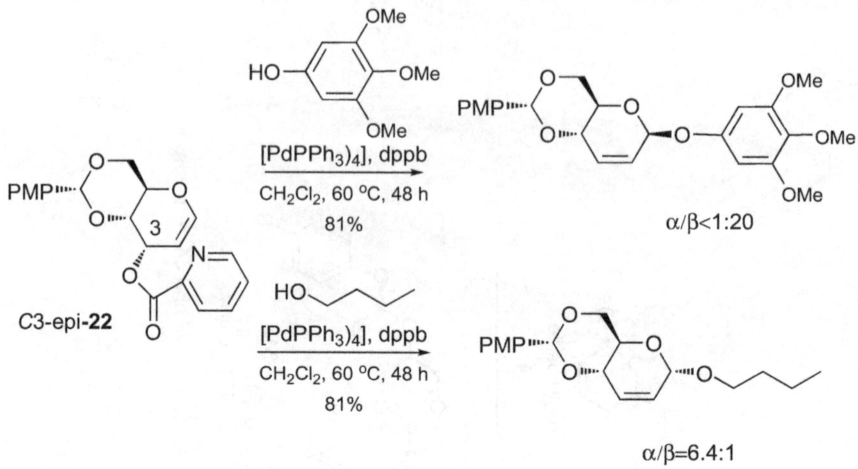

Scheme 10 Palladium-catalyzed *O*-glycosylation of alcohols and phenols with *C3-epi*-**22**.

Finally, Liu and coworkers have employed the decarboxylative allylation approach to the one-pot synthesis of allylic *N*-glycosyl imidazole analogues, *e.g.* **32**, by reaction of 3-OH glycals, *i.e.* **29**, with carbonyl di-imidazole derivatives, (**30**) in the presence of $Pd(PPh_3)_4$, which takes place *via in situ* formed intermediate allylic carbamates, *i.e.* **31** (Scheme 12).[14]

The reaction proceeded well with D-glycal substrates to give the corresponding *N*-glycosyl derivatives, *e.g.* **32c** and **32d**, respectively, and with 2-substituted carbonyl diimidazole analogues to give the desired allylic *N*-glycosyl derivatives, *e.g.* **32b**. Further examples involving purine and adenine type carbonyl derivatives also yielded the desired *N*-glycosyl heterocycles. On the other hand, the method proved to be stereoselective, and D-glycal derivative **33** furnished α-**32a** as the major compound, with a ratio $\alpha:\beta = 17:1$ (Scheme 13).[14] The method was also applied to the synthesis of *N*-furanosyl derivatives, which took place although with poor stereoselectivity.[14]

The palladium catalyzed **bis(*O*-arylation)** of 6-*O*-*tert*-butyldiphenylsilyl-3,4-di-*O*-isobutoxycarbonyl-D-glycal **34** with various phenols has been recently studied.[15] Thus, initial studies showed that dicarbonate **34** reacted with 2,5-dimethylphenol in the presence of $Pd_2(dba)_3$ as catalyst, in THF (60 °C) to give a mixture of 2,3-unsaturated glycoside **35a** and 3,4-unsaturated isomer **35b** (Scheme 14). The ratio of isomeric glycosides **35a** : **35b**, was found to be highly influenced by the ligand employed in the reaction. This tendency proved to be general with a variety of phenols, and dppe or dppb showed to be the best ligands in terms of overall yields. Better selectivity, however, in favor of the 2,3-unsaturated isomers was observed when dppe was used as ligand.

The proposed reaction pathway for this transformation invokes the reaction of two different Pd-π-allyl intermediates **36** and **38**, which react with the phenol nucleophiles (Scheme 15). The first reaction, leading to **37**, is regio- and stereoselective, whereas the phenol addition to **38** is stereoselective (Scheme 14). Thus, the attack of the nucleophile in **38** can occur at C-2 and C-4, which eventually leads to the formation of

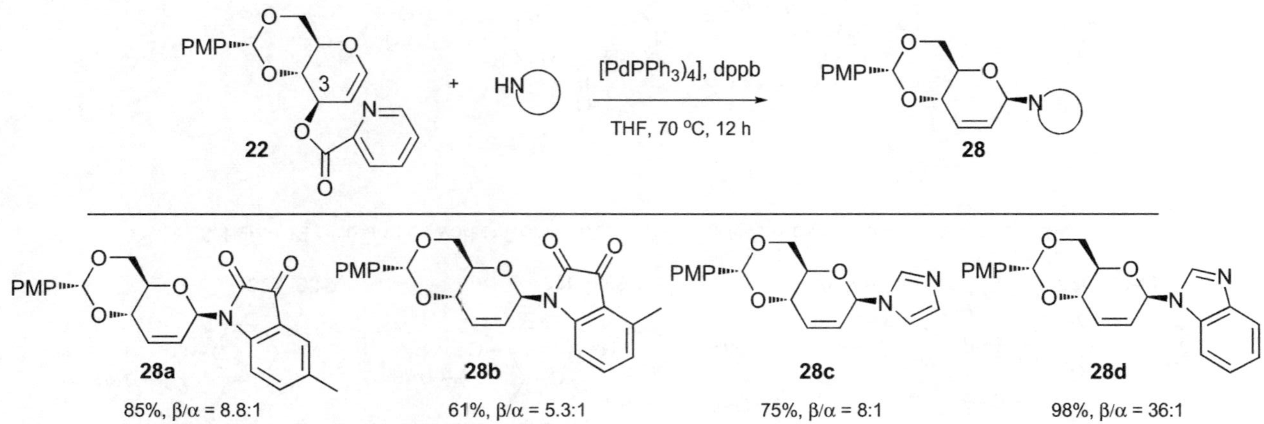

Scheme 11 Palladium-catalyzed *N*-glycosylation of N-heterocycles with C-3-*O*-picoloyl D-glycal **22**, leading to β-*N*-glycosyl derivatives **28**, in a stereoselective manner.

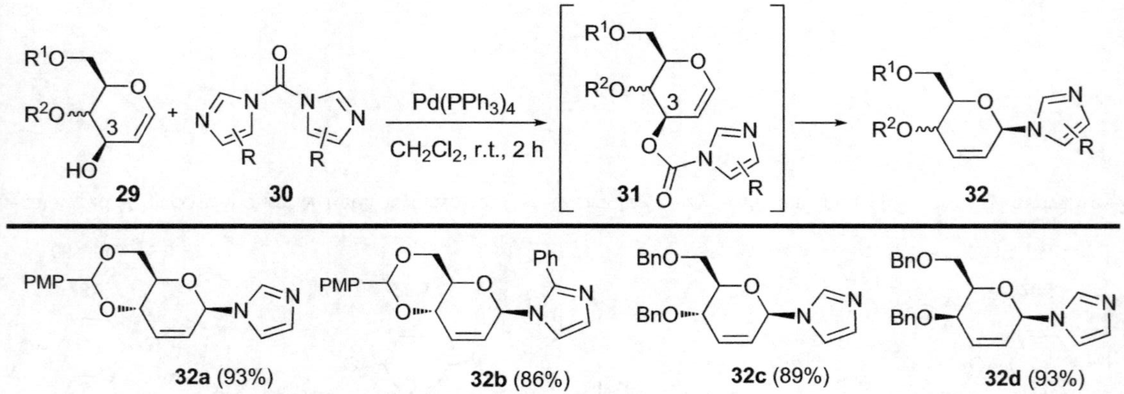

Scheme 12 Liu and coworkers' one-pot synthesis of *N*-glycosyl imidazole analogues **32**.

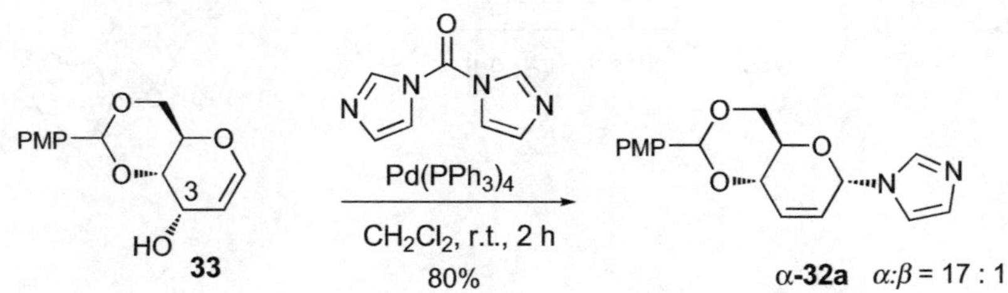

Scheme 13 Stereoselective synthesis of α-*N*-glycosyl imidazole α-**32a**.

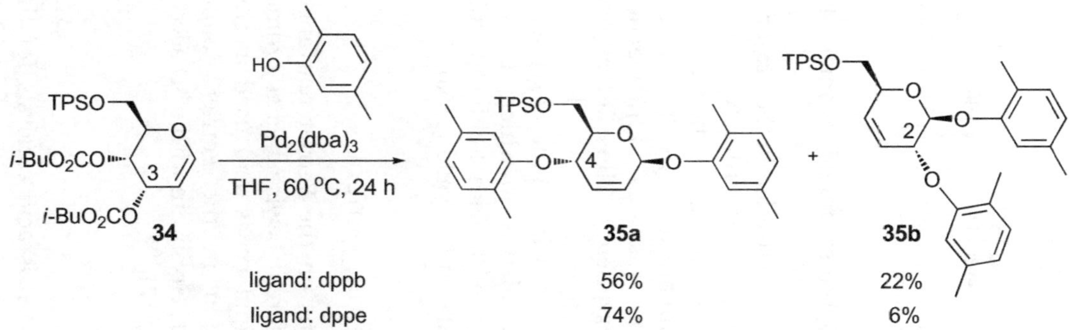

Scheme 14 Synthesis of 2,3- and 3,4-unsaturated *O*-glycosides **35a**,**b**.

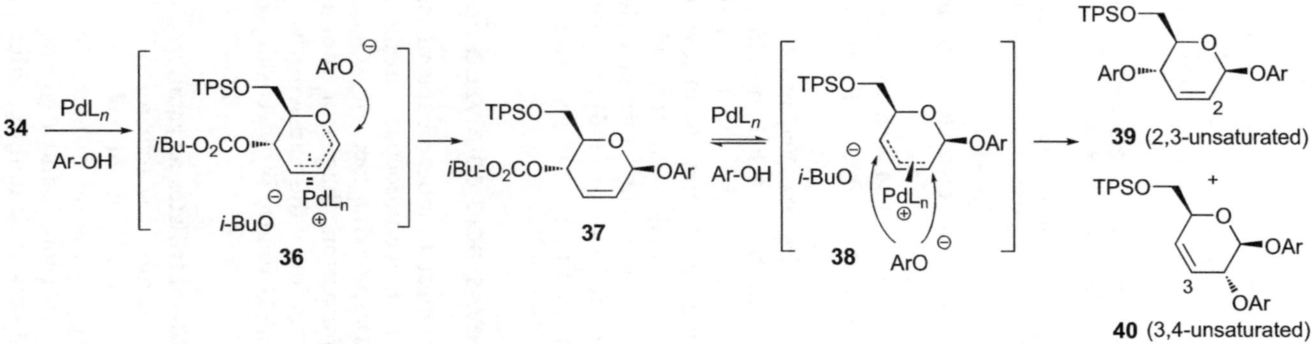

Scheme 15 Proposed reaction pathway for the formation of isomeric glycosides **39** and **40**.

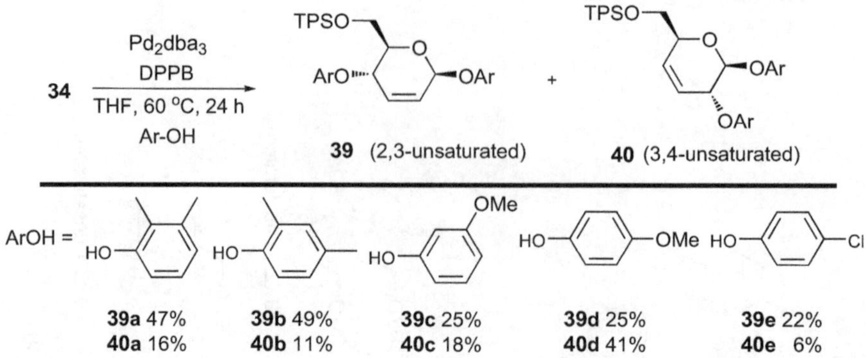

Scheme 16 Palladium catalyzed bis(*O*-arylation) reaction of D-glycal derivative **34** with different phenols (yields of isolated compounds are shown).

2,3-unsaturated glycoside **39** and its 3,4-unsaturated isomer **40**, respectively.[15]

The reaction of **34** with a variety of phenols was studied in the presence of the less discriminating ligand dppb (Scheme 16). Accordingly, reaction with 2,3-dimethylphenol and 2,4-dimethylphenol gave the isomeric glycoside pairs **39a**, **40a**, and **39b**, **40b**, in similar overall yields, 63% and 60%, respectively. A striking different behavior was noted in the reaction of **34** with 3-methoxyphenol and 4-methoxyphenol. Thus, 3-methoxyphenol gave **39c** and **40c** in low yield (43% combined yield), the former being the major isomer, whereas reaction with 4-methoxyphenol produced with good overall yield (66%) a mixture in which the 3,4-unsaturated isomer (**40d**) prevailed. Reaction of **34** with 4-chlorophenol furnished a mixture of glycosides **39e** and **40e** in low yield (28%).[15]

3 Lewis-acid catalyzed Ferrier rearrangement

The highest number of recent contributions on the Ferrier rearrangement had, again, focused on the use of different Lewis acids as promotors. Despite this fact, $BF_3 \cdot OEt_2$, the original Lewis acid used in this transformation, has still found application in the Ferrier double glycosylation of the acid-sensitive carotenoid zeaxanthin with tri-*O*-acetyl-D-glycal **41** to give bis(glycoside) **42** (Scheme 17).[16]

3.1 Metal triflates as catalysts

The continuous search for mild, efficient, Lewis acid catalysts for the Ferrier rearrangement has resulted in the screening of copper(II) triflate for *O*-glycosylation,[17] and zinc(II) triflate for *O*- and *S*-glycosylation,[18] *N*-glycosylation,[19] and *C*-glycosylation.[20] Additionally, the screening of several salts of yttrium, samarium, thulium, and gadolinium, have revealed that their corresponding triflates, $Y(OTf)_3$ (*O*- and *S*-glycosylation),[21] $Sm(OTf)_3$ (*O*- and *S*-glycosylation),[22] $Tm(OTf)_3$ (*O*-, *S*-, and *C*-glycosylation),[23] $Gd(OTf)_3$ (*O*-, *S*-, and *C*-glycosylation),[24] respectively, were the best reagents to effect the Ferrier rearrangement.

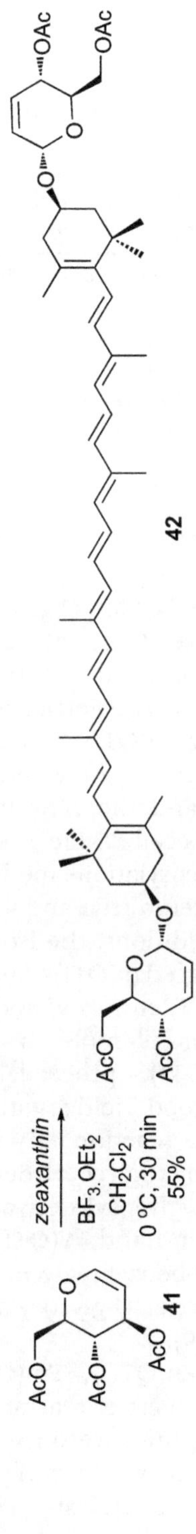

Scheme 17 Synthesis of a carotenoid glycoside, **42**, by Ferrier rearrangement.

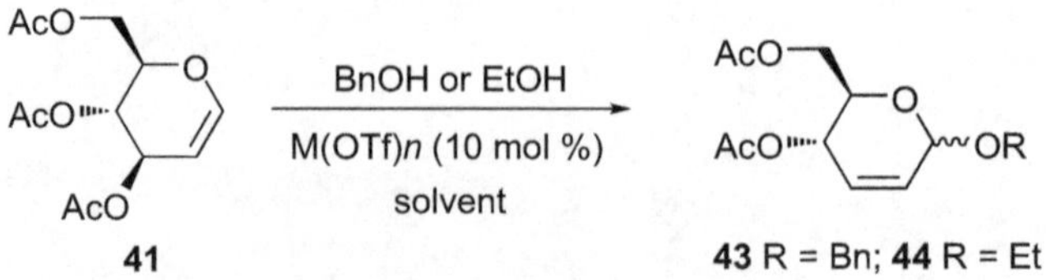

Scheme 18 Lewis-acid triflates as catalysts in the Ferrier glycosylation of **41**.

Table 1 A comparison of the catalytic effect of different Lewis-acid triflates (10 mol %) in the Ferrier rearrangement of tri-*O*-acetyl D-glycal **41**.

Entry	Catalyst	Alcohol	Solvent	Temperature	Time (min)	Yield	α : β Ratio	Ref.
i	$Cu(OTf)_2$	BnOH	CH_3CN	r.t.	10	98	7.3 : 1	17
ii	$Zn(OTf)_2$	BnOH	$ClCH_2CH_2Cl$	40 °C	240	98	7.3 : 1	18
iii	$Y(OTf)_3$	BnOH	CH_3CN	40 °C	10	85	9 : 1	21
iv	$Y(OTf)_3$	EtOH	CH_3CN	40 °C	15	91	17 : 1	21
v	$Sm(OTf)_3$	EtOH	CH_3CN	40 °C	10	88	33 : 1	22
vi	$Tm(OTf)_3$	EtOH	CH_3CN	40 °C	30	88	50 : 1	23
vii	$Gd(OTf)_3$	EtOH	CH_3CN	40 °C	30	91	8.1 : 1	24
viii	$Gd(OTf)_3$	BnOH	CH_3CN	40 °C	25	92	α-only	24

A comparative study on the Ferrier *O*-glycosylation of either benzyl alcohol or ethanol with tri-*O*-acetyl-D-glycal **41** mediated by these metal triflates (Scheme 18) has been compiled in Table 1. In all cases, the use 10 mol% catalyst was recommended. The Ferrier rearrangement of **41**, in the presence of BnOH, mediated by $Cu(OTf)_2$ or $Zn(OTf)_2$ gave similar results in terms of yield and α : β selectivity, however, $Zn(OTf)_2$ required a considerable higher temperature and more prolonged reaction time than $Cu(OTf)_2$ for the reaction to proceed (Table 1, entries *i*, *ii*). The best reaction conditions for Ferrier *O*-glycosylations mediated by $Y(OTf)_3$, $Sm(OTf)_3$, $Tm(OTf)_3$ and $Gd(OTf)_3$ were acetonitrile and 40 °C, respectively (Table 1, entries *iii*–*viii*). Under these conditions, the isolated glycoside yields were similar for all catalysts, although $Tm(OTf)_3$ and $Gd(OTf)_3$ appear to give highest α : β selectivities (Table 1, entries *vi* and *viii*).

Reaction of **41** with various phenols in the presence of $Y(OTf)_3$, $Sm(OTf)_3$, $Tm(OTf)_3$ or $Gd(OTf)_3$ proceeded to the corresponding *O*-glycosides, in moderate to good yields, with exclusive α-selectivity in most cases.[21–24] Conversely, the reaction **41** with 1-naphtol mediated by $Zn(OTf)_2$ gave an α,β-mixture of *C*-glycosyl derivatives (72%, 2 : 1 ratio), more likely by *O*-glycoside to-*C*-glycosyl derivative rearrangement in the reaction media.[6] On the other hand, $Y(OTf)_3$[21] and $Gd(OTf)_3$[24] were reported to activate 3,4,6-tri-*O*-benzyl-D-glycal **45**, less reactive than its acylated counterpart **41**, in Ferrier *O*-glycosylation processes leading to allylic glycosides **46** (Scheme 19).

Regarding *C*-glycosylation, $Zn(OTf)_2$, $Tm(OTf)_3$, and $Gd(OTf)_3$ have been reported to mediate on the Ferrier rearrangement in the presence of *C*-nucleophiles. Several nucleophiles were used in these studies including silanes and enolsilanes. A wider range of *C*-nucleophiles was screened in the case of $Zn(OTf)_2$ as catalyst.[20] For the sake of comparison,

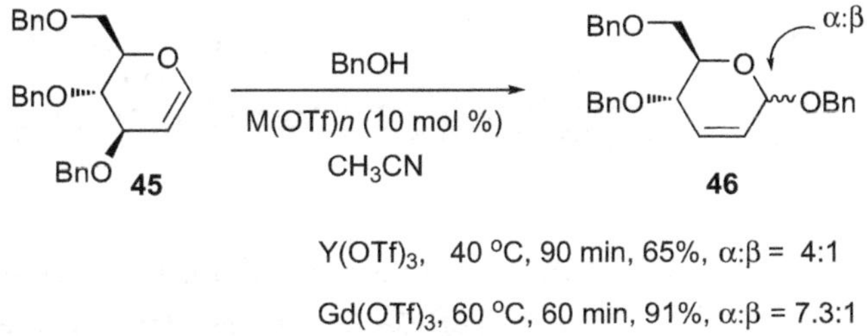

Scheme 19 Ferrier rearrangement of 3,4,6-tri-*O*-benzyl-D-glycal 45 catalyzed by $Y(OTf)_3$ or $Gd(OTf)_3$, in the synthesis of benzyl glucosides 46.

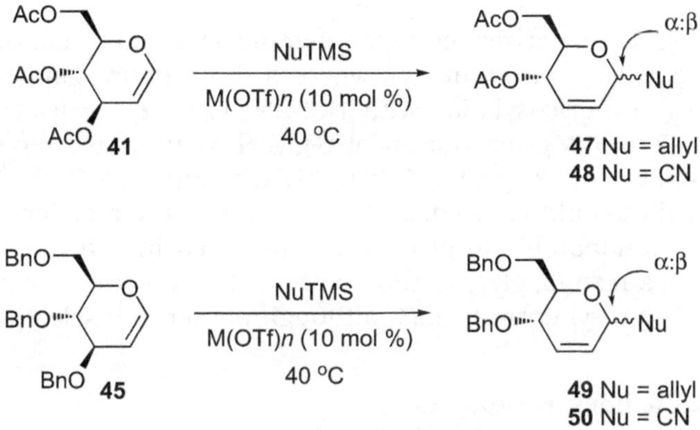

Scheme 20 *C*-Glycosylation of D-glycal derivatives 41 and 45, with allyltrimethylsilane and trimethylsilylcyanide.

Table 2 A comparison of the catalytic effect of different Lewis-acid triflates (10 mol %) in the Ferrier *C*-glycosylation of D-glycal derivatives 41 and 45.

Entry	Catalyst	Glucal	NuTMS	Solvent	Time (min)	Product, Yield, α : β ratio	Ref.
i	$Gd(OTf)_3$	**41**	AllylTMS	CH_3CN	40	**47**, 88%, α-only	24
ii	$Gd(OTf)_3$	**41**	TMSCN	CH_3CN	90	**48**, 85%, α-only	24
iii	$Tm(OTf)_3$	**41**	AllylTMS	CH_3CN	120	**47**, 88%, α-only	23
iv	$Tm(OTf)_3$	**41**	TMSCN	CH_3CN	120	**48**, 85%, α-only	23
v	$Zn(OTf)_2$	**41**	AllylTMS	$ClCH_2CH_2Cl$	60	**47**, 97%, α > 99%	20
vi	$Zn(OTf)_2$	**41**	TMSCN	$ClCH_2CH_2Cl$	60	**48**, 98%, 2 : 1	20
vii	$Gd(OTf)_3$	**45**	AllylTMS	CH_3CN	25	**49**, 68%, α-only	24
viii	$Gd(OTf)_3$	**45**	TMSCN	CH_3CN	130	**50**, 67%, 4 : 1	24
ix	$Zn(OTf)_2$	**45**	AllylTMS	$ClCH_2CH_2Cl$	120	**49**, 95%, α-only	20
x	$Zn(OTf)_2$	**45**	TMSCN	$ClCH_2CH_2Cl$	120	**50**, 98%, 2 : 1	20

the results of the reaction of D-glycal derivatives **41** and **45** with allyltrimethylsilane, and trimethylsilylcyanide leading to *C*-glycosyl derivatives **47–50** (Scheme 20), have been collected in Table 2. All the reactions were best carried out at 40 °C. Reactions with $Tm(OTf)_3$, and $Gd(OTf)_3$ were performed in acetonitrile, whereas 1,2-dichloroethane (DCE) was the solvent of choice when $Zn(OTf)_2$ was used as the catalyst. Higher

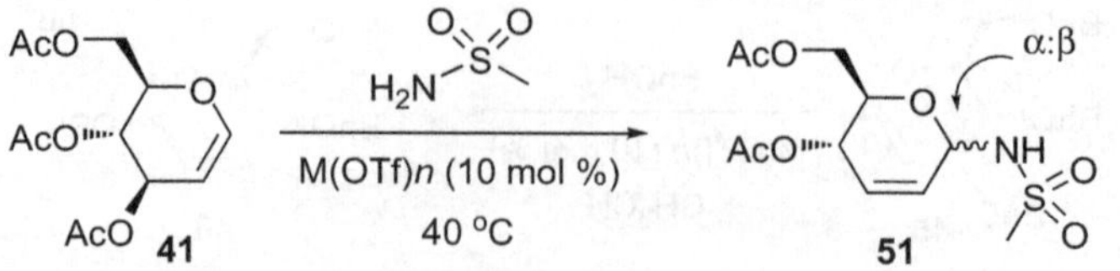

$Y(OTf)_3$, CH_3CN, 20 min, 85%, α-only

$Gd(OTf)_3$, CH_3CN, 84 min, 84%, α:β = 19:1

$Zn(OTf)_2$, $ClCH_2CH_2Cl$, 240 min, 92%, α:β = 3:1

Scheme 21 *N*-Glycosylation of methane sulphonamide with D-glycal **41**, under various metal triflate catalysts.

yields of *C*-glycosyl derivatives were generally obtained by use of $Zn(OTf)_2$ (Table 2, entries *v, vi, ix* and *x*), whereas this catalyst gave poorer α:β selectivities in *C*-glycosylation with TMSCN (Table 2, entries *vi* and *x*).

Finally, Ferrier *N*-glycosylation of D-glycal **41** was also reported with the assistance of $Zn(OTf)_2$,[19] $Tm(OTf)_3$,[23] and $Gd(OTf)_3$.[24] Several *N*-nucleophiles could be used, and the results of the reaction of **41** with methane sulphonamide to give **51** are displayed in Scheme 21. In line with previous results, glycosylation with $Zn(OTf)_2$ as catalyst gave better yields of *N*-glycosyl derivative **51**, although poorer α : β selectivities.

3.2 Metal halides as catalysts

The use of several metal halides has been reported in connection with Ferrier rearrangement processes. Among them, $RuCl_3$, $ZnBr_2$ and supported $FeCl_3$, have been used in *O*-glycosylation reactions, whereas *C*-glycosylation of acetylated glycals has been described in the presence of $RuCl_3$, $FeCl_3$ and SnX_4 (X = Cl, Br).

The use of $RuCl_3$ as catalyst in the Ferrier *O*-glycosylation has been independently reported by two research groups.[25,26] Chen and Lin described the use of $RuCl_3 \cdot 3H_2O$,[25] whereas non-hydrated $RuCl_3$ was employed by Krishna, Kashyap and coworkers,[26] which resulted in some differences in efficiency. Thus, reaction of benzyl alcohol with D-glycal **41** in acetonitrile, in the presence of $RuCl_3 \cdot 3H_2O$ (10 mol%) took place at 40 °C (10 min) to give benzyl glycoside **43** in 81% yield (α : β, 7 : 1) (Scheme 22a).[25] On the other hand, when $RuCl_3$ (10 mol%) was used as catalyst, the reaction was completed within five minutes at room temperature (r.t.). Further optimization concluded with the synthesis of benzyl glycoside **43** (98% yield, α : β = 6.7 : 1) with $RuCl_3$ (2 mol%) in acetonitrile at r.t. within 15 minutes (Scheme 22b).[26] Microwave-assisted $ZnBr_2$ catalyzed *O*-glycosylation has been applied to hindered phenols and alcohols.[27] Formation of benzyl glycoside **43** (93% yield, α : β, 3 : 1) under this reaction conditions was carried out in $CHCl_3$ under microwave (MW) irradiation (Scheme 22c).[27] Finally, recyclable $FeCl_3 \cdot 6H_2O$ supported on carbon ($FeCl_3 \cdot 6H_2O$/C) was used in the preparation of 2,3-unsaturated *O*- and *S*-glycosides.[28] The method when applied to BnOH as nucleophile furnished glycoside **43** in excellent yield (99%) and with good α : β selectivity (α : β, 8 : 1) (Scheme 21d).[28]

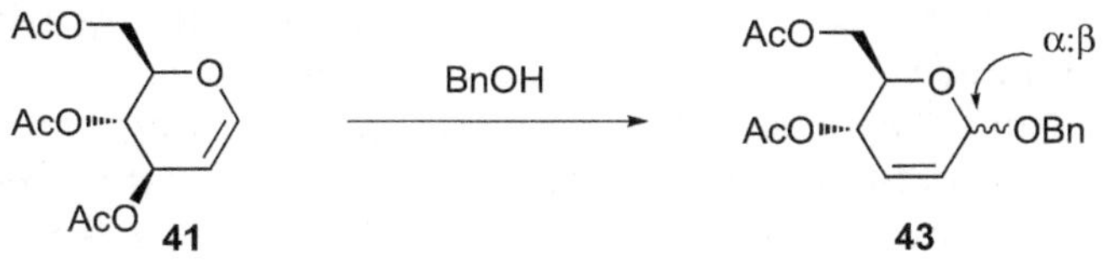

(a) $RuCl_3{\cdot}3H_2O$, 10 mol %, CH_3CN, 40 °C, 10 min, 81%, α:β = 7:1; Ref. 25
(b) $RuCl_3$, 2 mol %, CH_3CN, r.t., 15 min, 98%, α:β = 6.7:1; Ref. 26
(c) $ZnBr_2$, 10 mol %, $CHCl_3$, MW (400W), 72 °C, 25 min, 93%, α:β = 3:1; Ref. 27
(d) $FeCl_3{\cdot}6H_2O/C$, 5 mol %, CH_2Cl_2, r.t., 10 min, 99%, α:β = 8:1; Ref. 28

Scheme 22 Metal halides mediated Ferrier type *O*-glycosylation of benzyl alcohol with D-glycal **41**.

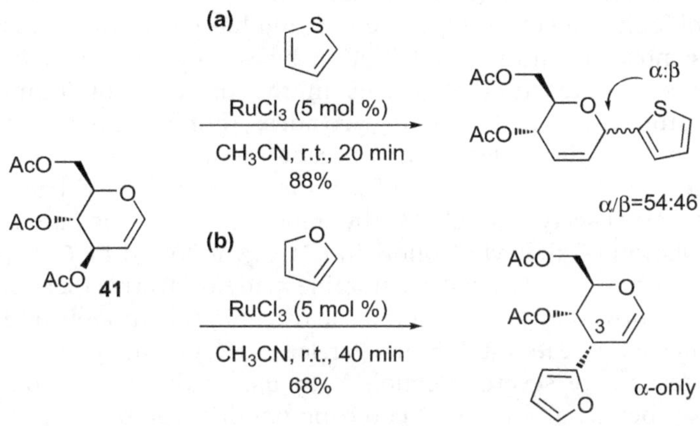

Scheme 23 $RuCl_3$ mediated Ferrier type *C*-glycosylation of thiophene and furan with D-glycal **41**.

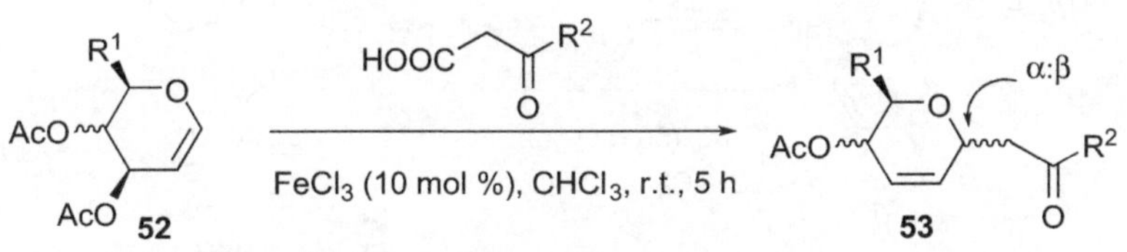

Scheme 24 Decarboxylative *C*-glycosylation of β-keto acids with acylated glycals.

C-Glycosylation of several acetylated and benzylated glycals has also been described with $RuCl_3$ (5 mol%) as catalyst (CH_3CN, r.t.).[29] Accordingly, various nucleophiles such as allyl trimethylsilane, trimethylsilyl cyanide, as well as thiophene furnished the expected *C*-glycosyl derivatives. Conversely, analogous reaction of **41** with furan resulted in the formation of a C-3 adduct, rather than the expected *C*-glycosylated compound at C-1 (Scheme 23).[29,30]

Liu and coworkers have described the use of $FeCl_3$ in an interesting decarboxylative *C*-glycosylation of glycals with β-keto acids.[31] The reaction was applied to various acetylated glycals, *i.e.* **52**, and gave moderate yields of anomeric mixtures of β-keto functionalized glycosides, *i.e.* **53** (Scheme 24).

Good α : β selectivities were only observed in the reaction of D-galactal derivatives with bulky β-keto acids bearing phenyl moieties.

Wieczoreck and Thiem have also reported an interesting transformation related to Ferrier-type *C*-glycosylation. They found that the reaction of tri-*O*-acetyl-D-glycal **41** with phenyl acetylene in the presence of $SnBr_4$ (CH_2Cl_2, −45 °C) produced the pure α-anomer **54** embodying a C–C linked (*E*)-2-bromo-2-phenylvinyl group as the sole reaction product (Scheme 25).[32] A related reaction with $SnCl_4$ provided the analogous derivative **55**.

3.3 Iodine and iodine-related promotors

Recognition that iodine could be used as a mild and effective Lewis acid to induce Ferrier rearrangement came in the mid-1990s.[33,34] Since then, the use of iodine in Ferrier-type glycosylation has been firmly established. In this context, an initial contribution by Saeeng's group had already shown that the reaction of glycals in the presence of iodine/cupric acetate produced 2-deoxy-2-iodoglycosides, rather than the expected 2,3-unsaturated glycosides.[35] In a recent report, the same group has described the one-pot synthesis of 2,3-unsaturated-glycosyl triazoles, *i.e.* **56**, from tri-*O*-acetyl glycal **41**, by tandem iodine-mediated Ferrier *O*-glycosylation of alkynyl alcohols in CH_2Cl_2, followed by CuI triggered click reaction, in CH_3CN, with a suitable azide derivative (Scheme 26).[36]

More recently, they have explored the use of iodine monobromide (IBr) as an alternative catalyst for the Ferrier rearrangement of D-glycal **41** in the presence of several alcohols.[37] By use of this methodology (IBr, 20 mol%), benzyl glycoside **43** could be obtained by reaction of benzyl

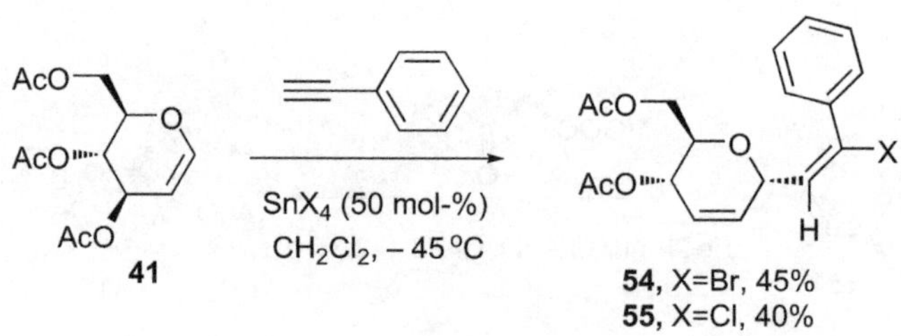

Scheme 25 *C*-Vinylation by Ferrier rearrangement.

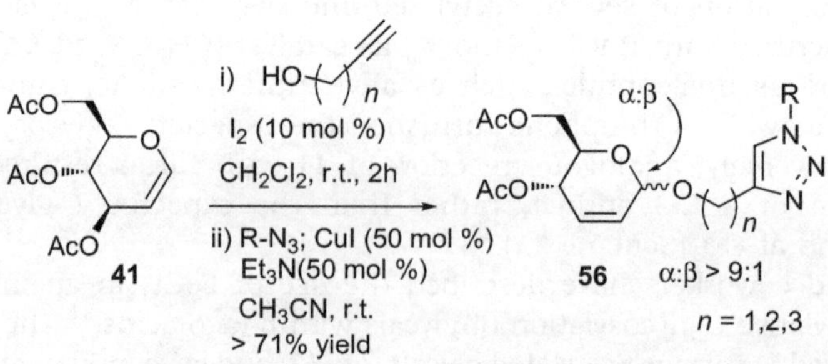

Scheme 26 Sequential *O*-glycosylation-Click reaction.

alcohol with tri-*O*-benzyl-D-glycal **41** in toluene/CH_2Cl_2 (1 : 1) (0–20 °C) in 89% yield (α : β = 4 : 1).

Iodine has been used in the *O*-glycosylation of Fmoc protected amino acid derivatives (serine, threonine and tyrosine) with acylated glycals in CH_2Cl_2, to give the desired *O*-glycosylated Fmoc amino acid derivatives in yields ranging from 33% to 89%, and with α : β selectivities higher than 9 : 1.[38] Iodine has also been reported as catalyst in the aza-Ferrier rearrangement of glycals with amides. Then, the use of equimolar amounts of iodine (1.0 equiv.), in CH_2Cl_2 at room temperature, was recommended to effect the Ferrier rearrangement of a series of carbamates, tosyl and mesyl amides with D-glycal derivatives **41** and **45**.[39]

4 Ferrier rearrangement mediated by acids

Zhang's group has reported in detail their previous findings on the use of sulfuric acid immobilized on silica gel (H_2SO_4–SiO_2) in the Ferrier rearrangement.[40] These studies have focused on the reaction of 3,4,6-tri-*O*-benzyl-D-glycal **45** with alcohols, thiols, and 2-naphtol as acceptors, in CH_2Cl_2.[41] Interestingly, the Ferrier rearrangement failed in the presence of phenol, however, the reaction with 2-naphtol took place to give a "rearranged" *C*-glycosyl naphtol (62%, α : β > 19 : 1).

Cyanuric chloride (TCT) has been found to promote the Ferrier rearrangement of D-glycal **41** in the presence of alcohols, in CH_2Cl_2 at room temperature (Scheme 27a).[42] The catalyst role in this process was assigned to HCl, which is formed along with cyanuric acid in the reaction of TCT with adventitious moisture.

Finally, 3,5-dinitrobenzoic acid (3,5-DNBA) was reported to catalyze the Ferrier rearrangement of acetylated glycals in the presence of alcohols and thiols in refluxing acetonitrile (Scheme 27b).[43]

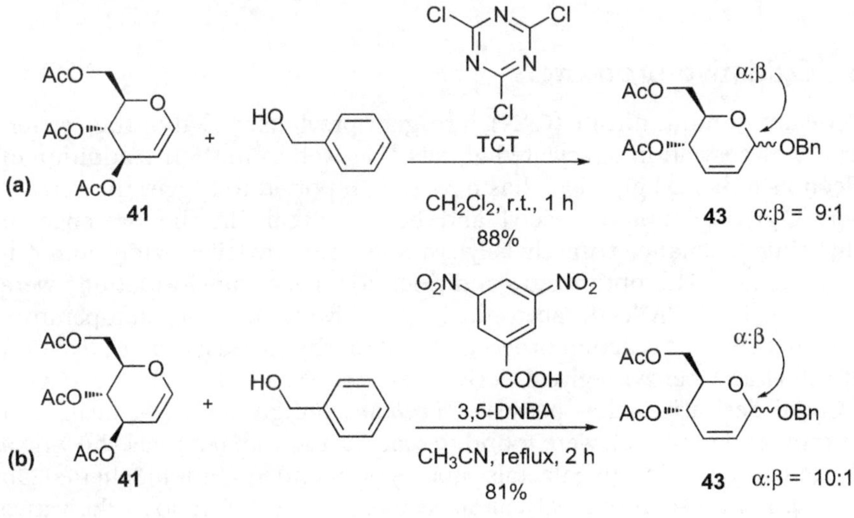

Scheme 27 Ferrier rearrangement of D-glycal **41** in the presence of benzyl alcohol.

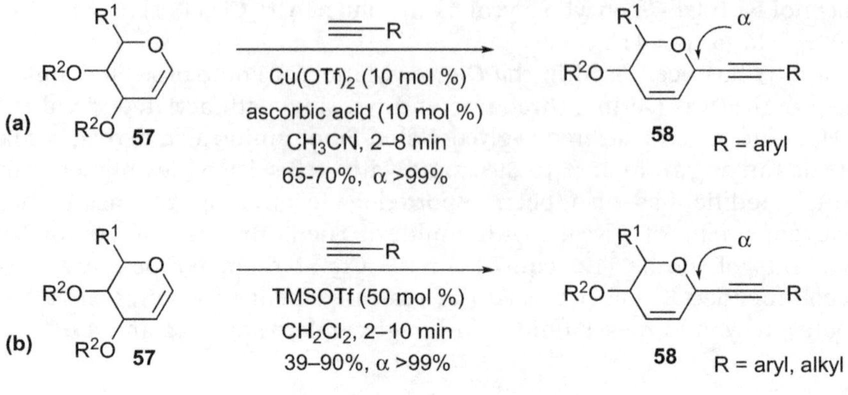

Scheme 28 *C*-Alkynylation of glycals.

4.1 *C*-alkynylation *via* Ferrier rearrangement

C-Alkynylation by Ferrier *C*-glycosylation has attracted considerable attention and deserves special attention. Two methods for α-selective *C*-alkynylation of glycals, *i.e.*, **57**→**58**, have recently been reported (Scheme 28). One of the methods involves the use of ascorbic acid and $Cu(OTf)_2$ (Scheme 28a),[44] whereas the second procedure takes place in the presence of TMSOTf (Scheme 28b).[45] Both methods involve the reaction of the free acetylene with the glycal, and the "active" catalyst appears to be, in both cases, TfOH. Regarding the nucleophilic species, a copper acetylene seems to operate under the first reaction conditions, while an *in situ* generated trimethylsilyl acetylene seems to be the nucleophile in the second method. Both protocols gave good *C*-glycosylation yields, *i.e.* yields for compounds type **58**, with a variety of glycals. Regarding the acetylene counterpart, the ascorbic acid/$Cu(OTf)_2$ method has only been applied to aryl acetylenes, whereas the TMSOTf method permitted the use of aliphatic and aromatic acetylenes.

5 Oxidative promoters

Ceric ammonium nitrate (CAN), a reagent previously used in the Ferrier-type *O*-glycosylation of acylated glycals,[46] as well as in the 1,2-addition of alcohols to benzyl glycals,[47] has now been reported to trigger the Ferrier-type *C*-glycosylation of acetyl and benzyl glycals in the presence of allyltrimethylsilane, trimethylsilylcyanide, trimethylsilyl azide, and triethylsilane.[48] The optimized conditions for these transformations were 1.0 equiv. of CAN, in acetonitrile as solvent at room temperature (Scheme 29). The reaction proceeded well for the stereoselective allylation of acetyl and benzyl D-glycal derivatives, *e.g.* **41, 45, 59** (Scheme 29a–c). D-Glycal derivatives that possess D-*arabino* configuration (D-glucals, not recommended name), were found to react faster than D-glycals showing a D-*lyxo* configuration (D-galactals, not recommended name) (Scheme 29b and c). Furanose derived glycal **60** reacted to give *C*-furanosyl derivative **62** (Scheme 29d). Reactions with TMSCN were less stereoselective than

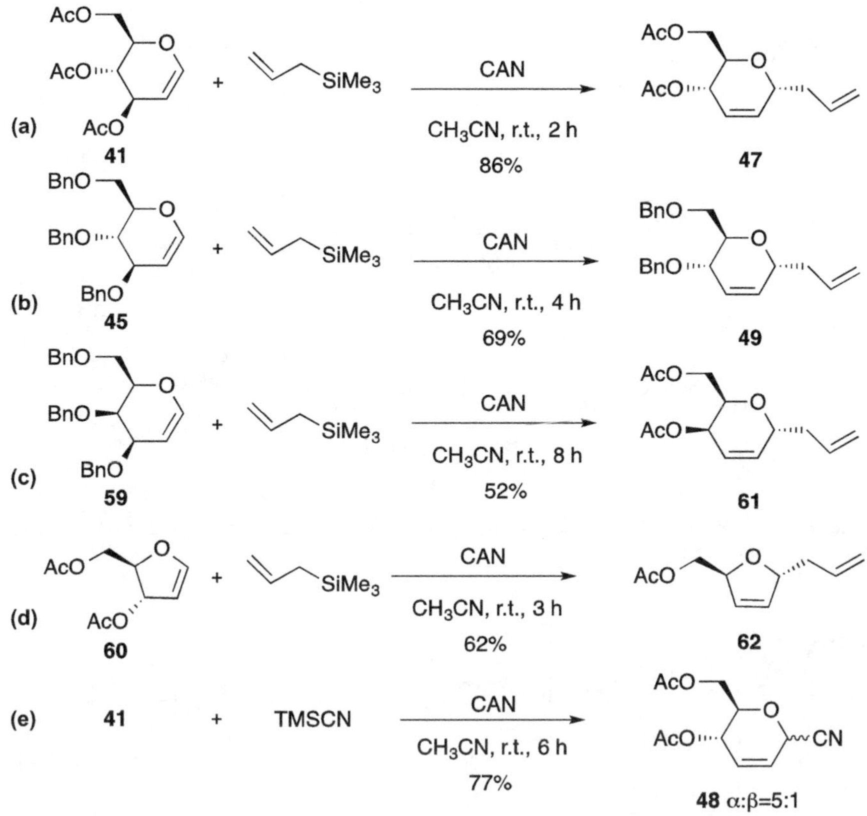

Scheme 29 Reaction of glycals with carbon nucleophiles in the presence of CAN.

those with allyltrimethylsilane, and D-glycal **41** furnished a 5 : 1 mixture of the α,β-anomers **48** (Scheme 29e).

6 Miscellaneous activation protocols

Ferrier rearrangement on a series of acetylated glycals in the presence of alcohols has been performed in deep eutectic solvents (DES).[49] Best results were obtained in a DES made from choline-malonic acid,[50] at 80 °C. Accordingly, allylic glycosides, *e.g.* **43**, were obtained in good yields although with poor α : β anomeric selectivity (Scheme 30a). Interestingly, the DES could be recycled up to four times without significant loss in activity.

A zirconium catalyst, generated electrochemically *in situ*, has shown to be effective in promoting Ferrier rearrangement of acetylated glycals in the presence of thiols, thiophenols, and allyltrimethylsilane, thus leading to *S*-glycosides and *C*-glycosyl derivatives, *e.g.* **63** and **47**, respectively (Scheme 30b and c).[51]

Two reports dealing with the Ferrier rearrangement of 3-OH glycals have appeared. In the first protocol, 4,6-di-*O*-benzyl-3-hydroxy-D-glycal **64** was reacted with diethyl phosphorochloridite in the presence of Et_3N (3.0 equiv.) to produce a transient 3-*O*-diethoxyphosphanyl-D-glycal

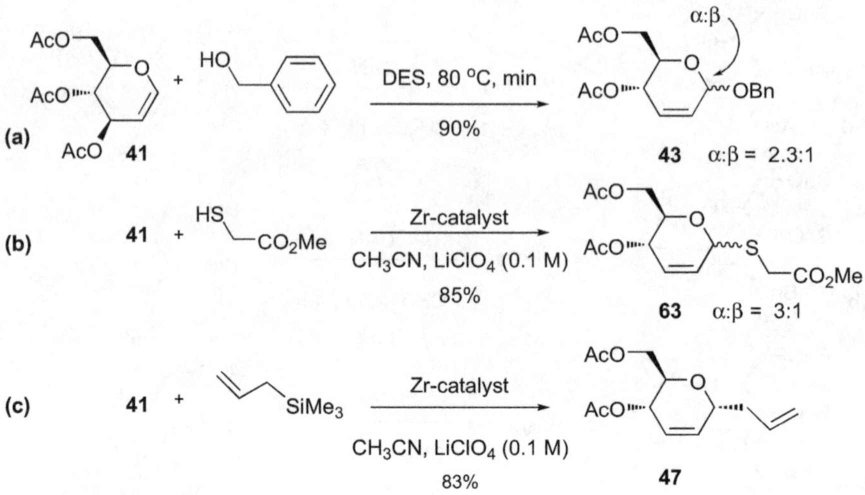

Scheme 30 Ferrier rearrangement of D-glycal **41** with D-*arabino* configuration (D-glucal, not recommended name).

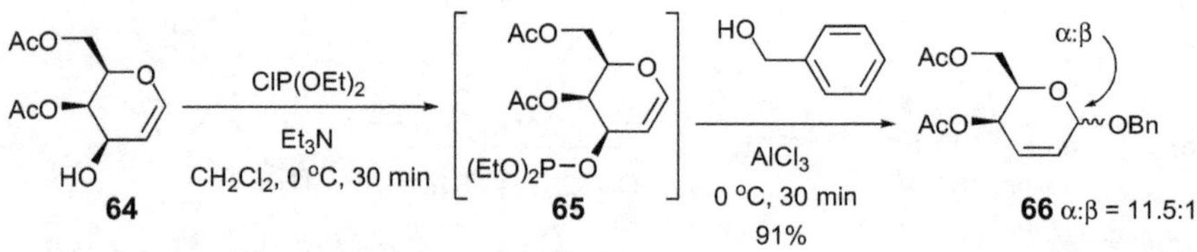

Scheme 31 Ferrier rearrangement of D-glycal **64** with D-*lyxo* configuration (D-galactal, not recommended name) *via* transient formation of a 3-*O*-diethoxyphosphanyl intermediate **65**.

intermediate, *i.e.* **65**, that experienced the Ferrier rearrangement with various alcohols in the presence of $AlCl_3$.[52] The two-step protocol was carried out in CH_2Cl_2 at 0 °C, in a one-pot operation to yield allylic glycosides, *e.g.* **66**, in good to excellent yields, and with excellent α-stereoselectivity (Scheme 31).

The second report dealt with a Ferrier-like process, the self-coupling of 3-OH glycal derivatives, *e.g.* **67**, under phase-transfer conditions. The reaction takes place through reactive 3-OTs intermediates, *e.g.* **68**, which experience the addition of the remaining 3-OH glycal to provide di-unsaturated disaccharides, *e.g.* **69** (Scheme 32).[53]

Nitrosyl tetrafluoroborate ($NOBF_4$), a catalyst introduced by Misra and coworkers for the Ferrier rearrangement,[54] has recently been employed in the sulfonamidoglycosylation of various glycals with *N*-(*O*-*tert*-butyldiphenylsilyl) hydroxysulfamide.[55] The reactions were carried out in refluxing CH_2Cl_2, or under microwave assisted heating. The ensuing *N*-glycosyl derivatives displayed good *in vitro* inhibitory activity against four human carbonic anhydrase isoforms.

Vankar and coworkers have introduced $AuCl_3$ in combination with phenylacetylene ($AuCl_3$/PhCCH) as an interesting new catalyst-system for the Ferrier rearrangement of glycals and 2-acetoxymethylglycals,

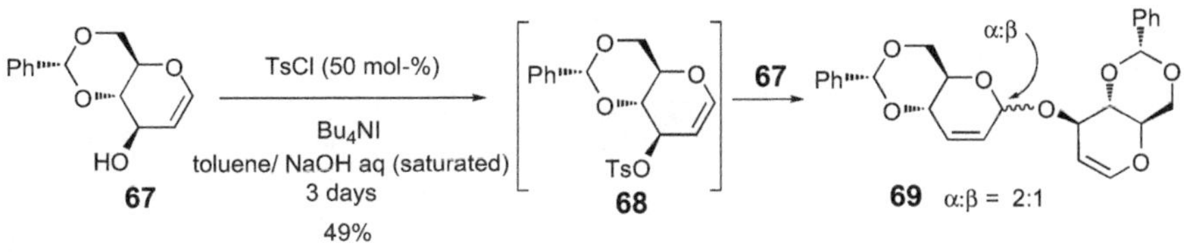

Scheme 32 Self-Ferrier rearrangement of 3-hydroxy glycal derivatives with TsCl under phase transfer conditions.

vide infra.[56] Based on previous knowledge on the activation of triple bonds by late transition metal cations, Vankar's group has found that phenylacetylene in addition to $AuCl_3$, was able to improve the overall outcome of the Ferrier rearrangement (reaction rate, yield, α/β selectivity) compared to when $AuCl_3$ had been used alone. For instance, reaction of tri-*O*-acetyl-D-glycal **41** in the presence of $AuCl_3$ (1 mol%) and phenylacetylene (1 mol%) in CH_2Cl_2 using allyl alcohol as the nucleophile was complete within 15 min (92%, $\alpha/\beta = 10:1$), to give glycoside **47**, compared to 9 h when $AuCl_3$ was used as catalyst (Scheme 33).[57] Reaction intermediates **I** and **II**, (Scheme 33) were tentatively suggested by the authors to describe the role of phenylacetylene in the catalytic cycle leading to allylic oxycarbenium intermediate **2** (Scheme 33). This catalyst-system was also used with sulfur, carbon, and nitrogen nucleophiles leading to the corresponding *S*-, *C*- and *N*-glycosyl derivatives, respectively.

7 Miscellaneous transformations of glycals

As a continuation of their interest in $Al(OTf)_3$ mediated transformations of glycals,[58] Williams' group has reported the preparation of several tricyclic derivatives, *e.g.* **73**, by $Al(OTf)_3$-triggered reaction of 3,4,6-tri-*O*-acetyl-D-glycal **70** in 1,2-dichloroethane (DCE), in the presence of various *para*-substituted phenols (Scheme 34).[59] For this transformation to occur, the authors proposed a reaction pathway involving an initial Ferrier rearrangement to give an allylic glycoside, *e.g.* **72**, which would evolve to the tricyclic derivative **73**, by a Friedel–Crafts C–C bond forming process. The authors were able to rule out the intermediacy of a 2-deoxy glycoside in the overall process, *i.e.* **70**→**73**, since heating of **71** in the presence of $Al(OTf)_3$ failed to produce the tricyclic benzopyran **73**. In this context, it is pertinent to mention that Booma and Balasubramanian had shown that 2-deoxy glycoside **71** could be transformed into **73** upon treatment with $BF_3 \cdot OEt_2$.[60]

A remarkable transformation, involving a Ferrier rearrangement, leading to pentacyclic benzopyran-fused pyranoquinolines has recently been reported by Kinfe's group.[61] They have described a three-component (glycal, aniline, salicylaldehyde) domino process that involves an initial Ferrier *C*-glycosylation of an *in situ* formed imine/enaminone, *e.g.* **74a,b**, with a glycal, *e.g.* **41**, leading to a 2,3-unsaturated *C*-glycosyl derivative, *e.g.* **75**, which undergoes a stepwise intramolecular cyclization, in the presence of the Lewis acid, to produce a pentacyclic structure, *e.g.* **76** (Scheme 35). The process, which is illustrated in

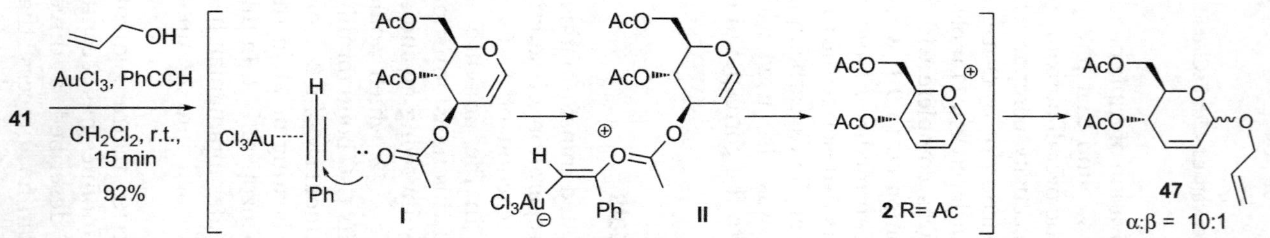

Scheme 33 Vankar's relay catalysis with the $AuCl_3$/PhCCH system in the Ferrier rearrangement of **41** in the presence of allyl alcohol.

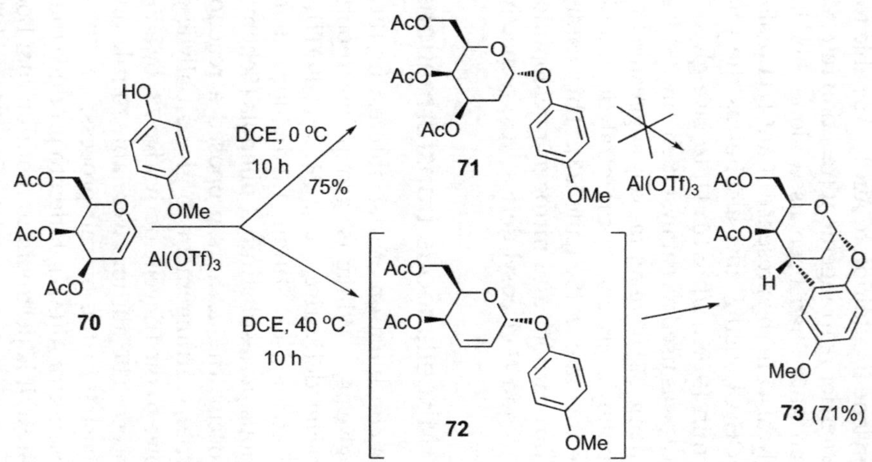

Scheme 34 Conversion of D-glycal **70** into chiral bridged benzopyran **73**, by sequential Ferrier rearrangement/Friedel–Crafts reactions.

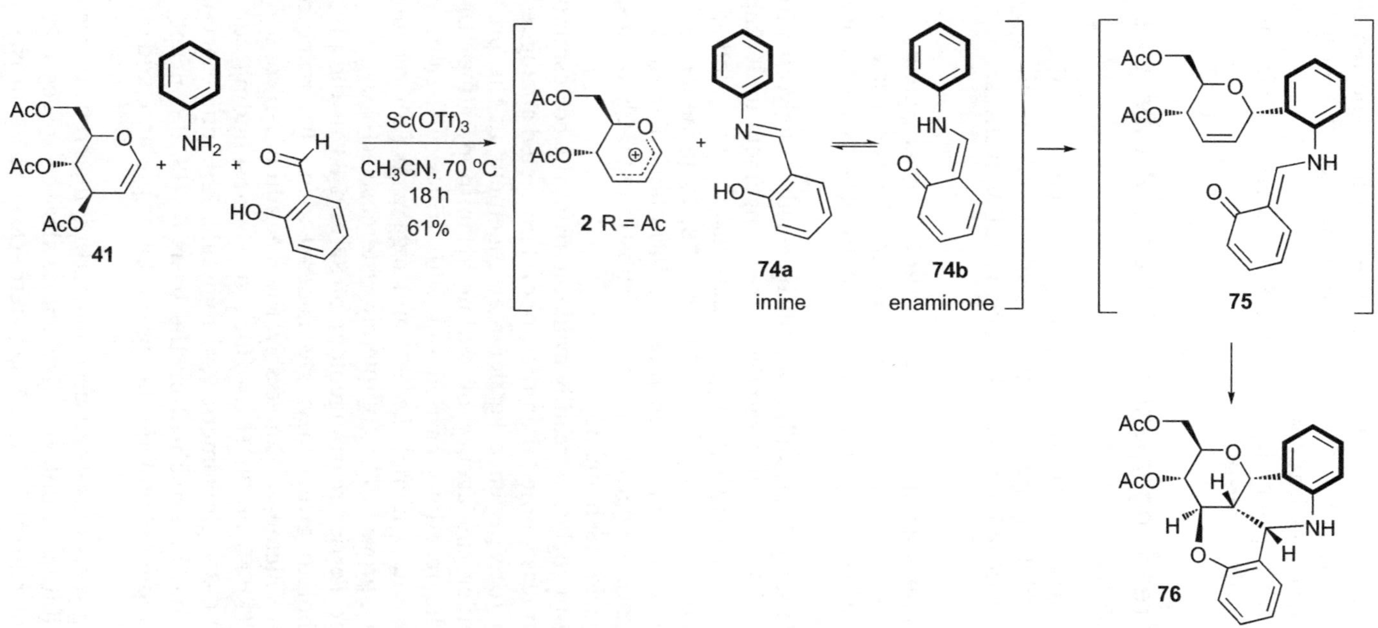

Scheme 35 Kinfe's three-component domino reaction of D-glycal **41**, aniline, and salicylaldehyde.

Scheme 35 with D-glycal **41**, was successfully applied to a variety of acetylated glycals in combination with differently substituted anilines, and salicylaldehydes. In this process four new stereocenters were generated in a completely stereoselective manner. The reaction proceeded with moderate to good yields with anilines or salicylaldehydes possessing electron-withdrawing groups. Conversely, the presence of electron-donating groups in the aromatic partners provided only traces of the pentacyclic derivatives.

8 Ferrier rearrangement on C-1 or C-2 substituted glycal systems

The incorporation of substituents at position C-1 or C-2 of glycals generates functionally rich substrates that are able to undergo Ferrier rearrangement to generate synthetically useful derivatives.

8.1 Ferrier rearrangement on C-2 substituted glycal systems

In the last two years, several reports on the Ferrier rearrangement of 2-halo, 2-nitro, 2-acetoxy and 2-acetoxymethyl glycals have appeared in the literature.

2-Iodoglycals, *e.g.* **77**, readily available by reaction of glycals with NIS in the presence of $AgNO_3$, undergo Ferrier rearrangement under the agency of $BF_3 \cdot OEt_2$ in CH_2Cl_2, to give 2,3-unsaturated 2-iodoglycosides (anomeric ratio 5 : 1), in good yields, *e.g.* **78, 79** (Scheme 36).[62] These derivatives can be engaged on inter- or intramolecular Heck coupling reactions to give the corresponding 2-*C*-substituted *O*-glycosides, *e.g.* **80, 81**, in good yields (Scheme 36).

Acylated 2-nitroglycals, readily available from the corresponding glycals[63] and far less studied than their benzyl protected analogues,[64] have been shown to experience Ferrier rearrangement in the presence of alcohols, under the agency of 4-dimethylaminopyridine (DMAP) in CH_2Cl_2.[65] The reactions carried out with tri-*O*-acetyl-2-nitro-D-glucal (not recommended name), *i.e.* **82**, and D-galactal (not recommended name), *i.e.* **83**, proved to be completely α-stereoselective with a series of alcohols. The Ferrier rearrangement of **82** and **83** in the presence of benzyl alcohol to give benzyl glycosides **84** and **85**, respectively, is illustrated in Scheme 37. Conversely, the reaction of 2-nitro D-glycals **82** or **83** with $TMSN_3$ or PhSH provided the corresponding equatorially oriented (β-) C-3 regioisomeric 2-nitroglycals. This difference in regioselectivity could be explained on the basis of HSAB concept,[66] where azide and thiophenol moieties being soft bases prefer to add on the soft acid center C-3.

2-Acetoxy glycals have been used as starting materials to access compounds exhibiting *N*-linked sugar enone derived glycosyl groups *via* Ferrier rearrangement.[67] Thus, 2,3,4,6-tetra-*O*-acetyl-2-hydroxy-D-glycal **86** reacted with a range of amines to give anomeric mixtures of *N*-glycosyl derivatives **88**, *via* intermediate enol-acetate **87** (Scheme 38). An excess (4.4 equiv.) of $BF_3 \cdot OEt_2$ proved to be the best promoter for this

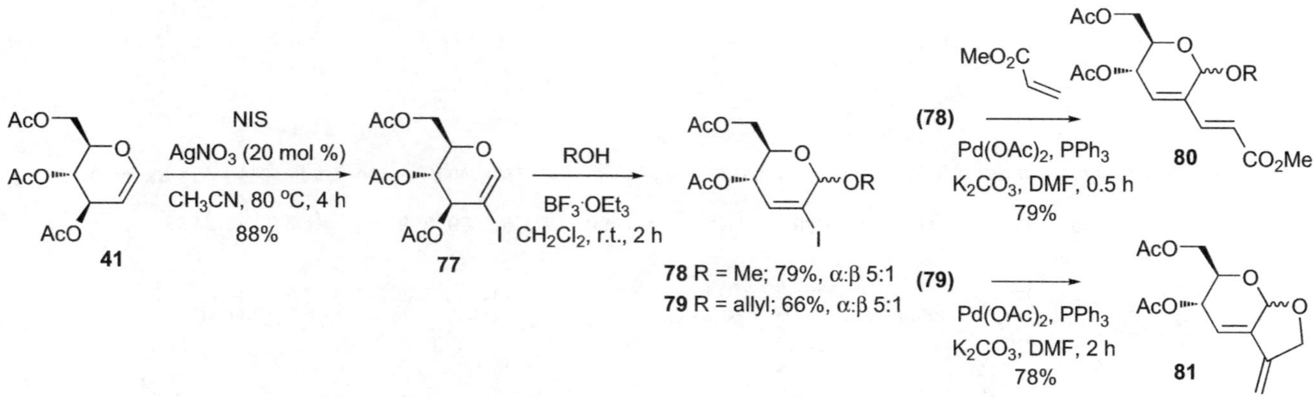

Scheme 36 Vankar's synthesis of 2-iodoglycals and their Ferrier rearrangement.

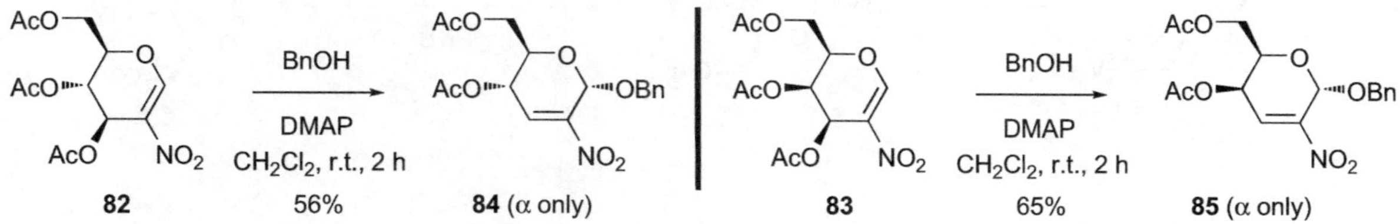

Scheme 37 Vankar's Ferrier rearrangement of 2-nitroglycals.

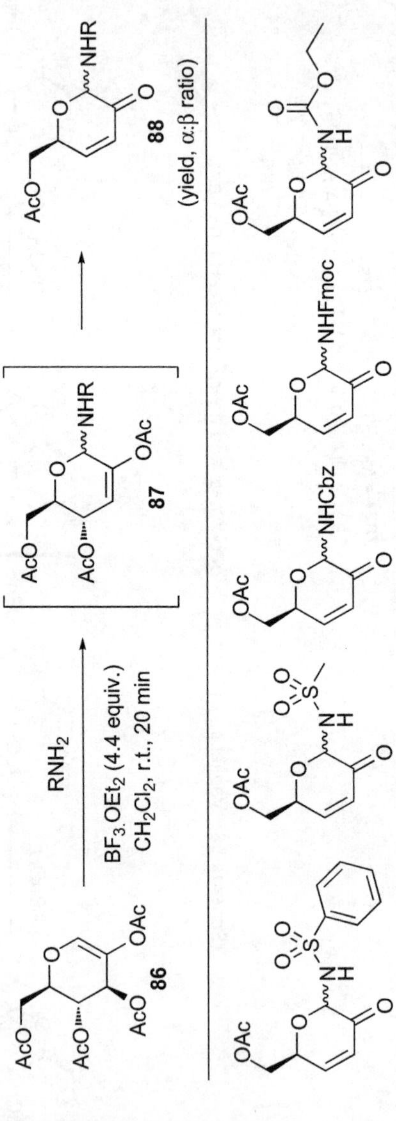

88a (73%, 5.2:1) **88b** (68%, 9:1) **88c** (54%, 3.4:1) **88d** (46%, 3:1) **88e** (55%, 9:1)

Scheme 38 Synthesis of *N*-glycosyl derivatives, embodying sugar enone derived glycosyl groups, by Ferrier rearrangement.

transformation (CH_2Cl_2, r.t.). Notably, less $BF_3 \cdot OEt_2$ loading led to the isolation of the Ferrier rearranged 2,3-unsaturated compounds type **87**, as the major products. This methodology has recently been applied in the synthesis of bicyclic glycosidase inhibitors.[68]

2-*C*-Acetoxymethyl glycals, *e.g.* **89**, first described by Booma and Balasubramanian in 1993,[69] are interesting substrates for Ferrier-like glycosylation. Their reaction is likely to proceed *via* an intermediate allylic oxycarbenium ion, *i.e.* **90**, generated by activation of the C-2′ allylic acetate moiety, which presents two reactive positions, C-1 and C-2′. Ferrier-like glycosylations with **89** have been reported to give allylic glycosides, *i.e.* **91**.[6,69] Along this line, Vankar's $AuCl_3$/PhCCH catalyst-system has been reported to activate glycals **89** in the presence of alcohols to give allylic glycosides **91**, in good yields (Scheme 39a).[56] Conversely, a Ferrier-type rearrangement of 2-*C*-acetoxymethyl glycals **89**, catalyzed by iodine, in the presence of *N*-aryl amides gave rise to 2-*C*-*N*-aryl-amidomethyl glycals, *e.g.* **92a–e** (Scheme 39b).[70]

8.2 Ferrier rearrangement on C-1 substituted glycal systems

2-Deoxy-2,3-dihydro-*N*-acetylneuraminic acid derivatives, *e.g.* **93**, are key components of cell surface glycoconjugates involved in recognition processes. These compounds can formally be considered as *C*-1 branched glycals (*C*-glycals), and their Ferrier rearrangement has been studied. In this context, Ikeda and coworkers have reported in full their previous work on the Ferrier rearrangement of a 4,5-oxazoline derivative of sialic acid, *i.e.* **93** (Scheme 40).[71] In fact, a catalytic amount of montmorillonite K-10 clay-supported $Bi(OTf)_3$ [20% w/w loading of $Bi(OTf)_3$], in acetonitrile at room temperature, proved to be the best promoter to produce a variety of 3,4-unsaturated sialic acid derivatives, *i.e.* **95**, both in terms of yield and α/β selectivity. The proposed reaction pathway involves the intermediacy of a cyclic allylic oxonium ion **94** that undergoes nucleophilic attack at C-2, preferably from the β-face (opposite to the α-*C*-7-OAc group) resulting in the predominant formation of β-*O*-alkyl glycosides, *e.g.* **95a–c** and β-*S*-alkyl glycoside **95d** (Scheme 40).

The behavior of hexacarbonyldicobalt (*C*-1)-alkynyl glycals, *e.g.* **96**, towards nucleophiles in the presence of $BF_3 \cdot OEt_2$ has been recently studied.[72] In this context, reaction of 6-*O*-triisopropylsilyl derivative **96a** with heteroaromatic nucleophiles ($BF_3 \cdot OEt_2$, −20 °C) furnished C-3 substituted derivatives **97** in a completely stereoselective manner (Scheme 41). The use of TMSCN as the nucleophile, however, resulted in the formation of **98**, embodying the CN group C–C linked to carbon 1. On the other hand, treatment of 6-OH derivative **96b** under similar reaction conditions resulted in the formation of a branched tetrahydrofuran derivative **100**, as the major reaction product. To account for this transformation, the authors invoked a reaction pathway that involved formation of an initial intramolecular Ferrier rearrangement adduct **99**, followed by hydration of the ensuing glycal double bond and ring opening.

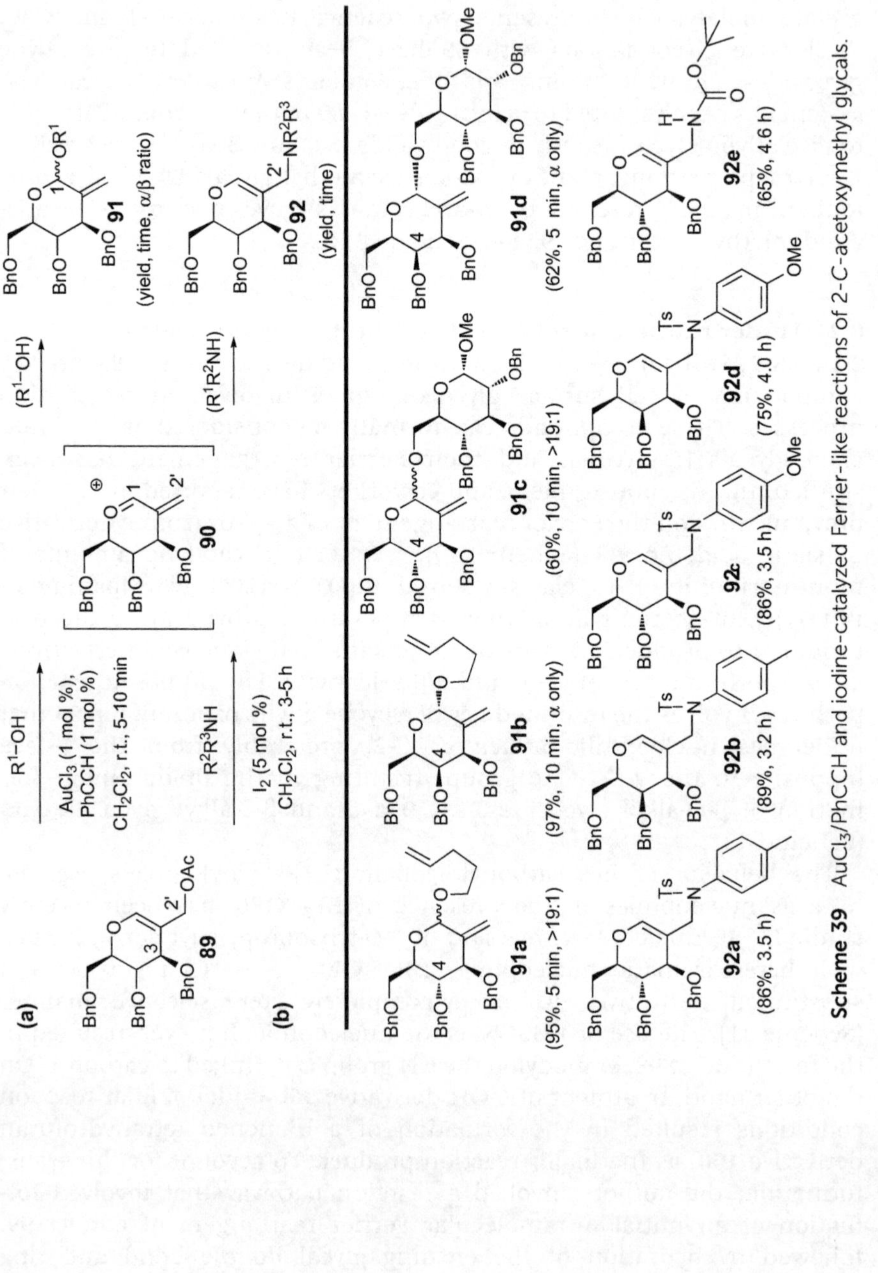

Scheme 39 AuCl$_3$/PhCCH and iodine-catalyzed Ferrier-like reactions of 2-*C*-acetoxymethyl glycals.

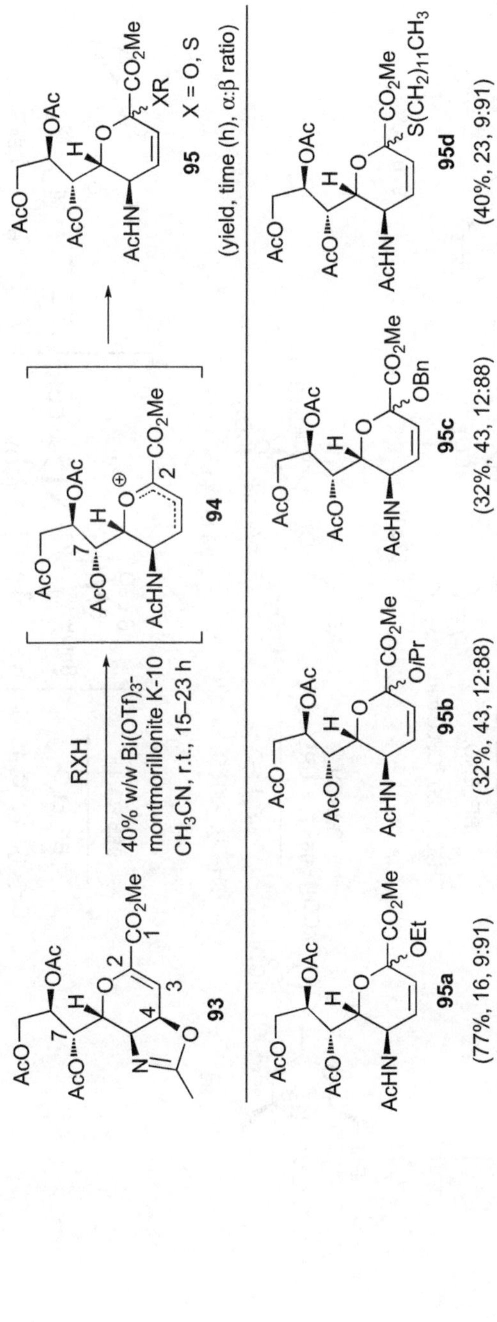

Scheme 40 Ferrier rearrangement of sialic acid derivative **93**.

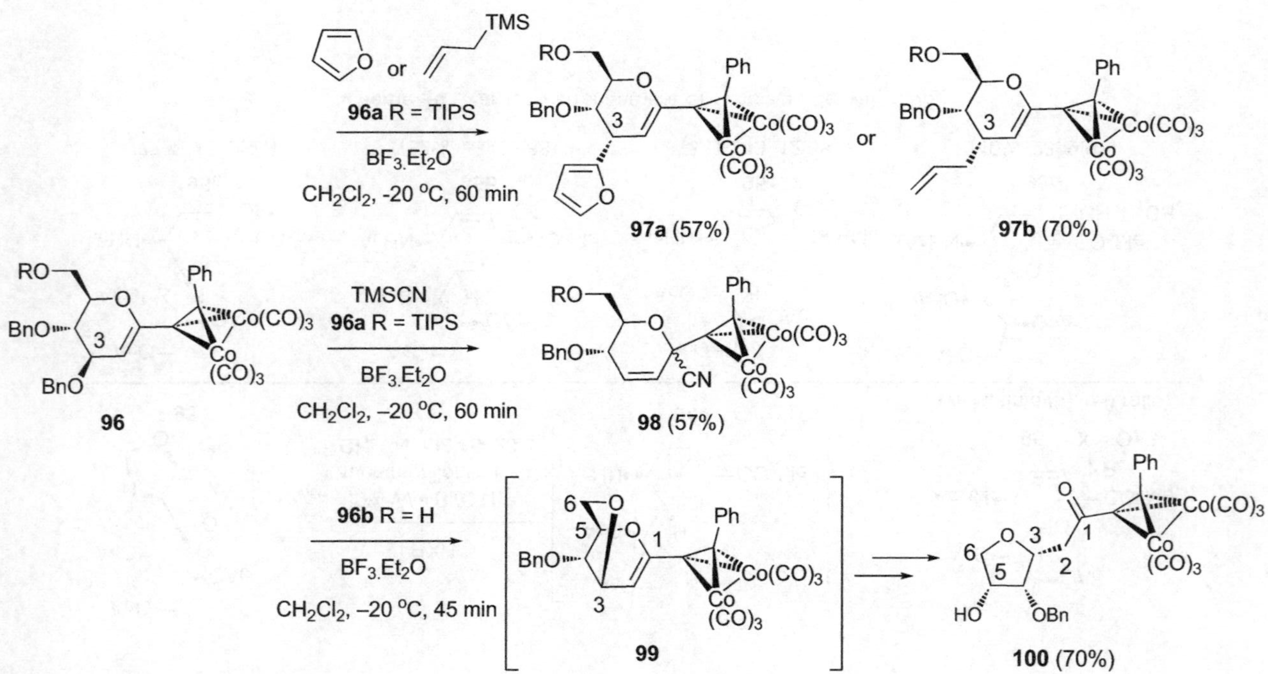

Scheme 41 Ferrier rearrangement of hexacarbonyldicobalt (*C*-1)-alkynyl glycals **96**.

9 Polarity inversion in the Ferrier rearrangement

The chemistry associated with the Ferrier rearrangement has, to date, been related with the interaction of the electrophilic Ferrier allylic oxycarbenium ion **2**, with nucleophiles. In a seminal contribution, Doisneau, Beau and coworkers have recently managed to widen the scope of the Ferrier rearrangement by generating nucleophilic allyl samarium species able to react with electrophiles.[73] Accordingly, direct treatment of glycals (equipped with a C-3 carbonate) with samarium diiodide (SmI_2) in THF at room temperature produced allylic samarium derivatives **101** that were able to react with external electrophiles (ketones) to generate C-3 branched compounds, *i.e.* **102** (Scheme 42).

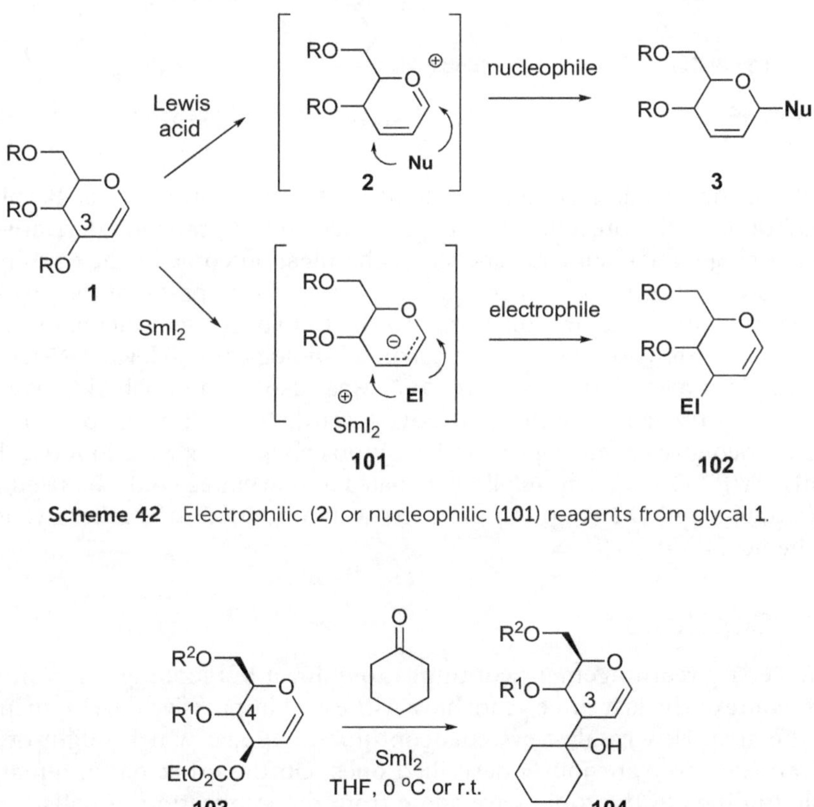

Scheme 42 Electrophilic (**2**) or nucleophilic (**101**) reagents from glycal **1**.

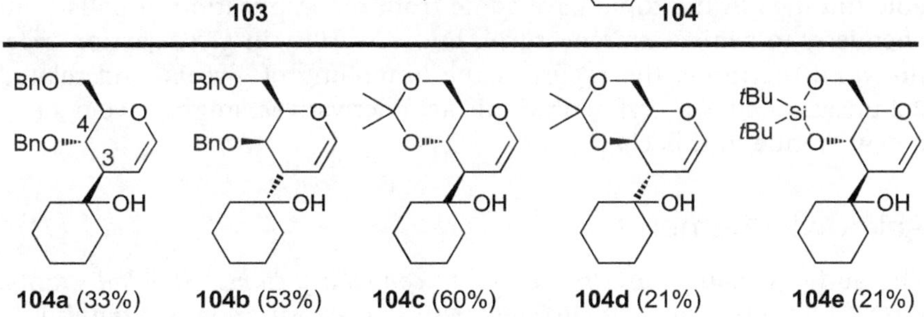

Scheme 43 SmI_2-induced umpolung of glycals **103**.

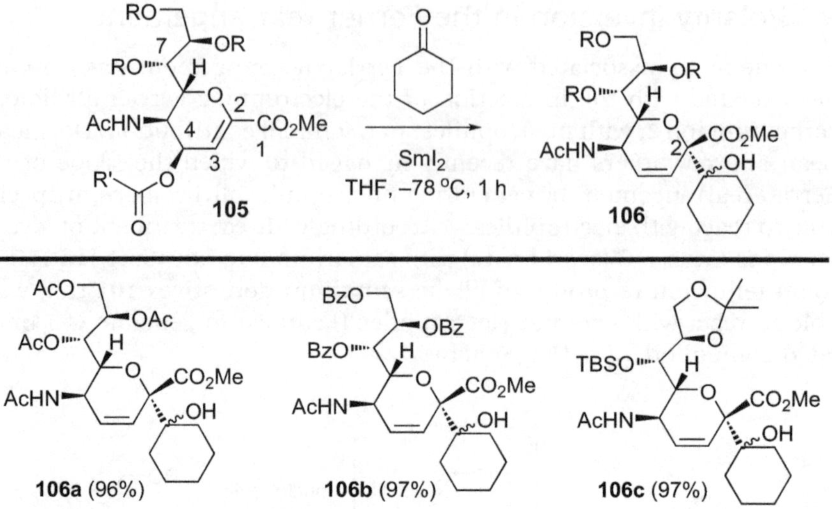

Scheme 44 SmI_2-induced umpolung of glycals from *N*-acetyl neuraminic acid derivatives, **105**.

Thus, the reaction of glycal carbonates **103** with SmI_2 under Barbier conditions in the presence of ketones provided C-3 branched derivatives **104** as single regioisomers. The stereochemical outcome of the coupling came dictated by the chiral center at C-4, with the trapping of the allylic-samarium-species taking place *anti* to the C-4 substituent (Scheme 43).

Further extension of this protocol to biologically relevant *N*-acetyl neuraminic acid derivatives, *e.g.* **105**, was also accomplished.[73] These derivatives proved to be more reactive towards SmI_2 than the previously mentioned glycals, and optimized reactions conditions were found to be SmI_2, THF, −78 °C, 1 h. Allylic benzoates and acetates could be used as substrates, and C-2 coupling products were obtained in excellent yields (Scheme 44).

10 Conclusion

The Ferrier rearrangement continues to enjoy a burgeoning interest. In this context, the last three years have witnessed impressive developments in the area. New catalyst systems continue to appear which might offer advantages over previously described ones. On the other hand, remarkable findings in the topic have come from the application of palladium chemistry to achieve stereocontrol, and versatility in glycosylation reactions. Furthermore, the report of the umpolung of glycals, and related 2,3-unsaturated *N*-acetyl neuraminic acid derivatives, might have opened a new avenue to this chemistry.

Acknowledgements

The authors thank the *Ministerio de Economía y Competitividad* grants: CTQ2012-32114 and CTQ2015-66702-R. Silvia Miranda is grateful to *Ministerio de Economia y Competitividad* for a predoctoral scholarship.

References

1 R. J. Ferrier, W. G. Overend and A. E. Ryan, *J. Chem. Soc. C*, 1962, 3667.

2 *See for example:* (a) J. J. Li, *Name Reactions*, Springer, Berlin, 3rd edn, 2006; (b) L. Kürti and B. Czabo, *Strategic Applications of Named Reactions in Organic Synthesis*, Elsevier Academic Press, Burlington, 2005.

3 The name Ferrier (I) reaction, is occasionally used to differentiate this transformation from another reaction, also described by Prof. Ferrier, in which hex-5-enopyranosyl derivatives are transformed into functionalized cyclohexanones, even though this process is also named *Ferrier carbocyclization*, or Ferrier (II) reaction: (a) R. J. Ferrier, *J. Chem. Soc., Perkin Trans. 1*, 1979, 1455; (b) R. J. Ferrier and S. Middleton, *Chem. Rev.*, 1993, **93**, 2779.

4 (a) R. J. Ferrier and J. O. Hoberg, *Adv. Carbohydr. Chem. Biochem.*, 2003, **58**, 55; (b) R. J. Ferrier and O. A. Zubkov, *Org. React.*, 2003, **62**, 569; (c) R. J. Ferrier, *Top. Curr. Chem.*, 2001, **215**, 153.

5 A. A. Ansari, R. Lahiri and Y. D. Vankar, *ARKIVOC*, 2013, **2**, 316.

6 A. M. Gomez, F. Lobo, C. Uriel and J. C. Lopez, *Eur. J. Org. Chem.*, 2013, 7221.

7 R. J. Ferrier and N. Prasad, *J. Chem. Soc. C*, 1969, 570.

8 J. Zeng, J. Ma, S. Xiang, S. Cai and X.-W. Liu, *Angew. Chem., Int. Ed.*, 2013, **52**, 5134.

9 (a) M. F. Carroll, *J. Chem. Soc.*, 1940, 704; (b) M. F. Carroll, *J. Chem. Soc.*, 1940, 1266; (c) M. F. Carroll, *J. Chem. Soc.*, 1941, 507; (d) W. Kimel and A. C. Cope, *J. Am. Chem. Soc.*, 1943, **65**, 1992.

10 S. Xiang, Z. Lu, J. He, K. L. M. Hoang, J. Zeng and X.-W. Liu, *Chem. – Eur. J.*, 2013, **19**, 14047.

11 S. Xiang, J. He, Y. J. Tan and X.-W. Liu, *J. Org. Chem.*, 2014, **79**, 11473.

12 K. L. M. Xiang, J. Hoang, Y. J. He, Tan and X.-W. Liu, *Angew. Chem., Int. Ed.*, 2015, **54**, 604.

13 L. Ji, S.-H. Xiang, W.-L. Leng, K. L. M. Hoang and X.-W. Liu, *Org. Lett.*, 2015, **17**, 1357.

14 S. Xiang, J. He, J. Ma and X.-W. Liu, *Chem. Commun.*, 2014, **50**, 4222.

15 A. Kubiak, R. Kolodziuk, S. Porwanski and A. Zawisza, *Carbohydr. Res.*, 2015, **417**, 34.

16 M. Hanaura, A. Agócs, K. Böddi, J. Deli and V. Nagy, *Tetrahedron Lett.*, 2014, **55**, 3635.

17 B. Srinivas, T. R. Reddy, P. R. Krishna and S. Kashyap, *Synlett*, 2014, **25**, 1325.

18 G. Narasimba, B. Srinivas, P. R. Krishna and S. Kashyap, *Synlett*, 2014, **25**, 523.

19 T. R. Reddy, S. K. Battina and S. Kashyap, *J. Carbohydr. Chem.*, 2015, **34**, 133.

20 T. R. Reddy, D. S. Rao and S. Kashyap, *RSC Adv.*, 2015, **5**, 28338.

21 P. Chen and S. Li, *Tetrahedron Lett.*, 2014, **55**, 5813.

22 P. Chen and D. Zhang, *Tetrahedron*, 2014, **70**, 8505.

23 P. Chen and B. Bi, *Tetrahedron Lett.*, 2015, **56**, 4895.

24 P. Chen and J. Su, *Tetrahedron*, 2016, **72**, 84.

25 P. Chen and L. Lin, *Tetrahedron*, 2013, **69**, 10045.

26 B. Srinivas, G. Narasimba, P. R. Krishna and S. Kashyap, *Synthesis*, 2014, **46**, 1191.

27 D. J. Bound, B. K. Bettadaiah and P. Srinivas, *Synth. Commun.*, 2014, **44**, 2565.

28 J. Zhou, H. Chen, J. Shan, J. Li, G. Yang, X. Chen, K. Xin, J. Zhang and J. Tang, *J. Carbohydr. Chem.*, 2014, **33**, 313.

29 B. Srinivas, T. R. Reddy and S. Kashyap, *Carbohydr. Res.*, 2015, **406**, 86.

30 J. S. Yadav, B. V. S. Reddy, J. V. Raman, N. Niranjan, S. K. Kumar and A. C. Kunwar, *Tetrahedron Lett.*, 2002, **43**, 2095.

31 H. Y. Tan, S. Xiang, W. L. Leng and X.-W. Liu, *RSC Adv.*, 2014, **4**, 34816.
32 E. Wieczoreck and J. Thiem, *J. Carbohydr. Chem.*, 2013, **32**, 502.
33 (a) T. A. Houston, E. Klemke and M. Koreeda, *203rd ACS National Meeting*, San Francisco, 1992, ORGN 533; (b) M. Koreeda, T. A. Houston, B. K. Shull, E. Klemke and R. J. Tuinman, *Synlett*, 1995, 90.
34 B. K. Banik, M. S. Manhas and A. K. Bose, *J. Org. Chem.*, 1994, **59**, 4714.
35 U. Sirion, S. Purintawarrakun, P. Sahakitpichan and R. Saeeng, *Carbohydr. Res.*, 2010, **345**, 2401.
36 W. Mangsang, U. Sirion and R. Saeeng, *Carbohydr. Res.*, 2013, **375**, 79.
37 R. Saeeng, O. Siripru and U. Sirion, *Heterocycles*, 2015, **91**, 849.
38 L. Ren, Y. Liu, G.-H. Yu, J.-Z.- Hao and M.-S. Cheng, *Chem. Pap.*, 2014, **68**, 525.
39 Z. Begum, C. Kishore, V. V. Reddy and B. V. S. Reddy, *Tetrahedron Lett.*, 2014, **55**, 6048.
40 (a) J. F. Zhou, X. Chen, Q. B. Wang, B. Zhang, L. Y. Zhang, A. Yusulf, Z. F. Wang, J. B. Zhang and J. Tang, *Chin. Chem. Lett.*, 2010, **21**, 922; (b) J. Zhou, B. Zhang, G. Yang, X. Chen, Q. Wang, Z. Wang, J. Zhang and J. Tang, *Synlett*, 2010, 893.
41 J. Zhang, B. Zhang, J. Zhou, H. Chen, J. Li, G. Yang, Z. Wang and J. Tang, *J. Carbohydr. Chem.*, 2013, **32**, 380.
42 X. Yang and N. Li, *Sci. World J.*, 2014, 307895, DOI: 10.1155/2014/307895.
43 N. Bodipati, S. R. Palla, V. Komera and R. K. Peddinti, *Tetrahedron Lett.*, 2014, **55**, 6878.
44 A. K. Kusunuru, M. Tatina, S. K. Yousuf and D. Mukherjee, *Chem. Commun.*, 2013, **49**, 10154.
45 S. Debari, M. Kumar, R. Deshidi, M. Rizvi and B. A. Shah, *Beilstein J. Org. Chem.*, 2014, **10**, 2649.
46 (a) J. S. Yadav, B. V. S. Reddy and S. K. Pandey, *New J. Chem.*, 2001, **25**, 538; (b) S. Yaya, S. Sorho, L. Cottier and G. Descotes, *Bull. Chem. Soc. Ethiop.*, 2005, **19**, 233.
47 K. Pachamuthu and Y. D. Vankar, *J. Org. Chem.*, 2001, **66**, 7511.
48 A. A. Ansari, Y. S. Reddy and Y. D. Vankar, *Beilstein J. Org. Chem.*, 2014, **10**, 300.
49 S. M. Rokade and P. M. Bhate, *Carbohydr. Res.*, 2015, **415**, 28.
50 (a) Q. Zhang, K. D. O. Vigier, S. Royer and F. Jérôme, *Chem. Soc. Rev.*, 2012, **41**, 7108; (b) E. L. Smith, A. P. Abbott and K. S. Ryder, *Chem. Rev.*, 2014, **114**, 11060.
51 D. Stevanovic, A. Pejovic, I. Damljanovic, A. Minic, G. A. Bogdanovic, M. Vukicevic, N. S. Radulovic and R. D. Vukicevic, *Carbohydr. Res.*, 2015, **407**, 111.
52 Y.-B. Chen, S.-I. Wang, Z.-P. Lin, C.-H. Lin, M.-T. Hsieh and H.-C. Lin, *Tetrahedron*, 2015, **71**, 350.
53 K. Komor, W. Szeja, T. Bieg, N. Kuznik, G. Pastuch-Gawołek and R. Komor, *Tetrahedron Lett.*, 2014, **55**, 3709.
54 (a) A. Sau, A. Santra and A. K. Misra, *Synlett*, 2012, 2341; (b) A. Santra, G. Guchhait and A. K. Misra, *Synlett*, 2013, 581.
55 (a) J. Omboumaa, D. Vullo, C. T. Supuran and J.-Y. Winum, *Bioorg. Med. Chem.*, 2014, **22**, 6353; (b) J. Ombouma, D. Vullo, P. Dumy, C. T. Supuran and J.-Y. Winum, *ACS Med. Chem. Lett.*, 2015, **6**, 819.
56 R. Roy, P. Rajasekaran, A. Mallick and Y. D. Vankar, *Eur. J. Org. Chem.*, 2014, 5564.
57 R. Balamurugan and S. R. Koppolu, *Tetrahedron*, 2009, **65**, 8139.
58 D. B. G. Williams, S. B. Simelanea and H. H. Kinfe, *Org. Biomol. Chem.*, 2012, **10**, 5636.

59 S. B. Simelane, H. H. Kinfe, A. Muller and D. B. G. Williams, *Org. Lett.*, 2014, **16**, 4543.
60 C. Booma and K. K. Balasubramanian, *Tetrahedron Lett.*, 1995, **36**, 5807.
61 P. T. Moshapo, M. Sokamisa, E. M. Mmutlane, R. M. Mampab and H. H. Kinfe, *Org. Biomol. Chem.*, 2016, **14**, 5627–5638. A correction to this manuscript has recently appeared: *Org. Biomol. Chem.*, 2016, **14**, 5861.
62 S. Dharum and Y. D. Vankar, *Org. Lett.*, 2014, **16**, 1172.
63 P. K. Kancharla, Y. S. Reddy, S. Dharuman and Y. D. Vankar, *J. Org. Chem.*, 2011, **76**, 5832.
64 W. Xue, J. Sun and B. Yu, *J. Org. Chem.*, 2009, **74**, 5079.
65 S. Dharuman, P. Gupta, P. K. Kancharla and Y. D. Vankar, *J. Org. Chem.*, 2013, **78**, 8442.
66 2-Acetoxymethyl glycals behaved in a similar manner, see: (a) P. Nagaraj and N. G. Ramesh, *Tetrahedron*, 2011, **67**, 9322; (b) W. Priebe and A. Zamojski, *Tetrahedron*, 1980, **36**, 287.
67 F. Ding, R. William and X.-W. Liu, *J. Org. Chem.*, 2013, **78**, 1293.
68 A. A. Ansari, P. Rajasekaran, M. M. Khan and Y. D. Vankar, *J. Org. Chem.*, 2014, **79**, 1960.
69 C. Booma and K. K. Balasubramanian, *J. Chem. Soc., Chem. Commun.*, 1993, 1394–1395.
70 J. S. Yadav, G. Narasimhulu, N. Umadevi, Y. V. Reddy and B. V. S. Reddy, *Tetrahedron Lett.*, 2013, **54**, 871.
71 (a) M. Oba, Y. Ueno, S. Kitani, T. Hayakawa, T. Takahashi, T. Suzuki, M. Sato and K. Ikeda, *Heterocycles*, 2014, **89**, 69; (b) K. Ikeda, Y. Ueno, S. Kitani, R. Nishino and M. Sato, *Synlett*, 2008, 1027.
72 (a) F. Lobo, A. M. Gomez, S. Miranda and J. C. Lopez, *Chem. Eur. – J.*, 2014, **20**, 10492; (b) J. C. Lopez, F. Lobo, S. Miranda, C. Uriel and A. M. Gomez, *Pure Appl. Chem.*, 2014, **86**, 1357.
73 T. X. Le, C. Papin, G. Doisneau and J.-M. Beau, *Angew. Chem., Int. Ed.*, 2014, **53**, 6184.

FimH antagonists – solubility *vs.* permeability

Lijuan Pang,† Jacqueline Bezençon,† Simon Kleeb,†
Said Rabbani, Anja Sigl, Martin Smiesko, Christoph P. Sager,
Deniz Eris, Oliver Schwardt and Beat Ernst*

DOI: 10.1039/9781782626657-00248

Urinary tract infections (UTIs) caused by uropathogenic *Escherichia coli* (UPEC) are among the most prevalent infections worldwide. Since frequent antibiotic treatment favors the emergence of antibiotic resistance, efficient non-antibiotic strategies are urgently needed. The first step of the pathogenesis of UTI is the bacterial adherence to urothelial host cells, a process mediated by the mannose-binding adhesin FimH located at the tip of bacterial pili. In a preliminary study, biphenyl α-D-mannopyranosides with an electron-withdrawing carboxylate on the aglycone were identified as potent FimH antagonists. Although passive permeability could be established by masking the carboxylate as an ester, insufficient solubility and fast hydrolysis did not allow to maintain the therapeutic concentration in the bladder for the requested period of time. By modifying the substitution pattern, molecular planarity and symmetry of the biphenyl aglycone could be disrupted leading to improved solubility. In addition, when heteroatoms were introduced to the aglycone, antagonists with further improved solubility, metabolic stability as well as passive permeability were obtained. The best representative, the pyrrolylphenyl mannoside **42f** exhibited therapeutic urine concentration for up to 6 h and is therefore a promising oral candidate for UTI prevention and/or treatment.

1 Introduction

Urinary tract infections (UTIs) – also known as acute cystitis or bladder infections – are among the most prevalent infectious diseases worldwide. UTIs affect millions of people every year and account for significant morbidity and high medical costs.[1] Since symptomatic UTIs require antibiotic treatment and recurrent antibiotic exposure leads to the ubiquitous problem of antimicrobial resistance, efficient non-antibiotic prevention and treatment strategies are urgently needed.[2] More than 70% of UTIs are caused by uropathogenic *Escherichia coli* (UPEC).[1a,3] In the first step of the infection, UPEC adhere to urothelial cells, which prevents them from being cleared by micturition but also triggers the invasion into host cells.[4] This initial contact is mediated by the bacterial adhesin FimH located at the tip of type 1 pili.[5] FimH consists of an N-terminal lectin domain and a C-terminal pilin domain. The carbohydrate recognition domain (CRD) of the lectin domain specifically recognizes oligmannosides being part of the glycoprotein uroplakin Ia located on the urinary bladder mucosa, whereas the pilin domain regulates the switch between the various affinity states of the CRD.[6] Blocking the FimH-CRD with carbohydrates or mimetics thereof prevents

Institute of Molecular Pharmacy, Pharmacenter, University of Basel, Klingelbergstr. 50, 4056 Basel, Switzerland. E-mail: beat.ernst@unibas.ch
†These authors equally contributed to this work.

the bacterial adhesion as well as the subsequent infection. This approach is therefore regarded as a potential therapy for prevention and/or treatment of UTIs.[7]

Over the last three decades, mannosides and oligomannosides with various aglycones, such as *n*-alkyl,[8d] phenyl,[8e] dioxocyclobutenylaminophenyl,[8f] umbelliferyl,[8e] biphenyl,[8g–k] indol(in)ylphenyl,[8l] triazolyl[8m] or thiazolylamino[8n] have been tested as potential antagonists for preventing type 1 pili-mediated bacterial adhesion.[8] In addition, different multivalent presentations of mannose derivatives have been explored.[9]

In general, when a daily therapy is required, oral administration is the standard care. To achieve oral availability as well as a therapeutic concentration in the bladder over an extended period of time, successful oral absorption (*i.e.* solubility and permeability), metabolic stability, and slow and prolonged renal excretion (*i.e.* renal reabsorption) are required. As previously described,[8h] the carboxylic acid moiety in biphenyl α-D-mannoside **1a** – its electron-withdrawing potential is essential for an enhanced π–π stacking interaction – impairs the membrane permeability and, as a consequence, the potential for oral absorption. Therefore, the polar carboxylate was masked as ester (→**1b**[8h]), exhibiting comparable electron withdrawing properties but improving passive permeability and renal reabsorption by increased lipophilicity. However, since the ester **1b** was characterized by low aqueous solubility (12 $\mu g\, mL^{-1}$) and fast metabolic cleavage ($t_{1/2} = 2.1$ min), the absorptive flux through the intestinal mucosa was limited and the renal excretion accelerated.[10] As a consequence, the therapeutic concentration in the bladder could not be maintained for a sufficient period of time.[8h] Therefore, emanating from antagonist **1b**, these pharmacokinetic drawbacks were addressed by structural modifications of the aglycone (Fig. 1).

2 Results and discussion

To improve relevant pharmacokinetic parameters (*i.e.* oral absorption, metabolism and renal excretion of FimH antagonist **1b**), we focused on three approaches. First, the molecular planarity and symmetry of the biphenyl aglycone was disrupted by modifying the substitution pattern[11] (Fig. 1a). Second, the hydrophobicity was improved by heterocyclic biaryl aglycones (Fig. 1b). Finally, oral availability was improved by replacing the carboxylic acid by the bioisosteric cyano group (Fig. 1c).[8o,11,12]

2.1 Synthesis of FimH antagonists

Biphenyl mannosides (Schemes 1 and 2). Compounds **1a,b**, and **2a,b** (Table 1) were synthesized as previously described.[8h] For the synthesis of **6a** and **6b**, iodide **3**[8l] was reacted with boronic acid **4** in a palladium-catalyzed Suzuki–Miyaura coupling to yield mannoside **5**. Subsequent deprotection afforded the test compounds **6a** and **6b** (Scheme 1).

Lewis acid-promoted glycosylation of phenol **8** with fluoride **7**[13] followed by a Suzuki–Miyaura coupling of bromide **9** with boronic acid **4** gave **10** (Scheme 2). Finally, deprotection yielded biphenyls **11a** and **11b**.

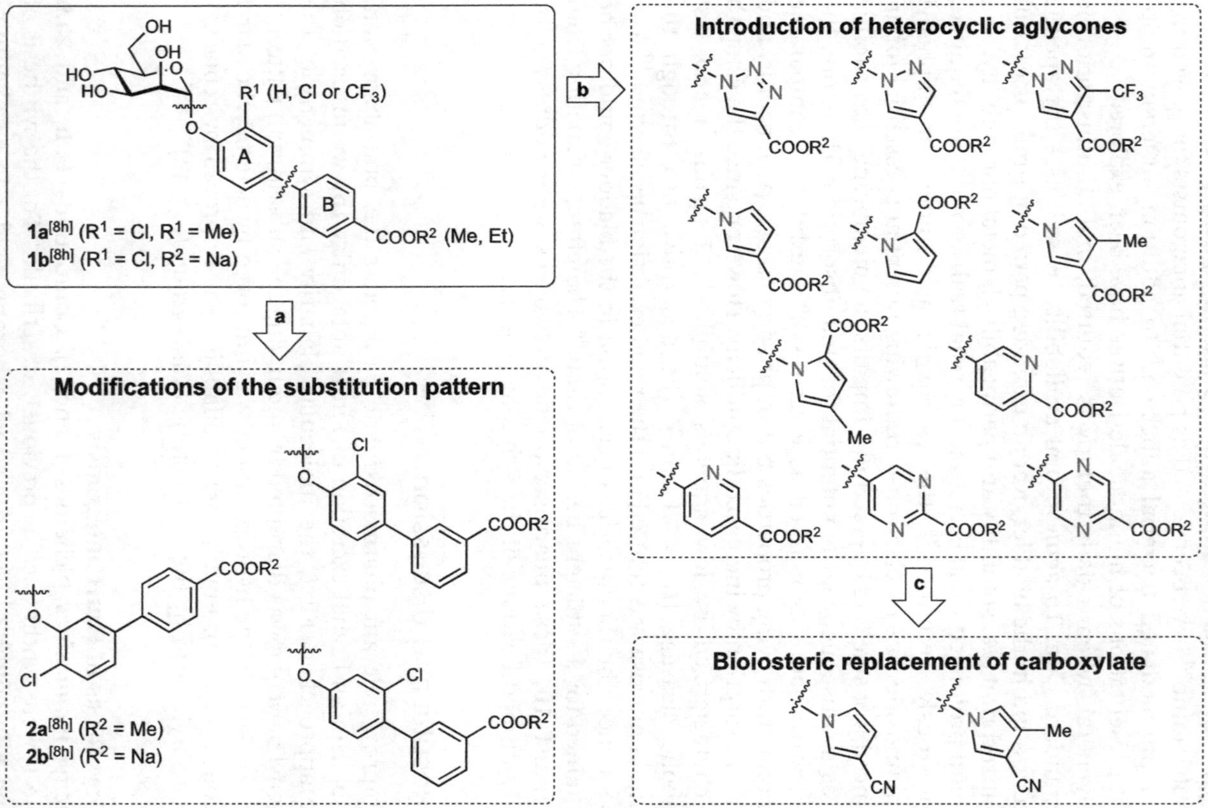

Fig. 1 Variations of the aglycone of FimH antagonists **1a** and **1b** by (**a**) modifying the substitution pattern, (**b**) introducing heteroaryl aglycones and (**c**) replacing the carboxylate moiety with a bioisosteric cyano group. Ring A is adjacent to the anomeric center, whereas ring B is in the terminal position.

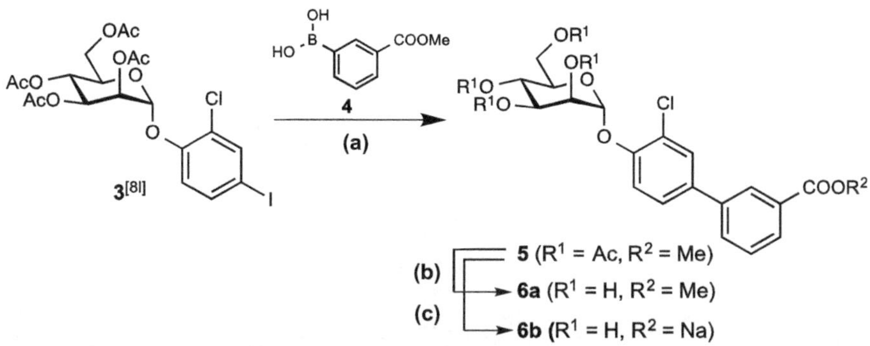

Scheme 1 (a) Pd(Cl$_2$)dppf·CH$_2$Cl$_2$, K$_3$PO$_4$, DMF, 80 °C, overnight (70%); (b) NaOMe, MeOH, rt, 4 h (quant.); (c) 0.2 N aq. NaOH, MeOH, rt, overnight (50%).

Heteroaromatic building blocks (Scheme 3). Starting with the commercial aminophenols **12a,b**, the azidophenols **13a,b** were obtained *via* a diazotransfer reaction using freshly prepared triflyl azide in pyridine and copper(II) sulfate as catalyst.[14] Because of low stability, **13a,b** were used without purification in a subsequent copper(I)-catalyzed Huisgen cycloaddition[15] with ethyl propiolate, yielding the triazolylphenols **14a,b** (Scheme 3A). By using an Ullmann-type copper-diamine-catalyzed *N*-arylation,[16] 1*H*-pyrazole-4-carboxylate was coupled with 4-iodoanisole (**15**) in *N*-methyl-2-pyrrolidone (NMP) to furnish **16a.** Because of the low reactivity of the ethyl 3-trifluoromethyl-1*H*-pyrazole-4-carboxylate, the coupling reaction was carried out under solvent-free conditions to yield **16b** in quantitative yield. Demethylation of **16** with $AlCl_3$ gave the pyrazolylphenol derivatives **17.** Due to the instability of **17b** under $AlCl_3$/nBu$_4$NI conditions, a solution of $AlCl_3$ in 1-dodecanethiol was used to accelerate the reaction and to suppress byproduct formation (Scheme 3B).[17] The pyrimidinyl derivative **20** was prepared *via* a *n*BuLi-mediated carboxylation with CO_2 followed by esterification (Scheme 3C).[18] To synthesize the cyano-substituted pyrroles **23a,b**, benzotriazol-1-ylmethyl isocyanide (**21**) was treated with the electron-deficient alkenes **22a,b** under basic heterocyclization conditions (Scheme 3D).[19]

Triazolylphenyl and pyrazolylphenyl mannosides (Scheme 4). Mannosylation of the phenols **14a,b** and **17a,b** (see Scheme 3A & B) with mannosyl fluoride **7**[13] and $BF_3 \cdot Et_2O$ as promoter, yielded exclusively the α-mannosides **24a,b** and **27a,b.** Deacetylation (→**25a–c** and **28a,b**) followed by ester hydrolysis gave the test compounds **26a,b** and **29a,b.**

Pyridinylphenyl, pyrazinylphenyl, and pyrimidinylphenyl mannosides (Scheme 5). Mannosyl fluoride **7**[13] was treated with 4-iodophenol or 4-bromo-2-trifluoromethylphenol in the presence of $BF_3 \cdot Et_2O$. The resulting iodide **30**[8k] and bromide **36**[8k] were transformed into the boronic acid pinacol esters **31**[8k] and **37** under Miyaura-borylation conditions. In a palladium-catalyzed Miyaura–Suzuki coupling[20] of the heteroaryl halides **20** (see Scheme 3C) and **32a–c** (commercially available) with boronic acid ester **31**, heteroarylphenyl mannosides **33a–d** were obtained in good to excellent yields. Similarly, mannoside **38** was prepared by coupling of ester **37** and pyridinylchoride **32a.** Deacetylation under Zemplén

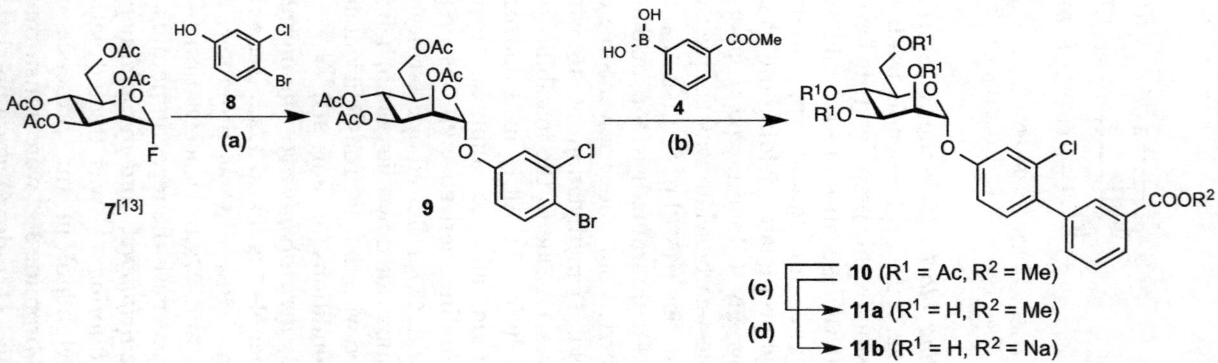

Scheme 2 (a) $BF_3 \cdot Et_2O$, DCM, mol. sieves 4 Å, 0 °C to rt, overnight (90%); (b) $Pd(Cl_2)dppf \cdot CH_2Cl_2$, K_3PO_4, DMF, 80 °C, overnight (76%); (c) NaOMe, MeOH, rt, 4 h (quant.); (d) 0.2 N aq. NaOH, MeOH, rt, overnight (61%).

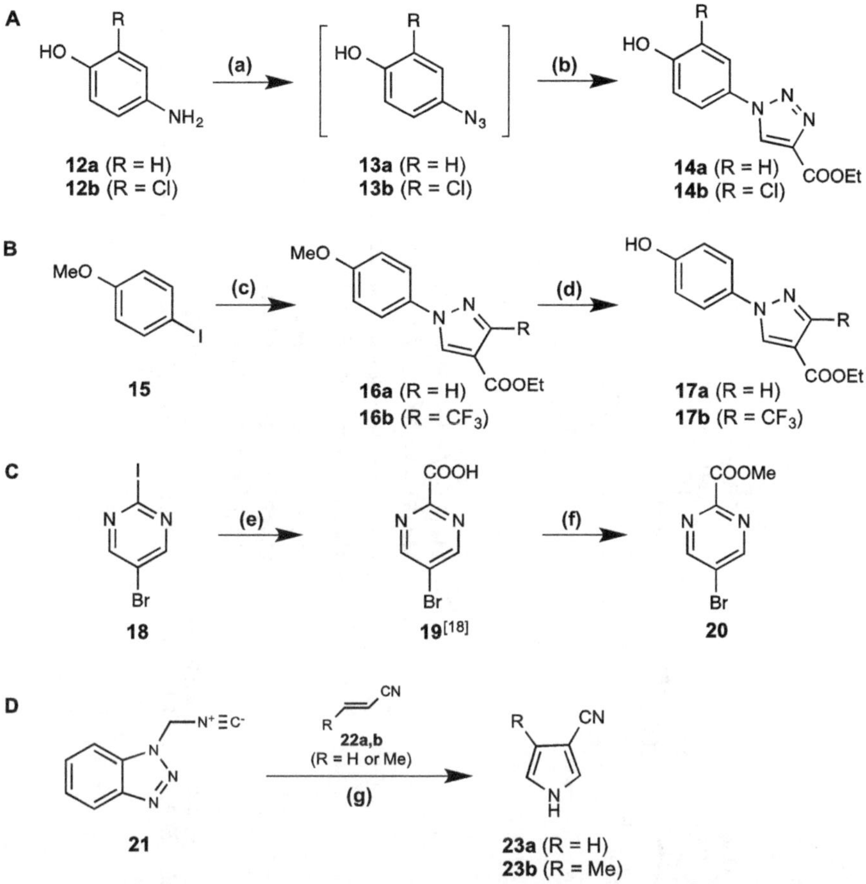

Scheme 3 (a) TfN_3, $CuSO_4$, triethylamine, pyridine, 0 °C to rt, 2 h; (b) ethyl propiolate, $CuSO_4 \cdot 5H_2O$, sodium ascorbate, *t*BuOH/H_2O (1 : 1), rt, 30 min (yield for two steps: 77% for **14a**, 48% for **14b**); (c) ethyl 1*H*-pyrazole-4-carboxylate or ethyl 3-trifluoromethyl-1*H*-pyrazole-4-carboxylate, CuI, *trans*-*N,N′*-dimethyl-1,2-cyclohexanediamine, K_2CO_3, NMP as solvent for **16a** and solvent free for **16b**, 110 °C, 24 h (80% for **16a**, quant. for **16b**); (d) $AlCl_3$, cat. *n*Bu_4NI, DCE (for **17a**), or $AlCl_3$ in 1-dodecanethiol (for **17b**), 0 °C to rt (60% for **17a**, 26% for **17b**); (e) i. *n*BuLi, hexane, toluene, −78 °C, 1 h; ii. CO_2 (g), −78 °C to rt, 7 h; (f) conc. H_2SO_4 (0.8 equiv.), MeOH, reflux, overnight (37% for two steps); (g) nitrile **22a,b**, *t*BuOK, THF, 0 °C to reflux, 2 h (60% for **23a**, 54% for **23b**).

conditions (→**34a–d, 39**) followed by saponification of the methyl ester yielded the sodium salts **35a–d** and **40.**

Pyrrolylphenyl mannosides (Schemes 6 and 7). In a copper catalyzed *N*-arylation, pyrroles **23a,b** (see Scheme 3D) and **23c–f** (commercial) were coupled with mannoside **3**[8l] (*ortho*-Cl) to yield the pyrrolylphenyl mannosides **41a–f** (Scheme 6).[17] Under similar conditions, mannosides **30** (without *ortho*-substituent) and **36** (*ortho*-CF_3) were coupled with pyrrole **23f** to yield **47** and **48** (Scheme 7). Because of partial deacetylation of the sugar moiety during *N*-arylation, the crude products were reacetylated to facilitate purification. Deacetylation of the mannose moiety (→**42a–f, 49** and **50**) followed by saponification of the alkyl esters gave the test compounds **43–46, 51** and **52.**

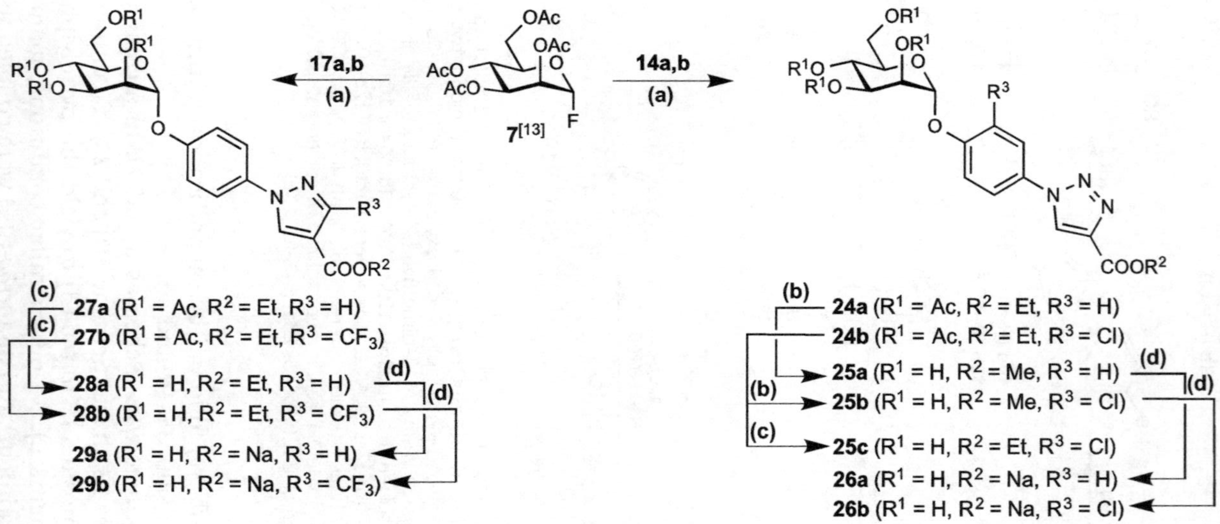

Scheme 4 (a) $BF_3 \cdot Et_2O$, DCM, mol. sieves 4 Å, 0 °C to rt, overnight (79% for **24a**, 76% for **24b**, 98% for **27a**, 64% for **27b**); (b) NaOMe, MeOH, rt, 4 h (74% for **25a**, 80% for **25b**); (c) NaOEt, EtOH, rt, overnight (74% for **25c**, 95% for **28a**, 82% for **28b**); (d) 0.2 N aq. NaOH, MeOH, rt, overnight (30% for **26a**, 90% for **26b**, 70% for **29a**, 79% for **29b**).

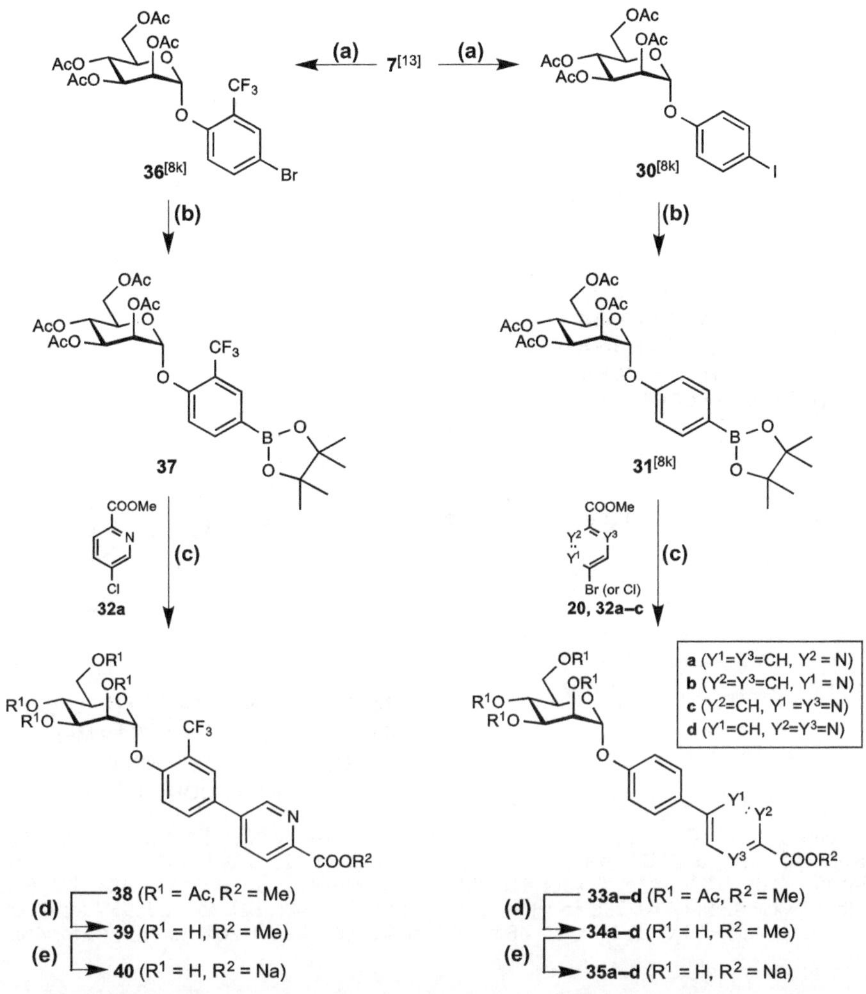

Scheme 5 (a) 4-Iodophenol or 4-bromo-2-trifluoromethylphenol, $BF_3 \cdot Et_2O$, DCM, mol. sieves 4 Å, 0 °C to rt, overnight (70%[8k] for **30**, 80%[8k] for **36**); (b) bis(pinacolato)diborone, $Pd(Cl_2)dppf \cdot CH_2Cl_2$, KOAc, DMF, 85 °C, overnight (80% for **31**, 83% for **37**); (c) $Pd(Cl_2)dppf \cdot CH_2Cl_2$, K_3PO_4, DMF, 85 °C, overnight (60% for **33a**, 80% for **33b**, 68% for **33c**, 40% for **33d**, 57% for **38**); (d) NaOMe, MeOH, rt, 4 h (36% for **34a**, 24% for **34b**, 36% for **34c**, 89% for **34d**, 60% for **39**); (e) 0.2 N aq. NaOH, MeOH, rt, overnight (32% for **35a**, 48% for **35b**, 44% for **35c**, 60% for **35d**, 90% for **40**).

2.2 Physicochemical properties and *in vitro* pharmacokinetics of FimH antagonists

For the assessment of the potential for intestinal absorption, physicochemical properties, *i.e.* lipophilicity ($\log D_{7.4}$), aqueous solubility, and permeability through an artificial membrane (PAMPA, $\log P_e$) as well as a Caco-2 cell monolayer (P_{app}) were determined (Table 1).[21–24] The following permeability thresholds for the effective permeability $\log P_e$ (low permeability is expected for $\log P_e < -6.3$ cm s^{-1}; moderate ≤ -5.7 cm s^{-1}; high > -5.7 cm s^{-1})[25] and for apparent permeability P_{app} (low permeability is expected for $P_{app} < 2 \times 10^{-6}$ cm s^{-1}; moderate $\leq 20 \times 10^{-6}$ cm s^{-1};

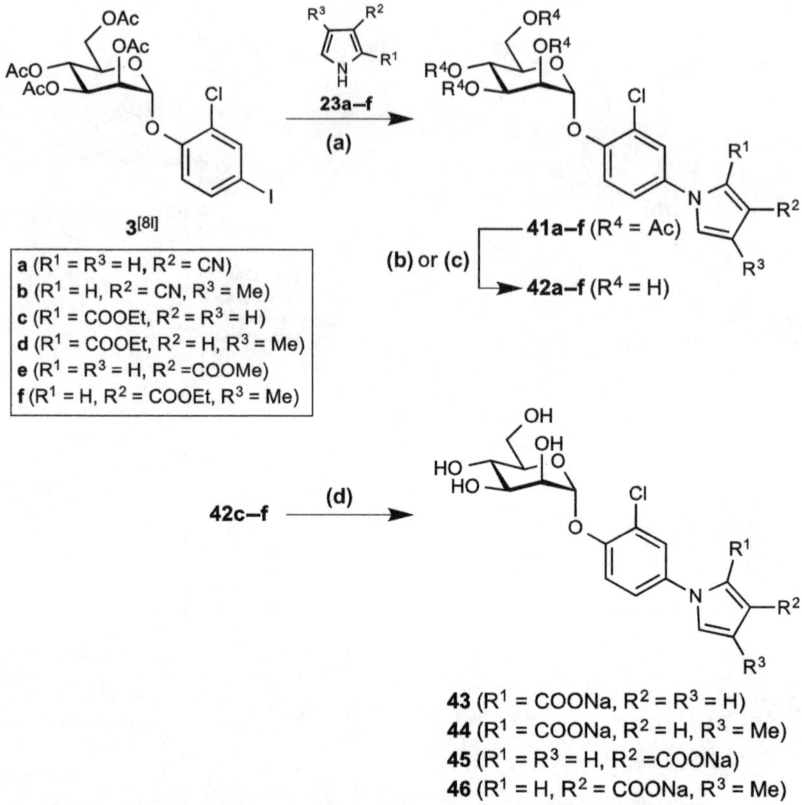

Scheme 6 (a) i. CuI, (±)-*trans*-1,2-diaminocyclohexane, K_3PO_4, 1,4-dioxane, 110 °C, overnight; ii. Ac_2O, DMAP, pyridine, rt, overnight (44% for **41a**, 92% for **41b**, 33% for **41c**, 64% for **41d**, 99% for **41e**, 77% for **41f**); (b) NaOMe, MeOH, rt, 4 h (65% for **42a**, 38% for **42b**, 83% for **42e**); (c) NaOEt, EtOH, rt, overnight (91% for **42c**, 61% for **42d**, 93% for **42f**); (d) NaOH, MeOH/H_2O (1 : 2), rt, 12–48 h (58% for **43**, 40% for **44**, 20% for **45**, 57% for **46**).

high > 20×10^{-6} cm s^{-1})[26] were used to classify the antagonists. As expected, all metabolites, *i.e.* the free carboxylates showed no or low permeability ($\log P_e < -6.3$ cm s^{-1}) and are not further discussed in this section. Furthermore, the esters were incubated with rat liver microsomes (RLM) for estimating their susceptibility to carboxylesterase (CES)-mediated hydrolysis.[27] Table 1 indicates the metabolic half lives ($t_{1/2}$) as determinants of the rate of bioconversion to the respective acid.

Solubility of biphenyl mannosides. As observed in our previous study,[8h] the biphenyl derivatives **1a** and **2a** (Table 1, entries 2 & 4) showed low aqueous solubility (≤24 μg mL^{-1}, Table 1), probably due to the symmetrical *para*-substitution leading to stacking effects. In order to disrupt this symmetry, the carboxylic acid moiety in **1a** was moved from the *para*- to the *meta-position* (→**6a**, entry 6). Since the dihedral angle remained unaffected (Fig. 2; global minima values calculated with MacroModel, version 9.9[30]) only moderately improved aqueous solubility (41 μg mL^{-1}, Table 1) resulted. However, when the chloro substituent on ring A was shifted from the *ortho*- to the *meta-position* (→**11a**, entry 8), the dihedral angle was enlarged from 39.6° in **6a** to 60.3° in **11a** (Fig. 2). The resulting

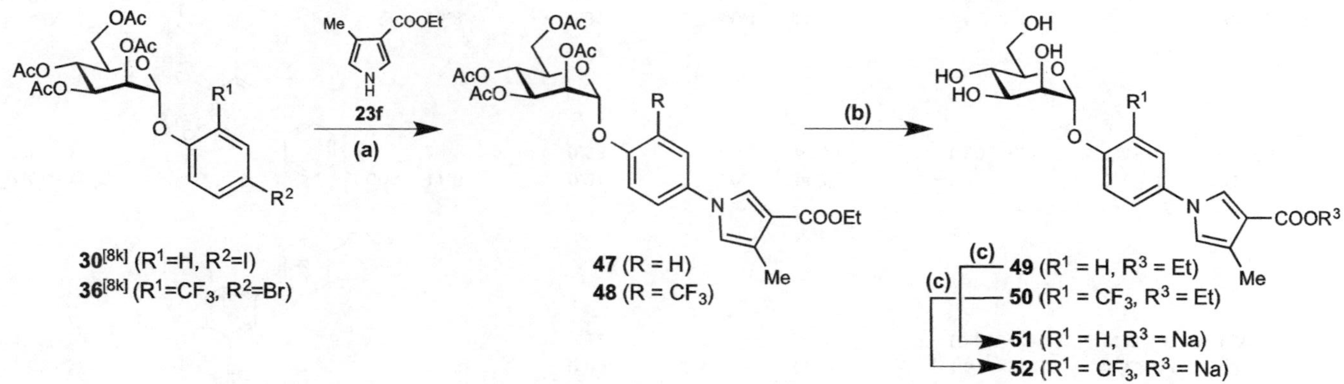

Scheme 7 (a) i. CuI, (±)-*trans*-1,2-diaminocyclohexane, K_3PO_4, 1,4-dioxane, 110 °C, overnight; ii. Ac_2O, DMAP, pyridine, rt, overnight (94% for **47**, 49% for **48**); (b) NaOEt, EtOH, rt, overnight (46% for **49**, 85% for **50**); (c) NaOH, MeOH/H_2O (1:2), rt, 48 h (99% for **51**, 35% for **52**).

Table 1 Physicochemical (log $D_{7.4}$, solubility, PAMPA, and Caco-2), *in vitro* pharmacokinetic (microsomal stability) and pharmacodynamics (IC_{50}) parameters of FimH antagonists; ring A is adjacent to the anomeric center, whereas ring B is in the terminal position.

Entry	Compound	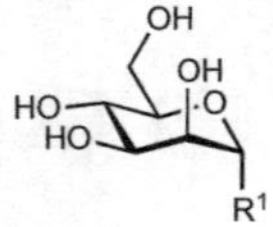	Binding Assay[a]					Caco-2 P_{app}[e] [10^{-6} cm s^{-1}]		Microsomal stability $t_{1/2}$[f] [min]
			IC_{50} [nM]	rIC_{50}	log $D_{7.4}$[b]	Solubility[c] [µg mL^{-1}]	PAMPA log P_e[d] [cm s^{-1}]	a→b	b→a	
1	**53**[8d]	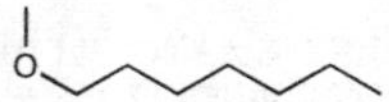	54.9	1	1.7	>3000[g]	−4.9	7.0 ± 0.6	9.4 ± 0.2	n.d.[h]
2	**1a**[8h] (R^2 = Me)	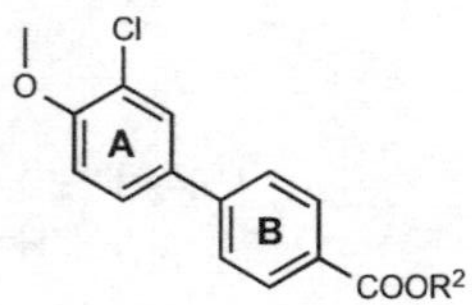	4.8	0.09	2.3	11.9[g]	−4.6	5.3 ± 0.6	17.5 ± 1.3	2.1
3	**1b**[8h] (R^2 = Na)		6.7	0.12	−0.8	>3000[g]	−10.0 ± 0.0	0.23 ± 0.03	0.38 ± 0.04	n.d.[h]
4	**2a**[8h] (R^2 = Me)	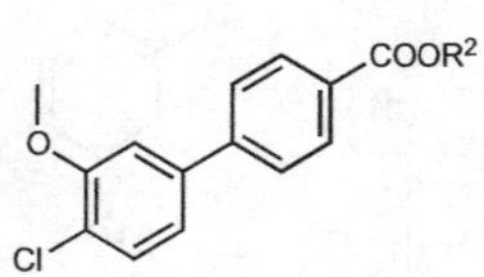	11.8	0.21	1.7 ± 0.1	24.3[g]	−4.7	6.1 ± 1.2	21.1 ± 1.2	22
5	**2b**[8h] (R^2 = Na)		29	0.53	< −1.5	>3000[g]	−10.0 ± 0.0	1.2 ± 0.4	1.2 ± 0.1	n.d.[h]
6	**6a** (R^2 = Me)	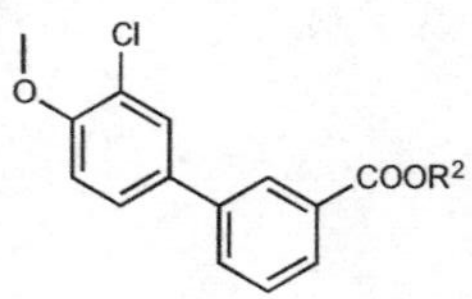	16.8	0.31	2.7 ± 0.1	41 ± 3	−4.6 ± 0.2	6.7 ± 0.4	20.7 ± 2.5	72
7	**6b** (R^2 = Na)		12	0.22	n.d.[h]	n.d.[h]	n.d.[h]	n.d.[h]	n.d.[h]	n.d.[h]
8	**11a** (R^2 = Me)	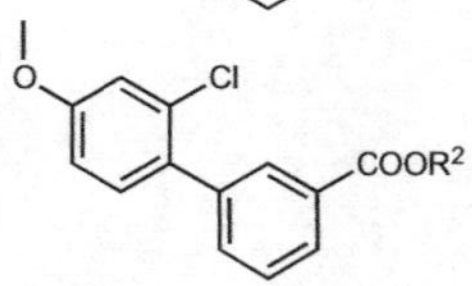	23.3	0.42	2.7 ± 0.1	134 ± 6	−4.5 ± 0.1	4.5 ± 0.3	10.8 ± 0.7	13
9	**11b** (R^2 = Na)		53	0.97	n.d.[h]	n.d.[h]	n.d.[h]	n.d.[h]	n.d.[h]	n.d.[h]

10	**25a** (R^2 = Me)	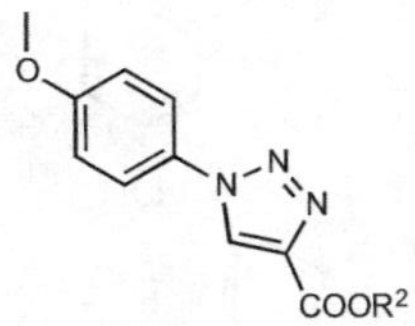	7.6	0.14	−0.6 ± 0.0	>180	−9.4 ± 0.3	n.d.[h]	n.d.[h]	38
11	**26a** (R^2 = Na)		16	0.29	n.d.[h]	>180	−10.0 ± 0.0	n.d.[h]	n.d.[h]	n.d.[h]
12	**25b** (R^2 = Me)	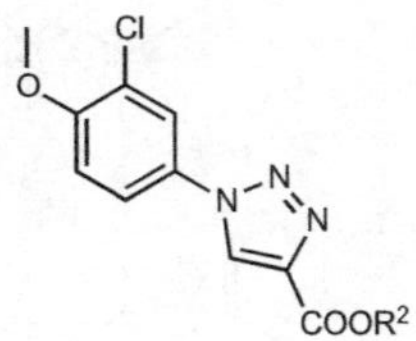	10.5	0.19	0.0 ± 0.0	>150	−9.1 ± 1.8	n.d.[h]	n.d.[h]	32
13	**25c** (R^2 = Et)		n.d.[h]	—	0.7 ± 0.0	>150	−10	n.d.[h]	n.d.[h]	42
14	**26b** (R^2 = Na)		21	0.38	< −1.5	>150	−6.7 ± 0.1	n.d.[h]	n.d.[h]	n.d.[h]
15	**28a** (R^2 = Et)	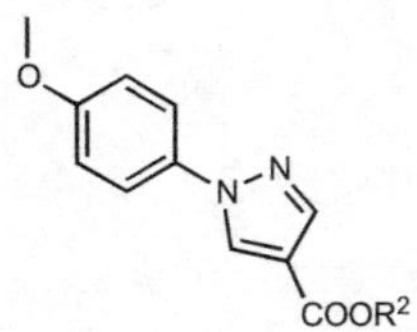	32.4	0.59	0.9 ± 0.0	>180	−6.6 ± 0.1	n.d.[h]	n.d.[h]	>120
16	**29a** (R^2 = Na)		111	2.02	n.d.[h]	>180	−10.0 ± 0.0	n.d.[h]	n.d.[h]	n.d.[h]
17	**28b** (R^2 = Et)	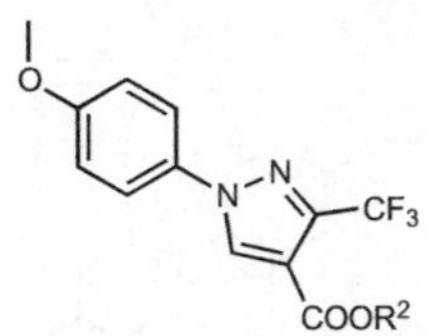	31.9	0.58	2.1 ± 0.0	>180	−5.7 ± 0.1	1.3 ± 0.1	12.4 ± 2.4	113
18	**29b** (R^2 = Na)		112	2.04	< −1.5	>180	−10.0 ± 0.0	n.d.[h]	n.d.[h]	n.d.[h]
19	**34a** (R^2 = Me)	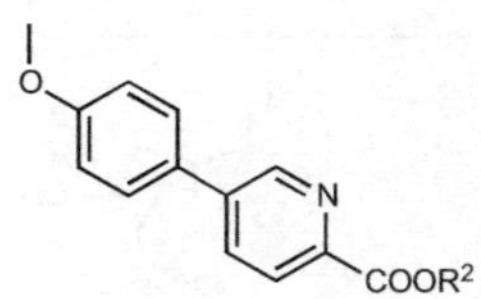	n.d.[h]	—	0.2 ± 0.0	>130	−7.5 ± 0.2	0.22 ± 0.05	2.3 ± 0.1	10
20	**35a** (R^2 = Na)		16	0.29	n.d.[h]	>180	−8.6 ± 1.6	n.d.[h]	n.d.[h]	n.d.[h]

Table 1 (*Continued*)

Entry	Compound	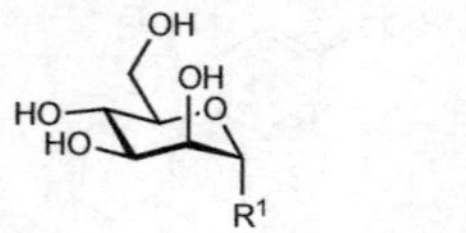	Binding Assay[a] IC$_{50}$ [nM]	rIC$_{50}$	$\log D_{7.4}$[b]	Solubility[c] [μg mL^{-1}]	PAMPA $\log P_e$[d] [cm s^{-1}]	Caco-2 P_{app}[e] [10^{-6} cm s^{-1}] a→b	b→a	Microsomal stability $t_{1/2}$[f] [min]
21	**34b** (R^2 = Me)	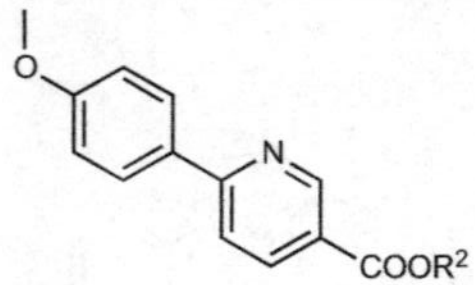	28.6	0.52	1.0 ± 0.0	59 ± 6	−6.3 ± 0.0	0.64 ± 0.06	8.3 ± 0.4	11
22	**35b** (R^2 = Na)		46	0.84	< −1.5	>180	−8.5 ± 1.8	n.d.[h]	n.d.[h]	n.d.[h]
23	**34c** (R^2 = Me)	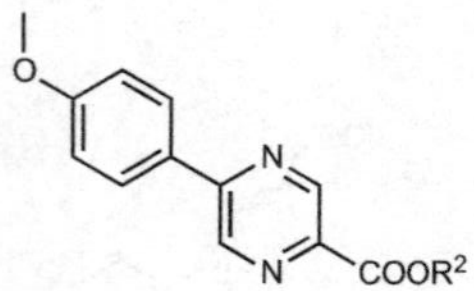	33.1	0.60	0.1 ± 0.1	>150	−7.6 ± 0.0	0.24 ± 0.01	1.8 ± 0.2	11
24	**35c** (R^2 = Na)		39	0.71	< −1.5	>180	−8.6 ± 1.6	n.d.[h]	n.d.[h]	n.d.[h]
25	**34d** (R^2 = Me)	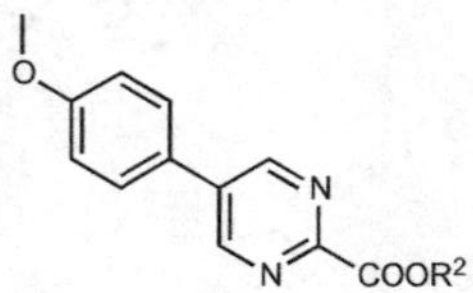	10.2	0.19	< −1.0	95 ± 6	−8.5 ± 0.1	0.16 ± 0.03	0.22 ± 0.05	24
26	**35d** (R^2 = Na)		35	0.64	n.d.[h]	n.d.[h]	−7.6 ± 1.6	n.d.[h]	n.d.[h]	n.d.[h]
27	**39** (R^2 = Me)	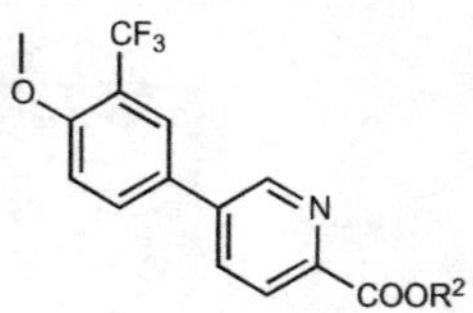	9.3	0.17	1.3 ± 0.1	>180	−8.6 ± 1.7	0.33 ± 0.04	7.2 ± 0.7	8.2
28	**40** (R^2 = Na)		20	0.36	< −1.5	>180	−9.3 ± 1.4	n.d.[h]	n.d.[h]	n.d.[h]
29	**42a**	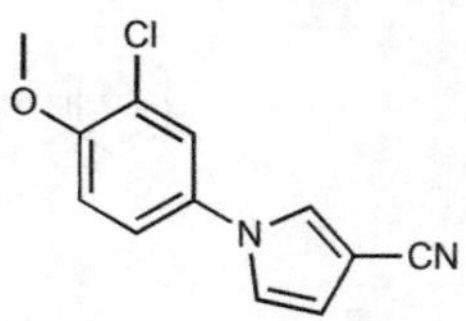	29	0.53	1.5 ± 0.1	>180	−8.8 ± 2.0	n.d.[h]	n.d.[h]	n.d.[h]

30	**42b**	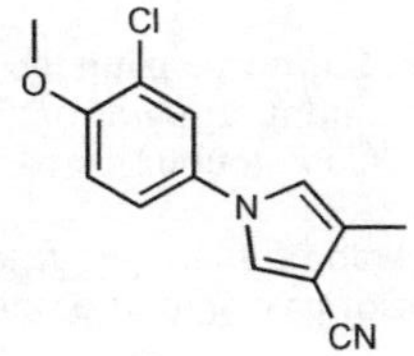	25	0.46	2.0 ± 0.1	69 ± 20	−6.3 ± 0.1	n.d.[h]	n.d.[h]	n.d.[h]
31	**42c** (R^2 = Et)	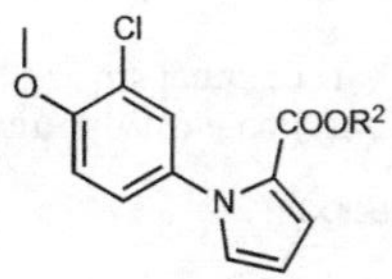	60.7	1.11	2.0 ± 0.0	>180	−5.2 ± 0.0	n.d.[h]	n.d.[h]	>120
32	**43** (R^2 = Na)		75	1.37	< −1.5	>180	−10.0 ± 0.0	n.d.[h]	n.d.[h]	n.d.[h]
33	**42d** (R^2 = Et)	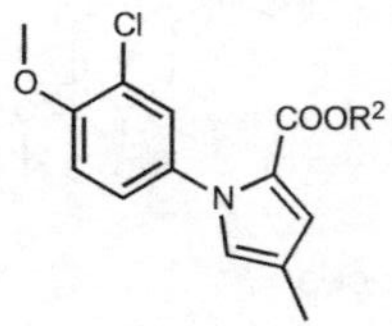	42.3	0.77	2.7 ± 0.0	34 ± 4	−4.8 ± 0.1	5.0 ± 0.2	35.6 ± 1.0	84
34	**44** (R^2 = Na)		23	0.42	0.7 ± 0.1	>180	−9.2 ± 1.7	n.d.[h]	n.d.[h]	n.d.[h]
35	**42e** (R^2 = Me)	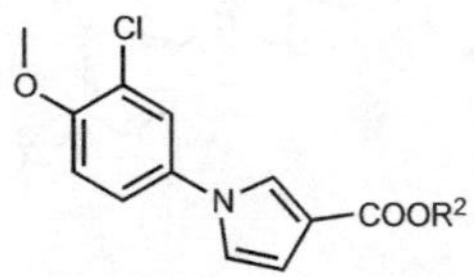	18.5	0.34	2.1 ± 0.2	>180	−6.0 ± 0.1	n.d.[h]	n.d.[h]	>120
36	**45** (R^2 = Na)		25	0.46	< −1.5	>180	−10.0 ± 0.0	n.d.[h]	n.d.[h]	n.d.[h]
37	**42f** (R^2 = Et)	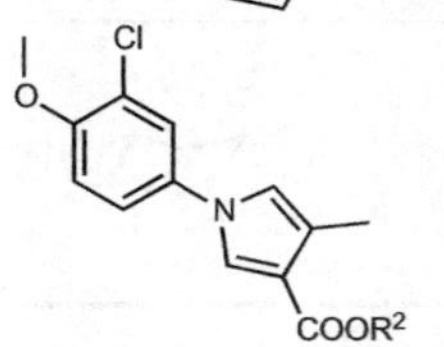	15.2	0.28	2.8 ± 0.1	>180	−4.8 ± 0.1	6.4 ± 0.7	30.0 ± 2.9	>120
38	**46** (R^2 = Na)		25	0.46	−0.5 ± 0.1	>180	−10.0 ± 0.0	n.d.[h]	n.d.[h]	n.d.[h]

Table 1 (*Continued*)

Entry	Compound	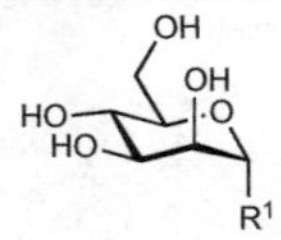	Binding Assay[a] IC50 [nM]	rIC50	$\log D_{7.4}$[b]	Solubility[c] [µg mL^{-1}]	PAMPA $\log P_e$[d] [cm s^{-1}]	Caco-2 P_{app}[e] [10^{-6} cm s^{-1}] a→b	b→a	Microsomal stability $t_{1/2}$[f] [min]
39	**49** (R^2 = Et)	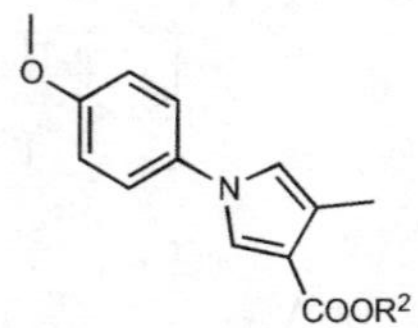	64.8	1.18	2.3 ± 0.0	>180	−5.1 ± 0.1	1.5 ± 0.5	17.2 ± 0.6	>120
40	**51** (R^2 = Na)		65	1.18	−1.0 ± 0.1	>180	−8.4 ± 1.8	n.d.[h]	n.d.[h]	n.d.[h]
41	**50** (R^2 = Et)	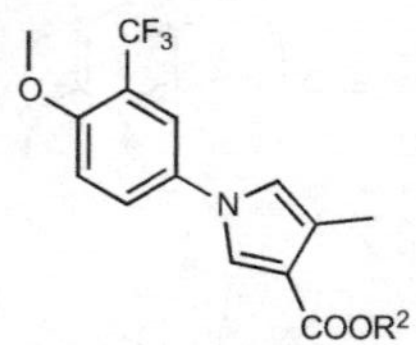	36.9	0.67	3.0 ± 0.1	135 ± 6	−5.0 ± 0.2	5.0 ± 0.3	26.1 ± 1.5	>120
42	**52** (R^2 = Na)		19	0.35	0.1 ± 0.2	>180	−8.6 ± 1.7	n.d.[h]	n.d.[h]	n.d.[h]

[a] The IC_{50} values were determined with a cell-free competitive binding assay.[28] The rIC_{50} values were calculated by dividing the IC_{50} of the compound of interest by the IC_{50} of reference compound **53**. This leads to rIC_{50} values below 1.0 for derivatives with higher affinity than reference **53** and rIC_{50} above 1.0 for compounds with lower affinity than **53**.
[b] Octanol–water distribution coefficients at pH 7.4 ($\log D_{7.4}$) were determined by a miniaturized shake flask procedure. The values are indicated as mean ± SD of sextuplicate determinations.[21]
[c] Kinetic solubility was measured in a 96-well format in triplicate at pH 7.4 using the µSOL Explorer solubility analyzer.[22]
[d] Permeation through an artificial membrane ($\log P_e$, effective permeability) was determined by PAMPA (parallel artificial membrane permeability assay) in quadruplicate at pH 7.4.[23,25]
[e] Permeation through a Caco-2 cell monolayer (P_{app}, apparent permeability) was assessed in the absorptive (a → b) and secretory (b → a) directions in triplicate.[24,29]
[f] Microsomal stability was determined in duplicate with pooled male rat liver microsomes (RLM, 0.125 mg mL^{-1}) at pH 7.4 and 37 °C.[27]
[g] Thermodynamic solubility obtained according to the procedure described in ref. 8*h*.
[h] n.d., not determined.

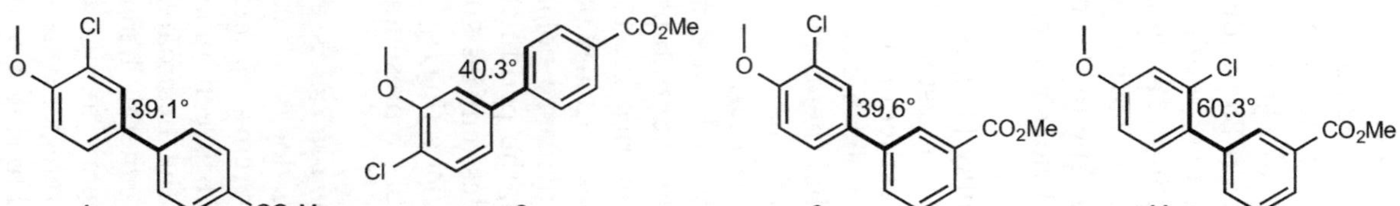

1a **2a** **6a** **11a**

Fig. 2 Dihedral angles between the aromatic rings of biphenyl aglycones of biaryl α-D-mannopyranosides (**1a**, **2a**, **6a**, and **11a**). The global minima values were calculated with MacroModel (version 9.9).[30]

disruption of the molecular planarity enhanced aqueous solubility up to 3-fold (from 41 to 134 $\mu g\,mL^{-1}$).

Permeability of biphenyl mannosides. Due to their high lipophilicity ($\log D_{7.4} \geq 1.7$), all biphenyl mannosides showed high effective ($\log P_e > -5.7\ cm\,s^{-1}$) and moderate apparent ($P_{app} > 2 \times 10^{-6}\,cm\,s^{-1}$) permeability.

Metabolic stability of biphenyl mannosides. Incubation with rat liver microsomes (RLM) induced a fast degradation of prodrug **1a** ($t_{1/2}$ 2.1 min, Table 1, entry 2). The esters of the biphenyl mannosides **2a** ($t_{1/2}$ 22 min, entry 4), **6a** ($t_{1/2}$ 72 min, entry 6), and **11a** ($t_{1/2}$ 13 min, entry 8) were less susceptible to the carboxylesterase (CES)-mediated metabolic turnover. The differing rates of hydrolysis may result from the change in the molecular geometry and therefore in the accessibility of the ester by the serine hydrolase CES.[31] We therefore attributed the rate differences of the CES-mediated hydrolysis primarily to the differing geometry of the aglycones, which, in case of **1a**, obviously orients the ester bond within the active site in an optimal position.

Overall, despite elevated solubility (134 $\mu g\,mL^{-1}$) and high effective permeability ($\log P_e\ -4.5\ cm\,s^{-1}$) of ester **11a**, fast metabolic ester hydrolysis ($t_{1/2} < 15$ min) reduces its therapeutic potential. Especially, the accelerated renal excretion of the polar metabolite (*i.e.* free carboxylate) impedes the maintenance of the therapeutic concentration in the bladder over an extended period of time.

Solubility of heteroaryl mannosides. In general, heterobiaryl mannosides (Table 1, entries 10–42) exhibited markedly higher aqueous solubility than the reference compound biphenyl α-D-mannoside **1a** (12 $\mu g\,mL^{-1}$). Triazole (→**25a–c**, entries 10, 12 & 13), pyrazole (→**28a**, entry 15), and pyridine derivatives (→**34a–d**, entries 19, 21, 23 & 25) exhibit higher solubility, due to increased polarity ($\log D_{7.4} < 1$). For the pyrazolylphenyl (→**28b**, entry 17), the pyridinylphenyl (→**39**, entry 27), and the pyrrolylphenyl mannosides (entries 29–42) high solubility was determined despite high lipophilicity ($\log D_{7.4} > 1$). As a trade-off, **42d** (entry 33) became the least soluble representative among all assessed heteroaryl mannosides (34 $\mu g\,mL^{-1}$).

Permeability of heteroaryl mannosides. As expected, the increase in polarity and solubility (*e.g.* →**25a**, $\log D_{7.4}\ -0.6$, entry 10) leads to a reduced permeability ($\log P_e -9.4\ cm\,s^{-1}$), *i.e.* poor oral absorption. Figure 3 illustrates the correlation between lipophilicity ($\log D_{7.4}$) and the effective permeability ($\log P_e$) of the heteroaryl mannoside esters and bioisosteres (entries 10–42). In general, heteroaryl mannosides with lipophilicity $\log D_{7.4} > 1$ have moderate to high effective permeability.

To enhance lipophilicity two strategies were explored: First, an *ortho*-chloro substituent was introduced to ring A of the biaryl aglycone of **25a** (→**25b**, entry 12) and, second, the methyl ester was replaced by an ethyl ester (→**25c**, entry 13). However, both strategies did not or only marginally influence the $\log P_e$ values and therewith the oral absorption potential. For the pyrazolylphenyl derivative **28a** (entry 15), although slightly more lipophilic than the triazolylphenyl **25c** (entry 13), low effective permeability ($\log P_e - 6.6\ cm\ s^{-1}$) was observed. By introducing a

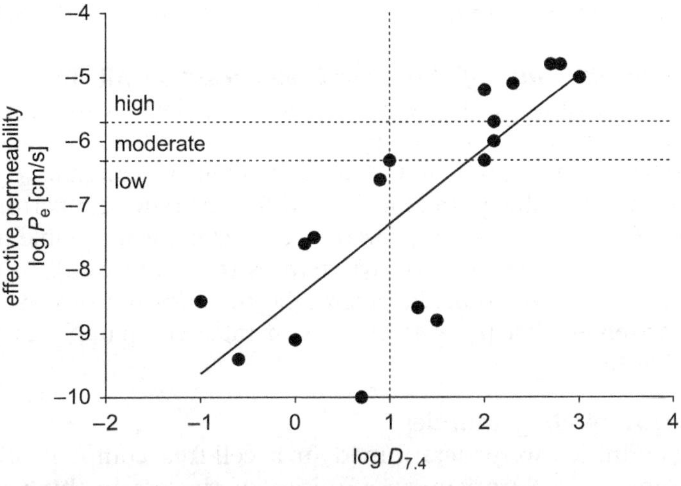

Fig. 3 Plot of $\log D_{7.4}$ *versus* effective permeability $\log P_e$ (cm s^{-1}; pH 7.4) for heteroaryl mannoside esters (entries 10–28, 31–42) and bioisosteres (entries 29 & 30). Data represent the mean only (Goodness of fit, R square = 0.64). Dashed lines on the *y*-axis indicate the thresholds between low, moderate, and high effective permeability (low < −6.3 cm s^{-1}, moderate $\log P_e \leq -5.7$ cm s^{-1} < high).[25] Dashed line on the *x*-axis represent the thresholds for lipophilicity ($\log D_{7.4} > 1$), where moderate effective permeability is reached.

trifluoromethyl substituent on the pyrazole moiety (→**28b**, entry 17), both lipophilicity (permeability) and solubility were improved, presumably sufficient for a successful intestinal uptake. Whereas introducing an *ortho*-trifluoromethyl substituent to ring A of the biaryl aglycone (→**39**, entry 27) exhibited despite increased lipophilicity only low effective permeability ($\log P_e$ −8.6 cm s^{-1}). By contrast, the pyrrolylphenyl mannosides **42a–f**, **49** and **50** (Table 1) with $\log D$ values of 1.5 to 3 and the $\log P_e$ values of −6.3 to −4.8 cm s^{-1} suggest modest to high permeability. Starting from antagonist **42c** (entry 31), the introduction of a 4-methyl substituent (→**42d**, entry 33) increased permeability, however, limited solubility to 34 μg mL^{-1}. In the case of the pyrrolylphenyl derivative **42e** (entry 35), both solubility and permeability could be successively enhanced by introducing a methyl substituent in the 4-position of the pyrrole moiety (→**49**, entry 39) and by modifying the *ortho*-substituent on ring A of the biaryl aglycone (→**50**, entry 41 & **42f**, entry 37). Especially, for the antagonists **42f** and **50**, the effective permeability (PAMPA, $\log P_e$ −4.8 cm s^{-1} and −5.0 cm s^{-1}, respectively) suggests a high oral absorption potential. Moreover, the absorptive flux (apical → basal) through the Caco-2 cell monolayer was moderate ($\log P_{app} > 2 \times 10^{-6}$ cm s^{-1}). Although the ratio $P_{app,b \to a}/P_{app,a \to b} > 2$ implies efflux–carrier activity,[29] we expected high systemic *in vivo* availability of **42f** and **50**, notably because efflux transporters at human intestines are considered to be easily saturable when compounds are administered at elevated doses (*e.g.* > 100 mg).[32] As expected, the bioisosteric replacement of the carboxylic moiety by a cyano group (→**42a,b**, entries 29 & 30) increased lipophilicity, but the PAMPA data still indicate low permeability for both derivatives ($\log P_e$ −8.8 cm s^{-1} and −6.3 cm s^{-1}, respectively).

Metabolic stability of heteroaryl mannosides. All five-membered heteroaryl derivatives with ester functions (triazoles→entries 10–14, pyrazoles→entries 15–18, and pyrroles→entries 29–42, Table 1) were found to be less susceptible to CES-mediated bioconversion ($t_{1/2}$ > 30 min) than the parent biphenyl mannoside **1a.** Surprisingly, the pyrrolylphenyl esters **42e,f, 49,** and **50** even proved to be metabolically stable and were only slowly hydrolyzed by carboxylesterases ($t_{1/2}$ > 120 min). In contrast, derivatives with six-membered heterocylic moieties (entries 19–28) were cleaved extremely fast ($t_{1/2}$ < 15 min) with rates comparable to biphenyl mannoside **1a.**

2.3 *In vitro* binding affinities

Binding affinities were determined in a cell-free competitive binding assay (Table 1).[28] Surprisingly, affinities of the esters (biphenyls and heteroaryl mannosides) were slightly improved compared to the corresponding acids, probably due to reduced desolvation costs.

Biphenyl mannosides. A comparison of antagonist **1b** (entry 3) with its regioisomers **2b, 6b,** and **11b** (entries 5, 7 & 9) indicates that changing the position of the carboxylic acid on the terminal ring B of the biphenyl aglycone as well as modifying the substitution pattern on ring A substantially reduced affinity. As previously reported, the *ortho*-chloro substituent present in the antagonists **1b** and **6b** provides additional van der Waals contacts leading to binding affinity in the low nanomolar range.[8k]

Heteroaryl mannosides. All heteroaryl mannosides (entries 10–42) were weaker binders than the biphenyl mannoside **1b,** however also showed IC_{50} values in the nanomolar range. *In silico* studies obtained with flexible docking (Glide software package[33]) to the FimH-CRD suggested a similar 'out-docking mode' (Fig. 4) for all heteroaryl mannosides (entries 10–42), establishing π–π-stacking with Tyr48 of the tyrosine gate.[8g]

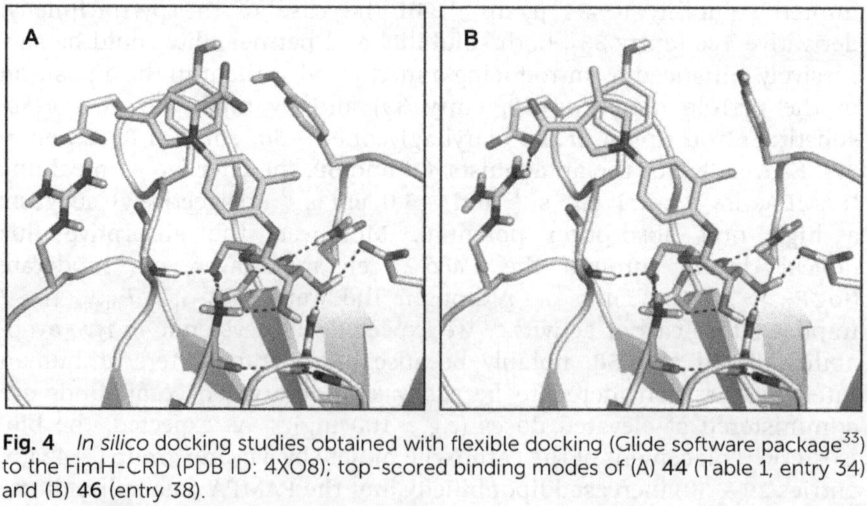

Fig. 4 *In silico* docking studies obtained with flexible docking (Glide software package[33]) to the FimH-CRD (PDB ID: 4XO8); top-scored binding modes of (A) **44** (Table 1, entry 34) and (B) **46** (entry 38).

Therefore, it is quite surprising that the triazolylphenyl mannosides (entries 10–14) and the pyrrolylphenyl mannosides (entries 29–42) showed approximately five-fold higher affinity than the pyrazolylphenyl analogues (entries 15–18). Furthermore, the substitution pattern had also an influence on the binding affinity as observed for the various pyrrolylphenyl mannosides (entries 29–42). In agreement with previous observations,[8i,k] an *ortho*-chloro or an *ortho*-trifluoromethyl substituent on ring A improved affinity approximately 3-fold (**51**, entry 40 *vs.* **46**, entry 38 & **52**, entry 42). Furthermore, the position of the electron-withdrawing carboxylic acid substituent in ring B affected the binding affinity as well. In the 3-position (→**45**, entry 36) a three-fold affinity increase compared to the 2-position (→**43**, entry 32) was observed. With an additional 4-methyl group, affinity was improved by a factor of three (**43**, entry 32 *vs.* **44**, entry 34), presumably due to sterically favored π–π stacking interactions between the heteroaromatic ring B and Tyr48 of the tyrosine gate.

2.4 *In vivo* pharmacokinetic study

Antagonist **42f** (entry 37, Table 1) exhibiting the best *in vitro* PK/PD profile was selected for an *in vivo* pharmacokinetic study. Since its solubility is not sufficient for the planned dose of 10 mg kg^{-1},[10a] 5% DMSO and 1% surfactant Tween 80 were used as solubilizer. The concentration–time profiles in urine and plasma are shown in Fig. 5.

Plasma concentrations of **42f** were low, with a C_{max} of only 0.04 μg mL^{-1} between 40 min and 1.5 h and dropped below the detection limit 3 h post application. In contrast, antagonist **42f** rapidly accumulates in the urine with a C_{max} ranging from 10 to 16 μg mL^{-1} at 1 to 3 h post application. After a stable concentration plateau, which is slightly shifted in time compared to plasma peak levels, **42f** could not be detected in urine 7 h post application.

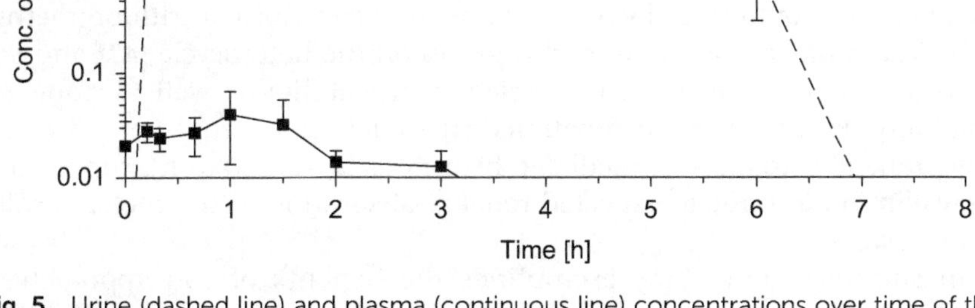

Fig. 5 Urine (dashed line) and plasma (continuous line) concentrations over time of the pyrrolylphenyl mannoside 42f after an application of a dose of 10 mg kg^{-1} (PBS containing 5% DMSO and 1% Tween 80). Shown are mean values with standard error of the mean for groups of three mice (C3H/HeN). The detection limit was at 0.02 μg mL^{-1}.

Accumulation in urine, resulting in a relatively constant plateau concentration over a time-period of about 3 h, can be related to several important interplaying mechanisms. Both, PAMPA[23] and transport through a Caco-2 cell monolayer predicted high (log P_e −4.8 cm s^{-1}) to moderate ($P_{app,a\text{-}b}$ 6.4 × 10^{-6} cm s^{-1}, Table 1) permeability for **42f**.[25,26] Although **42f** is a substrate of efflux transporters (Caco-2 cell experiment; b–a/a–b ratio > 2, Table 1),[29] the efflux transporters (*e.g.* P-gp) can be saturated by the applied high dose of 10 mg kg^{-1}.[32,34] Once in circulation, antagonist **42f** is filtered through the renal glomeruli in the kidneys. Based on its metabolic stability and a log $D_{7.4}$ value of 2.8, reabsorption from the filtrate in the proximal tubuli can be expected, leading to the observed delayed renal excretion.[10b,34,35] In summary, the observed PK profile of **42f** results from a good oral absorption (*i.e.* sufficient solubility and saturation of efflux transporter) with a delayed and prolonged elimination *via* the kidneys due to renal reabsorption.

3 Conclusions

Starting from ester **1a**, the present study aimed to optimize solubility and metabolic stability issues of biaryl mannosides in order to achieve high oral absorption of the ester and slow enzyme-mediated release of the polar metabolites.

First, our strategy to disrupt molecular planarity and symmetry of the biphenyl mannosides by modifying the substitution pattern, proved to be successful for improving solubility. Compared to antagonist **1a**, the ester **11a** showed a ten-fold improved solubility whereas membrane permeability remained high. However, since hepatic esterases rapidly convert ester **11a** to carboxylate **11b**, re-absorption from the tubuli is not possible leading to fast renal excretion. Furthermore, changing the positions of the substituents on the aglycone decreased affinity to the FimH-CRD, overriding the gain in the intestinal uptake potential (entries 4, 6 & 9).

In a second approach, the improvement of solubility and metabolic stability based on heterocyclic aglycones was studied. Thereby, triazole (**25a–c**, entries 10, 12 & 13), pyrazole (**28a,b**, entries 15 & 17), and six-membered heterocyclic moieties (**34a–d**, entries 19, 21, 23 & 25 and **39**, entry 27) proved highly beneficial to the aqueous solubility but in turn reduced lipophilicity and membrane permeability. Overall, only poor oral absorption could be observed. By contrast, the pyrrolylphenyl mannosides – optimized by the introduction of a chloro or trifluoromethyl substituent on ring A and a methyl group on the heterocycle (**42f** and **50**, entries 37 & 41) – exhibited sufficient permeability as well as aqueous solubility. Furthermore, incubations with rat liver microsomes, revealed low propensity to enzyme-mediated hydrolysis ($t_{1/2}$ > 120 min), supporting slow elimination due to expected renal re-absorption of the metabolically stable esters.

In summary, our study exemplifies the benefits of two approaches: Rearrangement of the substitution pattern to improve aqueous solubility and introduction of heteroaromatic aglycones to improve aqueous solubility and metabolic stability. For the ester **42f**, an optimal balance of pharmacodynamic, physicochemical and *in vitro* pharmacokinetic properties was realized. The measured high urine drug levels of **42f** (see Fig. 5) over an extended period of time considerably limit the dosing frequency and makes **42f** a promising candidate to be tested in a UTI disease model.

Abbreviations

Caco-2 cells	colorectal adenocarcinoma cells
CES	carboxylesterase
C_{max}	maximal concentration
CRD	carbohydrate recognition domain
D	octanol–water distribution coefficient
HPLC	high performance liquid chromatography
IC_{50}	half maximal inhibitory concentration
LC-MS	liquid chromatography mass spectrometry
PAMPA	parallel artificial membrane permeability assay
P_{app}	apparent permeability
P_e	effective permeability
RLM	rat liver microsomes
UPEC	uropathogenic *Escherichia coli*
UTI	urinary tract infection

Acknowledgements

The authors thank Prof. Dr med. Radek Skoda, Department of Biomedicine, University Hospital Basel, Switzerland, for giving us access to the animal facility. The financial support by the Swiss National Science Foundation (CS: SNF grant 200020_146202; DE: 31003A_144183; SK: 200020_129935) is gratefully acknowledged.

References

1 (a) T. M. Hooton and W. E. Stamm, Diagnosis and treatment of uncomplicated urinary tract infection, *Infect. Dis. Clin. North Am.*, 1997, **11**, 551–581; (b) T. J. Wiles, R. R. Kulesus and M. A. Mulvey, Origins and virulence mechanisms of uropathogenic *Escherichia coli*, *Exp. Mol. Pathol.*, 2008, **85**, 11–19; (c) S. D. Fihn, Acute uncomplicated urinary tract infection in women, *N. Engl. J. Med.*, 2003, **349**, 259–266.

2 (a) J. D. Schilling and S. J. Hultgren, Recent advances into the pathogenesis of recurrent urinary tract infections: the bladder as a reservoir for uropathogenic *Escherichia coli*, *Int. J. Antimicrob. Agents*, 2002, **19**, 457–460; (b) M. Blango and M. Mulvey, Persistence of uropathogenic *Escherichia coli* in the face of multiple antibiotics, *Antimicrob. Agents Chemother.*, 2010, **54**, 1855–1863.

3 C. Svanborg and G. Godaly, Bacterial virulence in urinary tract infection, *Infect. Dis. Clin. North Am.*, 1997, **11**, 513–529.

4 (a) M. A. Mulvey, Adhesion and entry of uropathogenic *Escherichia coli*, *Cell. Microbiol.*, 2002, **4**, 257–271; (b) D. S. Eto, T. A. Jones, J. L. Sundsbak and M. A. Mulvey, Integrin-mediated host cell invasion by type 1-piliated uropathogenic *Escherichia coli*, *PLoS Pathog.*, 2007, **3**, e100.

5 G. Capitani, O. Eidam, R. Glockshuber and M. G. Grütter, Structural and functional insights into the assembly of type 1 pili from *Escherichia coli*, *Microbes Infect.*, 2006, **8**, 2284–2290.
6 (a) I. Le Trong, P. Aprikian, B. A. Kidd, M. Forero-Shelton, V. Tchesnokova, P. Rajagopal, V. Rodriguez, G. Interlandi, R. Klevit, V. Vogel, R. E. Stenkamp, E. V. Sokurenko and W. E. Thomas, Structural basis for mechanical force regulation of the adhesin FimH via finger trap-like β sheet twisting, *Cell*, 2010, **141**, 645–655; (b) M. M. Sauer, R. P. Jakob, J. Eras, S. Baday, D. Eris, G. Navarra, S. Bernèche, B. Ernst, T. Maier and R. Glockshuber, Catch-bond mechanism of the bacterial adhesin FimH, *Nat. Commun.*, 2016, **7**, 10738.
7 N. Sharon, Carbohydrates as future anti-adhesion drugs for infectious diseases, *Biochim. Biophys. Acta*, 2006, **1760**, 527–537.
8 (a) N. Firon, I. Ofek and N. Sharon, Interaction of mannose-containing oligosaccharides with the fimbrial lectin of *Escherichia coli*, *Biochem. Biophys. Res. Commun.*, 1982, **105**, 1426–1432; (b) N. Firon, I. Ofek and N. Sharon, Carbohydrate specificity of the surface lectins of *Escherichia coli*, *Klebsiella pneumoniae*, and *Salmonella typhimurium*, *Carbohydr. Res.*, 1983, **120**, 235–249; (c) N. Sharon, Bacterial lectins, cell-cell recognition and infectious disease, *FEBS Lett.*, 1987, **217**, 145–157; (d) J. Bouckaert, J. Berglund, M. Schembri, E. De Genst, L. Cools, M. Wuhrer, C.-S. Hung, J. Pinkner, R. Slättegard, A. Zavialov, D. Choudhury, S. Langermann, S. J. Hultgren, L. Wyns, P. Klemm, S. Oscarson, S. D. Knight and H. De Greve, Receptor binding studies disclose a novel class of high-affinity inhibitors of the *Escherichia coli* FimH adhesin, *Mol. Microbiol.*, 2005, **55**, 441–455; (e) N. Firon, S. Ashkenazi, D. Mirelman, I. Ofek and N. Sharon, Aromatic alpha-glycosides of mannose are powerful inhibitors of the adherence of type 1 fimbriated *Escherichia coli* to yeast and intestinal epithelial cells, *Infect. Immun.*, 1987, **55**, 472–476; (f) O. Sperling, A. Fuchs and T. K. Lindhorst, Evaluation of the carbohydrate recognition domain of the bacterial adhesin FimH. Design, synthesis and binding properties of mannoside ligands, *Org. Biomol. Chem.*, 2006, **4**, 3913–3922; (g) Z. Han, J. S. Pinkner, B. Ford, R. Obermann, W. Nolan, S. A. Wildman, D. Hobbs, T. Ellenberger, C. K. Cusumano, S. J. Hultgren and J. W. Janetka, Structure-based drug design and optimization of mannoside bacterial FimH antagonists, *J. Med. Chem.*, 2010, **53**, 4779–4792; (h) T. Klein, D. Abgottspon, M. Wittwer, S. Rabbani, J. Herold, X. Jiang, S. Kleeb, C. Lüthi, M. Scharenberg, J. Bezençon, E. Gubler, L. Pang, M. Smiesko, B. Cutting, O. Schwardt and B. Ernst, FimH antagonists for the oral treatment of urinary tract infections: from design and synthesis to in vitro and in vivo evaluation, *J. Med. Chem.*, 2010, **53**, 8627–8641; (i) C. K. Cusumano, J. S. Pinkner, Z. Han, S. E. Greene, B. A. Ford, J. R. Crowley, J. P. Henderson, J. W. Janetka and S. J. Hultgren, Treatment and prevention of urinary tract infection with orally active FimH inhibitors, *Sci. Transl. Med.*, 2011, **3**, 109–115; (j) Z. Han, J. S. Pinkner, B. Ford, E. Chorell, J. M. Crowley, C. K. Cusumano, S. Campbell, J. P. Henderson, S. J. Hultgren and J. W. Janetka, Lead optimization studies on FimH antagonists: discovery of potent and orally bioavailable *ortho*-substituted biphenyl mannosides, *J. Med. Chem.*, 2012, **55**, 3945–3959; (k) L. Pang, S. Kleeb, K. Lemme, S. Rabbani, M. Scharenberg, A. Zalewski, F. Schädler, O. Schwardt and B. Ernst, FimH antagonists: structure-activity and structure-property relationships for biphenyl α-D-mannopyranosides, *ChemMedChem*, 2012, **7**, 1404–1422; (l) X. Jiang, D. Abgottspon, S. Kleeb, S. Rabbani, M. Scharenberg, M. Wittwer, M. Haug, O. Schwardt and B. Ernst, Anti-adhesion therapy for urinary tract infections – a balanced PK/PD profile

proved to be key for success, *J. Med. Chem.*, 2012, **55**, 4700–4713; (m) O. Schwardt, S. Rabbani, M. Hartmann, D. Abgottspon, M. Wittwer, S. Kleeb, A. Zalewski, M. Smiesko, B. Cutting and B. Ernst, Design, synthesis and biological evaluation of mannosyl triazoles as FimH antagonists, *Bioorg. Med. Chem.*, 2011, **19**, 6454–6473; (n) S. Brument, A. Sivignon, T. I. Dumych, N. Moreau, G. Roos, Y. Guérardel, T. Chalopin, D. Deniaud, R. O. Bilyy, A. Darfeuille-Michaud, J. Bouckaert and S. G. Gouin, Thiazolylaminomannosides as potent antiadhesives of type 1 piliated *Escherichia coli* isolated from Crohn's disease patients, *J. Med. Chem.*, 2013, **56**, 5395–5406; (o) S. Kleeb, L. Pang, K. Mayer, D. Eris, A. Sigl, R. C. Preston, P. Zihlmann, T. Sharpe, R. P. Jakob, D. Abgottspon, A. S. Hutter, M. Scharenberg, X. Jiang, G. Navarra, S. Rabbani, M. Smiesko, N. Lüdin, J. Bezençon, O. Schwardt, T. Maier and B. Ernst, FimH antagonists: bioisosteres to improve the in vitro and in vivo PK/PD profile, *J. Med. Chem.*, 2015, **58**, 2221–2239.

9 (a) T. K. Lindhorst, C. Kieburg and U. Krallmann-Wenzel, Inhibition of the type 1 fimbriae-mediated adhesion of *Escherichia coli* to erythrocytes by multiantennary D-mannosyl clusters: The effect of multivalency, *Glycoconjugate J.*, 1998, **15**, 605–613; (b) N. Nagahori, R. T. Lee, S.-L. Nishimura, S. Pagé, R. Roy and Y. C. Lee, Inhibition of adhesion of type 1 fimbriated *Escherichia coli* to highly mannosylated ligands, *ChemBioChem*, 2002, **3**, 836–844; (c) C. C. M. Appeldoorn, J. A. F. Joosten, F. A. Maate, U. Dobrindt, J. Hacker, R. M. J. Liskamp, A. S. Khan and R. J. Pieters, Novel multivalent mannose compounds and their inhibition of the adhesion of type 1 fimbriated uropathogenic *E. coli*, *Tetrahedron: Asymmetry*, 2005, **16**, 361–372; (d) A. Patel and T. K. Lindhorst, A modular approach for the synthesis of oligosaccharide mimetics, *Carbohydr. Res.*, 2006, **341**, 1657–1668; (e) M. Touaibia, A. Wellens, T. C. Shiao, Q. Wang, S. Sirois, J. Bouckaert and R. Roy, Mannosylated G(0) dendrimers with nanomolar affinities to *Escherichia coli* FimH, *ChemMedChem*, 2007, **2**, 1190–1201; (f) M. Durka, K. Buffet, J. Iehl, M. Holler, J.-F. Nierengarten, J. Taganna, J. Bouckaert and S. P. Vincent, The functional valency of dodecamannosylated fullerenes with *Escherichia coli* FimH – towards novel bacterial antiadhesives, *Chem. Commun.*, 2011, **47**, 1321–1323; (g) J. Bouckaert, Z. Li, C. Xavier, M. Almant, V. Caveliers, T. Lahoutte, S. D. Weeks, J. Kovensky and S. G. Gouin, Heptyl α-D-mannosides grafted on a β-cyclodextrin core to interfere with *Escherichia coli* adhesion: An *in vivo* multivalent effect, *Chem. – Eur. J.*, 2013, **19**, 7847–7855.

10 (a) C. A. Lipinski, Drug-like properties and the causes of poor solubility and poor permeability, *J. Pharmacol. Toxicol. Methods*, 2000, **44**, 235–249; (b) H. van de Waterbeemd, D. Smith, K. Beaumont and D. Walker, Property-based design: Optimization of drug absorption and pharmacokinetics, *J. Med. Chem.*, 2001, **44**, 1313–1333.

11 (a) M. Ishikawa and Y. Hashimoto, Improvement in aqueous solubility in small molecule drug discovery programs by disruption of molecular planarity and symmetry, *J. Med. Chem.*, 2011, **54**, 1539–1554; (b) A.-E. F. Nassar, A. M. Kamel and C. Clarimont, Improving the decision-making process in the structural modification of drug candidates: enhancing metabolic stability, *Drug Discovery Today*, 2004, **9**, 1020–1028.

12 N. A. Meanwell, Synopsis of some recent tactical application of bioisosteres in drug design, *J. Med. Chem.*, 2011, **54**, 2529–2591.

13 I. L. Scott, R. V. Market, R. J. DeOrazio, H. Meckler and T. P. Kogan, Stereospecific α-D-mannosylation, *Carbohydr. Res.*, 1999, **317**, 210–216.

14 A. Titz, Z. Radic, O. Schwardt and B. Ernst, A safe and convenient method for the preparation of triflyl azide, and its use in diazo transfer reactions to

primary amines, *Tetrahedron Lett.*, 2006, **47**, 2383–2385; R. Yan, F. Yang, Y. Wu, L. Zhang and X. Ye, An efficient and improved procedure for preparation of triflyl azide and application in catalytic diazotransfer reaction, *Tetrahedron Lett.*, 2005, **46**, 8993–8995.
15 V. Rostovtsev, L. Green, V. Fokin and K. Sharpless, A stepwise Huisgen cycloaddition process: Copper(I)-catalyzed regioselective "ligation" of azides and terminal alkynes, *Angew. Chem., Int. Ed.*, 2002, **41**, 2596–2599.
16 J. Antilla, J. Baskin, T. Barder and S. Buchwald, Copper-diamine-catalyzed *N*-arylation of pyrroles, pyrazoles, indazoles, imidazoles, and triazoles, *J. Org. Chem.*, 2004, **69**, 5578–5587.
17 (a) M. Node, K. Nishide, K. Fuji and E. Fujita, Hard acid and soft nucleophile system. 2. Demethylation of methyl esters of alcohol and phenol with an aluminum halide-thiol system, *J. Org. Chem.*, 1980, **45**, 4275–4277; (b) M. Node, K. Kumar, K. Nishide, S. Ohsugi and T. Miyamoto, Odorless substitutes for foul-smelling thiols: Syntheses and applications, *Tetrahedron Lett.*, 2001, **42**, 9207–9210.
18 B. Le Bourdonnec, R. Windh, L. Leister, Q. Zhou, C. Ajello, M. Gu, G. Chu, P. Tuthill, W. Barker, M. Koblish, D. Wiant, T. Graczyk, S. Belanger, J. Cassel, M. Feschenko, B. Brogdon, S. Smith, M. Derelanko, S. Kutz, P. Little, R. DeHaven, D. DeHaven-Hudkins and R. Dolle, Spirocyclic delta opioid receptor agonists for the treatment of pain: Discovery of *N,N*-diethyl-3-hydroxy-4-(spiro[chromene-2,4'-piperidine]-4-yl) benzamide (ADL5747), *J. Med. Chem.*, 2009, **52**, 5685–5702.
19 A. Katritzky, D. Cheng and R. Musgrave, Syntheses of imidazoles and pyrroles: BetMIC and TosMIC as complementary reagents, *Heterocycles*, 1997, **44**, 67–70.
20 M. Prieto, E. Zurita, E. Rosa, L. Munoz, P. Lloyd-Williams and E. Giralt, Arylboronic acids and arylpinacolboronate esters in Suzuki coupling reactions involving indoles. Partner role swapping and heterocycle protection, *J. Org. Chem.*, 2004, **69**, 6812–6820.
21 J. C. Dearden and G. M. Bresnen, The measurement of partition coefficients, *Quant. Struct. – Act. Relat.*, 1988, **7**, 133–144.
22 A. Avdeef, High-througput measurements of solubility profiles, in *Pharmacokinetic Optimization in Drug Research; Biological, Physicochemical and Computational Strategies*, ed. B. Testa, H. van de Waterbeemd, G. Folkers and R. Guy, Verlag Helvetica Chimica Acta, Zurich, 2001, pp. 305–326.
23 M. Kansy, F. Senner and K. Gubernator, Physicochemical high throughput screening: Parallel artificial membrane permeation assay in the description of passive absorption processes, *J. Med. Chem.*, 1998, **41**, 1007–1010.
24 P. Artursson and J. Karlsson, Correlation between oral-drug absorption in humans and apparent drug permeability coefficients in human intestinal epithelial (Caco-2) cells, *Biochem. Biophys. Res. Commun.*, 1991, **175**, 880–885.
25 A. Avdeef, S. Bendels, L. Di, B. Faller, M. Kansy, K. Sugano and Y. Yamauchi, PAMPA – critical factors for better predictions of absorption, *J. Pharm. Sci.*, 2007, **96**, 2893–2909.
26 T. Hou, J. Wang, W. Zhang and X. Xu, ADME evaluation in drug discovery. 7. Prediction of oral absorption by correlation and classification, *J. Chem. Inf. Model.*, 2007, **47**, 208–218.
27 M. Taketani, M. Shii, K. Ohura, S. Ninomiya and T. Imai, Carboxylesterase in the liver and small intestine of experimental animals and human, *Life Sci.*, 2007, **81**, 924–932.
28 S. Rabbani, X. H. Jiang, O. Schwardt and B. Ernst, Expression of the carbohydrate recognition domain of FimH and development of a competitive binding assay, *Anal. Biochem.*, 2010, **407**, 188–195.

29 I. Hubatsch, E. G. Ragnarsson and P. Artursson, Determination of drug permeability and prediction of drug absorption in Caco-2 monolayers, *Nat. Protoc.*, 2007, **2**, 2111–2119.
30 MacroModel, version 9.9, Schrödinger, LLC, New York, NY, 2012. Torsion angle measurements are based on the lowest-energy conformers.
31 (a) R. M. Wadkins, C. L. Morton, J. K. Weeks, L. Oliver, M. Wierdl, M. K. Danks and P. M. Potter, Structural constraints affect the metabolism of 7-ethyl-10-[4-(1-piperidino)-1-piperidino]carbonyloxycamptothecin (CPT-11) by carboxylesterases, *Mol. Pharmacol.*, 2001, **60**, 355–362; (b) J. M. Hatfield, M. Wierdl, R. M. Wadkins and P. M. Potter, Modifications of human carboxylesterase for improved prodrug activation, *Expert Opin. Drug Metab. Toxicol.*, 2008, **4**, 1153–1165; (c) G. Vistoli, A. Pedretti, A. Mazzolari and B. Testa, In silico prediction of human carboxylesterase-1 (hCES1) metabolism combining docking analyses and MD simulations, *Bioorg. Med. Chem.*, 2010, **18**, 320–329; T. Satoh and M. Hosokawa, Structure, function and regulation of carboxylesterases, *Chem. Biol. Interact.*, 2006, **162**, 195–211.
32 J. H. Lin, Drug-drug interaction mediated by inhibition and induction of P-glycoprotein, *Adv. Drug Delivery Rev.*, 2003, **21**, 53–81.
33 Glide, version 5.7, Schrödinger, LLC, New York, NY, 2011; 3D representations were made by the PyMOL Molecular Graphics System, Version 1.6 Schrödinger, LLC.
34 D. A. Smith, B. C. Jones and D. K. Walker, Design of drugs involving the concepts and theories of drug metabolism and pharmacokinetics, *Med. Res. Rev.*, 1996, **16**, 243–266.
35 M. V. Varma, B. Feng, R. S. Obach, M. D. Troutman, J. Chupka, H. R. Miller and A. El-Kattan, Physicochemical determinants of human renal clearance, *J. Med. Chem.*, 2009, **52**, 4844–4852.

Carbohydrate steroid hybrid architectures: the viewpoint of amphiphilicity and self-organisation

Zonglong Yang,[a] Rui Xu,[a] Stéphane Chambert,[a] Laurent Soulère,[a] Mohammed Ahmar,[a] Grahame Mackenzie,*[b] Stephen J. Cowling,[c] John W. Goodby*[c] and Yves Queneau*[a,b]

DOI: 10.1039/9781782626657-00274

Carbohydrate steroid hybrids (steroid glycosides, glycosyl steroids and related compounds) are ubiquitous natural products, with immense structural diversity. They also show a common feature of being amphiphilic in character, which will be the major subject reviewed in this chapter. It will be preceded by an introduction on the occurrence and properties of glycosteroids, and their classification into three main sub-families based on their overall molecular shape and polarity where systems have: (i) one saccharidic part and one steroidal part, possibly including a spacer chain; (ii) saccharidic connections at two different points of the steroid, and (iii) an additional appendage that can be either polar or non-polar.

1 Introduction

Glycosteroids are natural products showing a wide range of biological and pharmacological properties. Of prime importance is their specific combination of steroidal and saccharide units involving myriads of possibilities due to variations in the steroidal backbone and in the carbohydrate moiety. This complexity, results in a very diverse scope of roles and properties that provides fascinating challenges and opportunities for drug design. The occurrence in Nature, the biological roles and the pharmacological properties of glycosteroids have been reviewed in many reports, whereas others focused more on either structural aspects or synthetic approaches.[1–17]

Glycosteroids are hybrids in which two moieties with inverse polarity are connected; hence, they must be seen as amphiphilic systems, and indeed, some of their properties are connected with their ability to interact with other biomolecules within cell membranes and contribute to organized molecular and supramolecular systems. Amphiphilicity of such biomolecular systems often relates to the field of liquid crystals, and

[a]*Université de Lyon, INSA-Lyon, UMR 5246, CNRS, Université Lyon 1, CPE Lyon, ICBMS, Institut de Chimie et de Biochimie Moléculaires et Supramoléculaires, Bât. J. Verne, 20 avenue A. Einstein, 69621 Villeurbanne, France. E-mail: yves.queneau@insa-lyon.fr*

[b]*Department of Chemistry, University of Hull, Cottingham Road, Hull HU6 7RX, UK. E-mail: G.Mackenzie@hull.ac.uk*

[c]*Department of Chemistry, The University of York, University Way, Heslington, York YO10 5DD, UK. E-mail: john.goodby@york.ac.uk*

indeed, it is now established that the liquid crystalline state is widely encountered in biological systems.

Actually, the first observation of the liquid crystalline state was made when studying the melting behaviour of naturally occurring cholesteryl esters found in plants.[18,19] It was later found that many other families of biomolecules also exhibit the ability to exist in the liquid crystalline state, which is now understood to be based on the self-assembly and self-organization processes of elementary molecules.[20–28] Self-assembled or self-organized systems can result from auto-association of one single type of molecule, but in Nature systems are generally formed by more complex associations of different species. For example, within the regular lipidic bilayer organization of cell membranes, other components also establish low-energy interactions, such as glycosphingolipids that interact with cholesterol *via* non-covalent forces.[29] These so-called "lipid rafts",[30–36] are complex assemblies suggested to be the place where many biological functions occur. In such systems and others, glycolipids show a characteristic ability for existing in the liquid crystalline state.[37,38]

Having investigated glycolipids over the years[39–43] with the specific viewpoint of their self-assembly, self-organization and self-sorting properties, we later entered into the field of glycosteroids with the same curiosity.[44–46] Actually, the amphiphilicity of glycosteroids and their intriguing melting behaviour have been observed from the very beginning of investigations on naturally occurring plant sterol glycosides. For example, sitosteryl β-D-glucopyranoside **1** (Fig. 1) was studied in the early 20th century, at a time when the melting point was one of the key methods for characterizing new molecules, either extracted or newly synthesized by comparing them to authentic samples, and assessing their purity.[47] Another typical example is that of α-solanine, (**2**), a toxic molecule found in green potatoes, which was found to exhibit a fluid-like lamellar phase between the solid and liquid states.[19,48]

Diverse types of cholesterol-containing glycolipids can be found in Nature, which are referred to as steroid or steroidal glycosides, glycosyl steroids, glycosteroids, glycoalkaloids or terpene glycosides. Many of them belong to the family of steryl glucosides (SG), in which a glycosyl head group is attached to the 3-hydroxy group of a sterol; they are secondary metabolites in many plants, fungi, bacteria and animals.[1] One well known example is the family of saponins, some of which are thought to be the active compounds in traditional medicine.[3,13] The biological and pharmacological activities of saponins, such as cancer-related, anti-inflammatory, immunomodulatory, *etc.*, are abundantly documented.[2–5,8,9,13,15] Their family name "saponin" comes from their ability to stabilize foams and act as a soap (from the Latin "sapo"). These properties also result in possible uses of saponins such as foaming agents, emulsifiers, surfactants and adjuvants.

There are several other types of glycosteroid systems, which differ in their steroidal moiety, with variations in the hydroxylation pattern around the ABCD traditional steroid scaffold, in some the type of the additional E-ring with substituents on the E ring system, or its replacement by an open chain appendage, with the presence of double bonds, of

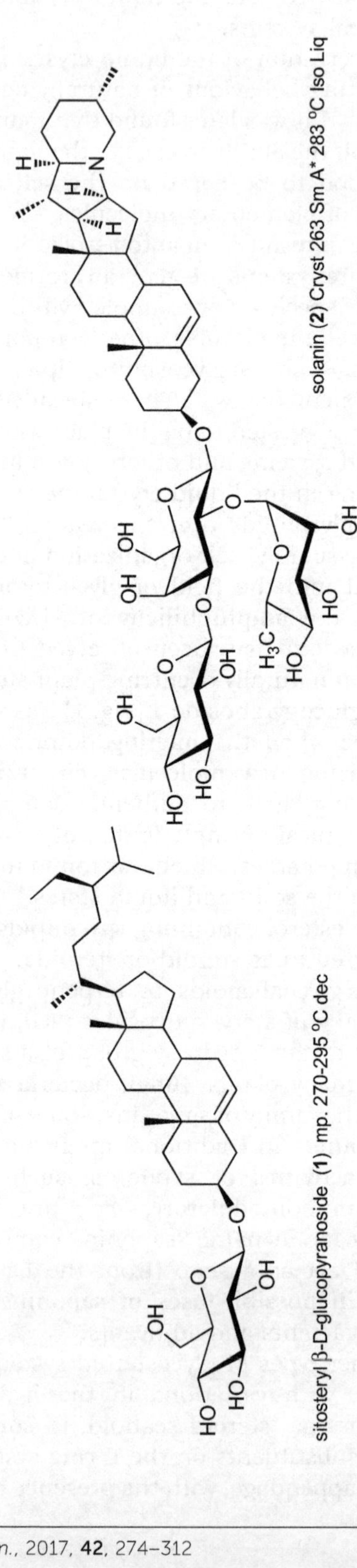

sitosteryl β-D-glucopyranoside (1) mp. 270-295 °C dec

solanin (2) Cryst 263 Sm A* 283 °C Iso Liq

Fig. 1 Strucutres of sistosteryl beta glucoside and of α-solanine.

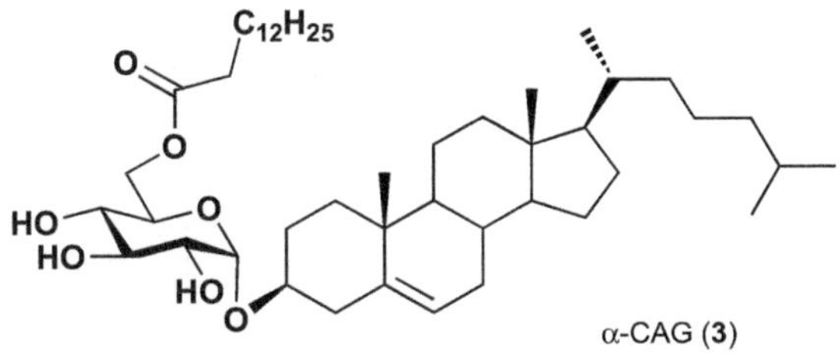

α-CAG (**3**)

Fig. 2 The structure of α-CAG, found in the membrane of *Helicobacter pylori*.

keto groups, *etc.* On the carbohydrate side, monosaccharidic, disaccharidic or more complex systems are encountered, including systems in which some saccharides link to the steroidal system at more than one site. Furthermore, these saccharides can be substituted with other groups, such as fatty acid esters in common with those found in the membranes of some bacteria, such as α-CAG (**3**), in *Helicobacter pylori* (Fig. 2). Part of the biological role of glycosteroids relates to their physicochemical behaviour; for example, it is thought α-CAG is implicated in the extraction of cholesterol from host membranes to those of bacteria.[49] Morphological changes and their consequences on membrane biophysics, notably the liquid crystalline behaviour, were considered as important parameters in the biological role or effects of compounds associated with the family of steroidal glycosides (SGs) or their acylated counterparts (ASGs).[1]

Thus far we have reviewed evidence in support of molecular shape being a key aspect of the construction of supramolecular assemblies. In the following section we provide a list of typical shapes that are encountered in the glycosteroid family, before focusing on reviewing the work reported on physicochemical investigations of the self-organisation behaviour of some of the family and model compounds.

2 The different classes of glycosteroids, classified by molecular shape

2.1 Main structural features and denomination system

There are several kinds of conjugates in which steroids are incorporated,[2] among them are: steroid–polyamine or polyether conjugates, such as squalamine (**4**) and the steroid (**5**) in which the steroid of each serves as a backbone for mimicking the overall structure of the cyclic peptide polymyxin B antibiotic; the steroid polyoxyethylenic diester (**6**), which shows activity similar to amphotericin B; steroid–amino acid conjugates, such as bufetoxin (**7**), which is isolated from the Chinese hoptoad and steroid–carbohydrate conjugates, such as digoxin (**8**), which is a naturally occurring cardiotonic glycoside (Fig. 3). Several types of hybrids combining cholesterol or bile acids with a biologically active moiety (β-lactam, hormone, porphyrin, nucleotides…) or with a second steroidal system (bis-steroidal conjugates) have also been reported. Some are multicomponent hybrids, for example having a steroid and both a

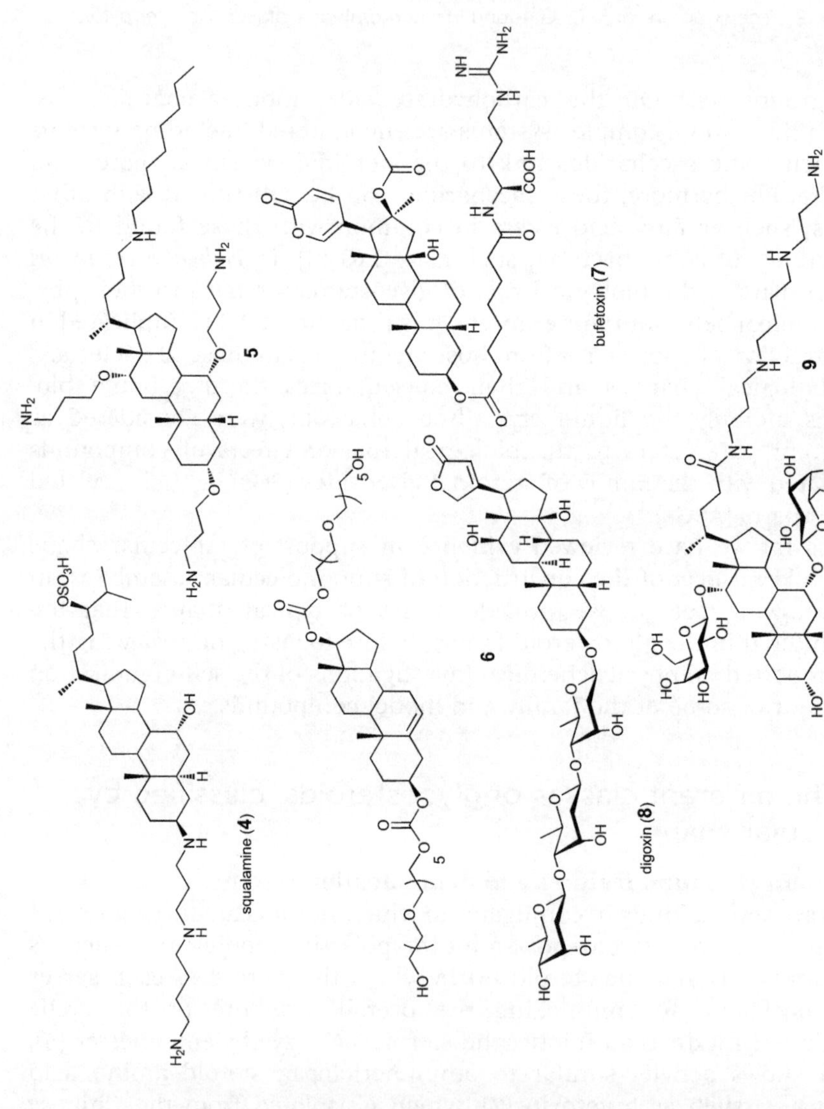

Fig. 3 Several types of steroidal conjugates.

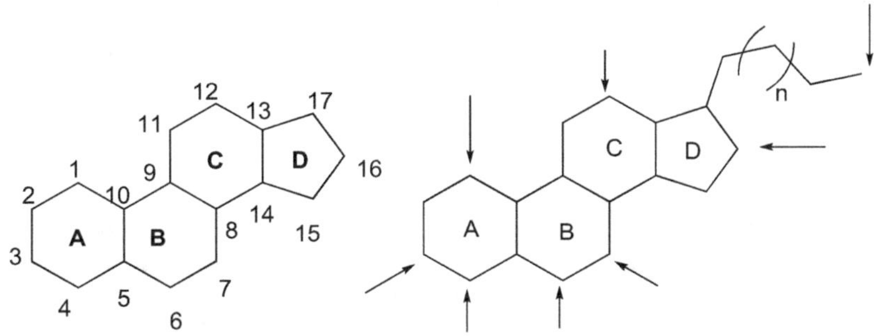

Fig. 4 Typical connective points (shown by arrows) between the steroidal and saccharidic moieties in most glycosteroids.

polyamine and a carbohydrate, such as bile acid conjugate (**9**), which exhibited promising properties for DNA transfection.[17]

In this chapter we focus only on steroid conjugates, which possess a carbohydrate moiety that can be either simple in construction or bear an additional appendage (often as a fatty acid ester group), which in some cases can involve a spacer, as for the most part in the case of synthetic analogues specifically designed for biological or physicochemical purposes.

Glycosteroids, in general, differ by: (i) the position at which connection is made with the carbohydrate moiety (which can be *via* either a glycosidic or ester linkage) (Fig. 4); (ii) the type of steroid (with the variability provided by: the number of cyclic systems; functional group pattern on the backbone or on external appendages; modifications on the backbone) (Fig. 5); (iii) the nature of the carbohydrate part (with several levels of complexity including: mono or oligosaccharides; specific nature of the monosaccharidic units; internal connectivity within the oligosaccharidic part; additional substitutions such as fatty acyl chains or ionic groups) (Fig. 6). Marrying the two families of saccharides and steroids, each of which show a fantastic level of structural variability, offers Nature one of the subtlest combination for molecular specificity.

In the following sections, we will use some code names for differentiating classes of compounds with respect to their overall molecular shape, using C for carbohydrate, S for steroid, F for fatty chain, *sp* for spacer and X for any other additional appendage, as schematized in Fig. 7.

2.2 "CS" and "C*sp*S" systems

The "CS" systems are those of simple carbohydrate–steroid hybrids, with one carbohydrate moiety connected to the steroidal backbone. Cholesteryl β-D-glucopyranoside is the simplest example. It was first synthesized in the early years of the 20th century along with a variety of glucosides to assist characterization of some natural vegetal extracts.[47] Since then, an immense number of different kinds of CS-type derivatives have been identified, mostly from plants, but also from fungi, bacteria and animals. Typical examples of triterpene glycosides including the saponins family

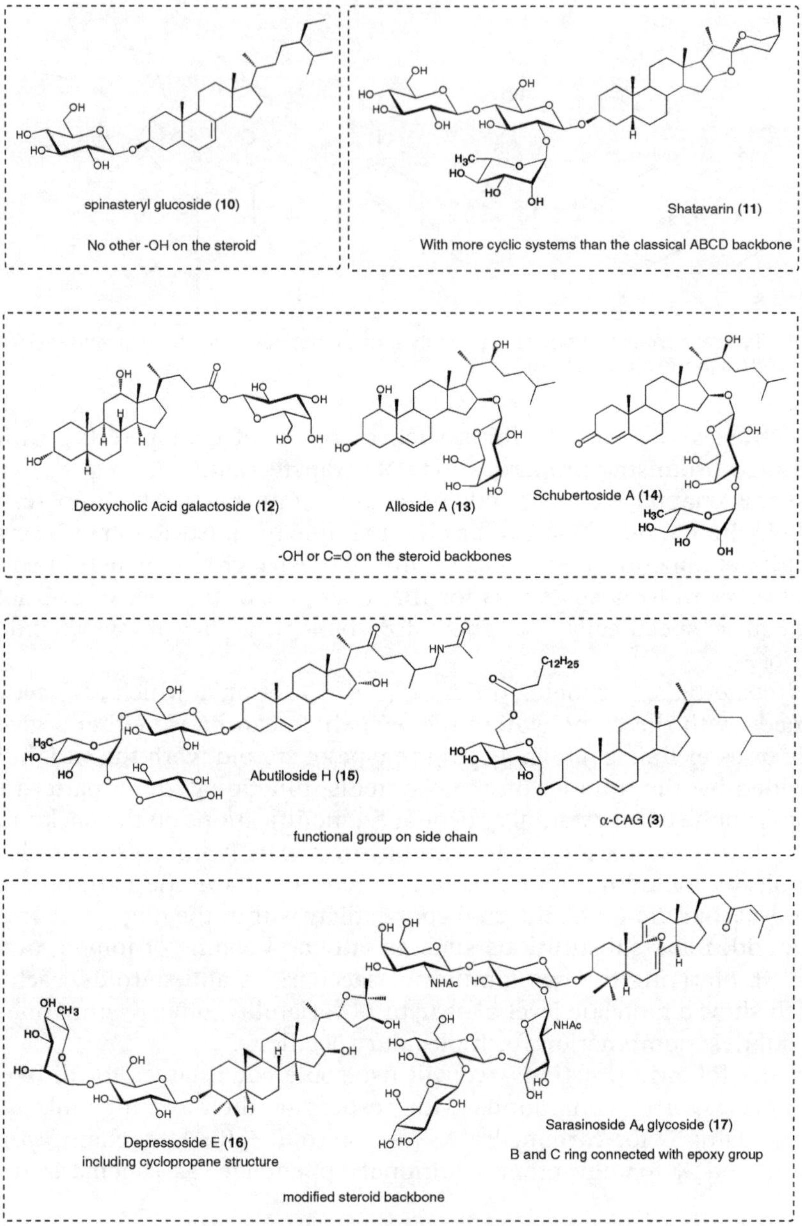

Fig. 5 Typical variations in the aglycon structure.[6,50–57]

and cardiac glycosides are shown in Fig. 8. For example, capsin (**33**) is a rather complex dammarane triterpene glycoside, extracted from the leaves of *Corchorus capsdaris*, identified by C. M. Hasan *et al.* as the 3-glucoside of 20, 24-epoxy-3β, 12β,25,30-tetrahydroxydammarane (capsugenin).[71] Other types of interesting systems are bile acids, such as ursodeoxycholic acid *N*-acetylglucosaminide (**34**), extracted from human urine, and identified by H.-U. Marschall *et al.* as a new type of bile acid conjugate.[72] Analogues galactosyl esters were also reported by

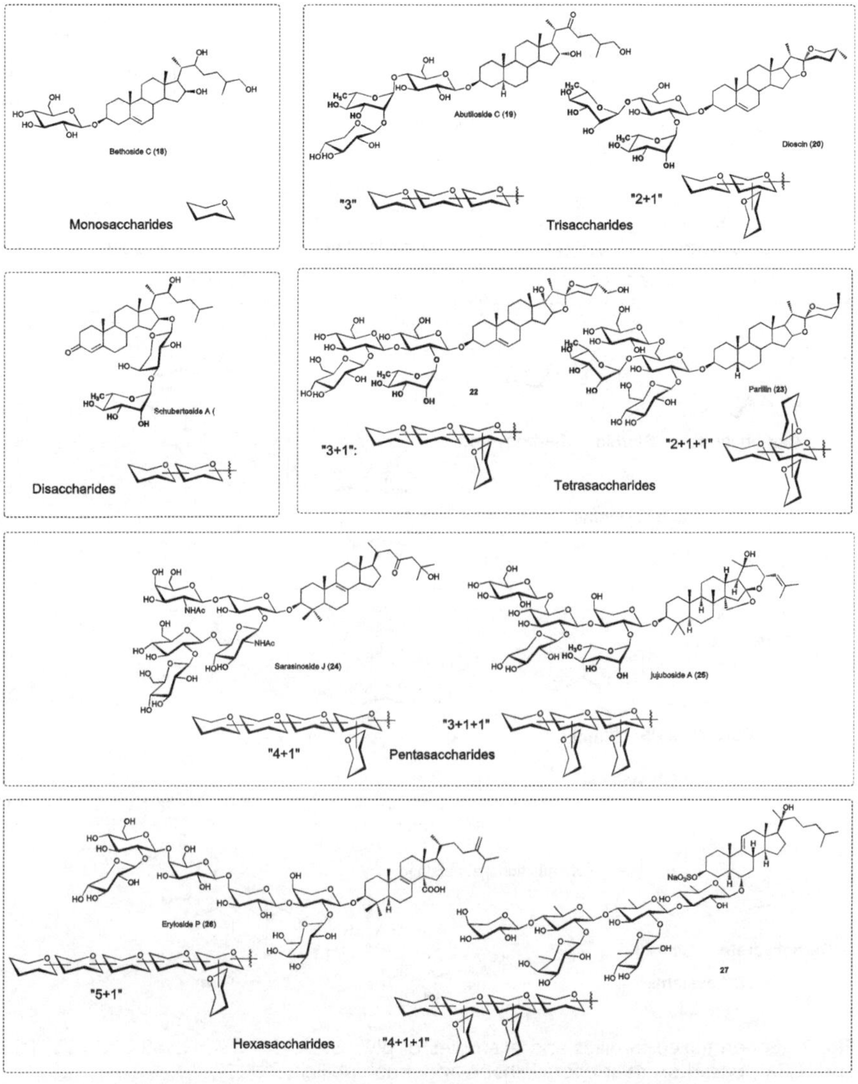

Fig. 6 Typical variations in the carbohydrate moiety.[58–66]

J. Goto and coworkers.[52] As with many other saponins, ginsenoside R_{g3} (**35**), having a simple β-2-*O*-glucosyl-β-glucose carbohydrate moiety, and found in the plant genus *Planax* (ginseng), exhibits potent antitumor activity.[5,73] Digitonin (**36**), a steroid conjugate with a more complex pentasaccharidic moiety, isolated from *Digitalis purpurea*,[74] is known for its ability to induce lysis in erythrocytes and other cells that contain cholesterol. The trisaccharidic digitoxin (**37**), is a widely prescribed drug for treating congestive heart failure and cardiac arrhythmia. It is a natural compound extracted from *Digitalis purpurea*, but can also be synthesized. For example, B. Yu and coworkers reported a very selective approach employing 3,4-di-*O*-*tert*-butyldiphenylsilyl-D-digitoxosyl *ortho*-cyclopropylethynylbenzoate as a donor and $Ph_3PAuOTf$ as catalyst, securing the formation of the β-digitoxosidic linkages in excellent yields

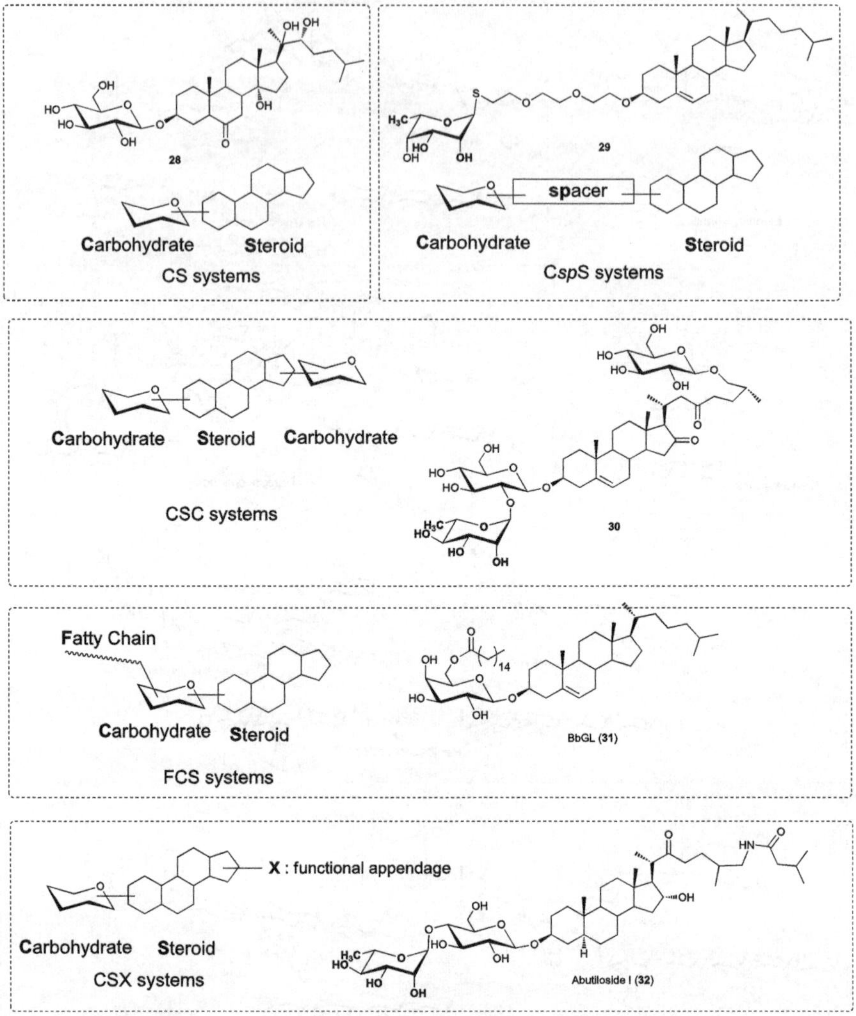

Fig. 7 Schematized families and examples of glycosteroids "CS", "C*sp*S", "CSC", "FCS" and "CSX", based on their overall shape and amphiphilicity.[55,67–70]

with β/α selectivity, and a 52% overall yield for the 9-step synthesis.[75] From marine organisms, an example is the complex pentasaccharidic holothurinoside A (**38**) isolated from the sea cucumber *Holothuria forskalii*, for which an interesting anticancer activity was found against P388 cells.[76] Fascinating synthetic work has been achieved by several groups either for synthesizing the natural glycosteroids in order to determine their biological properties, or analogues for varying the properties.[14,15,17] Some key contributions focus more on the way to perform the glycosylation efficiently to connect the steroid to the saccharide, either chemically or enzymatically.[16,77–79] Other papers have focused on the elaboration of either the steroid backbone,[14,15,80,81] or, specific modifications to the carbohydrate moiety.[82,83] "CS" mimics based on the estradiol backbone have also been reported.[84]

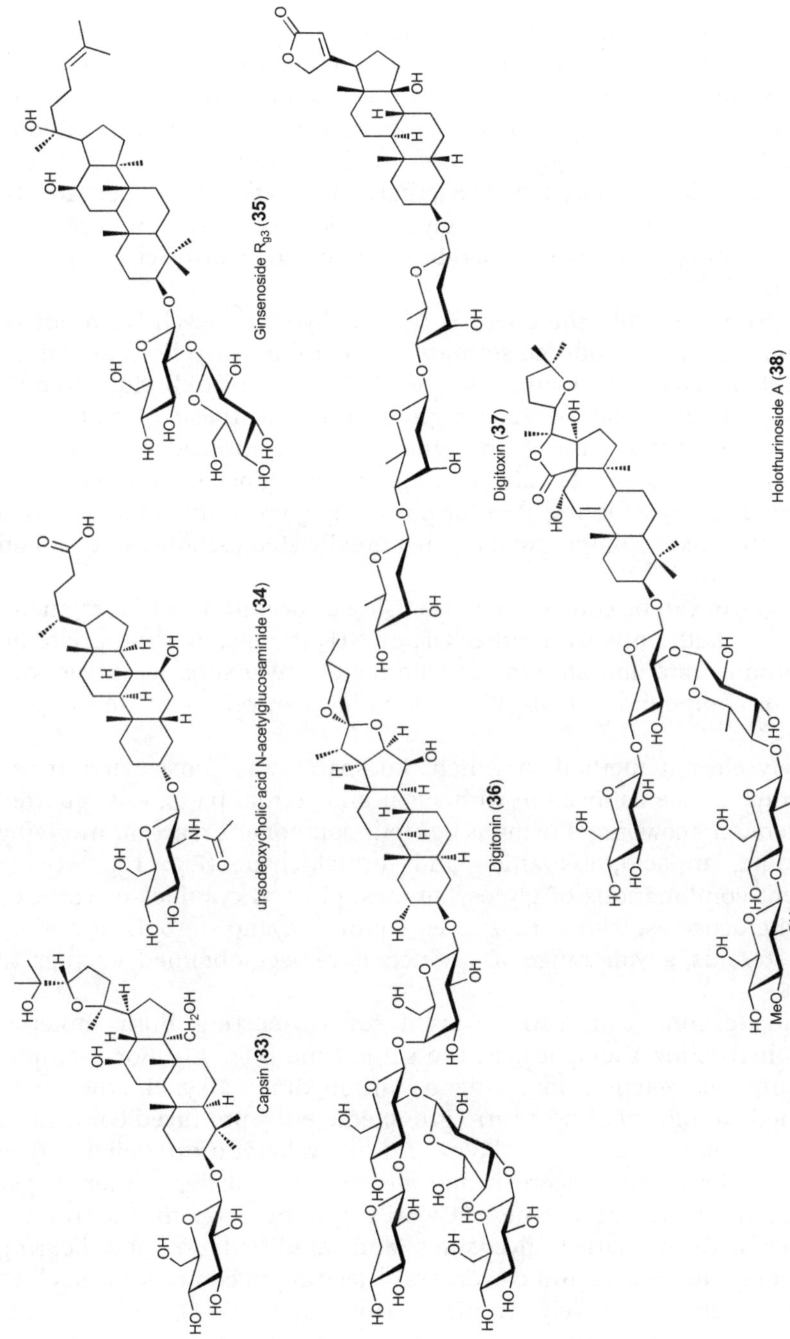

Fig. 8 Typical "CS" glycosteroids.[71–76]

The "C*sp*S" glycosteroids, which are all synthetic constructs, are also interesting compounds, because several have been targeted to possess self-assembling and/or specific surfactant properties, while others have been designed as mimics of naturally occurring glycosteroids. The spacer is often used as the connecting strategy to provide both (variable) elongation of the system and a reactive connective function. Different strategies for connecting the carbohydrate to the steroid on the spacer have been reported (Fig. 9). Faivre *et al.* reported the synthesis of compound **39**, obtained by glycosylation of the terminal OH of a steroid modified with a tetra(ethylenoxy)ether chain at *O*-3.[85] Alternatively, anomeric alkylation of various glycosylthiols was used to obtain a family of thioglucosides such as **40** designed for constructing specific liposomes.[68,86–90]

Our group reported the opening of carbohydrate bicyclic lactones by the amino group of modified steroids either aminodeoxycholesterol itself or the long chain analogues **41** and **42**.[44–46] A disaccharidic lactone opening was also used by Pucci's group in the synthesis of more elaborated derivatives such as **43** designed for their antioxidant properties (Fig. 10).[91] This lactone opening strategy was combined with a metathesis reaction as a key coupling step for preparing glycosteroids possessing a spacer, such as in compound **49**, using an allylated carbohydrate and an allylated steroid.[45,46]

The spacer can be constructed also using a succinic acid C-4 fragment, coupled at both ends with either OH or NH_2 groups, as appropriate on the carbohydrate and steroid (or triterpene) precursors, as in the synthesis of compounds **44–46** (Fig. 11), along with others made in these studies.[92–94]

A very elegant method in which the spacer was constructed at the same time as the connection with sugar and steroid parts, was reported by Rivera and coworkers using a Ugi four-component reaction involving an amine, an acid, a cyanide and formaldehyde (Fig. 12).[95] Using different combinations of glycosylamines, glycosyl cyanides or carboxymethyl glucosides, with carboxylated steroids, cyano steroids or aminodeoxy steroids, a wide range of products have been obtained, such as **47** and **48**.

Cycloadditions can also be used for connecting both moieties and constructing the spacer at the same time (Fig. 13). For example, a 1,3-dipolar reaction involving a carbohydrate alkynyl ether with modified steroids bearing a nitro group efficiently produced conjugates such as compound **50**.[96] Alkynyl–azido cycloaddition "click" type reactions have also been proposed recently, using either sugar azides and steroidal alkynes (Tiwari's group),[97] or in the reverse manner with propargyl glycosides and modified steroids bearing an azido group (Stasik and coworkers)[98] leading to compounds such as **51** and **52ab** respectively. A bis triazolyl system (**53**) was recently proposed by Tiwari and coworkers.[99] Other click type connections have been reported by Xie and coworkers and Somsák, Sun and coworkers.[100,101]

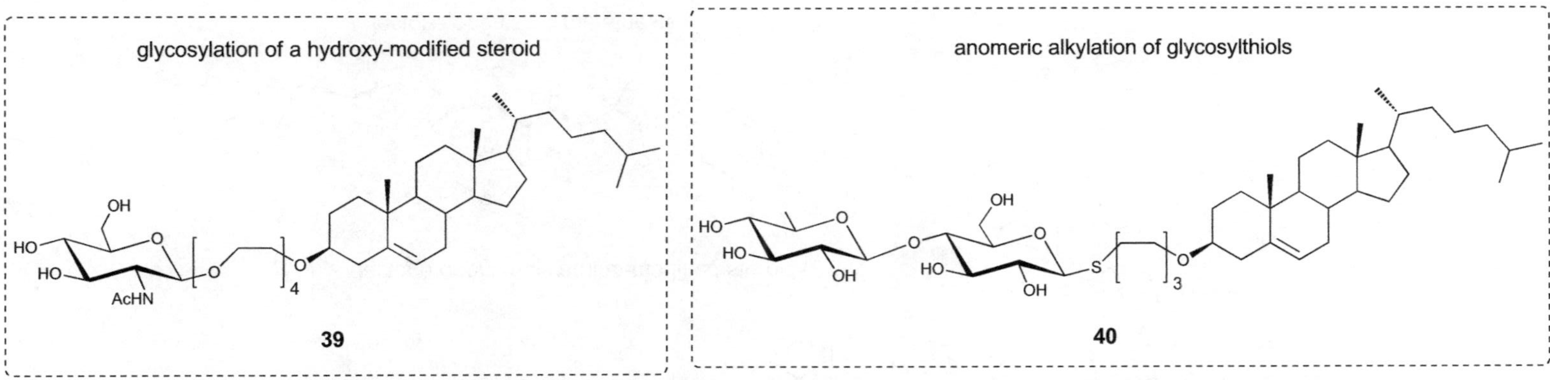

39 **40**

Fig. 9 Glycosylation strategies towards spacer-containing glycosteroid hybrids ("C*sp*S").[68,85–90]

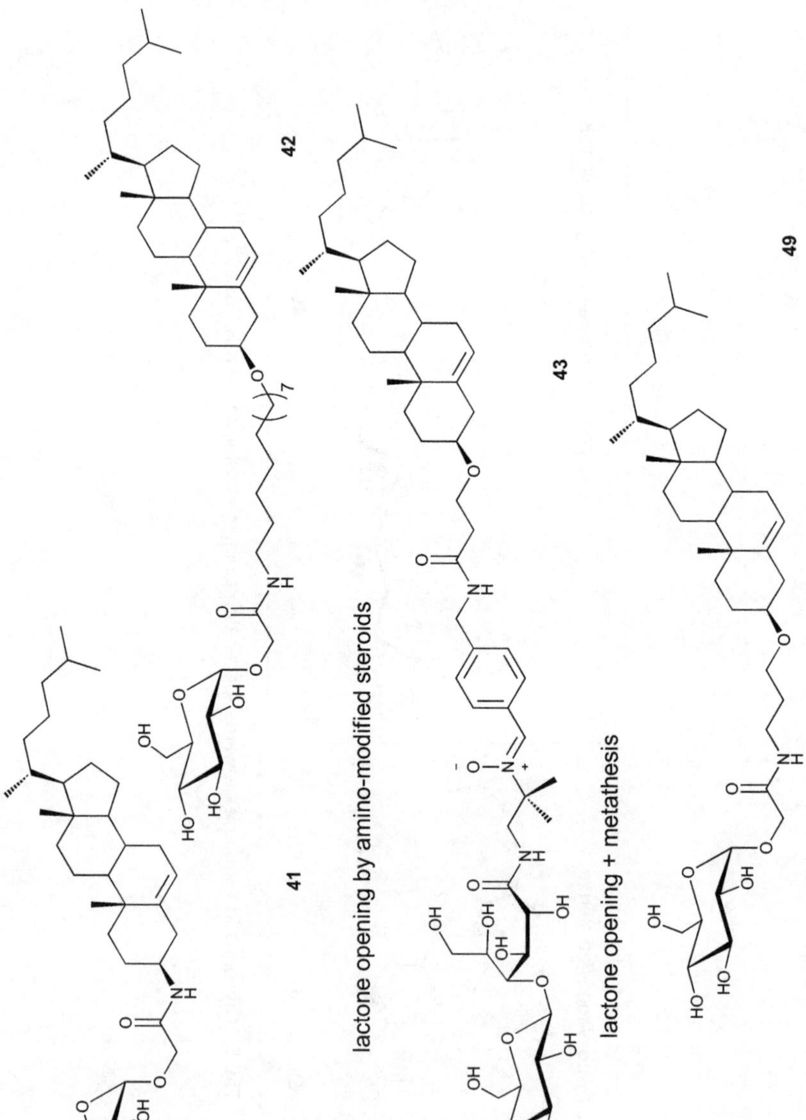

Fig. 10 Lactone opening strategies towards spacer-containing glycosteroids ("CspS").[44–46,91]

succinic diesters, bisamides or amidoesters

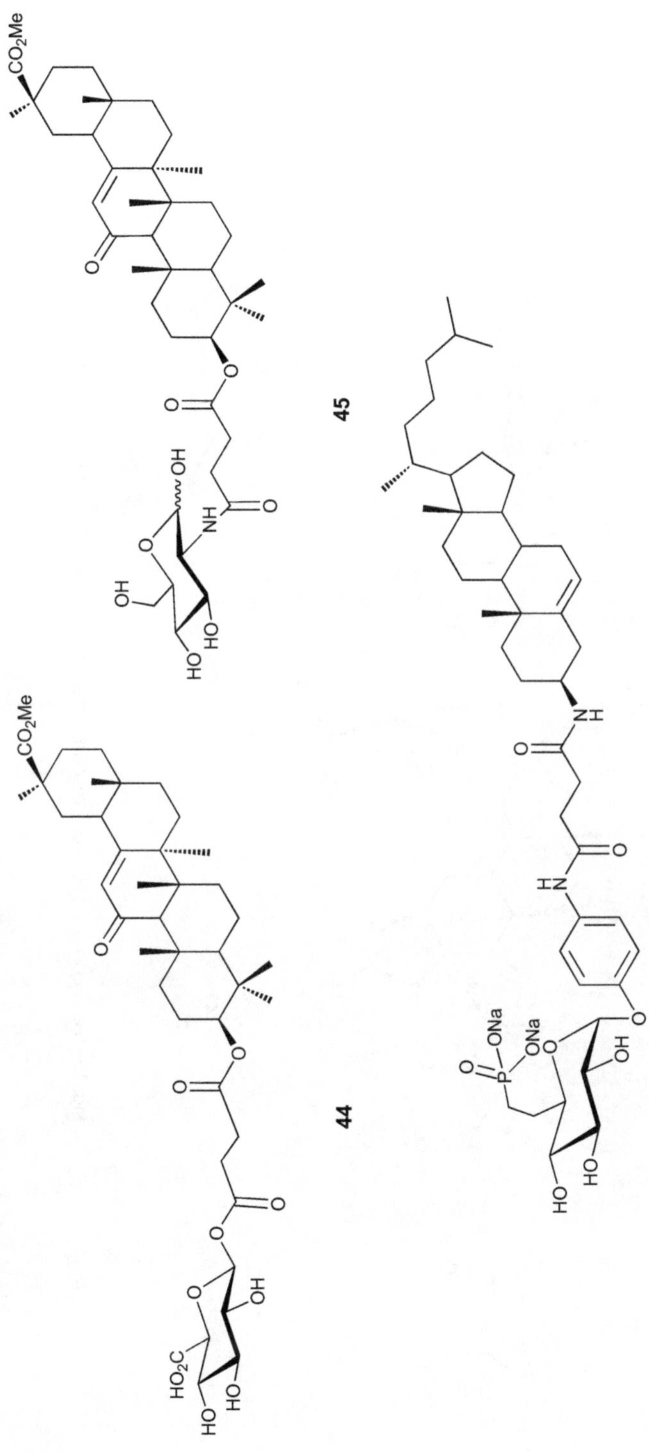

44

45

46

Fig. 11 Succinic spacer systems for ("*CspS*") glycosteroids.[92–94]

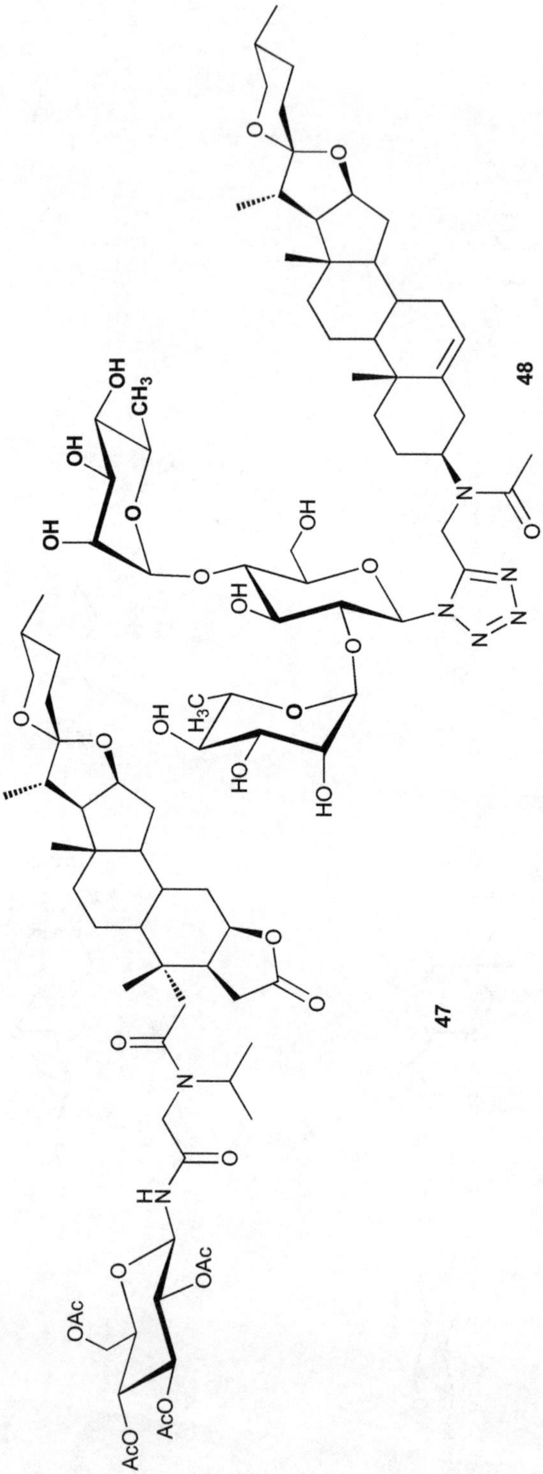

47

48

Fig. 12 Four component – Ugi reaction strategy towards "*CspS*" glycosteroids.[95]

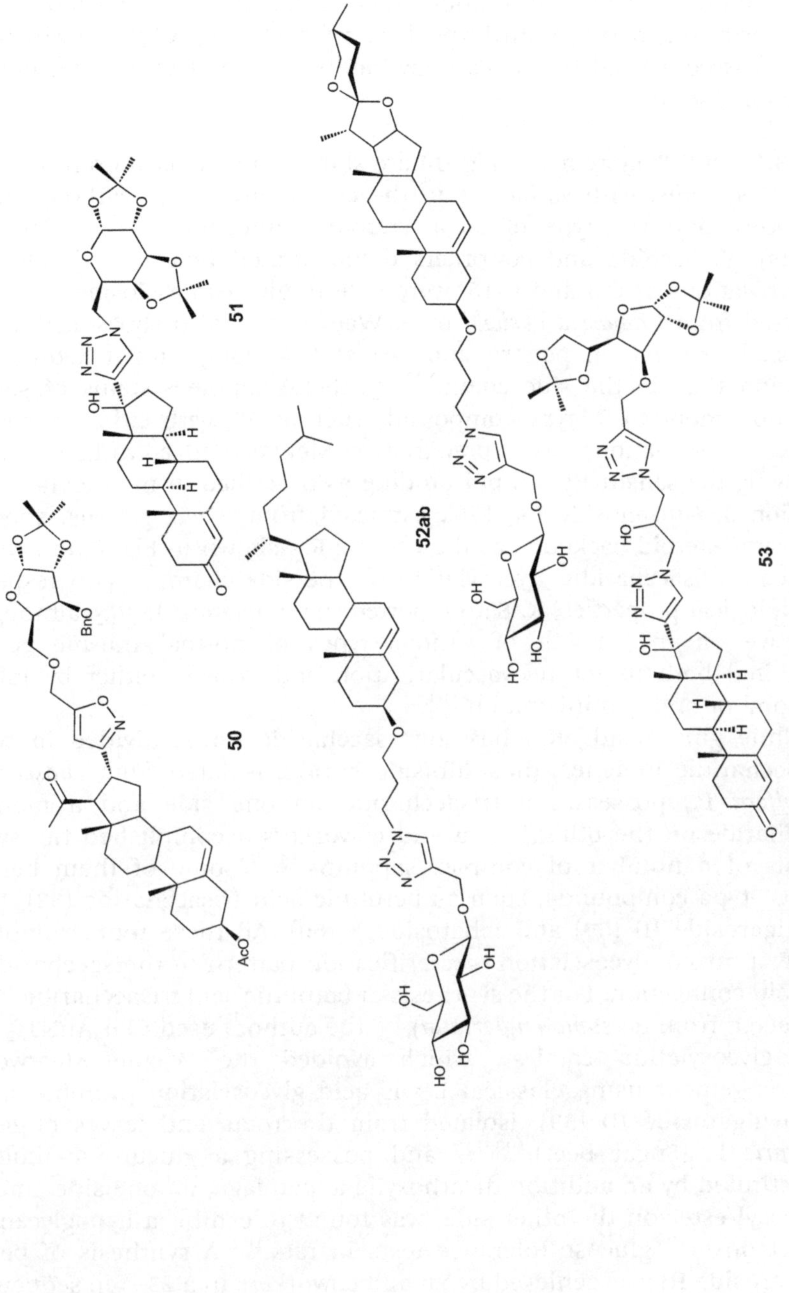

Fig. 13 Dipolar cycloadditions strategies towards spacer-containing glycosteroids.[96–99]

2.3 Bolaphilic systems: "CSC", "FCS" and "CSX"

Bolaphiles are systems in which two sub-structures with similar polarities are connected to another one with inverse polarity. We name "CSC" for those in which one steroid is connected to carbohydrate moieties on two different points of its backbone; thus, making a polar–non-polar–polar construction, and "FCS" systems in which the carbohydrate part possesses an additional hydrophobic appendage, which is in general a fatty chain. With respect to the combination of polar and non-polar moieties, it is worth considering a third type, here referred to as "CSX", in which a steroid is connected to one carbohydrate moiety and to another polar functional group.

2.3.1 "CSC" systems. In Fig. 14 are shown some typical examples of "CSC" systems, with variations in the complexity of the carbohydrate moieties, and the type of sugar–steroid connections (glycosides or esters). Y. Sashida and coworkers demonstrated how two simple β-glucosylations at C-3 and C-16, were able to give camassioside (**54**), as isolated from *Camassia cusickii* by S. Watson who also showed its very unusual oxygenation pattern and OH at C-7 along with a ketone at position C-23 in the side chain.[102] Synthetic simple systems of such "mono + mono" CSC-type compounds such as **55**, with either glucose, galactose or mannose were prepared by Menger and coworkers[103] for studying the sensitivity of lipid binding as described in more details in Section 3. Ginsenoside R_{b2} (**56**), extracted from *Panax ginseng*, shows the same steroid backbone as the CS-type R_{g3} shown in Fig. 8, but with a second disaccharidic glycosylation on the side chain.[73] With respect to biological properties, K. Sato reported that ginsenoside R_{b2}, although inactive on the growth of various types of normal murine cells, inhibited both tumor neovascularisation and growth, either by intra tumoral or oral administration.[104]

While ginsenoside R_{b2} has four saccharide units, divided in two disaccharidic moieties, diosbulbiside D (**57**), isolated from *Dioscorea bulbifera* L., possesses a trisaccharide on one side and a monosaccharide on the other.[105] Yu and coworkers accomplished the synthesis of a number of complex saponins,[14,15] some of them being "CSC"-type compounds, such as betulinic acid trisaccharide (**58**), betavulgaroside III (**59**) and lobatoside E (**60**). All these four examples show a mixed glycosylation – esterification pattern in the saccharide–steroid connection. For the synthesis of betulinic acid trisaccharide (**58**, extracted from *Bersama engleriana*),[106] the authors used PPh_3AuNTf_2 as the glycosylation catalyst, which avoided the Wagner–Meerwein rearrangement using classical Lewis acid glycosylation promoters.[107] Betavulgaroside III (**59**), isolated from the roots and leaves of *Beta vulgaris* L. (Sugar beet),[108–110] and possessing a glucuronic moiety substituted by an addition dicarboxylic appendage, on one side, and a glucosyl ester on the other side, was found to exhibit a hypoglycemic effect on oral glucose tolerance tests in rats.[109] A synthesis of betavulgaroside III was achieved by Yu and coworkers in a 23-step sequence and an 0.9% overall yield from D-glucose.[111]

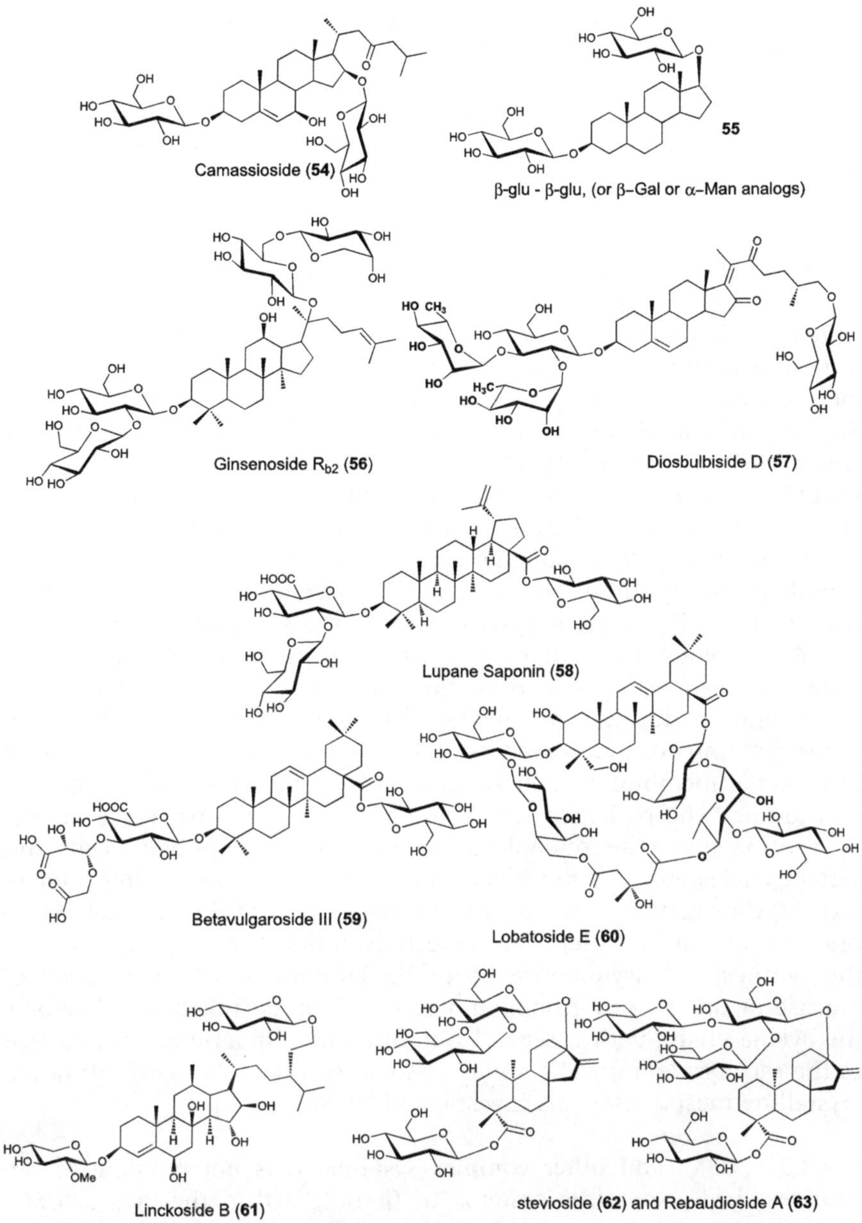

Fig. 14 Typical examples of CSC-type systems.[73,102–106,108–110,112,117]

Quite interesting also is the cyclic system found in lobatoside E (**60**), in which the two carbohydrate parts are connected together by a C5-diester motif. Lobatoside E, isolated from the herb of *Actinestemma lobatum* MAXlM. (Cucurbitaceae), was found to be the most potent antitumour active congener of a small group of the plant triterpene saponins named cyclic bisdesmosides.[112] A total synthesis of this very complex molecule was achieved by Yu and coworkers in a total of 73 steps and 1.2% overall yield.[113] Many other interesting "CSC" systems have been reported, among which was linckoside B (**61**) as reported by Ojika and coworkers[114] and synthesized by Y. Li and coworkers,[115] along with other compounds found in *Holboellia coriacea* Diels.[116]

Let us mention that in the popular sweetening products isolated from *Stevia rebaudiana*, the two components of principal interest for the sweetening properties are two CSC-type compounds, namely stevioside (**62**) and rebaudioside A (**63**).[117]

2.3.2 "FCS" systems. Acylated steroidal glycosides (ASG) are commonly found in many plants.[1,2,6,7,11,14] They have been identified as early as 1964 and since then have been intensely studied with respect to their occurrence and properties, in particular their interactions with membranes.[118–123] Examples of FCS-type compounds are shown in Fig. 15, such as daucosterol palmitate (**64**), which was isolated from *Linaria vulgaris* Mill by H. Hua *et al.*[124] Other types of "FCS" glycosteroids come from bacteria, such as cholesteryl 6-*O*-phosphatidyl-α-D-glucopyranoside and α-CAG from *Helicobacter pylori*, which is suggested to be related to gastric malignancies.[49] J. Gervay-Hague and coworkers completed the synthesis of α-CAG, in an overall yield of 45% starting from glucose.[125] The same group reported the synthesis of BbGL-1 (**31** and **65**),[126] which is another interesting FCS-type glycosteroid, isolated from *Borrelia burgdorferi* and is thought to be involved in developing host immunity during Lyme disease.[70] The role of these bacterial glycolipids remains not totally understood; however, their very high content in bacterial membranes is in keeping with them being major contributors towards the pathogenicity of those bacterial strains. For example, *H. pylori* is able to sense cholesterol rich regions of epithelial cells and extract cholesterol as cholesteryl glucoside (α-CG), α-CPG (**66**) and α-CAG (**3**) thus resulting in a change in the content of the host cell membranes that can be a step towards cell evolution.[49] Our group reported the synthesis of asymmetric bolaphile biomimics (**67**), composed of glucose or other head groups and steroidal units linked together by a methylene chain, with a second non-bridging aliphatic chain attached to the saccharide unit. The ability of these compounds to exhibit liquid crystalline mesophases was investigated (*cf.* Section 3).[45,46]

2.3.3 "CSX" and other complex systems. It is not possible to categorize all glycosteroidic systems to fit only within the "CS", "CSC" and "FCS" types of compounds. Indeed, other conjugations, which can exhibit polarity can occur, such as aminoacids or other functional groups, either in naturally occurring compounds or those obtained by synthesis. Some typical examples are depicted in Fig. 16, in addition to those already depicted above.

For example, Avenacin A-1 (**68**), isolated from the oat root, possesses a methylaminobenzoic ester that gives fluorescent properties to the molecule, which is a property that is extremely rare among saponins.[127] Avenacin A-1 was found to exhibit potent antifungal activity and confer resistance to a broad range of soilborne pathogens.[128] Chromosaponin I (CSI, **69**) is a conjugate of soyasaponin I and γ-pyrone, isolated from pea

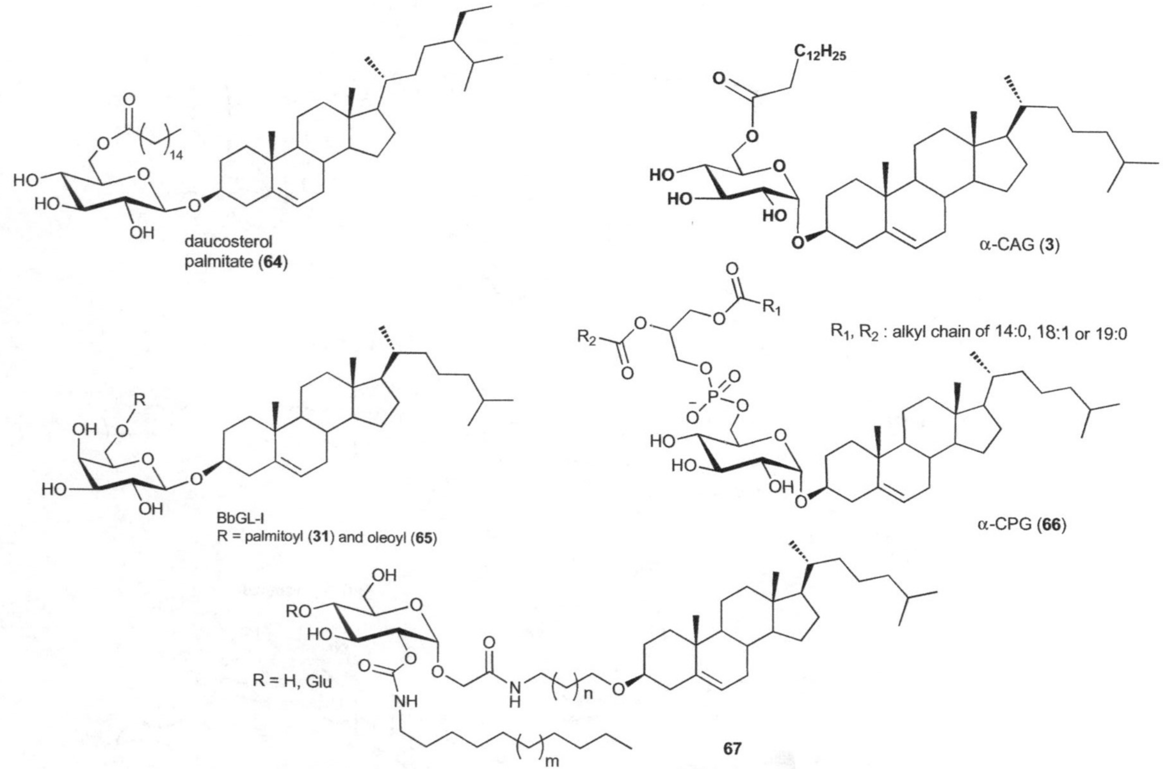

Fig. 15 Typical examples of FCS-type glycosteroids[45,46,49,70,124]

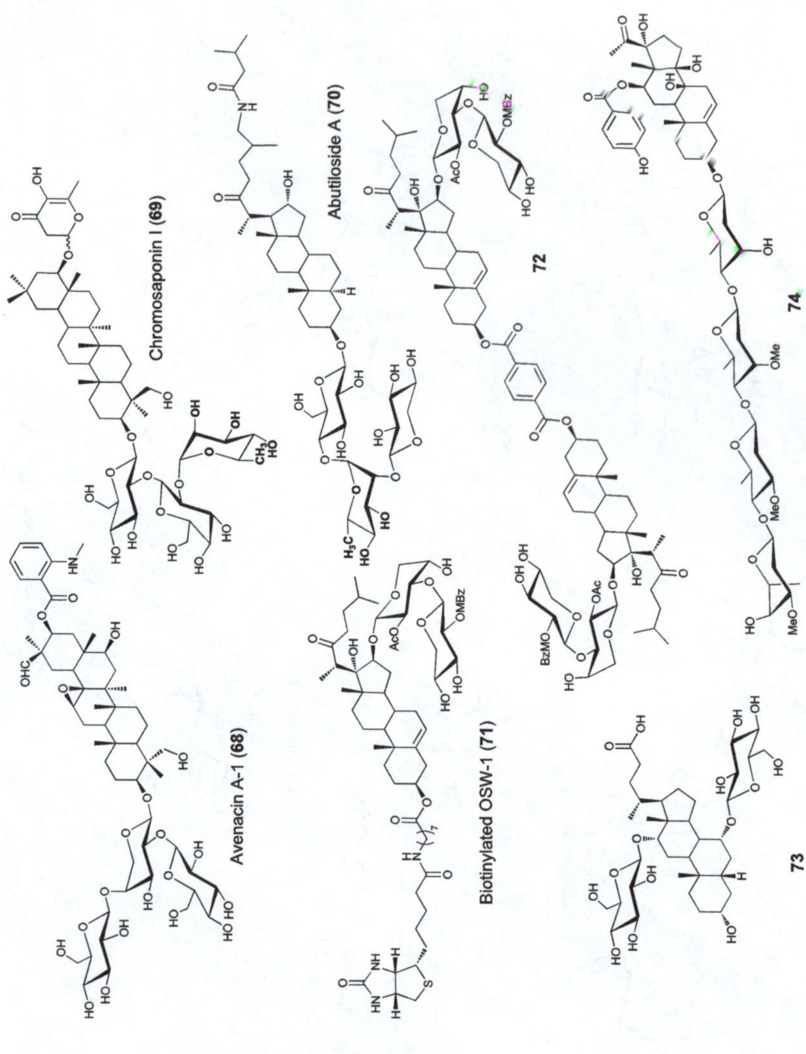

Fig. 16 Typical examples of "CSX" systems.

and other leguminous plants.[129] The role CSI is to stimulate the growth of roots by affecting cell elongation *via* inhibition of ethylene signaling.[130] Abutiloside A (**70**), which shows an amido side chain, was extracted from the roots of *Solanum abutiloides*, is regarded as a key intermediate in the biogenesis of steroidal alkaloids.[131] OSW-1 is a natural saponin of CS-type with potent antitumor activity,[10] which has been the stimulus for several syntheses to have been reported. Z. Jin and coworkers reported the synthesis of a biotinylated OSW-1 (**71**) conjugate, which was designed to determine the molecular target of OSW-1 in cells.[132] This "CSX" type biotinylated OSW-1 also exhibited inhibitory activity against cancer cells, indicating that the presence of biotin at C-3 does not prevent the interaction of OSW-1 with its putative targets. X. Ma *et al.* reported even more complex dimeric systems, such as the CS*sp*SC-type dimer (**72**) shown in Fig. 16. In this series, they demonstrated the importance of key structural features, namely the acylation pattern on the sugars and the configuration at C-16 and C-17 when assessing their antitumor activities.[133] In keeping with the strategy of mimicking natural amphiphilicity of glycosteroids, Bowe, Sofia, Axelrod, Kahne and coworkers looked for nonpeptidic facial amphiphiles as drug transport agents and synthesized a class of glycosylated bile acid derivatives (see Section 3), such as **73**. One key step was cholic acid glycosylation at the hindered C-7 and C-12 hydroxyl groups.[134–138]

As the key parameter is polarity, one can consider that some systems that we have included in the CS-type, in which the steroid backbone is highly oxygenated, might also come in the "CSX" subgroup, as for example compound **73** reported recently by Y. Bao.[139]

A few more complex systems should be mentioned. An interesting one is the very peculiar lipid A-type pentasaccharide **75** which has been identified in the *Bradyrhizobium* strain BTAi1 lipid content, showing a unique structure in which the lipopolysaccharide (LPS) lipid A is covalently substituted by an hopanoid structure. This compound is built on a pentasaccharidic backbone, with several fatty chains connected as amides on the aminodeoxy groups of two internal residues. One of the fatty chains is substituted at its other end with an hopanoid backbone through an ester bond. One galacturonic and one dimannosyl residue complete the structure, that we could consider as a SFCX system, for sure a very complex one. The discovery of this unique molecule by Molinaro and co-workers and the accompanying studies revealed the influence of structural variations in lipid A type compounds on the properties of cell membranes, notably in the context of plant–microbe symbioses.[140,141]

Let us briefly mention that macromolecular systems involving a combination of carbohydrates and steroids were also reported, such as compounds **76–79** constructed on chitin or chitosan backbones. Hydroxyethoxylation of the primary hydroxyl groups of the polysaccharidic chain possibly modulated the overall polarity and amphiphilicity of the polymers. These compounds were prepared with the aim of elaborating self-aggregated particles used in gene transfection or drug delivery applications (Fig. 17).[142–145]

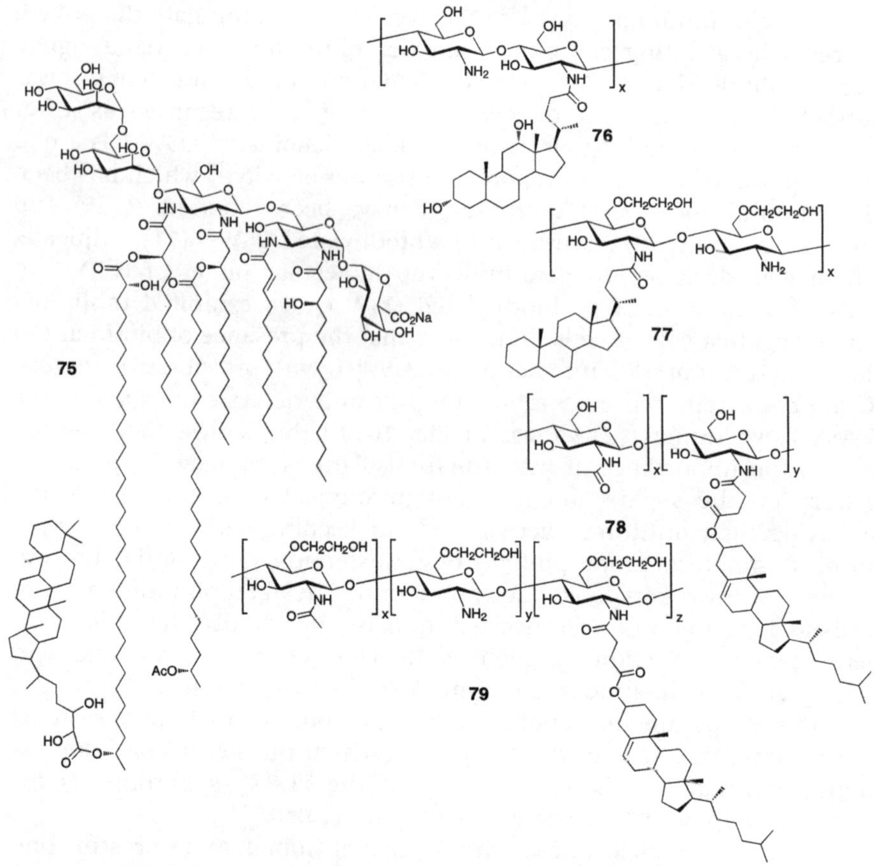

Fig. 17 Miscellaneous complex glycosteroidal systems.

3 Investigations on supramolecular systems made of glycosteroids

As early as 1913,[47] in an account aiming at clarifying the structure of naturally occurring phytosterols extracted from plants, Salway reported the anomalous melting of sitosteryl β-D-glucopyranoside ("the substance begins to soften at about 270 °C, and melts and decomposes at 295–300 °C") and cholesteryl β-D-glucopyranoside ("began to soften at 270 °C, and completely melted at 285 °C"). Since then, several other compounds have shown complex melting behaviour, and precise liquid–crystalline structures have been identified. These self-organising properties are also reflected in their solution behaviour, leading to self-assembling-organising supramolecular structures, such as liposomes or vesicles. In addition, the interactions with membranes reflect the specific self-assembling properties of amphiphilic glycosteroids, as well as some applications such as drug delivery agents and drug solubilizing agents. The following sections will focus mainly on liquid–crystalline properties of some glycosteroids; in addition, other properties will be briefly mentioned. Macromolecular systems are not included in this section, as their self-assembling properties were already reviewed.[8,146]

3.1 Glycosteroidic liquid crystals and materials

Although liquid crystal behaviour for derivatives of cholesterol and cholestanol have been abundantly documented, and despite the very early observations by Salway on the anomalous melting behaviour of cholesteryl glucosides, it was only in the mid 1990's that Vill and coworkers investigated the exact nature of the melting of α-solanin (**2**). α-Solanin is a toxic trisaccharidic alkaloid found in green potatoes, and was studied by Vill *et al.* with the aim of producing accurate analytical information necessary to improve characteristics of cultivars.[48] He found that the compound passes from the crystalline to the liquid crystal phase forming a smectic A phase at 263 °C, analogous to fatty alcohol glycosides, with a clearing point of 285 °C, which occurred with extensive decomposition. All other studies on naturally occurring glycosteroids, (*e.g.* by Menger on digitonin, Mezzenga on glycyrrhizic acid)[147,148] concerned the formation of vesicles or nanomaterials, which are discussed later in this section. When strictly speaking of liquid crystal investigations, all other reports concerned synthetic systems, which are briefly described below.

In 2007 Queneau, Goodby, Mackenzie and coworkers reported their first results on liquid crystalline glycosteroids.[44] The first series of compounds was prepared using the reaction of bicyclic 3,4,6-tri-*O*-acetylcarboxymethyl α-D-glycopyranoside 2-*O*-lactone,[149–152] which is a synthon able to conveniently deliver carbohydrate moieties for any nucleophilic functional group with aminodeoxy-cholesterol or -cholestanol, with either an α or β configuration at the steroid 3-position (Fig. 18).[44] High melting points were found for the four derivatives (in the range of 181–202 °C); however, only the cholestanyl derivatives (**81** and **82**) were found to be fluid enough for the formation of a smectic A phase, although occurring with rapid decomposition once the clearing point was reached (>230 °C). This series of compounds was further enriched by different structural variations: mono- or -disaccharides, adding a methylene spacer between the two moieties with various chain lengths (from C4 to C12), and presence of an additional fatty chain at *O*-2 of the sugar moiety for producing a "FCS" type structure.[45,46] All compounds in these series (**67**) were found to exhibit liquid–crystalline behaviour, for which transition temperatures are given in Table 1.

These variations gave evidence for significant shape changes depending on the length of the side chain and that of the methylene spacer. Once the side chain becomes too long to fill the space between the carbohydrate and the steroid units, the behaviour becomes more complex as two different mesophases are observed, first a smectic phase corresponding to a rod-like shape, then a columnar phase, corresponding to a wedge-shaped structure, thereby providing insights in the segregation between the two types of lipidic moieties. Models of the rod-like structure and the packing of the wedge-like structures are shown in Fig. 19.

Using a copper catalyzed cycloaddition of alkynyl sugars and azidopropyl steroids, Stasik and coworkers prepared a series of conjugates (**52ab** and **83ab–87ab**) of cholesterol and diosgenin (Fig. 20).[98] They found significant influence of the sugar, the glycosidic bond heteroatom, and of the steroid itself on the thermotropic behaviour (Table 2). Changing the saccharide moiety (rhamnose *vs.* glucose) led to significant differences depending

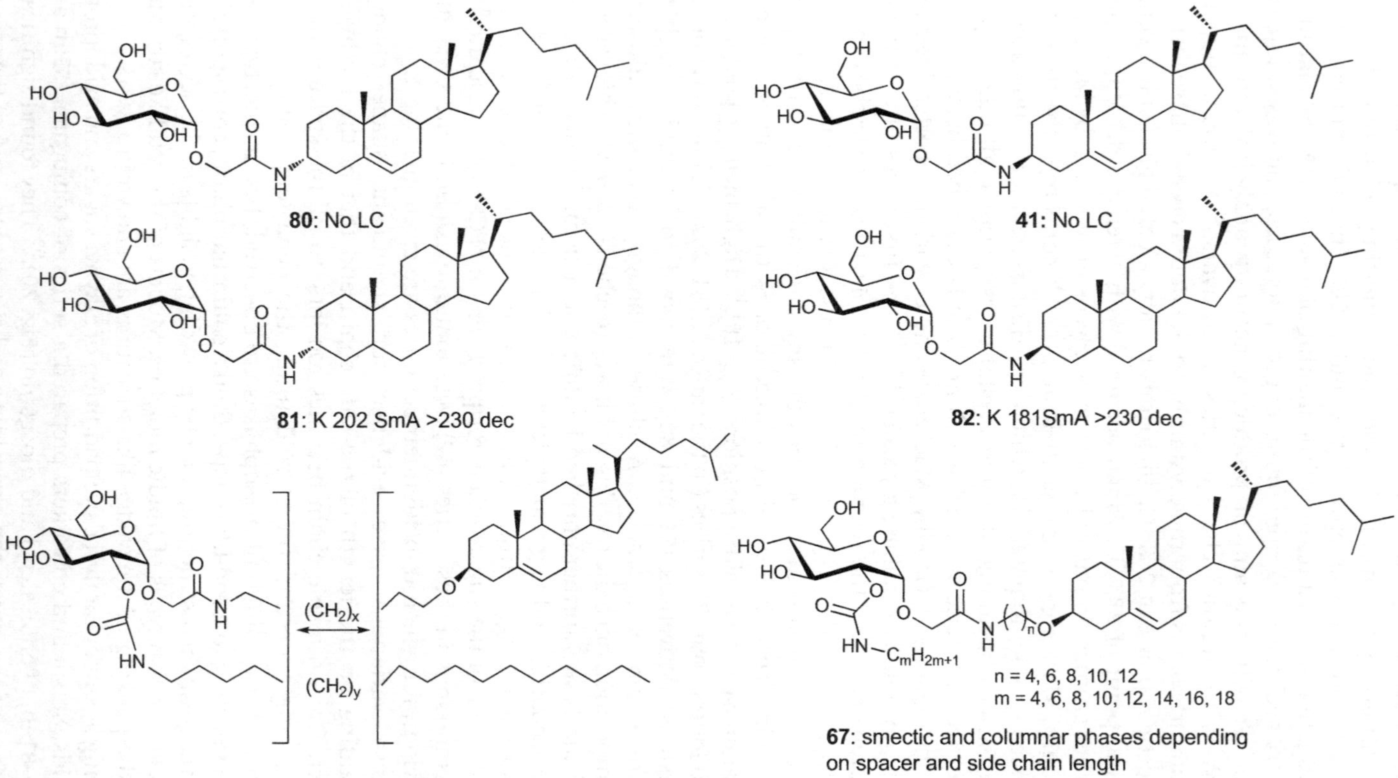

Fig. 18 Glycosteroid hybrids with spacers of different size, and possible 2-*O*-substitution with a fatty chain.[44,46]

Table 1 Influence of spacer and side chain on mesophase formation in compounds **67** (Fig. 18).[46]

			Transition temperatures (°C) and enthalpy of transition [kJ mol^{-1}]					
n	m	Cr		Lamellar		Col		Iso. Liq.
4	Nil	•	50 [127.60]	•	290 [6.48]	—	—	•
	4	•	96 [7.93]	•	251 [0.23]	—	—	•
	8	•	94 [17.91]	•	178 [0.99]	—	—	•
	10	•	115 [8.18]	•	164 [0.65]	—	—	•
	12	•	110 [23.63]	•	159 [0.62]	—	—	•
	14	•	128 [27.40]	•	142 [0.51]	•	165 [0.66]	•
	16	•	124 [13.35]	•	150 [1.43]	•	176 [1.09]	•
	18	•	103, 126 [9.13, 9.55]	•	146 [2.47]	•	183 [1.22]	•
6	Nil	•	50 [14.43]	•	269 [3.91]	—	—	•
	6	•	123 [26.57]	•	193 [2.21]	—	—	•
	8	•	109 [15.16]	•	178 [1.75]	—	—	•
	12	•	127 [24.11]	•	145 [0.37]	—	—	•
	14	•	132 [23.08]	•	126 [0.42]	•	155 [0.50]	•
	16	•	138 [29.12]	•	129 [0.30]	•	170 [1.04]	•
	18	•	138 [22.25]	•	138 [3.48]	•	180 [1.08]	•
8	Nil	•	60 [35.83]	•	244 [2.52]	—	—	•
	6	•	154 [32.03]	•	192 [3.28]	—	—	•
	8	•	126 [29.34]	•	188 [2.91]	—	—	•
	12	•	131 [15.99]	•	156 [0.81]	—	—	•
	14	•	123 [22.42]	—	—	•	144 [0.46]	•
	16	•	125 [18.42]	—	—	•	150 [0.89]	•
	18	•	131 [23.81]	—	—	•	170 [1.26]	•
10	Nil	•	75 [39.98]	•	—	—	246 [4.49]	•
	6	•	137 [30.56]	•	—	—	194 [4.54]	•
	8	•	152 [33.37]	•	—	—	181 [4.32]	•
	12	•	117 [20.94]	•	—	—	160 [2.30]	•
12	Nil	•	100 [10.66]	•	—	—	239 [5.00]	•
	6	•	136 [32.53]	•	—	—	192 [5.91]	•
	8	•	118 [22.00]	•	—	—	177 [5.50]	•
	12	•	118 [30.99]	•	—	—	159 [4.23]	•

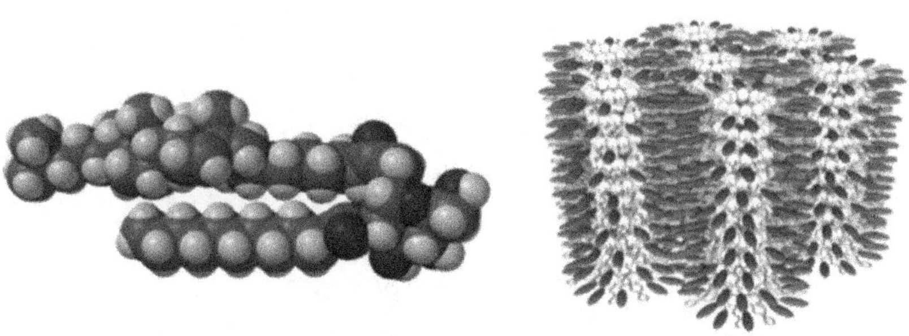

Fig. 19 Left: Model for the rod like structure of compound **67** with C_4 spacer and C_{12} side chain leading only to lamellar phase; Right: proposed structure for the hexagonal columnar phase of compound **67** with longer side chains (>C_{14}).[46]

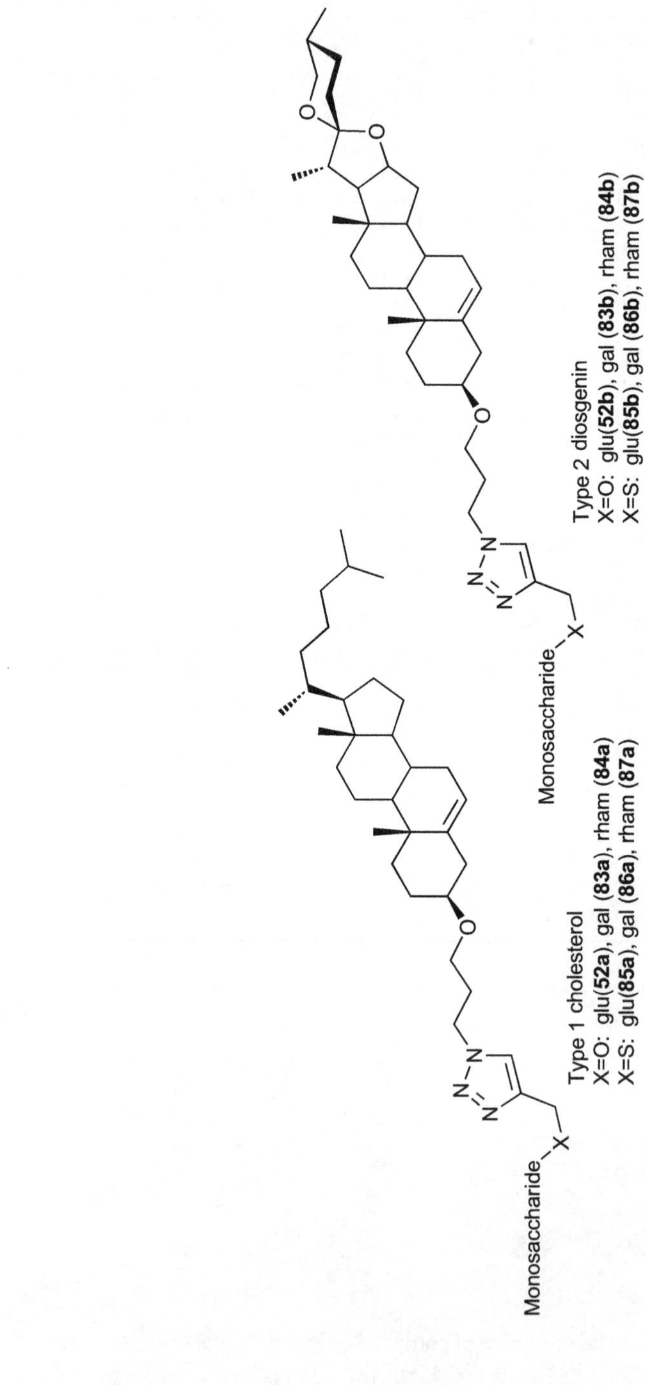

Fig. 20 Glycosidic and thioglycosidic triazolyl-tethered glycosteroids hybrids.[98]

Table 2 Influence of sugar and glycosidic bond type on mesophase formation in synthetic glycosteroids **52ab** and **83ab–87ab** (Fig. 15).[98]

X	Sugar		Cholesterol (type 1)		Diosgenin (type 2)
O	β-D-Glucose	**52a**	Tg 100 SmA 170 Iso. Liq. °C	**52b**	Tg 109 SmA 202 Iso. Liq. °C
	β-D-Galactose	**83a**	Tg 105 SmA 160 Iso. Liq. °C	**83b**	Tg 113 SmA 210 Iso. Liq. °C
	α-L-Rhamnose	**84a**	Tg 103 SmA 185 Iso. Liq. °C	**84b**	Tg 84 SmA 171 Iso. Liq. °C
S	β-D-Glucose	**85a**	Tg 85 SmA 238 Iso. Liq. °C	**85b**	Tg 143 SmA 240 Iso. Liq. °C
	β-D-Galactose	**86a**	Tg 87 SmA 195 Iso. Liq. °C	**86b**	Tg 141 SmA 220 Iso. Liq. °C
	α-L-Rhamnose	**87a**	Tg 86 SmA 183 Iso. Liq. °C	**87b**	Tg 108 SmA 180 Iso. Liq. °C

whether the steroid was cholesterol or diosgenin, thereby demonstrating that not only the number of OH groups in the carbohydrate should be considered. This series of compounds also permitted us to show that the type of atom used for the glycosidic bond, either oxygen or sulfur, has an influence on the phase transition temperatures with much higher clearing points occurring for thiogluco- and galactosides. This is discussed in terms of higher polarizability of the sulfur atom *vs.* oxygen. The C–O *vs.* C–S bond lengths might also be an important parameter in this respect.

Faivre and colleagues have reported the thermotropic and lyotropic behaviour of cholesteryl oligoethyleneglycol glycosides **39** (Fig. 21),[85] in the dry state and in the presence of various amounts of water, showing how much the hydration shell of the sugar significantly modifies the overall shape and system curvature. In 2012, Mezzenga and coworkers[153] investigated the self-assembly of sitosteryl glucoside (**1**) in bulk and in solution. In the bulk, they identified a lamellar phase, with strong microphase separation. In water, platelet-like structures were observed, with the sugar facing the solvent, and forming helical ribbons upon dilution, whereas in isopropanol, the steroid faced the solvent leading to the formation of much smaller spherical objects. The same group recently reported that the naturally occurring glycyrrhizic acid (**88**) was able to self-assemble in water, first forming long fibrils, then, at higher concentrations, nematic phases, and finally hydrogels.[148] Some of these assemblies were then used as scaffolds for designing new solid catalysts.

3.2 Glycosteroids in vesicles and liposomes, interactions with membranes, drug transport and solubilization

In addition to pure "liquid crystal" investigations, there have been a significant number of contributions dealing with the incorporation of glycosteroids and related compounds into liposomes and other types of materials. In 1978, Chabala and Shen and coworkers[68,86–88] reported several papers in which they demonstrated that synthetic thioglycosides connected to a steroid with a C-6 spacer chain such as **89** were able to be incorporated into liposomes (Fig. 22). Subsequently, Baldeschwieler and coworkers,[89,90] using the analogous 6-aminoglycosides **90–92** produced evidence for the recognition of surface groups of the lipid vesicles by specific cells and tissues. Maitani and coworkers also investigated the incorporation of steryl glucosides (derived from soybean) in liposomes

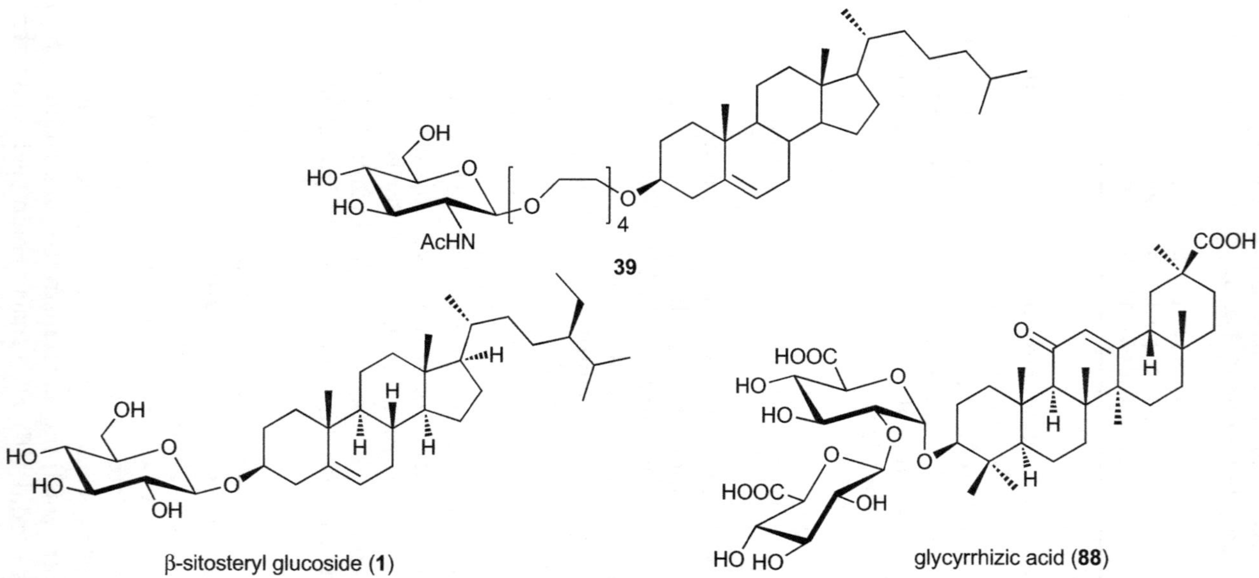

39

β-sitosteryl glucoside (**1**)

glycyrrhizic acid (**88**)

Fig. 21 Synthetic or natural glycosteroids used in studies of bulk and solution self-organization behaviour in Faivre and Mezzenga's contributions.[85,148,153]

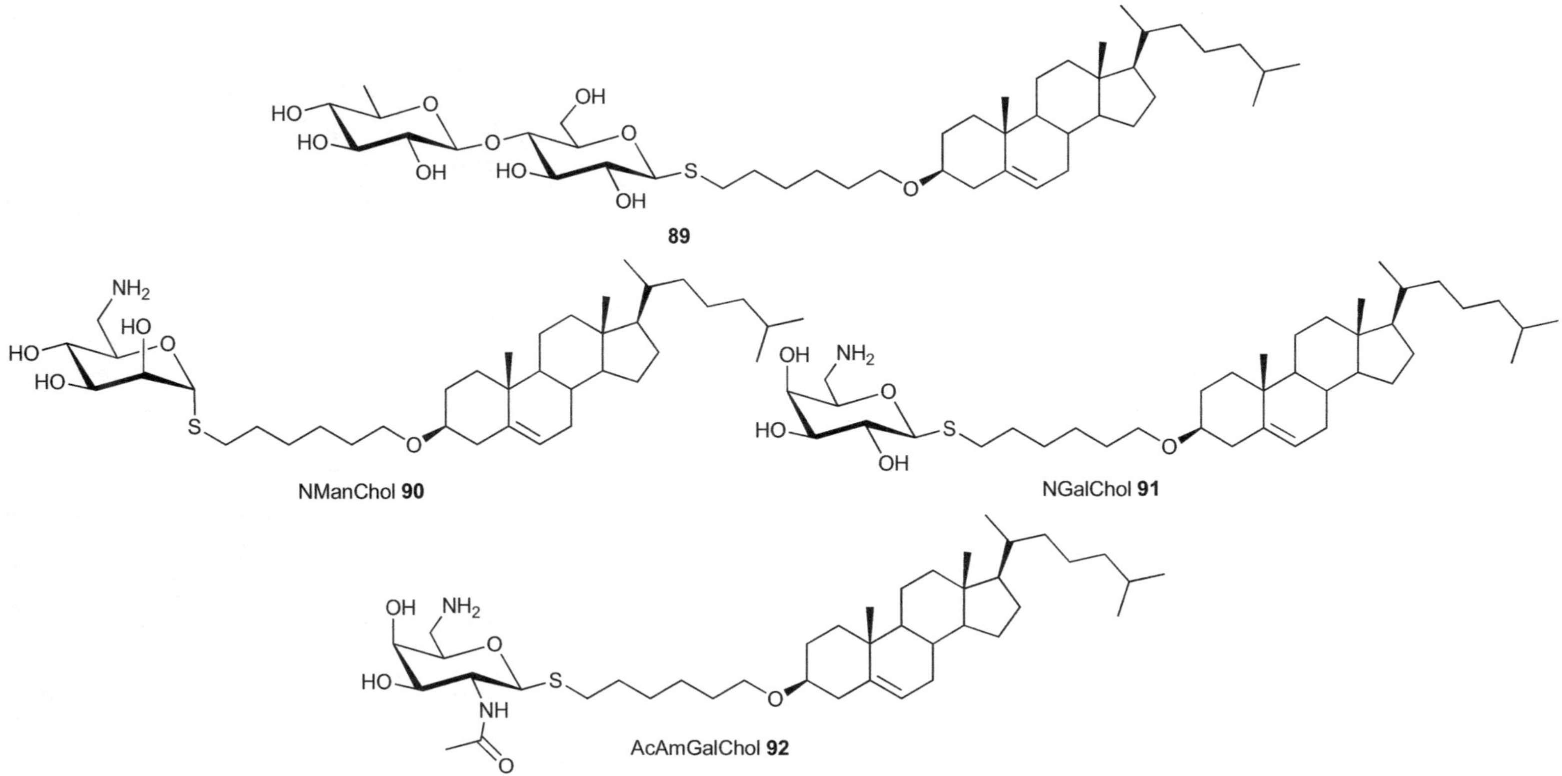

89

NManChol **90**

NGalChol **91**

AcAmGalChol **92**

Fig. 22 Synthetic glycosteroids able to establish (or interact with) liposomes or vesicles in Shen and Baldeschwieler contributions.[68,86–90]

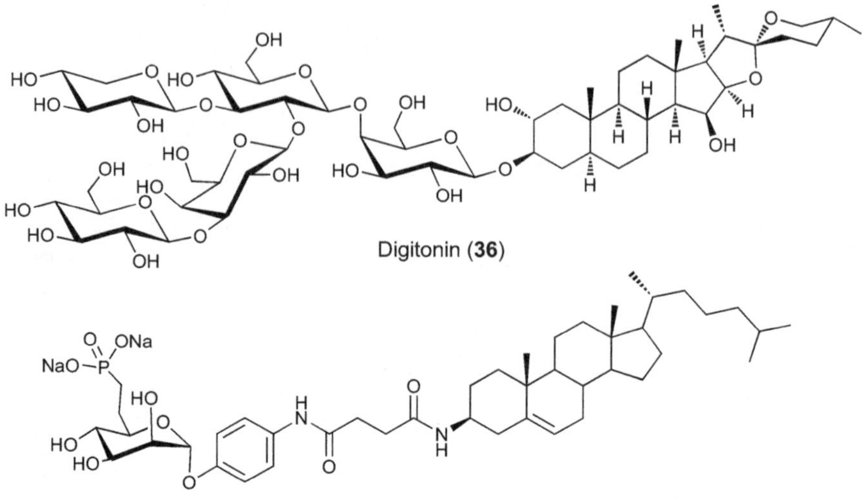

Digitonin (**36**)

mannose-6-phosphonate cholesterylamine (**46**)

Fig. 23 Naturally occurring or synthetic glycosteroids used in Menger and Barragan contributions.[94,147,156,157]

and found that the fluidity of the liposomal membrane was significantly influenced by the presence of the sugar moiety, as compared to the systems built with non-glycosidic sterols.[154,155]

Menger and coworkers investigated two types of "CS"-type sugar–steroid hybrids, digitonin (**36**) and synthetic amide **46** (Fig. 23).[147,156,157] In work aimed at investigating the effect of such a saponin with respect to its membrane disrupting properties, they exposed digitonin (**36**) to some giant vesicles made of stearoyl (or palmitoyl) phosphocholine, with or without membrane bound cholesterol. They found that only those containing cholesterol were disrupted by digitonin, indicating that specific complexes between cholesterol and digitonin take place within the bilayer resulting in destabilization of the system.[147] With Barragan and coworkers, they investigated the use of the non-hydrolysable sugar–steroid hybrid **46**[94] constructed with a semi-rigid spacer, for which they found an interesting ability to bind lipid bilayers and some cells, thereby being used as multivalent adhesives between cells and vesicle surfaces.[156]

The same researchers later focused their interest on CSC-type structures obtained by double glycosylation of 3,17-dihydroxy-androstane with glucose, galactose or mannose and studied their effect on membrane bilayers.[103] The bilayers were modified by intercalating the steroid–saccharide hybrids between the interdigitated DPPC molecules, and the nature of the saccharide led to significant changes in the melting temperatures, as a result of how much they separate the DPPC molecules (Fig. 24).

Steryl glycosides were also studied with respect to their ability to transport drugs, or at least, to help their solubilization. Liu reported the use of a combination of terpene glycosides as solubilizing agents,[158] in keeping with Tanaka's and Kouno's earlier reports.[159–162] A very nice work in this field was reported by Bowe, Sofia, Axelrod, Kahne and coworkers,[134–138] who designed "CSCX" type compounds (**73** and **93–96**, Fig. 25) on a bile

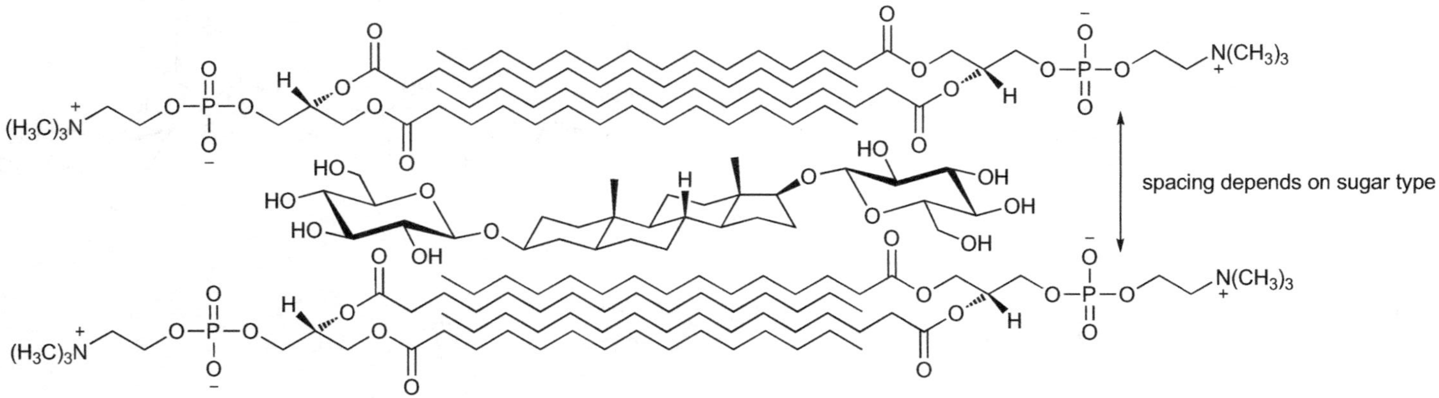

Fig. 24 Glucose–glucose steroid **55** interdigitated into a DPPC bilayer.[103]

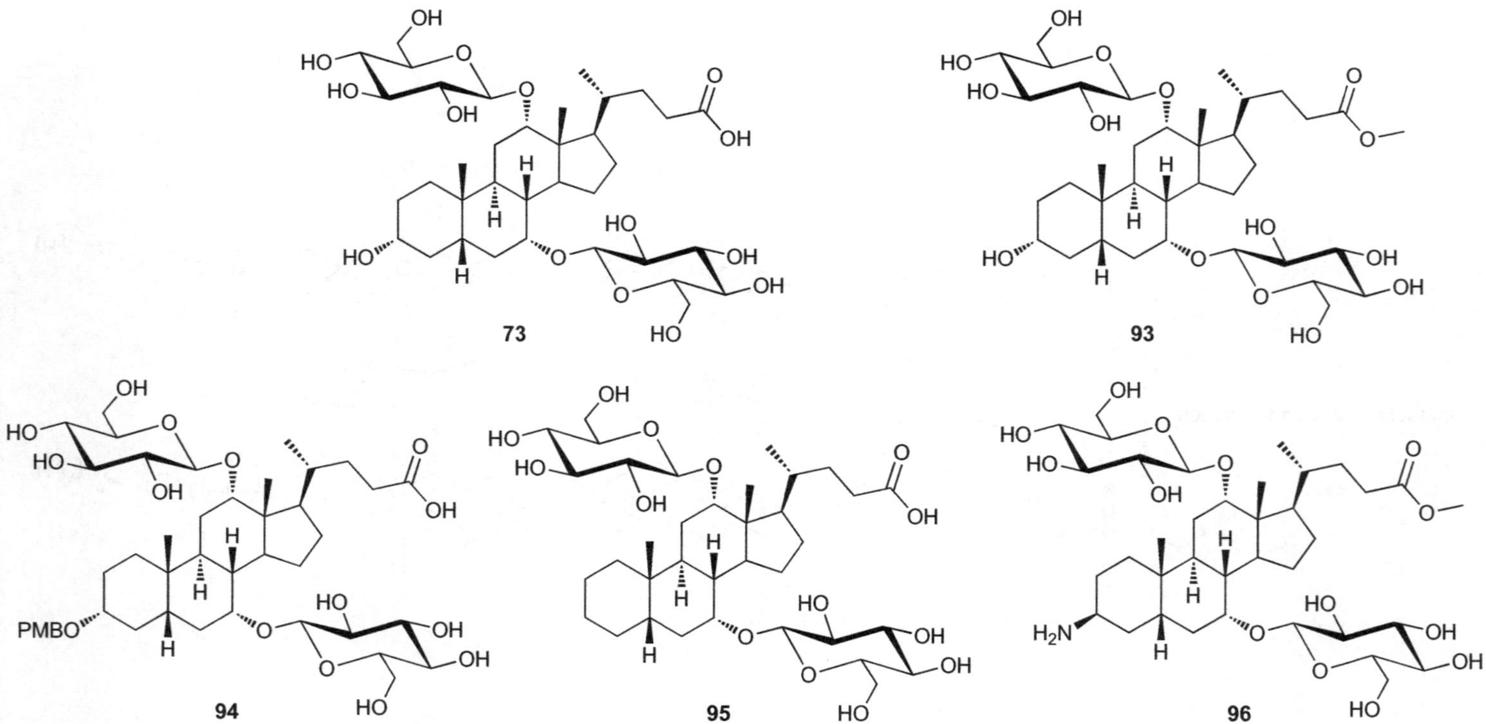

Fig. 25 Examples of glycosteroids used in drug transport or vesicle formation.[134–138]

acid backbone and for which they found a significant increase of intestinal absorption over a range of drugs, due to the hydrophobicity of the steroid nucleus. However, the overall hydrophobicity of the molecule was not predictive of the transport efficacy of bile acid-based enhancers.

4 Conclusion

In this account, our goal was to focus on the overall level of amphiphilicity of glycosteroids and related compounds, in order to guide the way we look at them, not only with their detailed structural patterns, but also foreseeing their potential self-assembly behaviour or their ability to interact with other amphiphilic biomolecules or modifying the properties of supramolecular biological architectures. This can aid the understanding of some of the known biological properties of saccharide–steroid hybrids, and orientate the design of new molecules of biological or therapeutic interest.

Acknowledgements

We want to associate colleagues who were involved in the original work in the field: F. Ali-Rachedi, E. J. Davis, A. Doutheau, N. M. Xavier. Financial support from CNRS and MENESR is also gratefully acknowledged, as well as grants from the China Scholarship Council to YZ and RX. The authors would also like to that the EPSRC for financial support for this work through the Platform Grant EP/D055261.

References

1 S. Grille, A. Zaslawski, S. Thiele, J. Plat and D. Warnecke, *Prog. Lipid Res.*, 2010, **49**, 262.
2 D. B. Salunke, B. G. Hazra and V. S. Pore, *Curr. Med. Chem.*, 2006, **13**, 813.
3 A. Sun, X. Xu, J. Lin, X. Cui and R. Xu, *Phytother. Res.*, 2015, **29**, 187.
4 A. Faizal and D. Geelen, *Phytochem. Rev.*, 2013, **12**, 877.
5 M. A. Lacaille-Dubois and H. Wagner, *Phytomedicine*, 1996, **2**, 363.
6 L. Nyström, A. Schär and A.-M. Lampi, *Eur. J. Lipid Sci. Technol.*, 2012, **114**, 656.
7 N. V. Ivanchina, A. A. Kicha and V. A. Stonik, *Steroids*, 2011, **76**, 425.
8 S. Schwarz, N. M. Xavier and A. P. Rauter, in *Carbohydrate Chemistry, Chemical and Biological Approaches*, ed. A. P. Rauter and T. K. Lindhorst, Specialist Periodical Reports, Royal Society of Chemistry, Cambridge, 2012, vol. 37, p. 326.
9 J. P. Munafo and T. J. Gianfagna, *Nat. Prod. Rep.*, 2015, **32**, 454.
10 Y. Tang, N. Li, J. A. Duan and W. Tao, *Chem. Rev.*, 2013, **113**, 5480.
11 T. Cheriet, I. Mancini, R. Seghiri, F. Benayache and S. Benayache, *Nat. Prod. Res.*, 2015, **29**, 1589.
12 W. Schoner, *Eur. J. Biochem.*, 2002, **269**, 2440.
13 V. L. Challinor and J. J. De Voss, *Nat. Prod. Rep.*, 2013, **30**, 429.
14 B. Yu, J. Sun and X. Yang, *Acc. Chem. Res.*, 2012, **45**, 1227.
15 Y. Yang, S. Laval and B. Yu, *Adv. Carbohydr. Chem. Biochem.*, 2014, **71**, 137.
16 N. H. Thuan and J. K. Sohng, *J. Ind. Microbiol. Biotechnol.*, 2013, **40**, 1329.
17 B. Heasley, *Chem. – Eur. J.*, 2012, **18**, 3092.

18 F. Reinitzer, *Monatsh. Chem.*, 1888, **9**, 421.
19 V. Vill, *Mol. Cryst. Liq. Cryst.*, 1992, **213**, 67.
20 G. T. Stewart, *Liq. Cryst.*, 2003, **30**, 541.
21 S. A. Jewell, *Liq. Cryst.*, 2011, **38**, 1699.
22 G. T. Stewart, *Liq. Cryst.*, 2004, **31**, 443.
23 J. W. Goodby, *Liq. Cryst.*, 1998, **24**, 25.
24 F. Ercole, M. R. Whittaker, J. F. Quinn and T. P. Davis, *Biomacromolecules*, 2015, **16**, 1886.
25 F. Xu, H. Wang, J. Zhao, X. Liu, D. Li, C. Chen and J. Ji, *Macromolecules*, 2013, **46**, 4235.
26 Y. Shen, X. Chen and J. Wen, *Mol. Cryst. Liq. Cryst.*, 2011, **537**, 76.
27 L. Schefer, A. Sánchez-Ferrer, J. Adamcik and R. Mezzenga, *Langmuir*, 2012, **28**, 5999.
28 P. Kirsch and T. Mergner, *J. Fluorine Chem.*, 2006, **127**, 146.
29 J. Fantini and F. J. Barrantes, *Biochim. Biophys. Acta*, 2009, **1788**, 2345.
30 M. Sugahara, M. Uragami, X. Yan and S. L. Regen, *J. Am. Chem. Soc.*, 2001, **123**, 7939.
31 H. Cao, N. Tokutake and S. L. Regen, *J. Am. Chem. Soc.*, 2003, **125**, 16182.
32 V. Rondelli, G. Fragneto, S. Motta, E. Del Favero, P. Brocca, S. Sonnino and L. Cantù, *Biochim. Biophys. Acta*, 2012, **1818**, 2860.
33 H. M. Kim, B. R. Kim, H. J. Choo, Y. G. Ko, S. J. Jeon and C. H. Kim, *ChemBioChem*, 2008, **9**, 2830.
34 H. M. Kim, H. J. Choo, S. Y. Jung, Y. G. Ko, W. H. Park, S.-J. Jeon, C. H. Kim, T. Joo and B. R. Cho, *ChemBioChem*, 2007, **8**, 553.
35 D. Lingwood and K. Simons, *Science*, 2010, **327**, 46.
36 S. Sonnino, A. Prinetti, L. Mauri, V. Chigorno and G. Tettamanti, *Chem. Rev.*, 2006, **106**, 2111.
37 J. W. Goodby, V. Gortz, S. J. Cowling, G. Mackenzie, P. Martin, D. Plusquellec, T. Benvegnu, P. Boullanger, D. Lafont, Y. Queneau, S. Chambert and J. Fitremann, *Chem. Soc. Rev.*, 2007, **36**, 1971.
38 J. W. Goodby, S. J. Cowling, E. J. Davis, Y. Queneau, in *Carbohydrate Chemistry, Chemical and Biological Approaches*, ed. A. P. Rauter, T. K. Lindhorst and Y. Queneau, Specialist Periodical Reports, Royal Society of Chemistry, Cambridge, 2014, vol. 40, p. 312.
39 M. K. Singh, R. Xu, S. Moebs, A. Kumar, Y. Queneau, S. J. Cowling and J. W. Goodby, *Chem. – Eur. J.*, 2013, **19**, 5041.
40 Y. Queneau, J. Gagnaire, J. J. West, G. Mackenzie and J. W. Goodby, *J. Mater. Chem.*, 2001, **11**, 2839.
41 V. Molinier, P. H. J. Kouwer, Y. Queneau, J. Fitremann, G. Mackenzie and J. W. Goodby, *Chem. Commun.*, 2003, 2860.
42 V. Molinier, P. J. J. Kouwer, J. Fitremann, A. Bouchu, G. Mackenzie, Y. Queneau and J. W. Goodby, *Chem. – Eur. J.*, 2006, **12**, 3547.
43 V. Molinier, P. H. J. Kouwer, J. Fitremann, A. Bouchu, G. Mackenzie, Y. Queneau and J. W. Goodby, *Chem. – Eur. J.*, 2007, **13**, 1763–1775.
44 S. Chambert, A. Doutheau, Y. Queneau, S. J. Cowling, J. W. Goodby and G. Mackenzie, *J. Carbohydr. Chem.*, 2007, **26**, 27.
45 F. Ali Rachedi, S. Chambert, F. Ferkous, Y. Queneau, S. J. Cowling and J. W. Goodby, *Chem. Commun.*, 2009, 6355.
46 R. Xu, F. Ali-Rachedi, N. M. Xavier, S. Chambert, F. Ferkous, Y. Queneau, S. J. Cowling, E. J. Davis and J. W. Goodby, *Org. Biomol. Chem.*, 2015, **13**, 783.
47 A. H. Salway, *J. Chem. Soc., Trans.*, 1913, **103**, 1022.
48 V. Grassert and V. Vill, *Liq. Cryst. Today*, 1994, **4**, 4.

49 H. J. Wang, W. C. Cheng, H. H. Cheng, C. H. Lai and W. C. Wang, *Mol. Microbiol.*, 2012, **83**, 67.
50 H. Park, T. H. Lee, F. Chang, H. J. Kwon, J. Kim and H. Kim, *Bull. Korean Chem. Soc.*, 2013, **34**, 1339.
51 M. Pise, J. Rudra, S. Bundale, D. Begde, N. Nashikkar and A. Upadhyay, *J. Med. Plants Res.*, 2011, **5**, 507.
52 T. Goto, A. Shibata, D. Sasaki, N. Suzuki, T. Hishinuma, G. Kakiyama, T. Iida, N. Mano and J. Goto, *Steroids*, 2005, **70**, 185–192.
53 Y. S. Vollerner, S. D. Kravets, A. S. Shashkov, B. Tashkhodzhaev, M. B. Gorovits, M. R. Yagudaev and N. K. Abubakirov, *Chem. Nat. Compd.*, 1991, **27**, 198.
54 K. Kawashima, Y. Mimaki and Y. Sashida, *Chem. Pharm. Bull.*, 1991, **39**, 2761.
55 H. Yoshimitsu, M. Nishida, M. Yoshida and T. Nohara, *Chem. Pharm. Bull.*, 2002, **50**, 284.
56 M. Zahid, A. Ali, O. Ishurd, A. Ahmed, Z. Ali, V. U. Ahmad and Y. Pan, *Helv. Chim. Acta*, 2002, **85**, 689.
57 E. A. Santalova, V. A. Denisenko, P. S. Dmitrenok, D. V. Berdyshev and V. A. Stonik, *Nat. Prod. Commun.*, 2006, **1**, 265.
58 V. L. Challinor, J. M. U. Stuthe, P. G. Parsons, L. K. Lambert, R. P. Lehmann, W. Kitching and J. J. De Voss, *J. Nat. Prod.*, 2012, **75**, 1469.
59 R. H. Tian, E. Ohmura, H. Yoshimitsu, T. Nohara and M. Matsui, *Chem. Pharm. Bull.*, 1996, **44**, 1119.
60 M. Sautour, A. C. Mitaine-Offer and M. A. Lacaille-Dubois, *J. Nat. Med.*, 2007, **61**, 91.
61 Y. Mimaki, Y. Sashida, O. Nakamura, T. Nikaido and T. Ohmoto, *Phytochemistry*, 1993, **33**, 675.
62 V. L. Challinor, P. G. Parsons, S. Chap, E. F. White, J. T. Blanchfield, R. P. Lehmann and J. J. De Voss, *Steroids*, 2012, **77**, 504.
63 H. F. Dai, R. A. Edrada, R. Ebel, M. Nimtz, V. Wray and P. Proksch, *J. Nat. Prod.*, 2005, **68**, 1231.
64 H. Otsuka, T. Akiyama, K. I. Kawai, S. Shibata, O. Inoue and Y. Ogihara, *Phytochemistry*, 1978, **17**, 1349.
65 A. S. Antonov, A. I. Kalinovsky, V. A. Stonik, S. S. Afiyatullov, D. L. Aminin, P. S. Dmitrenok, E. Mollo and G. Cimino, *J. Nat. Prod.*, 2007, **70**, 169.
66 S. W. Yang, T. M. Chan, A. Buevich, T. Priestley, J. Crona, J. Reed, A. E. Wright, M. Patel, V. Gullo, G. Chen, B. Pramanik and M. Chu, *Bioorg. Med. Chem. Lett.*, 2007, **17**, 5543.
67 Y. Mimaki, Y. Sashida and H. Shimomura, *Phytochemistry*, 1989, **28**, 3453.
68 M. M. Ponpipom, R. L. Bugianesi and T. Y. Shen, *J. Med. Chem.*, 1980, **23**, 1184.
69 X. Hou and F. Chen, *Yaoxue Xuebao*, 1998, **33**, 923.
70 G. Ben-Menachem, J. Kubler-Kielb, B. Coxon, A. Yergey and R. Schneerson, *Proc. Natl. Acad. Sci.*, 2003, **100**, 7913.
71 C. M. Hasan, A. Islam, M. Ahmed, M.-D. Ahmed and P. G. Waterman, *Phytochemistry*, 1984, **23**, 2583.
72 H. Marschall, B. Egestad and H. Matern, *J. Biol. Chem.*, 1989, **264**, 12989.
73 C. N. Tong, H. Matsuda and M. Kubo, *Yakugaku Zasshi*, 1992, **112**, 856.
74 E. W. McChesney, F. C. Nachod, M. E. Auerbach and F. O. Laquer, *J. Am. Pharm. Assoc., Sci. Ed.*, 1948, **37**, 364.
75 Y. Ma, Z. Li, H. Shi, J. Zhang and B. Yu, *J. Org. Chem.*, 2011, **76**, 9748.
76 J. Rodriguez, R. Castro and R. Riguera, *Tetrahedron*, 1991, **47**, 4753.
77 W. W. Win and R. W. Franck, *J. Org. Chem.*, 1997, **62**, 4510.

78 T. M. Beale and M. S. Taylor, *Org. Lett.*, 2013, **15**, 1358.
79 A. M. Szpilman and E. M. Carreira, *Org. Lett.*, 2009, **11**, 1305.
80 P. Gupta and G. Panda, *Eur. J. Org. Chem.*, 2014, 8004.
81 M. Tamiya, N. Isaka, T. Kitazawa and M. Ishiguro, *Asian J. Org. Chem.*, 2014, **3**, 264.
82 S. M. Wang, W. Z. Ge, H. M. Liu, D. P. Zou and X. B. Yan, *Steroids*, 2004, **69**, 599.
83 J. F. Li, S. J. Chen, Y. Zhao and J. X. Li, *Carbohydr. Res.*, 2009, **344**, 599.
84 F. S. Ekholm, G. Schneider, J. Wölfling and R. Leino, *Eur. J. Org. Chem.*, 2011, 1064.
85 V. Faivre, P. L. Bardonnet, P. Boullanger, H. Amenitsch, M. Ollivon and F. Falson, *Langmuir*, 2009, **25**, 9424.
86 J. C. Chabala and T. Y. Shen, *Carbohydr. Res.*, 1978, **67**, 55.
87 M. M. Pompipon, R. L. Bugianesi and T. Y. Shen, *Can. J. Chem.*, 1980, **58**, 214.
88 M. M. Ponpipom, M. S. Wu, J. C. Robbins and T. Y. Shen, *Carbohydr. Res.*, 1983, **118**, 47.
89 M. R. Mauk, R. C. Gamble and J. D. Baldeschwieler, *Science*, 1980, **207**, 309.
90 M. R. Mauk, R. C. Gamble and J. D. Baldeschwieler, *Proc. Natl. Acad. Sci.*, 1980, **77**, 4430.
91 F. Choteau, G. Durand, I. Ranchon-Cole, C. Cercy and B. Pucci, *Bioorg. Med. Chem. Lett.*, 2010, **20**, 7405.
92 P. Iveson and D. V. Parke, *J. Chem. Soc. C*, 1970, **15**, 2038.
93 R. M. Kondratenko, S. R. Mustafina, L. A. Baltina, F. Z. Galin and G. A. Tolstikov, *Chem. Nat. Compd.*, 2005, **41**, 7.
94 S. Vidal, A. Morère and J. L. Montero, *Heteroat. Chem.*, 2003, **14**, 241.
95 D. G. Rivera, K. Pérez-Labrada, L. Lambert, S. Dörner, B. Westermann and L. A. Wessjohann, *Carbohydr. Res.*, 2012, **359**, 102.
96 K. S. Wankhede, V. V. Vaidya, P. S. Sarang, M. M. Salunkhe and G. K. Trivedi, *Tetrahedron Lett.*, 2008, **49**, 2069.
97 D. Kumar, K. B. Mishra, B. B. Mishra, S. Mondal and V. K. Tiwari, *Steroids*, 2014, **80**, 71.
98 R. Beaulieu, S. Gottis, C. Meyer, E. Grand, V. Deveaux, J. Kovensky and I. Stasik, *Carbohydr. Res.*, 2015, **404**, 70.
99 K. B. Mishra, B. B. Mishra and V. K. Tiwari, *Carbohydr. Res.*, 2014, **399**, 2.
100 K. Cheng, J. Liu, X. Liu, H. Li, H. Sun and J. Xie, *Carbohydr. Res.*, 2009, **344**, 841.
101 K. Cheng, J. Liu, H. Sun, E. Bokor, K. Czifrák, B. Kónya, M. Tóth, T. Docsa, P. Gergely and L. Somsák, *New J. Chem.*, 2010, **34**, 1450.
102 Y. Mimaki, Y. Sashida and K. Kawashima, *Chem. Pharm. Bull.*, 1992, **40**, 148.
103 B. N. A. Mbadugha and F. M. Menger, *Org. Lett.*, 2003, **5**, 4041.
104 K. Sato, M. Mochizuki, I. Saiki, Y. Yoo, K. Samukawa and I. Azuma, *Biol. Pharm. Bull.*, 1994, **17**, 63.
105 H. Liu, G. X. Chou, J. M. Wang, L. L. Ji and Z. T. Wang, *Planta Med.*, 2011, **77**, 845.
106 A. L. Tapondjou, T. Miyamoto and M.-A. Lacaille-Dubois, *Phytochemistry*, 2006, **67**, 2126.
107 Y. Li, J. Sun and B. Yu, *Org. Lett.*, 2011, **13**, 5508.
108 Y. M. Yoshikawa, T. Murakami, M. Kadoya, H. Matsuda, O. Muraoka, J. Yamahara and N. Murakami, *Chem. Pharm. Bull.*, 1996, **44**, 1212.
109 M. Yoshikawa, T. Murakami, M. Kadoya, H. Matsuda, J. Yamahara, O. Muraoka and N. Murakami, *Heterocycles*, 1995, **8**, 1621.
110 M. Yoshikawa, T. Murakami, M. Inaduki, K. Hirano, J. Yamahara and H. Matsuda, *Chem. Pharm. Bull.*, 1997, **45**, 561.
111 S. Zhu, Y. Li and B. Yu, *J. Org. Chem.*, 2008, **73**, 4978.

112 T. Fujika, M. Iwamoto, Y. Iwase, S. Hachiyama, H. Okabe, T. Yamauchi and K. Mihashi, *Chem. Pharm. Bull.*, 1989, **37**, 2355.
113 C. Zhu, P. Tang and B. Yu, *J. Am. Chem. Soc.*, 2008, **130**, 5872.
114 J. H. Qi, M. Ojika and Y. Sakagami, *Bioorg. Med. Chem.*, 2002, **10**, 1961.
115 Q. Liu, Y. Yu, P. Wang and Y. Li, *New J. Chem.*, 2013, **37**, 3647.
116 Z. Zhu, T. Chen, H. F. Pi, H. L. Ruan, J. Z. Wu and P. Zhang, *Chem. Nat. Compd.*, 2015, **51**, 890.
117 C. Hellfritsch, A. Brockhoff, F. Stähler, W. Meyerhof and T. Hofmann, *J. Agric. Food Chem.*, 2012, **60**, 6782.
118 M. Lepage, *J. Lipid Res.*, 1964, **5**, 587.
119 C. Grunwald, *Plant Physiol.*, 1971, **48**, 653.
120 J. B. Mudd and T. T. McManus, *Plant Physiol.*, 1980, **65**, 78.
121 M. S. Webb, T. C. Irving and P. L. Steponkus, *Biochim. Biophys. Acta, Biomembr.*, 1995, **1239**, 226.
122 K. K. Halling, B. Ramstedt and J. P. Slotte, *Biochim. Biophys. Acta, Biomembr.*, 2008, **1778**, 1100.
123 K. Schrick, S. Shiva, J. C. Arpin, N. Delimont, G. Isaac, P. Tamura and R. Welti, *Lipids*, 2012, **47**, 185.
124 H. Hua, X. Li and H. Zhang, *J. Shenyang Pharm. Univ.*, 2000, **17**, 40.
125 R. A. Davis, C. H. Lin and J. Gervay-Hague, *Chem. Commun.*, 2012, **48**, 9083.
126 S. S. Kulkarni and J. Gervay-Hague, *Org. Lett.*, 2008, **10**, 4739.
127 K. Papadopoulou, R. E. Melton, M. Leggett, M. J. Daniels and A. E. Osbourn, *Proc. Natl. Acad. Sci.*, 1999, **96**, 12923.
128 P. Mylona, A. Owatworakit, K. Papadopoulou, H. Jenner, B. Qin, K. Findlay, L. Hill, X. Qi, S. Bakht, R. Melton and A. Osbourn, *Plant Cell*, 2008, **20**, 201.
129 S. Tsurumi and Y. Tsujino, *Physiol. Plant.*, 1995, **93**, 785.
130 A. Rahman, A. Ahamed, T. Amakawa, N. Goto and S. Tsurumi, *Plant Physiol.*, 2001, **125**, 990.
131 R. H. Tian, E. Ohmura, M. Matsui and T. Noharas, *Phytochemistry*, 1997, **44**, 723.
132 Y. Kang, C. Lou, K. B. R. Ahmed, P. Huang and Z. Jin, *Bioorg. Med. Chem. Lett.*, 2009, **19**, 5166.
133 X. Ma, B. Yu, Y. Hui, Z. Miao and J. Ding, *Bioorg. Med. Chem. Lett.*, 2001, **11**, 2153.
134 Y. Cheng, D. M. Ho, C. R. Gottlieb, D. Kahne and M. A. Bruck, *J. Am. Chem. Soc.*, 1992, **114**, 7319.
135 P. Venkatesan, Y. Cheng and D. Kahne, *J. Am. Chem. Soc.*, 1994, **116**, 6955.
136 C. L. Bowe, L. Mokhtarzadeh, P. Venkatesan, S. Babu, H. R. Axelrod, M. J. Sofia, R. Kakarla, T. Y. Chan, J. S. Kim, H. J. Lee, G. L. Amidon, S. Y. Choe, S. Walker and D. Kahne, *Proc. Natl. Acad. Sci.*, 1997, **94**, 12218.
137 M. J. Sofia, R. Kakarla, N. Kogan, R. Dulina, Y. W. Hui, N. T. Hatzenbuhler, D. Liu, A. Chen and T. Wagler, *Bioorg. Med. Chem. Lett.*, 1997, **7**, 2251.
138 H. R. Axelrod, J. S. Kim, C. B. Longley, E. Lipka, G. L. Amidon, R. Kakarla, Y. W. Hui, S. J. Weber, S. Choe and M. J. Sofia, *Pharm. Res.*, 1998, **15**, 1876.
139 D. Wang and Y. R. Bao, *Chem. Nat. Compd.*, 2015, **51**, 897.
140 A. Silipo, G. Vitiello, D. Gully, L. Sturiale, C. Chaintreuil, J. Fardoux, D. Gargani, H. I. Lee, G. Kulkarni, N. Busset, R. Marchetti, A. Palmigiano, H. Moll, R. Engel, R. Lanzetta, L. Paduano, M. Parrilli, W. S. Chang, O. Holst, D. K. Newman, D. Garozzo, G. D'Errico, E. Giraud and A. Molinaro, *Nat. Commun.*, 2014, **5**, 5106.
141 A. Molinaro, O. Holst, F. Di Lorenzo, M. Callaghan, A. Nurisso, G. D'Errico, A. Zamyatina, F. Peri, R. Berisio, R. Jerala, J. Jimenez-Barbero, A. Silipo and S. Martin-Santamaria, *Chem. – Eur. J.*, 2015, **21**, 500.

142 Y. H. Kim, S. H. Gihm, C. R. Park, K. Y. Lee, T. W. Kim, I. C. Kwon, H. Chung and S. Y. Jeong, *Bioconjugate Chem.*, 2001, **12**, 932.
143 S. Kwon, J. H. Park, H. Chung, I. C. Kwon and S. Y. Jeong, *Langmuir*, 2003, **19**, 10188.
144 X. B. Yuan, H. Li and Y. B. Yuan, *Carbohydr. Polym.*, 2006, **65**, 337.
145 J. M. Yu, Y. J. Li, L. Y. Qiu and Y. Jin, *J. Pharm. Pharmacol.*, 2009, **61**, 713.
146 M. P. Patel, R. R. Patel and J. K. Patel, *J. Pharm. Pharm. Sci.*, 2010, **13**, 536.
147 F. M. Menger and J. S. Keiper, *Angew. Chem., Int. Ed.*, 1998, **37**, 3433.
148 A. Saha, J. Adamcik, S. Bolisetty, S. Handschin and R. Mezzenga, *Angew. Chem., Int. Ed.*, 2015, **54**, 5408.
149 S. Trombotto, A. Bouchu, G. Descotes and Y. Queneau, *Tetrahedron Lett.*, 2000, **41**, 8273.
150 S. Trombotto, M. Danel, J. Fitremann, A. Bouchu and Y. Queneau, *J. Org. Chem.*, 2003, **68**, 6672.
151 A. Listkowski, P. Ing, R. Cheaib, S. Chambert, A. Doutheau and Y. Queneau, *Tetrahedron: Asymmetry*, 2007, **18**, 2201.
152 R. Cheaib, A. Listkowski, S. Chambert, A. Doutheau and Y. Queneau, *Tetrahedron: Asymmetry*, 2008, **19**, 1919.
153 A. Sánchez-Ferrer, J. Adamcik and R. Mezzenga, *Soft Matter*, 2012, **8**, 149.
154 K. Muramatsu, T. Masumizu, Y. Maitani, S. H. Hwang, M. Kohno, K. Takayama and T. Nagai, *Chem. Pharm. Bull.*, 2000, **48**, 610.
155 Y. Maitani, K. Nakamura and K. Kawano, *Curr. Pharm. Biotechnol.*, 2005, **6**, 81.
156 V. Barragan, F. M. Menger, K. L. Caran, C. Vidil, A. Morère and J.-L. Montero, *Chem. Commun.*, 2001, 85.
157 F. M. Menger, *C. R. Chim.*, 2009, **12**, 54.
158 Z. Liu, US20120121696 A1, 2012.
159 H. Kimata, N. Sumida, N. Matsufuji, T. Morita, K. Ito, N. Yata and O. Tanaka, *Chem. Pharm. Bull.*, 1985, **33**, 2849.
160 Y. Sasaki and O. Tanaka, *Chem. Pharm. Bull.*, 1988, **36**, 3491.
161 T. Tanaka, M. Kataoka, N. Tsuboi and I. Kouno, *Chem. Pharm. Bull.*, 2000, **48**, 201.
162 T. Tanaka, H. Zhang, Z. H. Jiang and I. Kouno, *Chem. Pharm. Bull.*, 1997, **45**, 1891.

Recent advances in the synthesis of imino sugars. An insight into the cascade addition of Grignard reagents to halonitriles/cyclization

Michał Malik* and Sławomir Jarosz*
DOI: 10.1039/9781782626657-00313

This review is focused on the recent advances in the synthesis of imino sugars. It is divided into the following sections: S_N2 cyclizations, reactions of cyclic imines, reactions of cyclic nitrones, and ring-closing metathesis. In a separate category, a cascade addition of Grignard reagents to halonitriles followed by the S_N2-type cyclization is discussed – a rarely used, but convenient approach to the synthesis of five- and six-membered heterocycles containing an endocyclic nitrogen atom.

1 Introduction

Imino sugars represent a broad group of organic compounds widely present in Nature. These compounds, closely related to sugars, with the endocyclic nitrogen atom instead of the oxygen atom,[1] possess interesting biological properties.[2] For example, *Miglitol* is used as an anti-diabetic drug,[2c] whereas *Miglustat* is applied against type 1 Gaucher disease.[2d] Both are analogs of deoxynojirimycin (DNJ; Fig. 1), probably the most recognized imino sugar.[3]

First synthesis of DNJ was reported by Paulsen in 1966.[4] A decade later, this imino sugar was isolated from *Bacillus* bacteria and its potential as an anti-diabetic drug was recognized by the scientists from the Bayer company.[5] Soon after, bicyclic imino sugars were isolated from natural sources;[6] they were proven to possess strong anti-tumour[7a,7b] and anti-glycosidase[7c] activity.

Rapid progress in the art of the total synthesis allowed for the development of numerous synthetic routes leading to naturally occurring imino sugars. The creativity of researchers led also to many novel structures, not occurring in Nature.[8]

There are at least two approaches, by which the existing literature reports may be categorized. The most straightforward and intuitive way is to sort the data by the type of the alkaloid group to which the target compound belongs, *i.e.* indolizidines, quinolizidines, pyrrolizidines, piperidines, pyrrolidines, *etc.*

However, from a purely synthetic point of view, it is more interesting to learn at first glance how the target compound was obtained. Therefore, the material collected in this report is based on the type of the reaction leading to the desired imino sugar scaffold. Despite the fact that such organization may provoke some ambiguities, we believe that it will

Institute of Organic Chemistry, Polish Academy of Sciences, ul. Kasprzaka 44/52, 01-224 Warsaw, Poland. E-mail: michal.malik@icho.edu.pl; slawomir.jarosz@icho.edu.pl

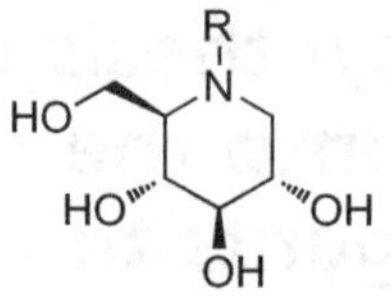

Fig. 1 Deoxynojirimycin (DNJ) and its well-known analogs.

simplify this presentation, especially for synthetic organic chemists, and will allow for an easier discussion.

The following main categories cover the vast majority of the existing literature data on the synthesis of imino sugars: the S_N2-type cyclizations, reactions of cyclic imines (including reductive aminations), reactions of cyclic nitrones, and ring-closing metathesis (RCM). Recently, we have proposed a useful methodology for the preparation of 2-allyl- and 2,2-diallyl-substituted piperidines and pyrrolidines based on the cascade addition of Grignard reagents to halonitriles and subsequent S_N2 cyclization.[9] Although this methodology is still not well-explored (only several precedents exist in the literature), its evident applicability to the synthesis of imino sugars convinced us to discuss it as a separate category. Few other approaches, not fitting into any of the above-mentioned categories, such as the aza-Achmatowicz reaction and cycloadditions, will also be discussed.

Since the synthesis of imino sugars is reported in many excellent books and reviews,[1,3,8b,10] we have decided to focus on the recent literature. Moreover, we have focused on the approaches arising from the chiral pool, which is the most popular methodology in the total synthesis of natural products. Nevertheless, some papers based on the less common *de novo* asymmetric approach (*i.e.* starting from achiral substrates) are also discussed.

2 S_N2 cyclizations

The S_N2-type cyclizations are among the most common methodologies leading to imino sugars. The earliest total syntheses of mono- and bicyclic compounds of this class were often accomplished *via* an intramolecular displacement of a leaving group by a nitrogen nucleophile as demonstrated, for example, in the first total synthesis of (+)-castanospermine.[11] Due to the reliability of this approach, many novel imino sugars are still synthesized on this way.

The synthetic routes in this methodology are usually long, since they require tedious manipulations with the protecting groups, a major drawback of this approach. However, this is not always the case, as is demonstrated by some of the following examples.

Vankar's group reported the synthesis of a novel family of bicyclic imino sugars, which can be regarded as conformationally locked pyrrolidines (Scheme 1).[12] *Syn*-dihydroxylation of D-glucal-derived compound **1** gave diol **2** as an inseparable mixture of isomers ($dr = 3{:}2$). After tritylation, both diastereoisomers (**3** and **4**) were easily separated by chromatography and the major diastereoisomer **3** was converted into the bicyclic imino sugar **6**

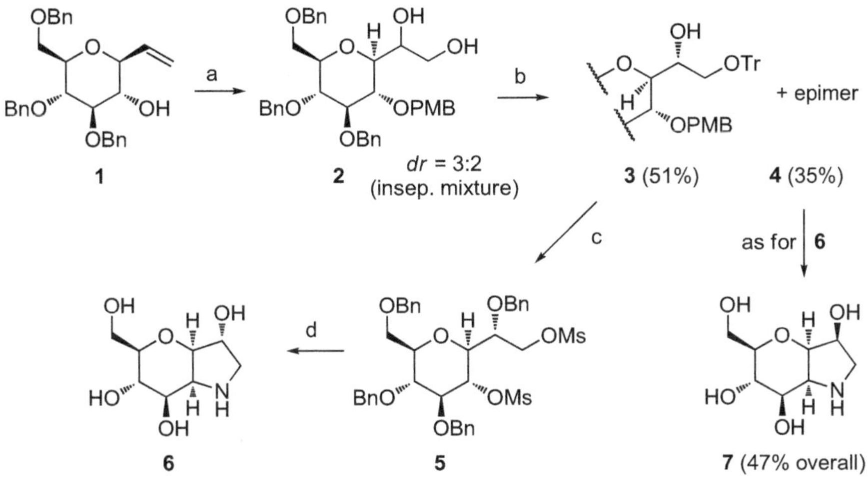

Scheme 1 Reagents and conditions: (a) i. PMBCl, DMF, NaH, 93%, ii. OsO_4 (cat), NMO, 89%; (b) TrCl, DCM, Et_3N, **3** (51%) + **4** (35%); (c) i. BnCl, DMF, NaH, ii. $TrBF_4$, DCM, iii. MsCl, DMAP, Et_3N, 56% (3 steps); (d) i. $BnNH_2$, 120 °C, ii. H_2, $Pd(OH)_2/C$, EtOH, 75% (2 steps).

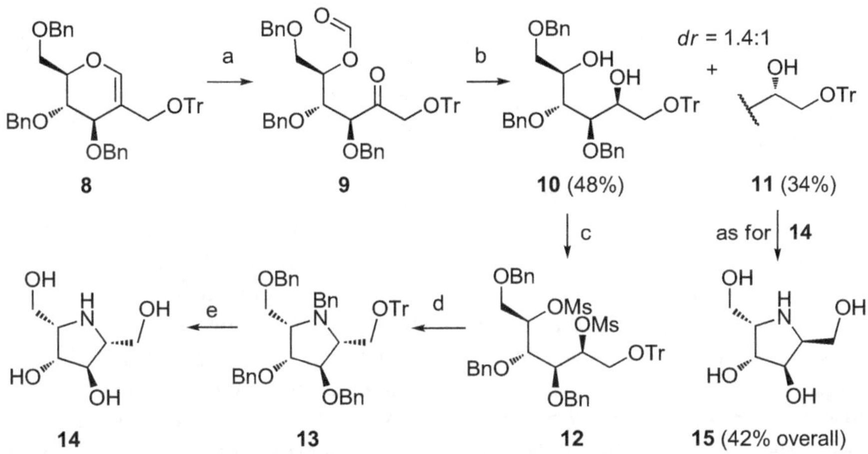

Scheme 2 Reagents and conditions: (a) i. OsO_4 (cat.), NMO, ii. $NaIO_4$, 74% (2 steps); (b) $NaBH_4$, MeOH, **10** (48%) + **11** (34%); (c) MsCl, Et_3N, DMAP, DCM, 89%; (d) $BnNH_2$, 140 °C, 78%; (e) i. TFA, DCM, 85%, ii. H_2, $Pd(OH)_2/C$, MeOH, HCl, 62%.

by a standard methodology shown in Scheme 1. Similar sequence of reactions performed on its epimer **4** provided compound **7**. Both derivatives: **6** and **7**, are good inhibitors of α-galactosidase ($IC_{50} \approx 1$ mM).

In a more recent paper,[13] Vankar and Ansari presented an approach to dihydroxymethyl dihydroxypiperidines. The representative synthesis (shown in Scheme 2) was initiated from derivative **8**, prepared from the protected D-glucal. Oxidative cleavage of the double led to dicarbonyl compound **9** which, after reduction, provided diastereomeric diols **10** and **11** in the ratio 1.4 : 1. Isomer **10** was transformed into the corresponding pyrrolidine **13** in a sequence of reactions involving mesylation and S_N2

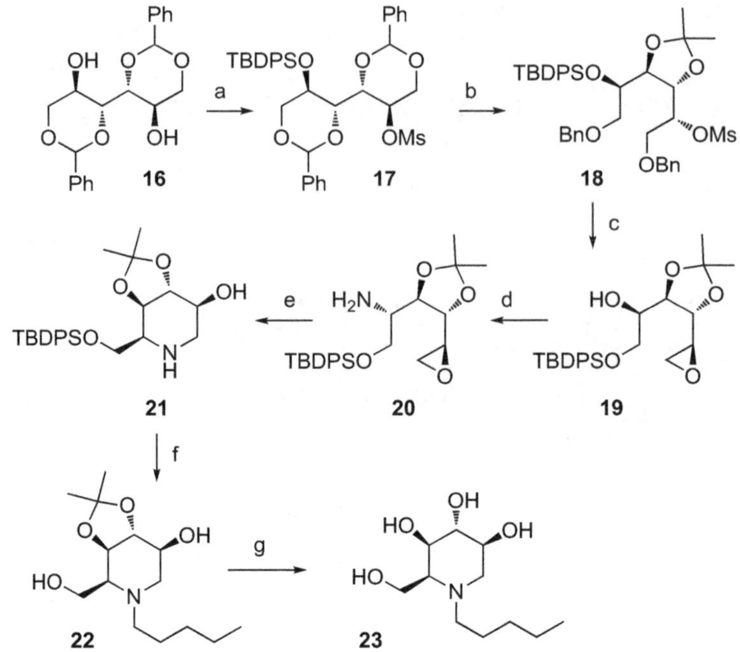

Scheme 3 Reagents and conditions: (a) i. TBDPSCl, imid., DMF, 69%, ii. MsCl, Et_3N, DCM, 94%; (b) i. $BF_3 \cdot Et_2O$, Et_3SiH, DCM, 85%, ii. 2,2-DMP, acetone, CSA (cat.), 80%; (c) i. H_2, Pd/C, EtOH, 95%, ii. K_2CO_3, MeOH, 60%; (d) i. Tf_2O, py, ii. NaN_3, DMF, 90% (2 steps), iii. H_2, Pd/$CaCO_3$, MeOH, 87%; (e) MeOH, reflux, 93%; (f) i. *n*-pentanal, $NaBH_3CN$, AcOH (cat.), MeOH, 81%, ii. TBAF, THF, 95%; (g) Dowex H, MeOH, 90%.

cyclization with $BnNH_2$. After detritylation and hydrogenolysis, the final product **14** was obtained in 43% overall yield (from **10**). The same sequence of reactions performed on diastereoisomer **11** afforded derivative **15** in 42% overall yield. Both imino sugars showed good inhibitory activity against α- and β-gluco-sidases (0.02 mM < IC_{50} < 0.15 mM).

Baskaran and coworkers devised a synthetic route to piperidine imino sugar **23** (Scheme 3),[14] a potent anticancer agent. Selective monosilylation of D-mannitol derived compound **16** followed by mesylation, gave derivative **17**. It was then subjected to regioselective, reductive cleavage of the benzylidene acetals, followed by protection of the released secondary hydroxyl groups with 2,2-dimethoxypropane (2,2-DMP), giving compound **18**. After removal of the benzyl groups, the diol was treated with base. Formation of the oxirane ring was accompanied by a migration of the silyl ether, which finally provided compound **19**. The remaining free hydroxyl group was transformed into triflate, which was then treated with NaN_3. The resulting azide was reduced (H_2/Pd) to afford amine **20**. Refluxing of this amine in MeOH yielded piperidine **21** in high yield. Its reaction with *n*-pentanal, under the reductive-amination conditions, followed by a cyclization, afforded *N*-alkylated piperidine **22**. Removal of the silyl and isopropylidene protecting groups yielded the final imino sugar **23**.

In another example, the Behr's group reported a new synthetic route to the analogs of natural alkaloid (−)-codonopsinine (Scheme 4).[15] This

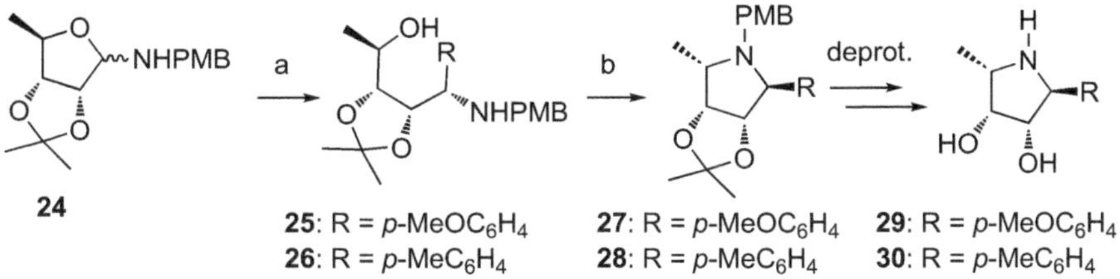

Scheme 4 Reagents and conditions: (a) RMgCl, THF, 52% (**25**) or 52% (**26**); (b) MsCl, py, 63% (**27**) or 78% (**28**).

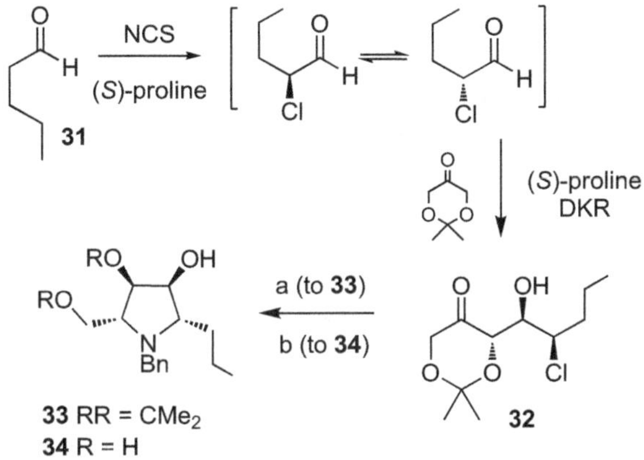

Scheme 5 Reagents and conditions: (a) i. $BnNH_2$, AcOH, MS 4 Å, THF, then $NaBH_3CN$, ii. $NaHCO_3$, toluene, 105 °C; (b) i. $BnNH_2$, AcOH, MS 4 Å, THF, then $NaBH_3CN$, ii. MeOH, 120 °C, microwaves.

quite concise approach was inititiated from the D-ribose-derived hemiaminal **24**. Highly diastereoselective Grignard addition (*de*>99 : 1) of arylmagnesium chlorides to **24** afforded either **25** or **26**; both underwent the S_N2 cyclization providing **27** or **28** respectively. Final deprotection afforded the free imino sugars: **29** and **30** (Scheme 4); both were exceptionally good inhibitors of fucosidase (IC_{50} ~ 6 nM).

Very recently, a concise *de novo* asymmetric approach to the pyrrolidine-based imino sugars was presented by Britton and co-workers.[16] Starting from a wide range of aldehydes, they used a tandem α-chlorination/DKR aldol reaction with the protected dihydroxyacetone catalyzed by (*S*)-proline, to obtain linear γ-haloketones in good yields and high *ee* values (>90%). In a representative example (Scheme 5), aldehyde **31** was transformed into γ-haloketone **32** which underwent a reductive amination reaction yielding the corresponding linear amine. The subsequent intra-molecular S_N2 cyclization conducted in toluene gave the protected pyrrolidine **33**, while the same process in methanol, with the assistance of microwaves, afforded deprotected derivative **34**. Several of the obtained pyrrolidines were transformed into the bicyclic analogs of (+)-hyacinthacine and (−)-steviamine. The wide range of the compounds obtained, in addition to the high diastereo- and enantioselectivity of the developed process, makes the methodology proposed

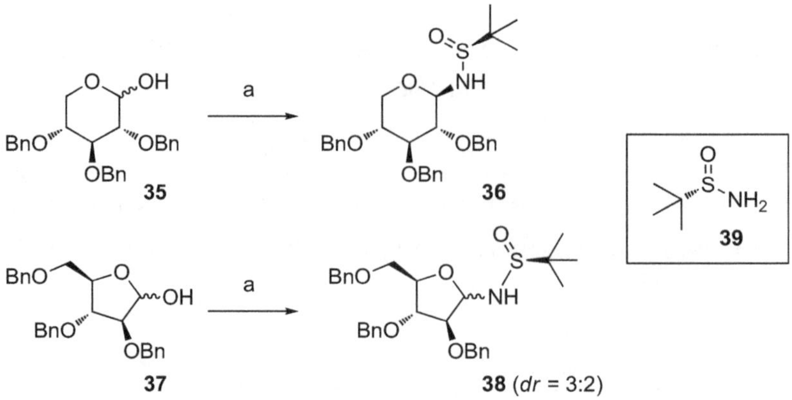

Scheme 6 Reagents and conditions: (a) **39** (2 equiv.), MS 4 Å, $Ti(OEt)_4$, toluene, 70 °C, 45% (route leading to **36**) or 71% (to **38**).

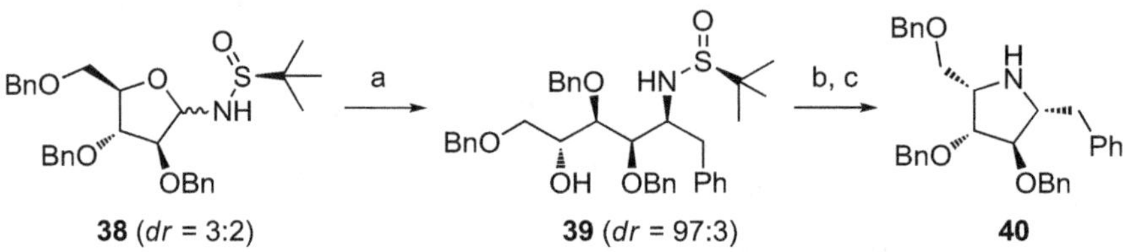

Scheme 7 Reagents and conditions: (a) BnMgCl, THF, LiCl, −60 to −20 °C, 94%; (b) i. Et_3N, MsCl, THF, ii. *t*-BuOK, THF, 50%; (c) i. HCl, MeOH, ii. Amberlite OH, quant.

by Britton very well suited for the development of large libraries of imino sugars.

Martin and coworkers showed a convenient route towards *N*-*tert*-butanesulfinyl aldimines and ketimines, such as **36** and **38** (Scheme 6).[17]

As the authors point out, reports on the stability and reactivity of such compounds are very limited. Therefore, they have decided to investigate the addition of organo-metallic species to the obtained derivatives. Reaction with the Grignard reagents turned out to be good-yielding, whereas diastereoselectivities ranged from poor to good. Some of the resulting iminoalditols were transformed into derivatives of pyrrolidine, such as **40** (Scheme 7). Gratifyingly, despite the use of **38** as a mixture of anomers, derivative **39** was obtained as practically single isomer. This fact was explained on the basis of equilibrium between the *N*-metallated anomers.

The group of Sas developed a novel synthetic route to polyhydroxylated indolizi-dines (Scheme 8).[18] Protected α-D-mannofuranoside **41** was transformed into derivative **43** by reaction with hydroxylamine **42** (nitrone formation) followed by intramolecular 1,3-dipolar cycloaddition. This compound was transformed into ammonium salt **44**, which was then cleaved under the basic conditions, giving the desired indolizidine scaffold **45**. Alternatively, compound **43** was transformed into the mesyl ester **46**, which cyclized upon treatment with K_2CO_3, providing the bicyclic compound **47**.

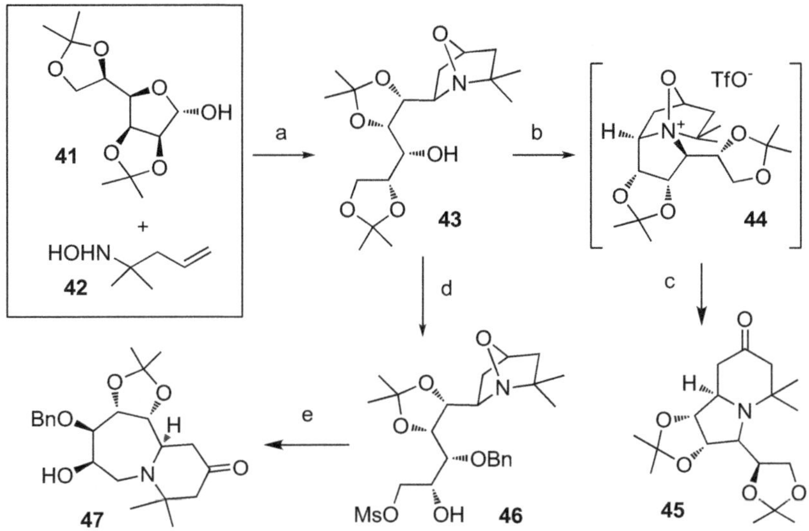

Scheme 8 Reagents and conditions: (a) toluene, 100 °C, 57%; (b) Tf_2O, py; (c) NaN_3, Bu_4NBr, DMF, 54% (2 steps); (d) i. NaH, DMF, BnBr, 94%, ii. TFA, $CHCl_3$, 83%, iii. MsCl, TEA, DCM, Bu_2SnO (cat.), 90%; (e) K_2CO_3, EtOH.

3 Reactions of cyclic imines

Another general approach towards imino sugars is based on the formation of cyclic imines and their subsequent reactions. The versatility of imines results from a number of possible transformations they can undergo; reduction to amines,[19] nucleophilic addition,[20] and aza-Diels–Alder cycloadditions[21] are among the most common.

Especially useful, in the context of the synthesis of imino sugars, are cyclic imines derived from the chiral pool. However, they are usually unstable and thus have to be used *in situ*, preferentially in a one-pot procedure. Among the transformations of such imines, the most known and straightforward is the reductive amination. This classical approach opens a convenient route to many naturally occurring imino sugars and their analogs. For example, the first, famous synthesis of DNJ was accomplished *via* a reductive amination process.[4]

This approach is also quite common in the recent literature. For example, Majewski and Palyam presented a useful organocatalytic approach to derivatives of DNJ, based on the reaction of dioxanone **48** with (*S*)-isoserinal hydrate **49** catalyzed by (*S*)-proline.[22] This process led to linear ketone **50** which was converted into *N*-isopropyl-L-ido-DNJ **51** (Scheme 9).

O'Doherty reported *de novo* asymmetric synthesis of 8a-*epi*-swainsonine.[23] The synthesis was initiated from furfural, which – in a few straightforward steps – was converted into the racemic alcohol (+/−)-**53** (Scheme 10).

Oxidation of the hydroxyl group followed by the highly enantioselective asymmetric reduction of the resulting ketone (in the presence of the Noyori's catalyst **52**) yielded alcohol (*R*)-**54** (*ee*>96%). Subsequent

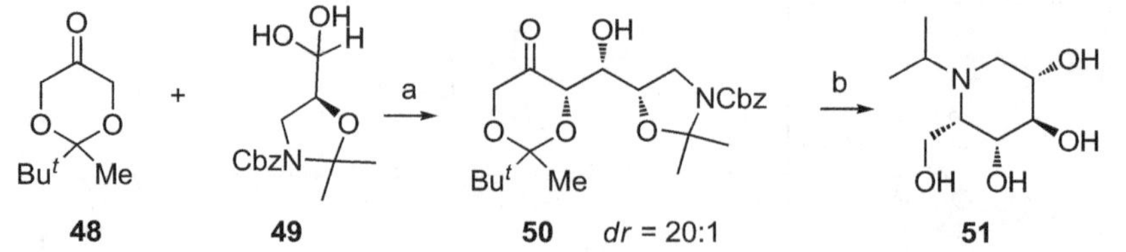

Scheme 9 Reagents and conditions: (a) (*S*)-proline, DMSO, LiCl, 69%; (b) H_2, Pd/C, MeOH, HCl, 82%.

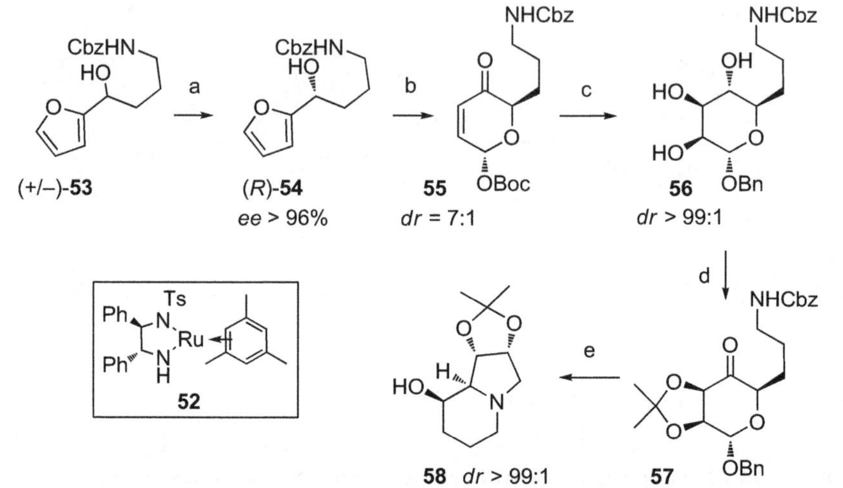

Scheme 10 Reagents and conditions: (a) i. MnO_2, THF, 90%, ii. cat. **52** (2 mol%), HCOOH, Et_3N, 91%; (b) i. NBS, AcONa, THF/H_2O, 92%, ii. Boc_2O, DCM, DMAP (cat.), 80%; (c) i. $Pd_2(dba)_3 \cdot CHCl_3$ (2.5 mol%), BnOH, DCM, Ph_3P (5 mol%), 88%, ii. $NaBH_4/CeCl_3$, MeOH, 87%; iii. OsO_4 (1 mol%), *t*-BuOH/acetone, NMO, 89%; (d) i. *p*-TsOH, 2,2-DMP, acetone, 87%, ii. Swern ox., 92%; (e) H_2, $Pd(OH)_2/C$, EtOH/THF, 76%.

Achmatowicz reaction and protection of the free hydroxyl group with Boc gave predominantly diastereoisomer **55**. This derivative was then converted into triol **56** (exclusively one isomer), in a synthetic route consisting of the Tsuji–Trost reaction with BnOH, subsequent reduction under the Luche conditions, and osmylation. After protection of both *syn*-hydroxyl groups with 2,2-DMP and the oxidation of the remaining OH, the resulting ketone **57** was subjected to the reductive conditions, under which the desired bicyclic scaffold **58** was formed. The probable course of this step involved removal of the benzyl and Cbz moieties followed by two consecutive reductive aminations. Finally, removal of the isopropylidene group under the acidic conditions gave the desired product. In a similar approach, O'Doherty obtained another diastereoisomer of swainsonine.[24]

Rauter and Queneau developed a methodology leading to imino sugar derivatives containing an α, β-unsaturated lactone moiety. Their approach started from known azido-derivative **59**,[25] which was transformed *via* a Wittig olefination into compound **60** (Scheme 11). This derivative was subjected to reductive conditions (H_2, Pd/C), after which an unstable, prone to dehydration hemiaminal was formed. Then, the

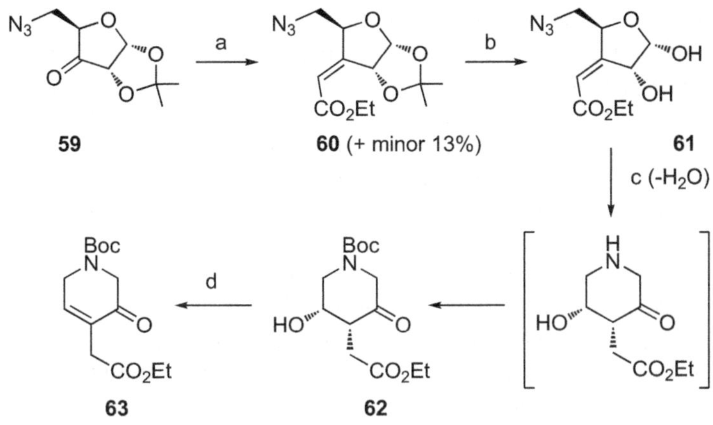

Scheme 11 Reagents and conditions: (a) $Ph_3P{=}CHCO_2Et$, $CHCl_3$, 74% (major isomer), 13% (minor); (b) 60% aq. TFA, 94%; (c) i. H_2, Pd/C, EtOH, then Boc_2O; (c) Ac_2O, py, 42% (3 steps).

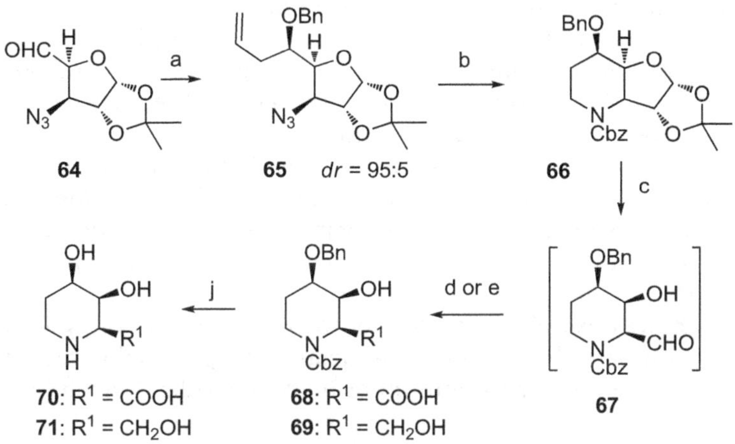

Scheme 12 Reagents and conditions: (a) i. Allyl bromide, Zn, THF, 61%, ii. BnBr, NaH, THF, 76%; (b) i. O_3, DCM, then Ph_3P, ii. $NaBH_3CN$, AcOH (cat.), iii. CbzCl, $NaHCO_3$, 25% (3 steps); (c) i. TFA/H_2O, ii. $NaIO_4$, acetone/H_2O; (d) NaH_2PO_4, $NaClO_2$, 30% H_2O_2, 67% (**68**; 3 steps from **66**); (e) $NaBH_4$, MeOH/H_2O, 40% (**69**; 3 steps from **66**); (f) H_2, Pd/C, MeOH, 92% (**70**) or 93% (**71**).

intermediate derivative was treated with Boc_2O, giving compound **62**. Acetylation of the latter led to **63**, a convenient precursor for the synthesis of novel imino sugars.

Apart from the previously mentioned approaches, the synthesis of cyclic imines can be also accomplished by a Staudinger/aza-Wittig transformation. The resulting unstable iminophosphorane can be either hydrolysed to an amine or reacted with a carbonyl compound to yield an imine. The utility of this approach was proven, for example, by Chattopadhyay, who devised a divergent approach to the analogs of fagomine and hydroxypipecolic acid (Scheme 12).[26] The synthesis was initiated from aldehyde **64**, easily prepared from D-glucose.[27] Highly diastereoselective Barbier reaction of **64** with allyl bromide ($dr = 95{:}5$), followed by benzylation provided the corresponding homoallylic acohol in which the newly formed hydroxyl group was protected as benzyl ether (**65**). Ozonolysis of the double bond followed by treatment with Ph_3P gave the cyclic imine (the product of the aza-Wittig reaction), which was reduced with $NaBH_3CN$ to the corresponding piperidine, finally protected with the Cbz group. Compound **66** was further transformed, *via* aldehyde **67**, into the analogs of either pipecolic acid (**70**) or fagomine (**71**).

The group of Blèriot used the solid-supported triphenylphosphane in a Staudinger/aza-Wittig tandem process to transform azide **72** into bicyclic derivative **73**.[28] The addition of various Grignard reagents to **73** provided a wide range of polyhydroxylated azepanes, *e.g.* the allyl derivative **74**. This compound could be easily converted into the interesting piperidine scaffold, such as **77** (Scheme 13).

As already stated previously, a useful synthetic approach to cyclic imines consists in the *N*-halogenation of a cyclic amine followed by dehydrohalogenation. This methodology may pose a serious problem, since it may lead to a mixture of region-isomeric imines. Therefore, the *C*-2 symmetrical precursors are usually preferred.

Recently, Yang and Shao exploited this methodology in the synthesis of lentiginosine and steviamine analogs.[29] Starting from the chiral pool, they obtained a family of various five-membered cyclic aldimines (Scheme 14). These compounds were subsequently subjected to the ytterbium(III) triflate-catalyzed aza-Diels–Alder reaction with the Danishefsky's diene, which provided a family of polyhydroxylated indolizidines. Some of them were easily transformed into imino sugars: (+)- and (−)-lentiginosine, and (−)-2-*epi*-steviamine. An example of such transformation is shown in Scheme 15.

Another, less exploited, possibility for the synthesis of cyclic imines consists in a partial reduction of a lactam group. Usually, reduction of amides is achieved with aluminum or boron hydrides.[30] Such harsh and

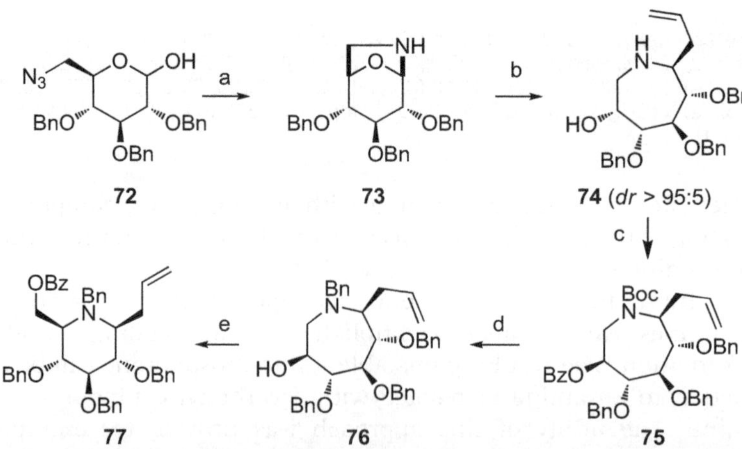

Scheme 13 Reagents and conditions: (a) supp. Ph_3P, THF, 62%; (b) allyl-MgBr, THF, 58%; (c) i. Boc_2O, AcOEt, H_2O, 83%, ii. DEAD, Ph_3P, *p*-nitrobenzoic acid, 93%; (d) i. K_2CO_3, THF/MeOH, 93%, ii. TFA, DCM, iii. DMF, BnBr, K_2CO_3, 82% (2 steps); (e) DEAD, Ph_3P, *p*-nitrobenzoic acid, 77%.

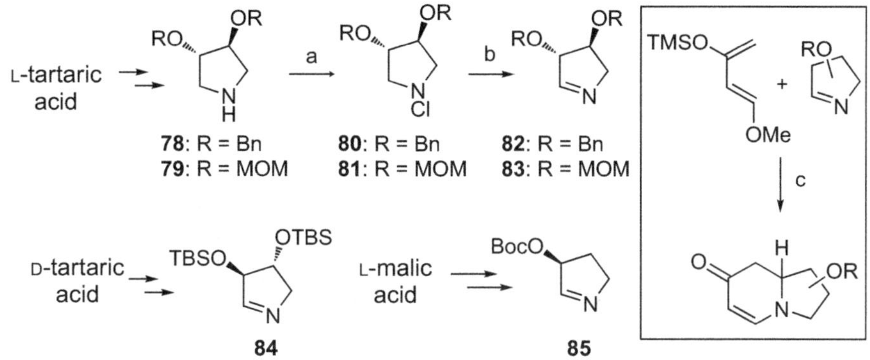

Scheme 14 Reagents and conditions: (a) NCS, Et_2O, 90% (**80**) or 88% (**81**); (b) DBU, Et_2O, 78% (**82**) or 75% (**83**); (c) $Yb(OTf)_3$, DCM.

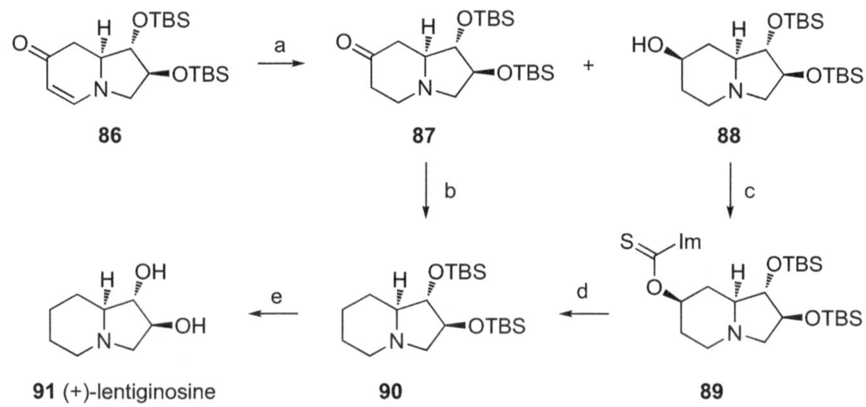

Scheme 15 Reagents and conditions: (a) H_2, Pd/C, AcOEt, 15% (**87**), 68% (**88**); (b) $TsNHNH_2$, MeOH, 55%; (c) TCDI, THF, 88%; (d) n-Bu_3SnH, AIBN, toluene, 75%; (e) MeCN, aq. HCl, then Dowex OH, 80%.

hard to control conditions, lead eventually to amines. However, some novel protocols were reported, in which aldimines and ketimines were obtained from the corresponding amides in good yields using Tf_2O/Et_3SiH[31] or the Schwartz's reagent.[32] Furman and coworkers used the latter to obtain a series of sugar-derived five- and six-membered imines and transformed them, *via* a nucleophilic addition of n-Bu_3Sn-allyl in a one-pot procedure, into various pyrrolidine and piperidine imino sugars (Scheme 16).[33] Yields of the products are usually high, but the diastereoselectivity of the process range from poor to excellent. Other nucleophiles, such as TMSCN, PhMgBr, or silyl enol ethers react similarly, providing differently substituted imino sugars.

Another approach proposed by Furman, similar to the methodology of Yang and Shao already shown in Scheme 14, describes the one-pot procedure in which the cyclic imine is reacted *in situ* with the Danishefsky's diene. This approach led to a series of bicyclic imino sugars.[34]

In one of the most recent papers from the Furman group, an elegant methodology for the preparation of peptidomimetics is proposed.[35] It is

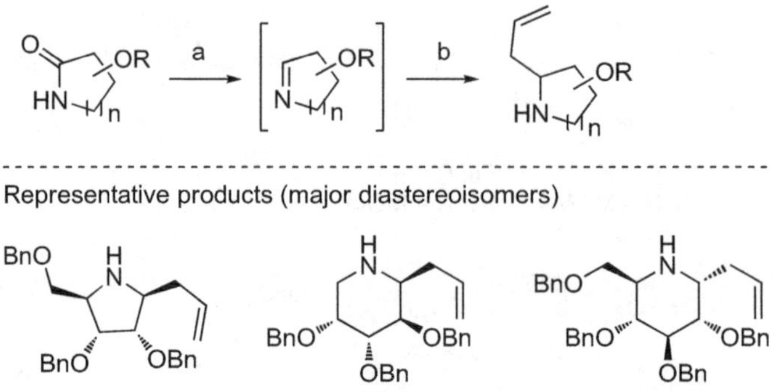

92 (69%)
dr = 4:1

93 (55%)
dr = 3:2

94 (84%)
dr = 9:1

Scheme 16 Reagents and conditions: (a) $Cp_2Zr(H)Cl$, THF; (b) *n*-Bu_3Sn-allyl, $Yb(OTf)_3$, THF.

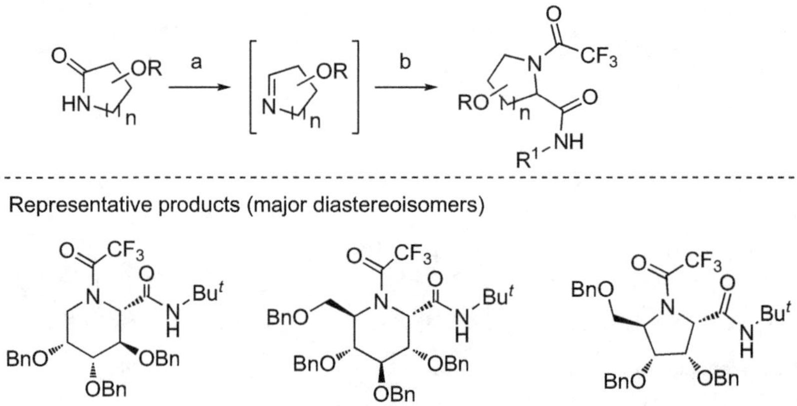

95 (76%)
dr = 3:1

96 (69%)
dr > 95:5

97 (72%)
dr = 5.3:1

Scheme 17 Reagents and conditions: (a) $Cp_2Zr(H)Cl$, THF; (b) TFA, isocyanide, THF.

based on the reduction of lactams with the Schwartz's reagent followed, in a one-pot manner, by the Joullié–Ugi reaction (Scheme 17).

However, despite the versatility of the Furman's approach, the methodology suffers from the necessity of the stoichiometric use of zirconium-based Schwartz's complex, a relatively expensive reagent.

4 Cascade addition of Grignard reagents to halonitriles/cyclization

It is well-established, that the addition of the Grignard reagents to nitriles results in a formation of the imine salts which, upon hydrolysis, yield ketones.[36] However, if a good leaving group, for example a halide or a mesyl ester, is present in the molecule, the reaction can proceed *via* a different route. Namely, an intramolecular S_N2 reaction takes place providing a cyclic imine.[37] Despite the simplicity and the synthetic potential

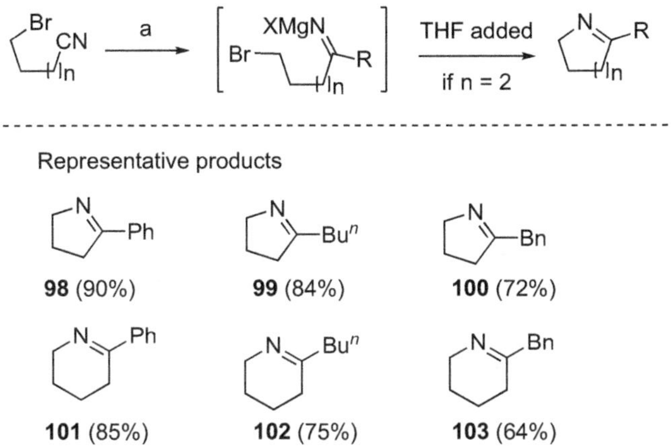

Scheme 18 Reagents and conditions: (a) RMgX, benzene/Et_2O.

of this approach, the reports on its application to the synthesis of nitrogen heterocycles are very limited.

In a seminal paper, Fry and Dieter described an addition of alkyl, vinyl, and phenyl magnesium halides to simple ω-bromonitriles.[37a] A typical procedure involved the addition of a solution of a Grignard reagent in diethyl ether to a solution of a substrate in benzene. According to this report, the five-membered imines are easily obtained, whereas the formation of the six-membered rings is much slower (Scheme 18). Addition of THF, after the initial nucleophilic addition, was crucial to accelerate the latter process and allows to obtain the six-membered imines in good yields. It should be pointed out, that the addition of allylmagnesium chloride to ω-bromonitriles was unsuccessful – the desired imine was not isolated by the authors. Instead, the authors observed the formation of diallylpiperidine in 17% yield, resulting from the addition of the second equivalent of the Grignard reagent to the cyclic imine. Reactions with vinylmagnesium bromide and *tert*-butyl magnesium chloride were also unsuccessful. The former gave a complicated mixture of products, whereas the latter left the starting material almost intact.

In the next paper,[37b] Fry and Dieter extended the scope of substrates to trimethyl-silyloxy-substituted ω-bromonitriles which allowed to accomplish total synthesis of (+/−)-pseudoconhydrine (Scheme 19).

In a more recent paper,[38] optically pure 5-bromo-2-hydroxypentanenitrile silyl ether **108** was treated with several different arylmagnesium bromides to afford, after reduction with $NaBH_4$ (the transitional cyclic imine was unstable, thus it had to be reduced *in situ*), the desired piperidine derivatives as practically single diastereo-isomers (Scheme 20).

The aliphatic Grignard reagents (butyl- and butenyl-magnesium halides) were also tested, but the products were obtained in moderate yields only. Since nitriles are susceptible to the α-deprotonation,[39] the loss of enantiomeric purity in the products was expected by the authors. However, no racemization occurred – the *ee* of the products was the same as in the starting material **108**.

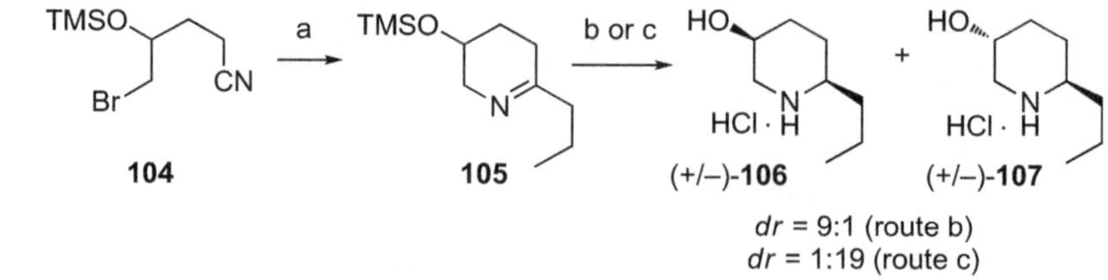

Scheme 19 Reagents and conditions: (a) *n*-propyl-MgCl, benzene/Et_2O, then THF, 74%; (b) i. H_2/Pt, EtOH, ii. conc. HCl, 95%; (c) i. $LiAlH_4$, THF, reflux, ii. conc. HCl, 86%.

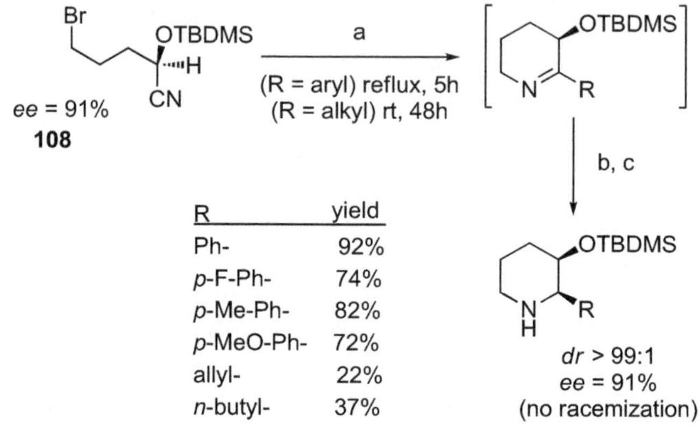

Scheme 20 Reagents and conditions: (a) RMgBr, THF; (b) $NaBH_4$, MeOH; (c) CbzCl, DCM/H_2O, Na_2CO_3.

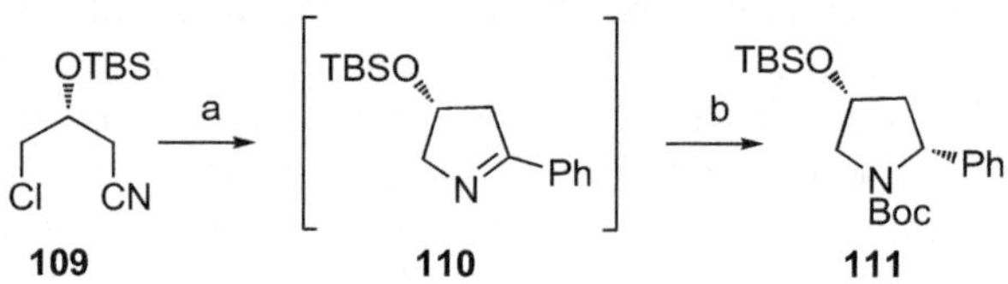

Scheme 21 Reagents and conditions: (a) PhMgBr, MTBE, 15 min, then THF or DME; (b) H_2, Pt/C, then Boc_2O, 89% (3 steps).

In the next report, the Maeda's group described an approach to 3,5-disubstituted pyrrolidine **111** (Scheme 21).[40] A structure of this type is found in some carbapenem antibiotics, *e.g.* in meropenem. The synthesis started from the readily available chloronitrile **109**, to which PhMgBr was added. In order to ensure a successful addition to the nitrile group, a relatively non-polar methyl-*tert*-butyl ether (MTBE) had to be used as a solvent. However, for the cyclization step, the addition of more polar THF or DME was crucial. The resulting cyclic imine **110** was subsequently hydrogenated to afford the desired pyrrolidine scaffold **111**.

The Behr's group developed a convenient methodology to the analogs of poly-hydroxylated ketimines, a family of naturally occurring unsaturated imino sugars (Scheme 22).[41]

They have established, that the reaction proceeds best in toluene at elevated temperature. According to their results, proper selection of the solvent plays a very important role in the successful outcome of the

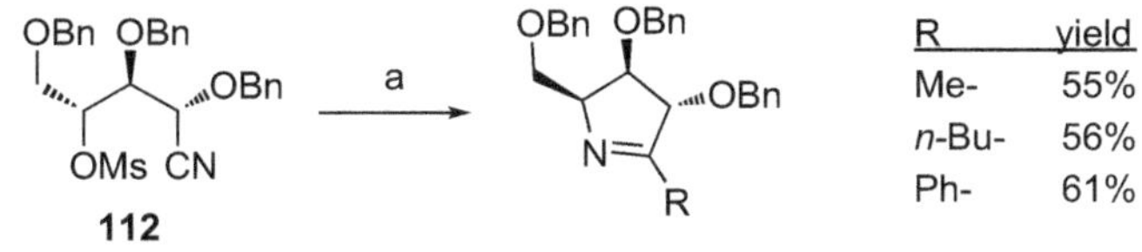

Scheme 22 Reagents and conditions: (a) RMgBr, toluene, 70 °C.

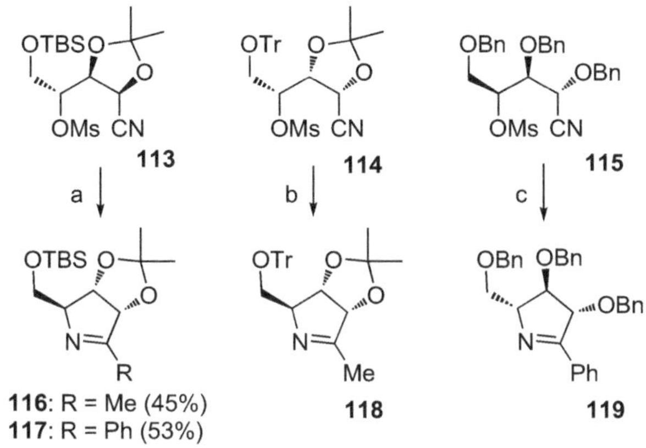

Scheme 23 Reagents and conditions: (a) RMgBr, toluene, 70 °C; (b) MeMgBr, toluene, 70 °C, 88%; (c) PhMgBr, toluene, 70 °C, then THF, 50%.

reaction. For example, the process of addition of a Grignard reagent to the nitrile group is much slower in the presence of even small amounts of THF. To investigate the tolerance of the protecting groups, the authors synthesized some other sugar-like substrates (Scheme 23) and tested them for the addition of the Grignard reagents.

Surprisingly, when derivative **115** was used as a substrate, under exactly the same conditions as for **113** and **114**, only the linear product was formed. The addition of THF to the reaction mixture after the Grignard reaction ensured the cyclization to imine **119**. In general, although the presence of THF before the addition of a Grignard reagent seems to slow down the reaction, it may benefit from the subsequent cyclization step in case of some substrates.

Our group reported that it is possible to perform the cascade addition of the Grignard reagents to halonitriles/cyclization with allylmagnesium bromide to obtain 2-allylsubstituted piperidine **121** (Scheme 24).[9a] In this approach, the Grignard reagent is added to a solution of ω-bromonitrile **120** in toluene, followed by the addition of $NaBH_4$ and MeOH. The reaction suffered, however, from the formation of 2,2-diallylsubstituted product **122**. Therefore, only a slight excess of All-MgBr had to be used to obtain **121** selectively.

Addition of All-MgBr to **120** yielded piperidine **121** as practically single stereo-isomer. We assumed, that the attack of the hydride anion on the imine moiety occurred from the *re* face and proceeded through a preferred chair-like transition state, rather than through a twisted boat-like transition state (Fig. 2).[42] Similar high stereoselectivity in the

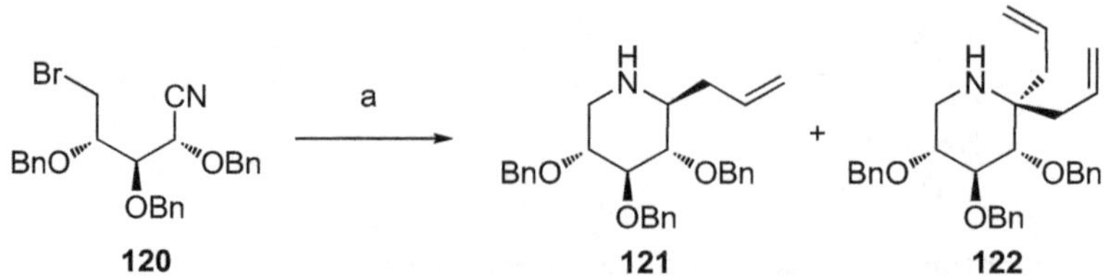

solvent	yield (121 / 122)
THF	28% / 13%
toluene	74% / 4%

Scheme 24 Reagents and conditions: (a) Allyl-MgBr, solvent, 0 °C, then MeOH, $NaBH_4$.

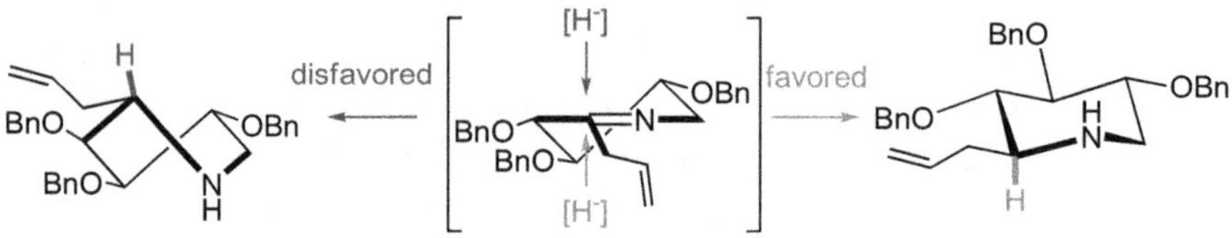

Fig. 2 Twisted boat-like transition state is disfavored and, therefore, **121** is the only observed diastereoisomer.

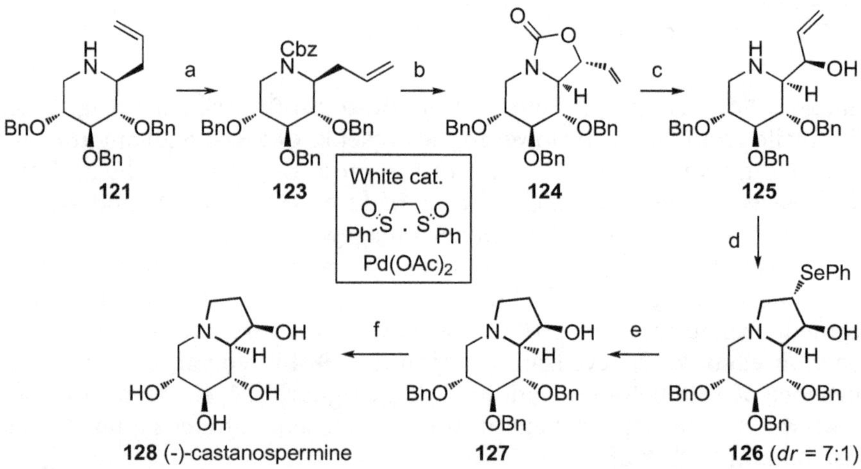

Scheme 25 Reagents and conditions: (a) benzyl chloroformate, MeCN, K_2CO_3, 95%; (b) White cat. (10 mol%), $Yb(OTf)_3$ (10 mol%), BQ, dioxane, 71%; (c) KOH, MeOH, 85%; (d) PhSeBr, DCM/py, 81% (dr = 7:1); (e) $NaBH_4$, $NiCl_2 \cdot 6H_2O$, MeOH/THF, 2 h, 60%; (f) i. H_2, $Pd(OH)_2/C$, MeOH, ii. Ac_2O, DMAP, py, 79% (2 steps), iii. MeOH, MeONa, then Amberlyst 15, quant.

nucleophilic addition to the endocyclic C=N bond in the six-membered rings has been described previously.[43]

Piperidine **121** served as a precursor in the preparation of several bicyclic imino sugars as demonstrated in a short six step synthesis of unnatural (−)-castanospermine **128** (Scheme 25).[44] Protection of **121** with the Cbz group gave derivative **123**, which – upon treatment with benzoquinone, White's Pd catalyst,[45,46] and $Yb(OTf)_3$ – was converted into oxazolidinone **124**, practically as a single diastereoisomer. Subsequent

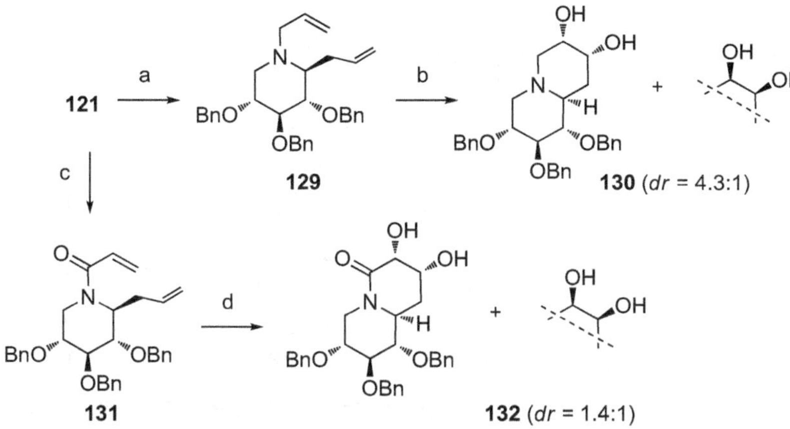

Scheme 26 Reagents and conditions: (a) allyl bromide, K_2CO_3, MeCN, 89%; (b) Grubbs II cat. (5 mol%), TFA, toluene/DCM, then evaporation, then $NaIO_4$, $CeCl_3 \cdot H_2O$ (20 mol%), MeCN/AcOEt/H_2O, 69%; (c) acryloyl chloride, Et_3N, DCM, 30 min, 91%; (b) Grubbs II cat. (5 mol%), toluene, then evaporation, then $NaIO_4$, $CeCl_3 \cdot H_2O$ (20 mol%), MeCN/AcOEt/H_2O, 74%.

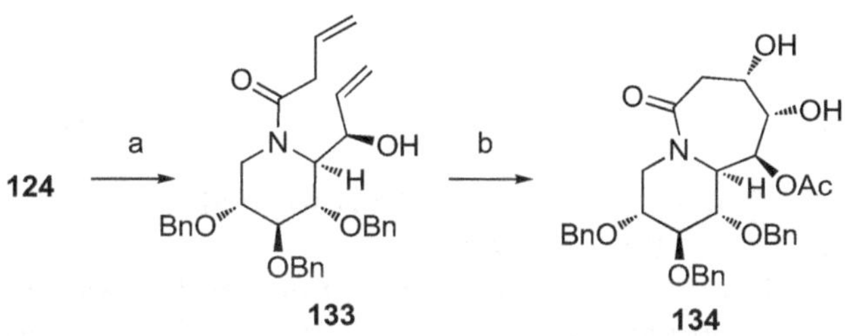

Scheme 27 Reagents and conditions: (a) allyl-MgBr, THF, 84%; (b) i. Ac_2O, DMAP, DCM/py, 92%, ii. Grubbs-II cat. (10 mol%), toluene, then evaporation, then $NaIO_4$, $CeCl_3 \cdot H_2O$ (20 mol%), MeCN/AcOEt/H_2O, 66%.

methanolysis gave aminoalcohol **125**, which was then treated with PhSeBr. This electrophile-induced a nucleophilic cyclization providing the bicyclic derivative **126.** Deselenylation and hydrogenolysis afforded the final alkaloid **128.**

Our group used also piperidine **121** for the preparation of other bicyclic structures such as **130, 132** (Scheme 26),[9a] and **134**[47] (Scheme 27). The key step in these syntheses consisted in the one pot ring-closing metathesis followed by the reuse of the ruthenium catalyst in the subsequent *syn*-dihydroxylation.[48]

Recently, we have expanded the scope of the transformation to shorter ω-bromonitriles (Scheme 28).[9b] The previously described sequential mode of the addition (first All-MgBr, then $NaBH_4$) failed in the case of bromonitriles with a shorter chain than in **120**; a mixture of 2-allyl- and 2,2-diallylsubstituted pyrrolidines was formed (1.5 : 1). Therefore, another approach has been proposed. To a solution of the corresponding ω-bromonitrile in toluene, a mild reducing agent – $Zn(BH_4)_2$[49] – was introduced *prior* to the addition of allylmagnesium bromide. Such relatively simple manipulation allowed us to obtain the desired monosubstituted

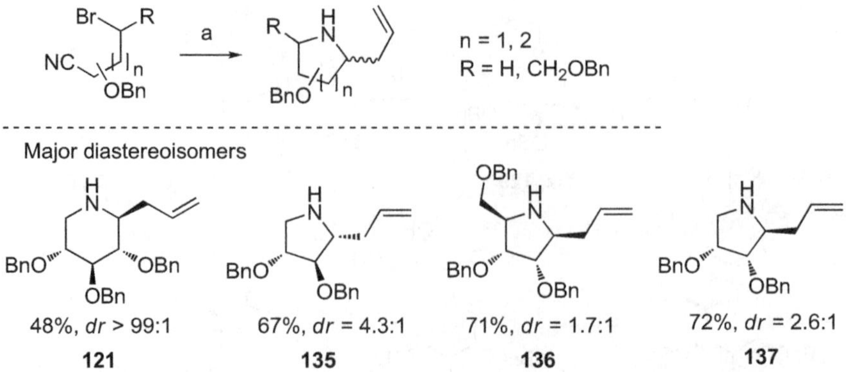

Scheme 28 Reagents and conditions: (a) toluene, $Zn(BH_4)_2$, 0 °C, allyl-MgBr.

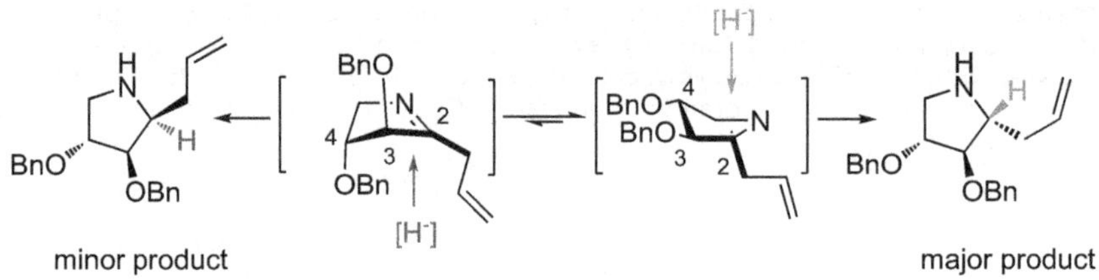

Fig. 3 Possible course of the addition of hydride anion to the transitional cyclic imine.

five-membered products **135**, **136**, and **137** in good yields, with only traces (<5%) of the disubstituted derivatives.

These pyrrolidines were formed with poor and moderate diastereoselectivities. By comparison of these results with those described by the Woerpel's group,[50] we have established, that the substituent at the C-3 position plays a directing role in the nucleophilic addition step. Thus, the allyl group at the C-2 position in the major isomers is consistently in the *anti*-relation to the benzyloxy group at the position C-3 (Fig. 3). The impact of the benzyloxy group at the C-4 seems to be negligible. Interestingly, in the Woerpel's model, the substituent at the C-4 has the strongest influence on the course of the addition. This fact can be attributed to the interaction between the alkoxy substituent at the C-4 and the positively charged oxocarbenium or iminium ions.[33–35,50,51] The reaction described by us is, however, carried out under the basic conditions, so such interaction is not applicable.

Pyrrolidine **135** served as a precursor in the synthesis of the alkaloid (−)-lentiginosine (Scheme 29). First, **135** (as a mixture of diastereoisomers) was treated with allyl bromide which provided the *N*-allyl derivatives **138** and **139**, easily separated by chromatography. The major isomer **139**, after treatment with trifluoroacetic acid, was subjected to the ring-closing metathesis affording the bicyclic derivative **140** in very good yield. Subsequent hydrogenation gave the final compound **141**.

We also proposed a useful approach enabling the selective synthesis of 2,2-diallylsubstituted derivatives **122**,[9a] **142**, and **143** (Scheme 30).[9b] The excess of All-MgBr (5 equiv.) was added to a solution of ω-bromonitrile in THF/DMPU in order to obtain good yields.

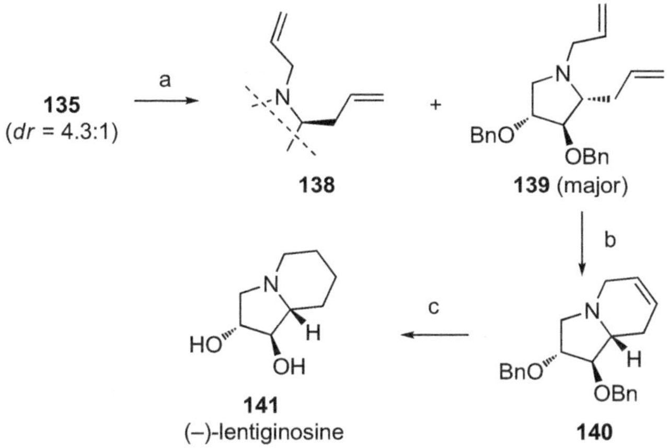

Scheme 29 Reagents and conditions: (a) allyl bromide, MeCN, K_2CO_3, 55% (**138**), 13% (**139**); (b) TFA, Grubbs-Hoveyda II cat. (5 mol%), toluene, 94%; (c) H_2, $Pd(OH)_2/C$, MeOH, 91%.

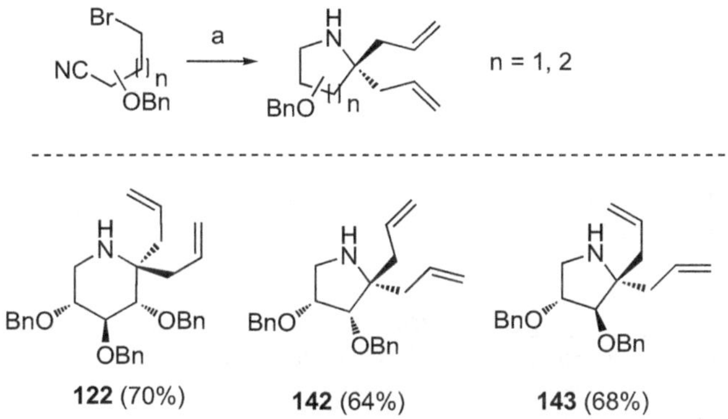

Scheme 30 Reagents and conditions: (a) allyl-MgBr, THF/DMPU.

In a stereodivergent and highly stereoselective approach (Scheme 31), piperidine **122** was transformed into either diol **149** (with a nitrogen atom masked as trifluoroacetate during the catalytic osmylation), or its diastereoisomer **148** (directing role of the nitrogen atom in the formation of osmate **146**).[9a]

5 Reactions of cyclic nitrones

Nitrones can undergo a variety of synthetically useful reactions: 1,3-dipolar cycloaddition with alkenes and alkynes to isoxazolidines,[52] nucleophilic addition which provide hydroxylamines,[53] or SmI_2-mediated cross couplings with carbonyl compounds affording vicinal amino alcohols.[54] Recently, special attention is focused on the preparation and application of the optically pure cyclic nitrones, being used in the synthesis of biologically important molecules.[55]

In addition to this high synthetic versatility, cyclic nitrones are relatively easily available, even from complex, sugar-derived substrates, since

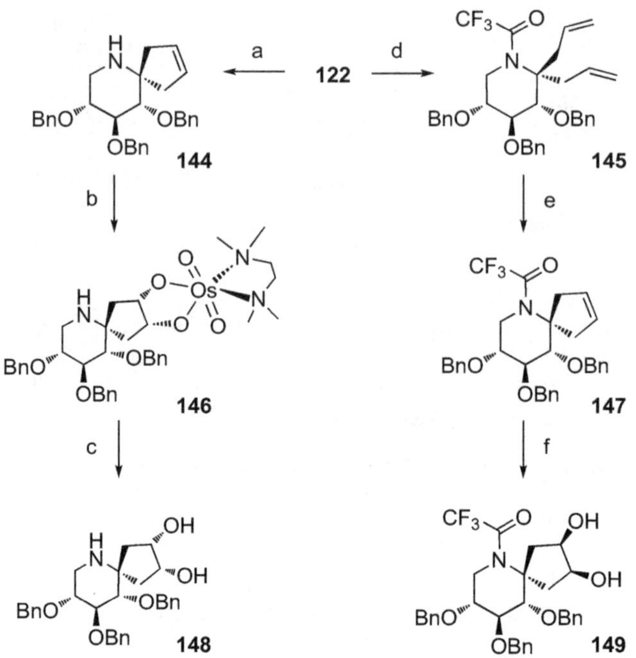

Scheme 31 Reagents and conditions: (a) Grubbs II cat. (10 mol%), HCl, toluene/CPME, 87%; (b) OsO_4, TMEDA, DCM/*t*-BuOH, 89%; (c) ethylenediamine, DCM, 81%; (d) TFAA, DMAP, py, 94%; (e) Grubbs II cat. (1 mol%), DCM, 95%; (f) OsO_4 (5 mol%), NMO, THF/*t*-BuOH/H_2O, 82%.

the protected hydroxyl groups are usually tolerated. The first possibility consists in the oxidation of hydroxylamines, imines, or amines.[56] Although straightforward, this approach lacks regioselectivity. Another way to obtain cyclic nitrones lays in an intramolecular S_N2 *N*-alkylation of oximes.[55] This is probably the easiest route to polyhydroxylated, cyclic (*i.e.* sugar-like) nitrones which are ideally suited for the preparation of imino sugars.

In the approach of Goti and Cardona to (+)-hyacinthacine A_1 and its analogs, D-ribofuranose-derived nitrone **150** was subjected to a 1,3-dipolar cycloaddition with *tert*-butylacrylate which provided a mixture of diastereomeric oxazolidines **151** and **152** ($dr = 1.5:1$; Scheme 32).[57]

The subsequent one-pot cleavage of the N–O bond in **151**, followed by a closure of the ring, gave lactam **153** in high yield. Reduction of this lactam with $LiAlH_4$ and removal of the benzyl groups gave the desired bicyclic compound **155**.

Yu and coworkers presented an elegant approach to (−)-hyacinthacine C_5 **164** (Scheme 33).[58] It was based on the nucleophilic addition of 2-lithio-1,3-dithiane derivative (+/−)-**157** to cyclic, sugar-derived nitrone **156**, followed by a Cope-House cyclization, which provided derivatives **158** and **159** as a separable mixture of diastereoisomers ($dr = 1:1$). Their reduction (performed independently for each isomer) with Zn/AcOH gave thioacetals **160** and **161**. Subsequent removal of the dithiane moiety in **160**, followed by a highly stereoselective reduction of the carbonyl group

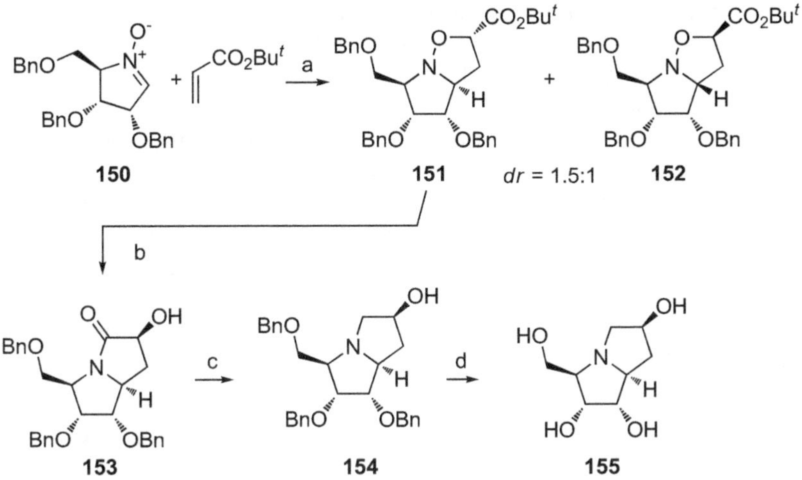

Scheme 32 Reagents and conditions: (a) DCM, 88%; (b) i. Zn, AcOH/H_2O, ii. Ambersep 900 OH, MeOH, 90% (2 steps); (c) $LiAlH_4$, THF, 82%; (d) H_2, Pd/C, MeOH, HCl, then DOWEX-OH, quant.

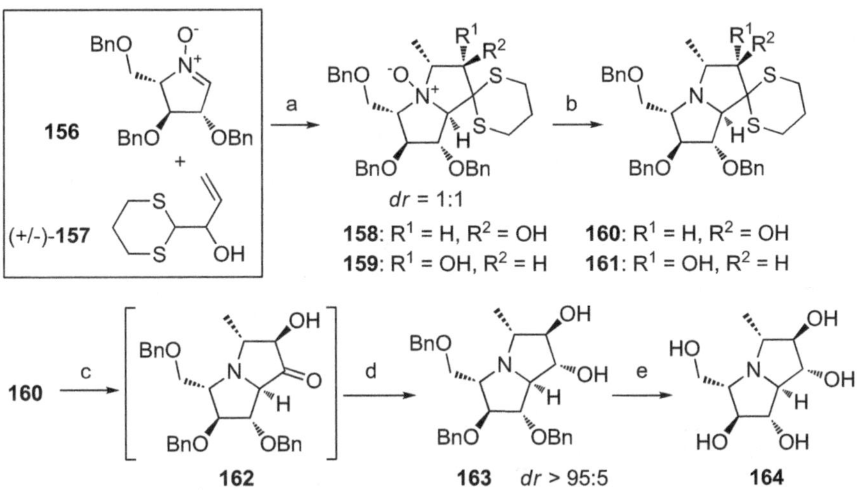

Scheme 33 Reagents and conditions: (a) i. first **157**, *n*-BuLi, TMEDA, THF, then **156**; ii. $CHCl_3$, 27% (**158**), 28% (**159**) (2 steps); (b) Zn, AcOH, **160** (98%) or **161** (98%); (c) NBS, $AgNO_3$, MeCN/H_2O; (d) $NaBH_4$, MeOH, 31% (2 steps); (e) H_2, Pd/C, MeOH, HCl, 93%.

in **162** ($dr > 95:5$), provided diol **163** in reasonable yield (31% over 2 steps) which, after deprotection, afforded the desired imino sugar **164**.

Recently, Michalak and coworkers presented the nucleophilic addition of various monosubstituted acetylenes to the optically pure nitrones, catalyzed by the NHC–Cu(I) complexes, leading to a family of 2-propargylhydroxylamines in high yields and very high diastereoselectivities (in most cases, $dr > 95:5$).[59] Such transformations usually require the use of the stoichiometric amounts of organometallic species under strictly anhydrous conditions. However, in the Michalak's methodology water is not only tolerable, but is also a preferred solvent. Interestingly, this is an unprecedented behaviour of Cu(I) acetylides in the combination with nitrones, since usually, under such conditions, the Kinugasa reaction takes place.[60] By this method the (−)-lentiginosine scaffold **167** was obtained (Scheme 34).

Recently, Py and Gilles reported the total synthesis of (+)-australine.[61] The key step consisted in a SmI_2-mediated cross-coupling reaction between sugar-derived cyclic nitrone **168** and β-silyl ethyl acrylate **169** which provided (not isolated) hydroxylamine **170**. Subsequent reduction/cyclization sequence gave lactam **171**, isolated as a practically single diastereoisomer. Finally, reduction of the lactam group, followed by a Tamao-Fleming oxidation and deprotection, afforded the target alkaloid **173** (Scheme 35).

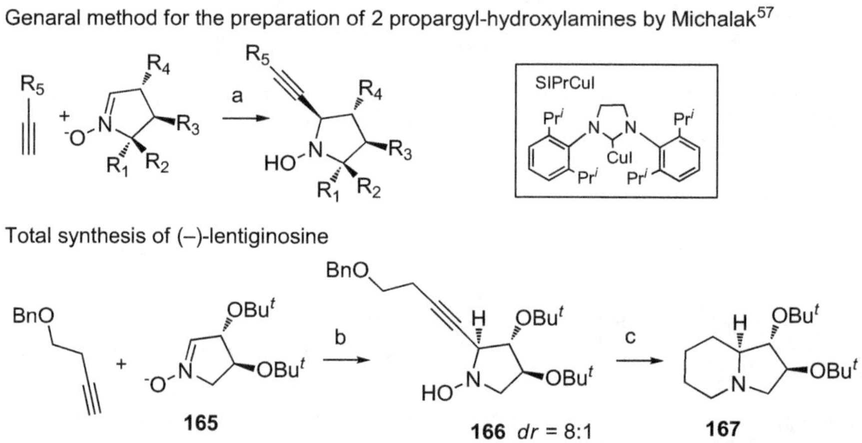

Scheme 34 Reagents and conditions: (a) SIPrCuI (5 mol%), Et_3N, H_2O; (b) SIPrCuI (10 mol%), TMG, H_2O, 66%; (c) i. H_2, Pd/C, EtOH, ii. Ph_3P (polymer-bound), CCl_4, DCM, 65% (2 steps).

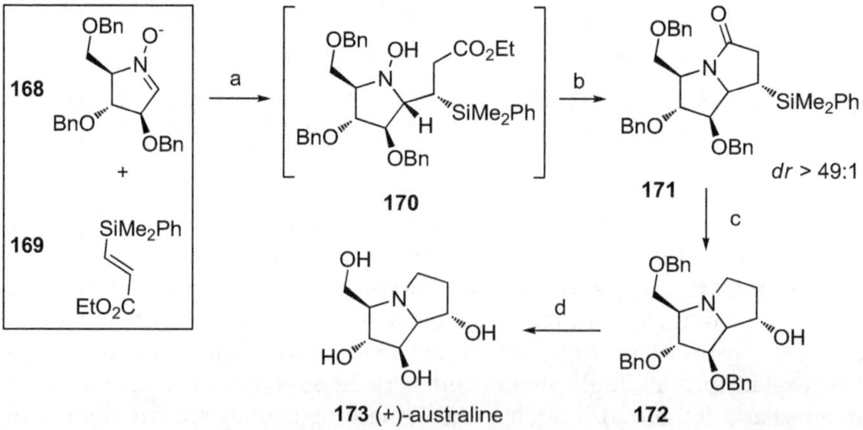

Scheme 35 Reagents and conditions: (a) SmI_2, HFIP, LiBr, THF; (b) Zn, AcOH,))), 64% (2 steps); (c) i. $BH_3 \cdot Me_2S$, THF, 73%, ii. t-BuOOH, KH, TBAF, DMF, 89%; (d) H_2, Pd/C, HCl, THF/MeOH, quant.

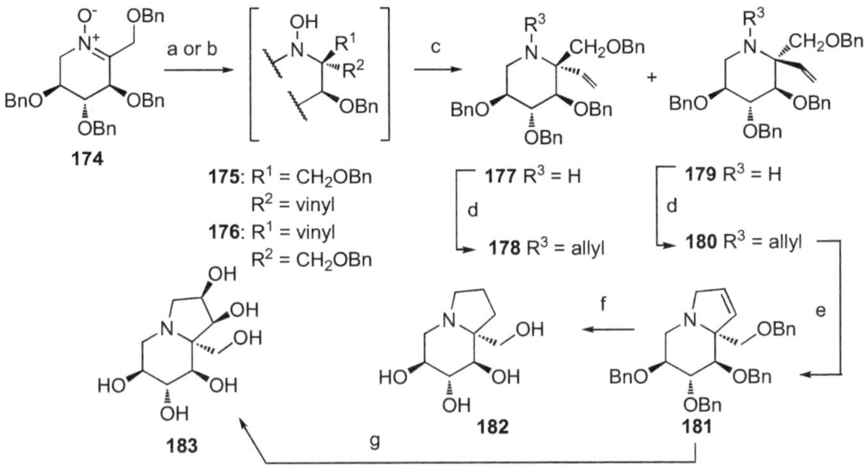

Scheme 36 Reagents and conditions: (a) vinyl-MgBr, DCM; (b) vinyl-MgBr, $ZnCl_2$, DCM; (c) Zn, AcOH/EtOH,))), [**177** : **179** = 15 : 85, 91%, route (a)] or [**177** : **179** = 98 : 2, 86%, route (b)]; (d) allyl-Br, MeCN, K_2CO_3, KI, 82% (**178**) or 81% (**180**); (e) Grubbs II cat. (5 mol%), DCM, 93%; (f) H_2, $Pd(OH)_2/C$, HCl, MeOH, 89%; (g) i. OsO_4, TMEDA, DCM, 62%, ii. H_2, $Pd(OH)_2/C$, HCl, MeOH, 91%.

Py and coworkers presented also a convenient route to 2,2-disubstituted piperidines. This strategy involved a highly selective, stereodivergent nucleophilic addition of vinylmagnesium bromide to L-sorbose-derived nitrone **174** which provided hydroxylamines **175** and **176** (Scheme 36).[62] Crude products were reduced and the resulting amines **177** and **179** were separated by chromatography. Subsequently, *N*-allyl-substituted piperidines **178** and **180** were obtained, and transformed into a series of bicyclic derivatives, for example **182** and **183**.

These compounds were quite potent inhibitors of α-glucosidase ($IC_{50} \approx 0.05$ μM for **182** and 1.5 μM for **183**).

6 Ring-closing metathesis

Soon after its discovery, the ruthenium-based olefin metathesis became widely applied in organic synthesis. This potent, versatile reaction offers an atom-economy approach, relatively stable catalysts, and the simplicity of use.[63]

Not surprisingly that the olefin metathesis, in its ring-closing variant (RCM), was also used with great success in the total synthesis of imino sugars.[64] This approach is fast (at least in terms of a number of synthetic steps involved) and reliable as was proven in numerous examples.

However, the RCM methodology has also several drawbacks. First, the basic amino group in the substrate, if present, deactivates the catalyst.[65] Therefore, before the reaction, it should be reversibly masked as, for example, a carbamate, amide, or ammonium salt. Second, the metathesis catalysts are still very expensive – a fact, that cannot be ignored when planning the synthesis on a larger scale. Moreover, the pollution of the products with ruthenium, even many steps after the metathesis reaction, is usually significant. This poses serious problems in the drug development, where the use of heavy metals is avoided.[66]

The key-step in the Madsen's synthesis of the nine-membered nitrogen heterocycle **187** was based on the RCM approach.[67] Unsaturated aldehyde **185**, readily obtained from the iodo derivative **184** by the Vasella reaction, was converted into the corresponding diene by a reductive amination with homoallylamine. The free nitrogen atom was protected as trifluoroacetate and the resulting amine **186** was subjected to the RCM reaction under the high-dilution conditions (<1 mM). This process afforded the desired macrocycle **187** in good yield, together with small amounts (*ca.* 7%) of compound **188** resulting from the isomerization of the terminal double bond *prior* to the RCM reaction. The main isomer **187** was oxidized to the epoxide, which was then converted into (+)-castanospermine **191** *via* an intramolecular opening of the oxirane ring (Scheme 37).

An elegant *de novo* total asymmetric synthesis of (+)-DNJ, based on the RCM approach, was proposed by Poisson (Scheme 38).[68] It was initiated from oxirane **193** (*ee*>99%), obtained by the Sharpless' asymmetric epoxidation of alcohol **192** with subsequent Payne rearrangement.

This oxirane was converted into oxazolidinone **194** which subsequently was transformed into benzylidene acetal **195** and alkylated with allyl iodide **196** to diene **197**. Ring closing metathesis reaction of this diene followed by hydroboration-oxidation afforded derivative **198** which, upon treatment with HCl, gave the target: (+)-deoxynojirimycin.

The Overkleeft synthesis of derivatives of DNJ was initiated from the enantiomerically pure TBS-protected cyanohydrin **199**. A sequence of a DIBAL-H reduction and subsequent reductive amination with amines (*S*)-**200** and (*R*)-**200** provided dienes **201** and **202**, respectively. Protection of the amine function with the Boc group and subsequent RCM reaction catalysed with the Grubbs I cat. afforded piperidines **203** and **203**; their *syn*-dihydroxylation and subsequent deprotection gave the final imino sugars (Scheme 39).[69]

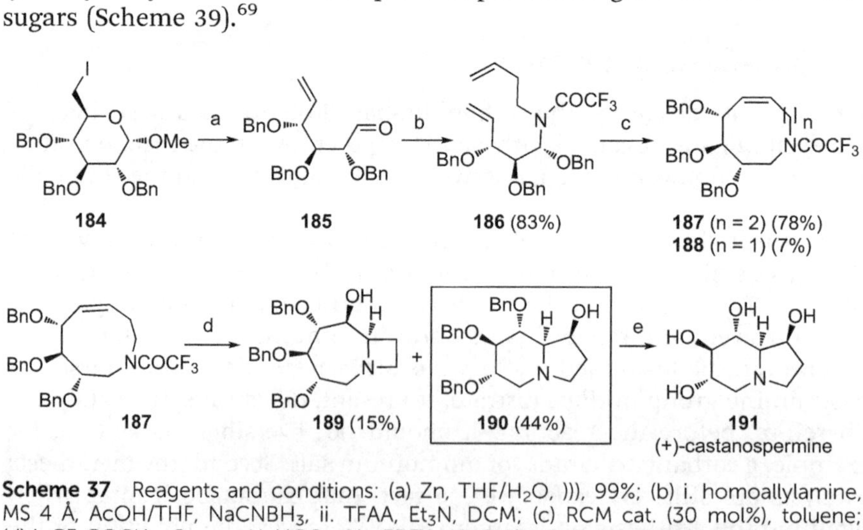

Scheme 37 Reagents and conditions: (a) Zn, THF/H_2O,))), 99%; (b) i. homoallylamine, MS 4 Å, AcOH/THF, $NaCNBH_3$, ii. TFAA, Et_3N, DCM; (c) RCM cat. (30 mol%), toluene; (d) i. CF_3COCH_3, Oxone, $NaHCO_3$, Na_2EDTA, MeCN/H_2O, ii. *t*-BuOK, Et_2O, H_2O; (e) H_2, Pd/C, 94%.

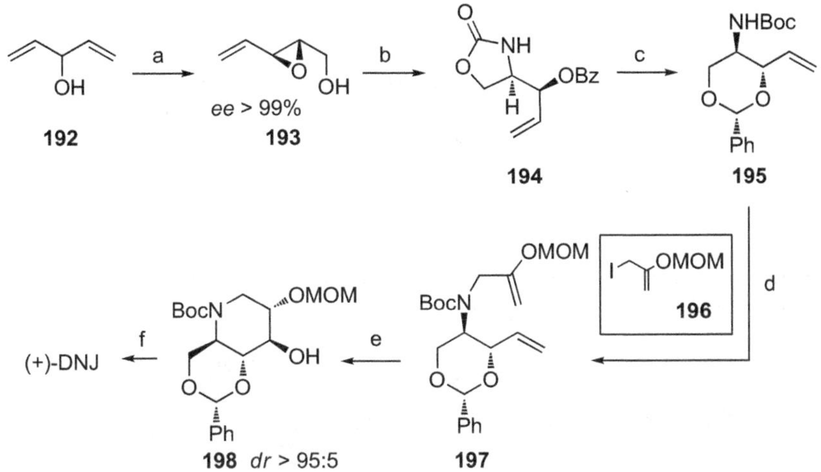

Scheme 38 Reagents and conditions: (a) Sharpless asymm. epox. and Payne rearr.; (b) i. BzNCO, Et_2O, 96%, ii. K_2CO_3, DCM/H_2O, $C_{12}H_{25}NMe_3Cl^+$ (cat.), 80%; (c) i. Boc_2O, Et_3N, DMAP, DCM, 91%, ii. EtONa, EtOH, 93%, iii. $PhCH(OMe)_2$, CSA, DCM, 91%; (d) **196**, NaH, DMF, 82%; (e) i. Hoveyda-Grubbs II cat. (10 mol%), BQ (10 mol%), toluene, reflux; ii. $BH_3 \cdot Me_2S$, THF, then $NaBO_3 \cdot 4H_2O$, 70% (2 steps); (f) EtOH, HCl, quant.

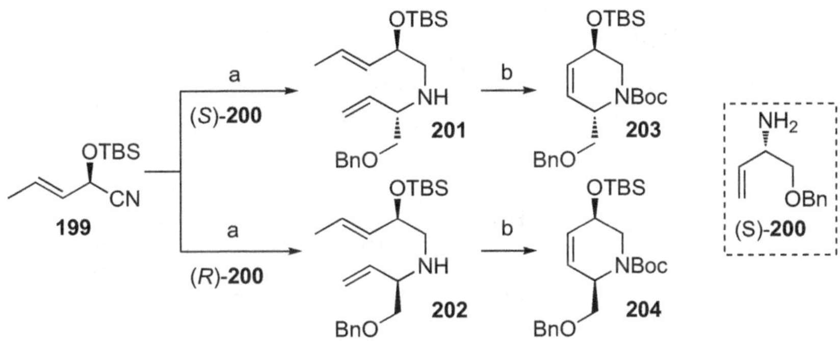

Scheme 39 Reagents and conditions: (a) DIBAL-H, Et_2O, then MeOH, (*S*)- or (*R*)-**200**, then $NaBH_4$, 80% (**201**) or 78% (**202**); (b) i. Boc_2O, THF, TEA, ii. Grubbs I cat. (3.5 mol%), DCM, 99% (**203**) or 95% (**204**) (2 steps).

An interesting approach to imino sugars was presented by Behr and coworkers.[70] Using an indium-mediated allylation of a carbonyl group in the unprotected pentosylamines such as **194**, they prepared the linear diene **195** with very high diastereoselectivity ($dr > 95:5$; Scheme 40). The latter compound was then *in situ* protected with *p*-toluenesulfonic acid (as ammonium salt) and subjected to the RCM reaction with the Hoveyda-Grubbs II cat. which afforded derivative **196** with a long, polyhydroxylated side-chain.

Very recently, Compain proposed an approach to spirocyclic imino sugars (Scheme 41).[71] The synthetic route started from known cyclobutanol **208**,[72] which was transformed into carbamate **209**. Rhodium-catalyzed intramolecular C–H amination performed on that compound led to oxazolidinone **211**. Subsequently, this compound was *N*-allylated and subjected to the ring-closing metathesis with the Grubbs II cat. As a

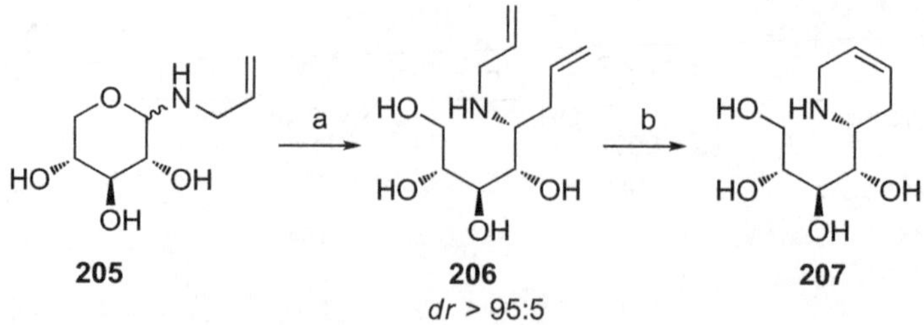

Scheme 40 Reagents and conditions: (a) In, allyl bromide, MeOH, 79%; (b) TsOH, DCM, Hoveyda-Grubbs II cat. (2.5 mol%), 55%.

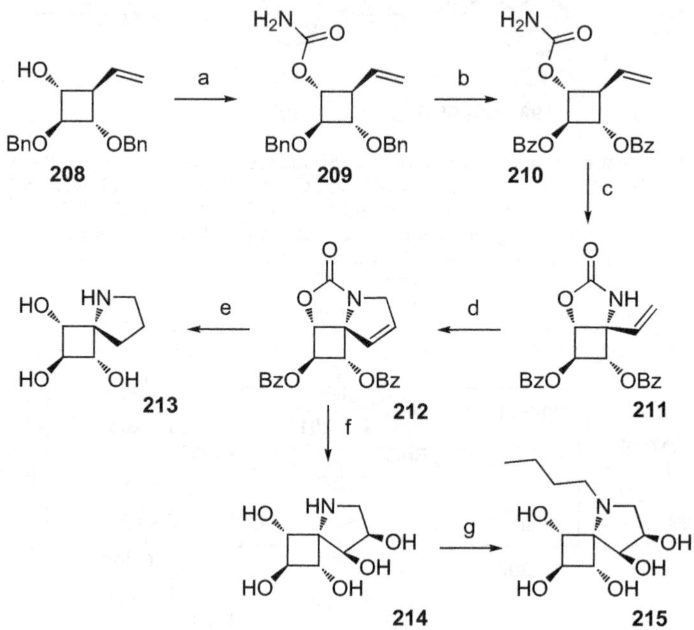

Scheme 41 Reagents and conditions: (a) i. $Cl_3C(O)NCO$, DCM, ii. K_2CO_3, MeOH, 94% (2 steps); (b) i. BCl_3, DCM, −60 °C, 84%, ii. BzCl, py, 0 °C, 87%; (c) $Rh_2(esp)_2$ (20 mol%), DCM, $PhI(OAc)_2$, MgO, reflux, 40%; (d) i. NaH, allyl-Br, DMF, 55%, ii. Grubbs II cat., reflux, 89%; (e) i. H_2, Pd/C, EtOH, AcOEt, quant., ii. DOWEX-OH, MeOH, H_2O, 70 °C, 35%; (f) i. OsO_4, NMO, *t*-BuOH, acetone, H_2O, 86%, ii. KOH, MeOH, quant.; (g) *n*-PrCHO, $NaBH_3CN$, MeOH, 58%.

result, derivative **212** was formed, which was then either reduced and deprotected giving **213**, or subjected to the *syn*-dihydroxylation and subsequent deprotection giving compound **214** as practically single diastereoisomer.

The latter was then reacted with *n*-butanal in the presence of $NaBH_3CN$, giving *N*-butyl derivative **215**, that can be regarded as a constrained analog of Miglustat.

7 Miscellaneous

Apart from the previously described, well-established methodologies, few different, less common approaches were also reported.

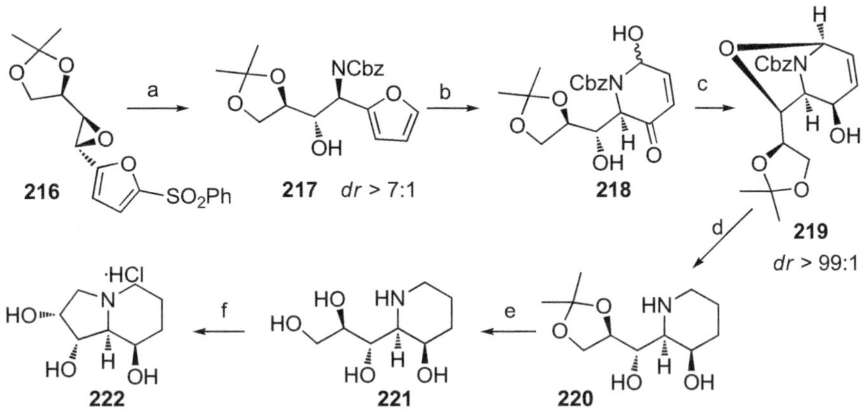

Scheme 42 Reagents and conditions: (a) i. aq. NH_3, MeOH, 79%, ii. Na/Hg, MeOH, 92%, iii. CbzCl, DCM/H_2O, $NaHCO_3$, 98%; (b) *m*-CPBA, DCM, 72%; (c) i. *p*-TsOH · H_2O, toluene, MS 4 Å, 80%, ii. $CeCl_3 \cdot 7H_2O$, $NaBH_4$, MeOH, 95%; (d) H_2, Pd/C, MeOH,67%; (e) *p*-TsOH · H_2O, MeOH, 93%; (f) DMF, Ph_3P, CCl_4, Et_3N, 80%.

Aggarwal and Bi proposed the synthesis of 8a-*epi*-swainsonine (**222**) from hemiaminal **218** which was prepared from the protected amine **217** in the aza-Achmatowicz[73] reaction as shown in Scheme 42.[74] The heterocyclic derivative **218** was, under the acidic conditions, transformed into the acetal; the carbonyl group was then reduced, under the Luche conditions, providing derivative **219** as practically single isomer. Reduction of the double bond with simultaneous removal of the Cbz group and cleavage of the *N*,*O*-acetal afforded piperidine **220** which was easily transformed, *via* **221**, into 8a-*epi*-swainsonine **222**.

Although the aza-variant of the Achmatowicz reaction is well established in the context of the synthesis of nitrogen heterocycles,[75] its application in the field of imino sugars is still limited, especially in recent literature. However, some earlier reports can be found.[76]

Another less common approach to the bicyclic imino sugars was presented by Jarosz and Magdycz.[77] It was based on the application of D-glucose-derived (*E*)-dienoaldehyde **223**, readily obtained by a controlled fragmentation of the corresponding allyltin derivative.[78] Reaction of this dienoaldehyde with hydroxylamine afforded oxime **224** which underwent, not the expected hetero Diels–Alder reaction, but the 1,3-dipolar cycloaddition providing the vinyl-substituted [5.5.0]-bicyclic oxazolidine **226** as a single stereoisomer. Allylation of the nitrogen atom followed by the RCM reaction catalysed with the Grubbs I cat. afforded **227**, easily transformed into [5.7.0]-bicylic derivative **228** (Scheme 43).

Alternatively, reaction of **223** with tosylamine provided *N*-tosyl imine **225**, which spontaneously underwent the expected aza-Diels–Alder reaction providing the desired [5.6.0]-bicylic compound **229** in a highly diastereoselective manner ($dr > 99:1$).[79] *cis*-Dihydroxylation of this compound led to diol **230** as a single stereoisomer, while *trans*-dihydroxylation afforded a mixture of stereoisomers with diol **231** highly predominating (Scheme 43).

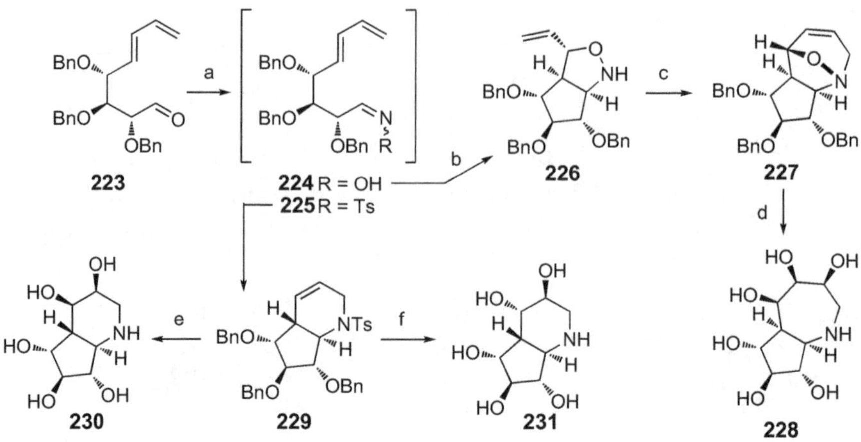

Scheme 43 Reagents and conditions: (a) $NH_2OH \cdot HCl$, EtOH/py, for **224** or $TsNH_2$, *p*-TsOH, DCM, for **225** (spont. cyclizes to **229**); (b) benzene/toluene, 10 kbar; (c) i. allyl-Br, DMF, K_2CO_3, 65%, ii. Grubbs I cat. (10 mol%), DCM, 97%; (d) i. OsO_4, NMO $\cdot$ H_2O, THF/*t*-BuOH/H_2O, 70%, ii. MeOH, NH_4Cl, Zn, 70%, iii. Na/NH_3, THF, 80%; (e) i. OsO_4, NMO $\cdot$ H_2O, THF/*t*-BuOH/H_2O, 85%; ii. Na/NH_3, THF, 57%; (f) i. HCO_2H/30% H_2O_2, ii. MeOH, 10% NaOH, 55% (+ 9% of the opposite *trans*-diol), iii. Na/NH_3, THF, 67%.

8 Conclusions

The most recent reports concerning the synthesis of imino sugars are summarized. Due to the promising biological activity they might have, these compounds represent an interesting synthetic target and, having usually many stereogenic centers, not a trivial one. Moreover, the total syntheses of imino sugars often serve as a challenging testing ground during the development of new synthetic tools.

The addition of the Grignard reagents to ω-halonitriles followed by an intramolecular displacement of a halide appears to be a convenient, but challenging way leading to the five- and six-membered cyclic imines, compounds of high synthetic utility, *e.g.* in the context of total synthesis of imino sugars. However, this approach has not been yet extensively studied – only several reports can be found in the literature.

Acknowledgements

The support from Grant: POIG.01.01.02-14-102/09 (part-financed by the European Union within the European Regional Development Fund) is acknowledged.

References

1 P. Compain and O. R. Martin, in *Imino sugars: From Synthesis to Therapeutic Applications*, ed. P. Compain and O. R. Martin, Wiley-VCH, Weinheim, 2007.
2 (a) A. A. Watson, G. W. J. Fleet, N. Asano, R. J. Molyneux and R. J. Nash, *Phytochemistry*, 2001, **56**, 265–295; (b) G. Horne, F. X. Wilson, J. Tinsley, D. H. Williams and R. Storer, *Drug Discov. Today*, 2011, **16**, 107–118; (c) P. C. Tyler and B. G. Winchester, in *Imino Sugars as Glycosidase Inhibitors: Nojirimycin and Beyond*, ed. A. E. Stutz, Wiley-VCH, Weinheim, 1999; (d) T. Cox,

R. Lachmann, C. Hollak *et al.*, *Lancet*, 2000, **355**, 1481–1485; (e) A. Banaś-Biela, F. Oulaïdi, S. Front, E. Gallienne, K. Ikeda-Obatake, N. Asano, D. A. Wenger and O. R. Martin, *ChemMedChem*, 2014, **9**, 2647–2652; (f) S. Front, E. Gallienne, J. Charollais-Thoenig, S. Demotz and O. R. Martin, *ChemMedChem*, 2016, **11**, 133–141.
3 M. S. M. Pearson, M. Mathe-Allainmat, V. Fargeas and J. Lebreton, *Eur. J. Org. Chem.*, 2005, 2159–2191.
4 H. Paulsen, *Angew. Chem., Int. Ed. Engl.*, 1966, **5**, 495–510.
5 H. Paulsen, in *Imino Sugars as Glycosidase Inhibitors: Nojirimycin and Beyond*, ed. A. E. Stuetz, Wiley-VCH, Weinheim, 1999.
6 L. D. Hohenschutz, E. A. Bell, P. J. Jewess, D. R. Leworthy, R. J. Pryce, E. Arnold and J. Clardy, *Phytochemistry*, 1981, **20**, 811–814.
7 (a) M. J. Humphries, K. Matsumoto, S. L. White and K. Olden, *Cancer Res.*, 1986, **46**, 5215–5222; (b) M. A. Spearman, B. C. Ballon, J. M. Gerrard, A. H. Greenberg and J. A. Wright, *Cancer Lett.*, 1991, **60**, 185–191; (c) B. G. Winchester, I. Cenci di Bello, A. C. Richardson, R. J. Nash, N. G. Ramsden and G. Fleet, *Biochem. J.*, 1990, **269**, 227–231.
8 (a) V. H. Lillelund, H. H. Jensen and M. Bols, *Chem. Rev.*, 2002, **102**, 515–553; (b) R. Lahiri, A. A. Ansari and Y. D. Vankar, *Chem. Soc. Rev.*, 2013, **42**, 5102–5118.
9 (a) M. Malik, G. Witkowski, M. Ceborska and S. Jarosz, *Org. Lett.*, 2013, **15**, 6214–6217; (b) M. Malik and S. Jarosz, *Org. Biomol. Chem.*, 2016, **14**, 1764–1776.
10 J. P. Michael, *Nat. Prod. Rep.*, 2008, **25**, 139–165.
11 R. C. Bernotas and B. Ganem, *Tetrahedron Lett.*, 1984, **25**, 165–168.
12 V. R. Doddi, H. P. Kokatla, A. P. J. Pal, R. K. Basak and Y. D. Vankar, *Eur. J. Org. Chem.*, 2008, 5731–5739.
13 A. A. Ansari and Y. D. Vankar, *J. Org. Chem.*, 2013, **78**, 9383–9395.
14 A. Aravind, M. G. Sankar, B. Varghese and S. Baskaran, *J. Org. Chem.*, 2009, **74**, 2858–2861.
15 A. Kotland, F. Accadbled, K. Robeyns and J.-B. Behr, *J. Org. Chem.*, 2011, **76**, 4094–4098.
16 M. Brlek-Bergeron, M. Meanwell and R. Britton, *Nat. Commun.*, 2015, **6**, 6903.
17 C. Cocaud, C. Nicolas, A. Bayle, T. Poisson, X. Pannecoucke and O. R. Martin, *Eur. J. Org. Chem.*, 2015, 4330–4334.
18 (a) E. Mironiuk-Puchalska, T. Rowicki, W. Sas and M. Koszytkowska-Stawińska, *Tetrahedron*, 2013, **69**, 9826–9831; (b) E. Mironiuk-Puchalska, M. Koszytkowska-Stawińska, M. Wielechowska and W. Sas, *Tetrahedron Lett.*, 2016, **57**, 199–202.
19 C. Claver and E. Fernandez, in *Modern Reduction Methods*, ed. P. G. Andersson and I. J. Munslow, Wiley VCH, Weinheim, 2008.
20 (a) B. W. Gung, in *Organic Reactions*, ed. L. E. Overman, Wiley VCH, Weinheim, 2004, vol. 64; (b) A. B. Charette, in *Chiral Amine Synthesis*, ed. T. C. Nugent, Wiley VCH, Weinheim, 2010.
21 L. F. Tietze and G. Kettschau, *Top. Curr. Chem.*, 1997, **189**, 1–120.
22 M. Palyam and M. Majewski, *J. Org. Chem.*, 2009, **74**, 4390–4392.
23 J. N. Abrams, R. S. Babu, H. Guo, D. Le, J. Le, J. N. Osbourn and G. A. O'Doherty, *J. Org. Chem.*, 2008, **73**, 1935–1940.
24 H. Guo and G. A. O'Doherty, *Tetrahedron*, 2008, **64**, 304–313.
25 S. D. Lucas, A. P. Rauter and H. P. Wessel, *J. Carbohydr. Chem.*, 2008, **27**, 172–187.
26 K. S. A. Kumar, J. S. Rathee, M. Subramanian and S. Chattopadhyay, *J. Org. Chem.*, 2013, **78**, 7406–7413.

27 J. M. J. Tronchet, B. Gentile, J. Ojha-Poncet, G. Moret, D. Schwarzanbach and F. Barblat-Ray, *Carbohydr. Res.*, 1977, **59**, 87–93.
28 M. Mondon, N. Fontelle, J. Désiré, F. Lecornué, J. Guillard, J. Marrot and Y. Blériot, *Org. Lett.*, 2012, **14**, 870–873.
29 J. Shao and J.-S. Yang, *J. Org. Chem.*, 2012, **77**, 7891–7900.
30 J. Seyden-Penne, *Reductions by the Alumino and Borohydrides in Organic Synthesis*, Wiley VCH, New York, 2 edn, 1997.
31 G. Pelletier, W. S. Bechara and A. B. Charette, *J. Am. Chem. Soc.*, 2010, **132**, 12817–12819.
32 B. Ganem and R. R. Franke, *J. Org. Chem.*, 2007, **72**, 3981–3987.
33 P. Szcześniak, S. Stecko, O. Staszewska-Krajewska and B. Furman, *Tetrahedron*, 2014, **70**, 1880–1888.
34 P. Szcześniak, S. Stecko, E. Maziarz, O. Staszewska-Krajewska and B. Furman, *J. Org. Chem.*, 2014, **79**, 10487–10503.
35 P. Szcześniak, E. Maziarz, S. Stecko and B. Furman, *J. Org. Chem.*, 2015, **80**, 3621–3633.
36 (a) F. F. Blicke and E.-P. Tsao, *J. Am. Chem. Soc.*, 1953, **75**, 5587–5590; (b) F. J. Weiberth and S. S. Hall, *J. Org. Chem.*, 1987, **52**, 3901–3904; (c) L. Miginiac, in *Handbook of Grignard Reagent*, ed. G. S. Silverman and P. E. Rakita, CRC Press, 1996.
37 (a) D. F. Fry, C. B. Fowler and R. K. Dieter, *Synlett*, 1994, 836–838; (b) D. F. Fry, C. B. Fowler, M. Brown, J. C. McDonald and R. K. Dieter, *Tetrahedron Lett.*, 1996, **37**, 6227–6230.
38 M. I. Monterde, R. Brieva and V. Gotor, *Tetrahedron: Asymmetry*, 2001, **12**, 525–528.
39 (a) F. F. Fleming, Y. Wie, W. Liu and Z. Zhang, *Org. Lett.*, 2007, **9**, 2733–2736; (b) G. Barker, M. R. Alshawish, M. C. Skilbeck and I. Coldham, *Angew. Chem.*, 2013, **125**, 7854–7857.
40 K. Maeda, Y. Yamamoto, K. Tomimoto and T. Mase, *Synlett*, 2001, **11**, 1808–1810.
41 J.-B. Behr, A. Kalla, D. Harakat and R. Plantier-Royon, *J. Org. Chem.*, 2008, **73**, 3612–3615.
42 M. A. T. Maughan, I. G. Davies, T. D. W. Claridge, S. Courtney, P. Hay and B. G. Davis, *Angew. Chem. Int. Ed.*, 2003, **42**, 3788–3792.
43 T.-H. Chan, Y.-F. Chang, J.-J. Hsu and W. C. Cheng, *Eur. J. Org. Chem.*, 2010, 5555–5559.
44 M. Malik, G. Witkowski and S. Jarosz, *Org. Lett.*, 2014, **16**, 3816–3819.
45 M. S. Chen, N. Prabagaran, N. A. Labenz and M. C. White, *J. Am. Chem. Soc.*, 2005, **127**, 6970–6971.
46 T. J. Osberger and M. C. White, *J. Am. Chem. Soc.*, 2014, **136**, 11176–11181.
47 M. Malik, M. Ceborska, G. Witkowski and S. Jarosz, *Tetrahedron: Asymmetry*, 2015, **26**, 29–34.
48 (a) A. A. Scholte, M. H. An and M. L. Snapper, *Org. Lett.*, 2006, **8**, 4759–4762; (b) S. Beligny, S. Eibauer, S. Maechling and S. Blechert, *Angew. Chem., Int. Ed.*, 2006, **45**, 1900–1903.
49 (a) S. Narasimhan and R. Balakumar, *Aldrichim. Acta*, 1998, **31**, 19–26; (b) S. Jarosz, *Carbohydr. Res.*, 1988, **183**, 201–207.
50 (a) C. H. Larsen, B. H. Ridgway, J. T. Shaw and K. A. Woerpel, *J. Am. Chem. Soc.*, 1999, **121**, 12208–12209; (b) C. H. Larsen, B. H. Ridgway, J. T. Shaw, D. M. Smith and K. A. Woerpel, *J. Am. Chem. Soc.*, 2005, **127**, 10879–10884.
51 E. R. van Rijssel, T. P. M. Goumans, G. Lodder, H. S. Overkleeft, G. A. van der Marel and J. D. C. Codée, *Org. Lett.*, 2013, **15**, 3026–3029.
52 K. V. Gothelf and K. A. Jørgensen, *Chem. Rev.*, 1998, **98**, 863–910.

53 R. Bloch, *Chem. Rev.*, 1998, **98**, 1407–1438.
54 F. Cardona and A. Goti, *Angew. Chem., Int. Ed.*, 2005, **44**, 7832–7835.
55 J. Revuelta, S. Cicchi, A. Goti and A. Brandi, *Synthesis*, 2007, **4**, 485–504.
56 I. A. Grigor'ev, in *Nitrile Oxides, Nitrones, And Nitronates In Organic Synthesis*, ed. H. Feuer, Wiley-VCH, Hoboken, 2008.
57 G. D'Adamio, A. Goti, C. Parmeggiani, E. Moreno-Clavijo, I. Robina and F. Cardona, *Eur. J. Org. Chem.*, 2011, 7155–7162.
58 W. Zhang, K. Sato, A. Kato, J. Yue-Mei, X.-G. Hu, F. X. Wilson, R. van Well, G. Horne, G. W. J. Fleet, R. J. Nash and C.-Y. Yu, *Org. Lett.*, 2011, **13**, 4414–4417.
59 Ł. Woźniak, O. Staszewska-Krajewska and M. Michalak, *Chem. Commun.*, 2015, **51**, 1933–1936.
60 B. Grzeszczyk, K. Poławska, Y. M. Shaker, S. Stecko, A. Mames, M. Woźnica, M. Chmielewski and B. Furman, *Tetrahedron*, 2012, **68**, 10633–10639.
61 P. Gilles and S. Py, *Org. Lett.*, 2012, **14**, 1042–1045.
62 J. Boisson, A. Thomasset, E. Racine, P. Cividino, T. Banchelin Sainte-Luce, J.-F. Poisson, J.-B. Behr and S. Py, *Org. Lett.*, 2015, **17**, 3662–3665.
63 D. Astruc, in *Olefin Metathesis: Theory and Practice*, ed. K. Grela, Wiley-VCH, Weinheim, 2014.
64 I. Dragutan, V. Dragutan, C. Mitan, H. C. M. Vosloo, L. Delaude and A. Demonceau, *Beilstein J. Org. Chem.*, 2011, **7**, 699–716.
65 P. Compain, *Adv. Synth. Catal.*, 2007, **349**, 1829.
66 K. R. Fandrick, J. Savoie, N. Yee, J. J. Song, C. H. Senanayake, in *Olefin Metathesis: Theory and Practice*, ed. K. Grela, Wiley-VCH, Weinheim, 2014.
67 T. Jensen, M. Mikkelsen, A. Lauritsen, T. L. Andresen, C. H. Gotfredsen and R. Madsen, *J. Org. Chem.*, 2009, **74**, 8886–8889.
68 G. Danoun, J. Ceccon, A. E. Greene and J.-F. Poisson, *Eur. J. Org. Chem.*, 2009, 4221–4224.
69 A. M. C. H. Nieuwendijk, M. Ruben, S. E. Engelsma, M. D. P. Risseeuw, R. J. B. H. N. van den Berg, R. G. Boot, J. M. Aerts, J. Brussee, G. A. van der Marel and H. S. Overkleeft, *Org. Lett.*, 2010, **12**, 3957–3959.
70 J.-B. Behr, A. Hottin and A. Ndoye, *Org. Lett.*, 2012, **14**, 1536–1539.
71 P.-A. Nocquet, R. Hensienne, J. Wencel-Delord, E. Wimmer, D. Hazelard and P. Compain, *Org. Biomol. Chem.*, 2015, **13**, 9176–9180.
72 P.-A. Nocquet, D. Hazelard, G. Gruntz and P. Compain, *J. Org. Chem.*, 2013, **78**, 6751–6757.
73 (a) O. Achmatowicz Jr., P. Bukowski, B. Szechner, Z. Zwierzchowska and A. Zamojski, *Tetrahedron*, 1971, **27**, 1973–1996; (b) M. A. Ciufolini, C. Y. W. Hermann, Q. Dong, T. Shimizu, S. Swaminathan and N. Xi, *Synlett*, 1998, 105–114.
74 J. Bi and V. K. Aggarwal, *Chem. Commun.*, 2008, 120–122.
75 F. van der Pijl, F. L. van Delft and F. P. J. T. Rutjes, *Eur. J. Org. Chem.*, 2015, 4811–4829.
76 (a) Y. M. Xu and W. S. Zhou, *Chin. J. Chem.*, 1998, **16**, 34–44; (b) J. Ostrowski, H. J. Altenbach, R. Wischnat and D. J. Brauer, *Eur. J. Org. Chem.*, 2003, 1104–1110; (c) A. Kennedy, A. Nelson and A. Perry, *Chem. Commun.*, 2005, 1646–1648.
77 M. Magdycz, P. Cmoch and S. Jarosz, *Heterocycles*, 2010, **80**, 1303–1318.
78 S. Jarosz, *Curr. Org. Chem.*, 2008, **12**, 985–994.
79 M. Magdycz and S. Jarosz, *Tetrahedron: Asymmetry*, 2013, **24**, 1412–1416.

Recent examples of novel synthetic approaches to diverse amino sugars

Zbigniew J. Witczak* and Roman Bielski
DOI: 10.1039/9781782626657-00344

This brief review discusses selected syntheses of amino sugars published in this century. The specific examples are divided into categories based on the location of the amino group in the final product. The review puts emphasis on diastereoselectivity and enantioselectivity effects. It considers mechanisms of the reactions introducing the amino group and applications of synthesized amino products. Potential and already implemented medical applications of presented compounds are also highlighted.

1 Introduction

Amino sugars are one of the most abundant classes of naturally occurring carbohydrates. Arguably, they are the most diverse category of natural compounds. The number of amino sugars identified in living organisms is greater than that of any other functionalized sugars. The most common ones are D-glucosamine, D-galactosamine and sialic acid. However, it must be emphasized that the major source of various amino sugars are bacteria. Amino sugars have been found in polysaccharides such as chitin and chitosan, in nucleic acids, in various medicinal compounds and, what is particularly important, they are components of tremendously diverse glycoconjugates including glycoproteins and glycopeptides. The medical importance of amino sugars cannot be overestimated. They act as inhibitors of glycosidases, are conjugated to various anticancer vaccines, comprise a plethora of key antibiotics, *etc.* Examples of carbohydrate drugs or carbohydrate-derived drugs containing an amino sugar unit are shown below and in Scheme 1.

Name	Trade name	Activity
Streptozotocin	Zanosar	Anticancer
Zanamivir	Relenza	Neuroaminidase inhibitor
Olsletamivir	Tamiflu	Neuroaminidase inhibitor
Topiramate	Topamax	Antiepilepsy
Enoxaparin	Lovenox	Thrombosis

In our opinion there are two reasonable ways to pursue discussing compounds structurally related to the given examples. One option is to consider the final structure of the synthetic target. Alternatively, one can

Department of Pharmaceutical Sciences, Nesbitt School of Pharmacy, Wilkes University, 84 W. South Street, Wilkes-Barre, PA 18766, USA. E-mail: zbigniew.witczak@wilkes.edu

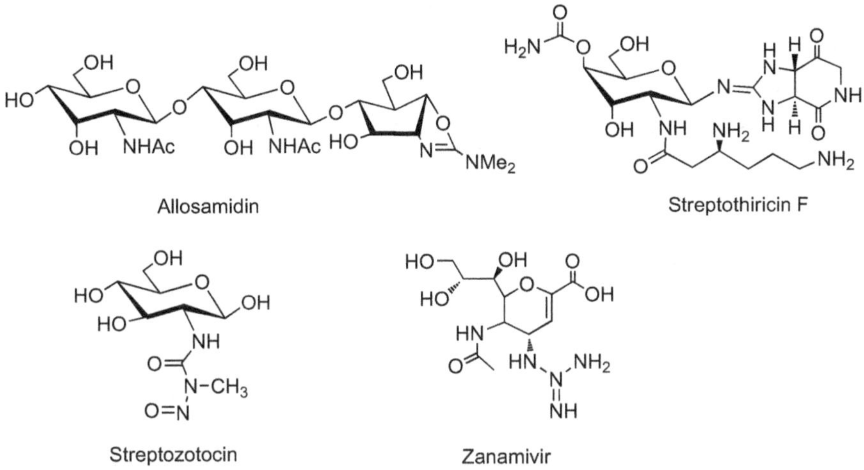

Scheme 1 Structures of medicinally relevant amino sugars and currently approved drugs with amino sugar motif.

look at the mechanism of the step introducing the amino group at a given position. We decided to employ the former. However, the other approach seems to be equally valid to us. Thus, let us briefly list the most common reactions leading to the formation of the carbon–amino bond in amino sugars. They are:

- Reduction of compounds already containing a carbon–nitrogen bond such as oximes, Schiff bases, nitro-compounds, *etc.*
- Nucleophilic substitution in compounds containing a good leaving group such as tosylate, mesylate, triflate, *etc.* The opening of cyclic compounds such as epoxides or cyclic sulfates belongs to this category. The nucleophiles used include azide, phthalate, ammonia, hydrazine, *etc.*
- Addition to double bonds such as an azide addition in the presence of CAN.
- Intramolecular reactions including rearrangements such as the Overman rearrangement or rearrangement of allylic azides, and reactions of non-allylic trichloro-(fluoro)acetimidates, carbamates, *etc.*

During the last decade and a half or so, many new amino sugars as components of naturally occurring oligosaccharides, polysaccharides and complex carbohydrates have been isolated from various sources and their structure has been determined. The topic of isolation and structure elucidation of natural amino sugars has been reviewed covering amino sugar importance and applications. Moreover, their validity to the multiple synthetic targets was assessed.[1–5] It is worth noting that, the recent classification[6] of rare bacterial amino sugars is based on their occurrence in nature as those sugars form key components of a variety of bacterial oligosaccharides, glycoproteins and glycoconjugates defined as zwitterionic polysaccharides. The following five types of rare amino sugars were highlighted: 2-acetamido-4-amino-2,4,6-trideoxy-D-galactose (AAT), bacillosamine (Bac), *N*-acetylfucosamine (FucNAc), 2,4-diacetamido-2,4,6-trideoxy-D-galactose (DATDG), and D-*xylo*-6-deoxy-4-keto-hexosamine (DKH).

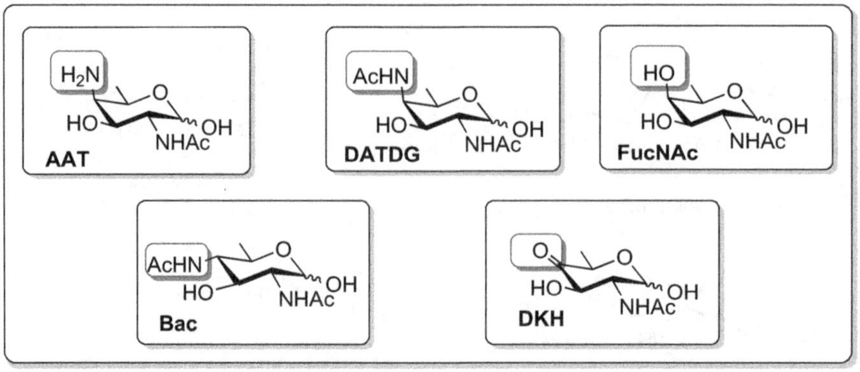

Scheme 2 The bacterial atypical amino sugars as key components of glycoconjugates.

They all share common structural characteristics having functional groups such as the acetamido group at C-2, hydroxy group at C-3 and methyl group at C-5. All these substituents are equatorial. Additionally, most of these amino sugars carry an axial substituent at C-4 (amino group in AAT, acetamido group in DATDG and hydroxyl group in FucNAc). Only bacillosamine (Bac) has an equatorial –NHAc group at C-4. D-xylo-6-deoxy-4-ketohexosamine (DKH) is equipped with a keto functional group at C-4. Those important structural functionalities are depicted in Scheme 2.

2 Synthetic strategies for the preparation of C-1 amino sugars

This category of amino sugars includes such important compounds as nucleosides and nucleotides. However, since synthetic strategies applicable to components of nucleic acid are usually very different from strategies used in the synthesis of other amino sugars, they are not included in this review.

One of the new and important applications of the Burgess reagent in the stereoselective synthesis of α- and β-glycosylamines was reported by the Nicolaou group.[7] This strategy is utilizing 3,4,6-protected lactols with exposed –OH groups at C-1 and C-2, to form sulfamidate ring. The following step of the stereoselective opening of sulfamidate ring with a variety of heteroatomic nucleophiles produces 1,2-*trans*-difunctionalized glycosylamine derivatives. A variety of diols based on carbohydrate synthons derived from D-glucose, D-galactose, and L-rhamnose were converted into α-disposed sulfamidates (Scheme 3).

The Imperiali group[8] reported a simple, one-step procedure transforming the hydroxy group at C-1 of chitobiose into the amino group with ammonium carbamate in methanol as depicted in Scheme 4.

The new selective amination protocol of unprotected sugar derivatives was also reported by Likhosherstov and coworkers[9] who used the same reagent, but with the major advantage of an easy isolation of the reaction product. Most importantly, the *N*-glycosylamines formed by the reaction with ammonia are precipitated as carbamic acid salts and this prevents

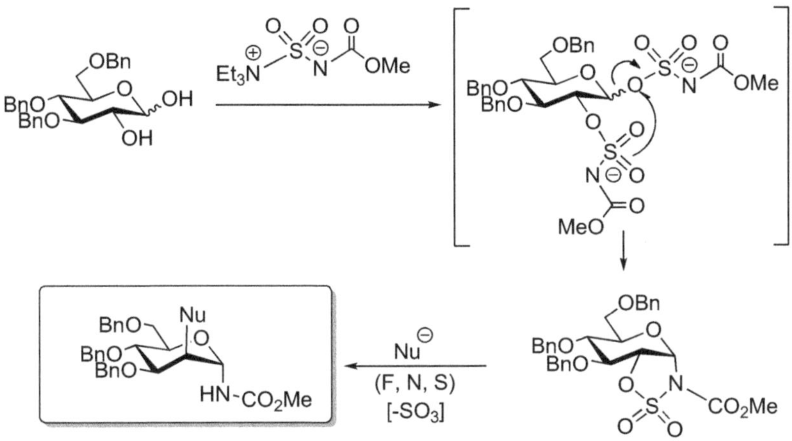

Scheme 3 Nicolaou's stereoselective synthesis of *N*-glucosamines using Burgess reagent.

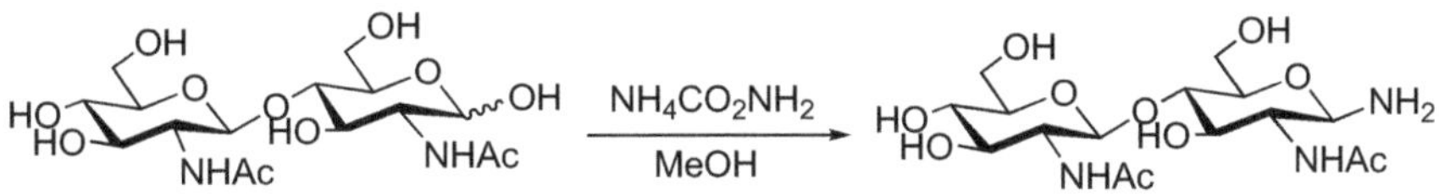

Scheme 4 Imperiali's synthesis of *N*-glycosylamines *via* amination with ammonium carbamate.

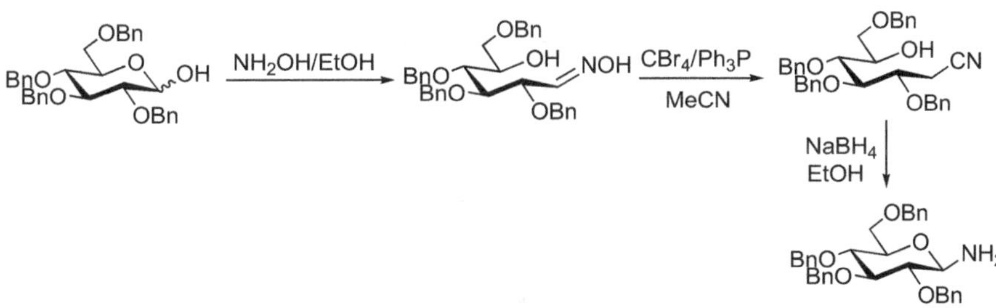

Scheme 5 Tranuner's synthesis on *N*-glucosylamines *via* borohydride reductive amination.

the hydrolysis and glycosylamine dimers formation. The free amine is isolated by base treatment under high vacuum.

The borohydride reductive cyclization of δ-hydroxy nitriles prepared from C-1 oximes followed by dehydration using a carbon tetrabromide/triphenylphosphane system was also explored as a convenient amination strategy to produce glycosylamines. The method, depicted in Scheme 5, was reported by Trauner and coworkers.[10] The methodology was further explored in the synthesis of *gluco-*, *galacto-*, *and manno-*configurational isomers. Interestingly, the strategy is not applicable to synthesizing furanosylamines.

Chagnault and coworkers reported[11] a convenient one-step approach to *N*-glucosylamines enabling the installation of (benzyl, nonyl and allyl)amine moieties at C-1. The synthetic approach starts from benzyl protected lactol as depicted in Scheme 6.

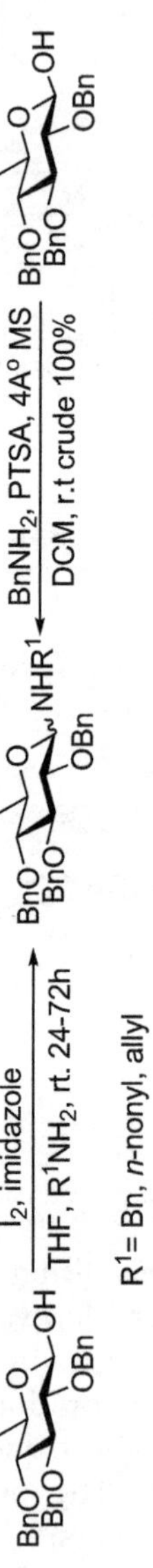

Scheme 6 Chagnault's synthesis of *N*-glucosylamines *via* iodine/imidazole catalyzed amination.

The combination of iodine/imidazole reagent in tetrahydrofuran as a solvent produced glycosylamines in quantitative yields. It is worth mentioning that *N*-benzylglucosylamine was produced even faster under this reaction condition than when using the Martin protocol.[12,13]

Liu and coworkers[14] have reported a mild and efficient protocol for the stereoselective synthesis of *N*-glycosylamines of enone sugars. Certain mechanistic studies required a stereoselective installation of the –NHR functionality at the anomeric center of the formed enone. Product formation is efficiently promoted by $BF_3 \cdot Et_2O$. A conventional reduction of the keto functionality at C-2 with sodium borohydride ($NaBH_4$) in the presence of cerium trichloride ($CeCl_3$) proceeded with a remarkable diastereoselectivity probably due to a steric hindrance imposed by the anomeric substituent adjacent to a carbonyl group at C-2. The subsequent dihydroxylation of the double bond with osmium tetroxide (OsO_4) in the presence of a catalytic amount of morpholine *N*-oxide (NMO) in aqueous *tert*-butanol proceeded by a diastereofacial addition of OsO_4 to the double bond from the same side of the ring as the existing allylic hydroxy group to produce *N*-sulfamidotalose as depicted in Scheme 7.

Motavia and coworkers[15] have described a novel acid-catalyzed rearrangement of glycosyl trichloroacetimidates leading to *N*-glycosylamines in moderate yields. The synthetic approach using trimethylsilyl triflate in dichloromethane solution in the presence of 4 Å molecular sieves, followed by reductive removal of the trichloroacetyl group produces *N*-glycosylamines in high yields as depicted below (Scheme 8).

3 Synthetic strategies for the preparation of C-2 amino sugars

There are relatively many stereoselective synthetic methodologies for synthesizing these compounds. The one reported by Bundle and coworkers[16] takes advantage of the synthetic utility of functionalized 1,2-oxazolines (Scheme 9). The synthetic sequence requires a stereoselective opening of 1,2-oxazoline with the formation of methyl 2-aminoglucoside in the pyranose form. The C-3 mesylated derivative after deprotection of 5,6-diisopropylidene functionality can be easily converted to epoxide, which can also be stereoselectively opened with an azide anion to form C-4 axial azide. C-6 mesylation of azide, followed by its reduction with sodium borohydride/nickel chloride produces methyl 2-amino-4-acetamidogulopyranoside.

Additionally, the opening of the epoxide with Amberlite IR-120 resin (H^+) produces methyl 2-acetamidogulopyranoside in 64% yield.

Another interesting synthesis of orthogonally protected bacterial rare sugars and D-glucosamine building blocks was recently offered by Emmadi and Kulkarni.[17] All precursors with axial and equatorial –OH at C-2 and C-4 positions (Scheme 10) were prepared from 1-thiophenyl glycoside. The functionalization of those important positions with the azido group produced orthogonally protected analogs, which can be conveniently converted into amino sugars.

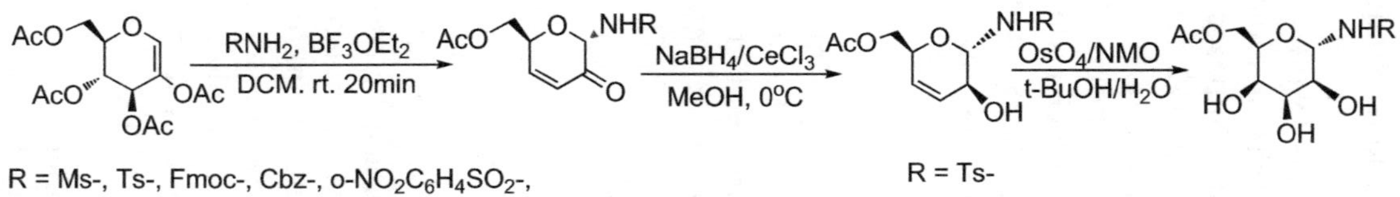

Scheme 7 Liu's stereoselective synthesis of *N*-glycosylamines from sugar enones.

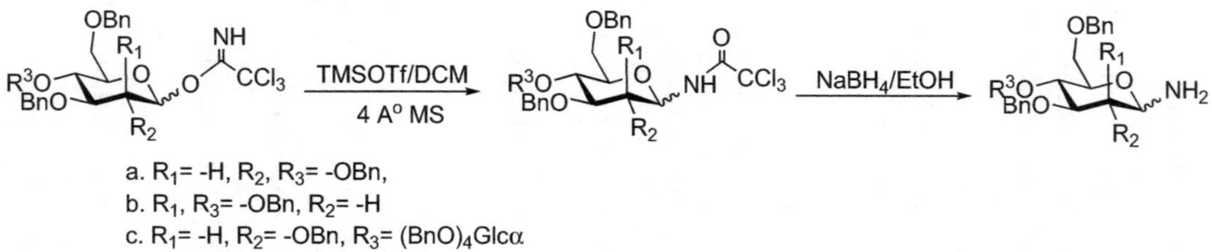

Scheme 8 Motavia's synthesis of *N*-glycosylamines *via* acid-catalyzed rearrangement.

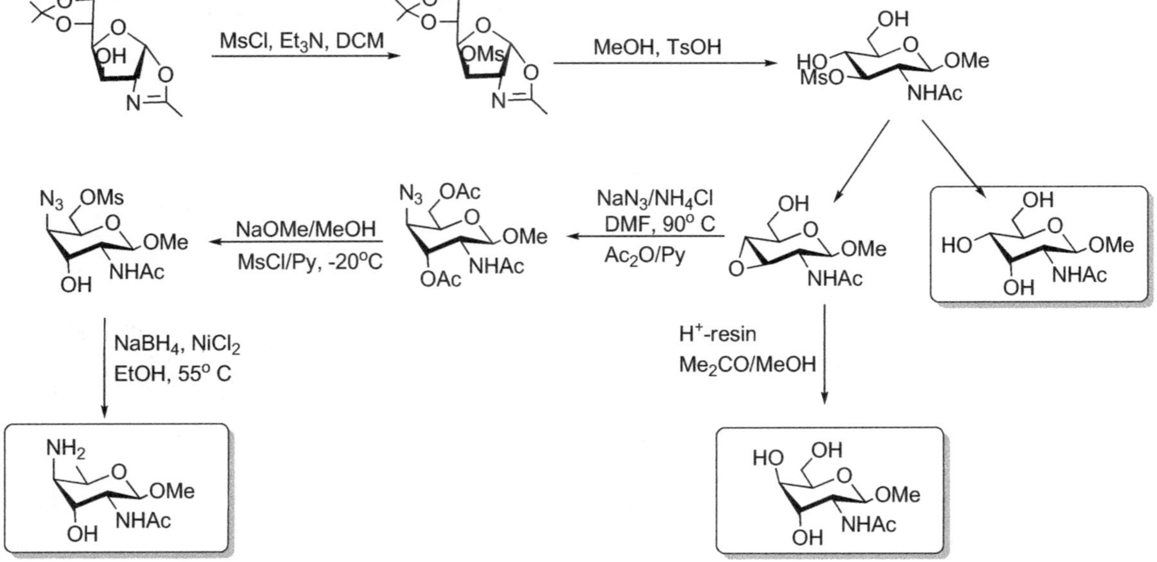

Scheme 9 Bundle's synthesis of 2-amino sugars *via* 1,2-oxazoline route.

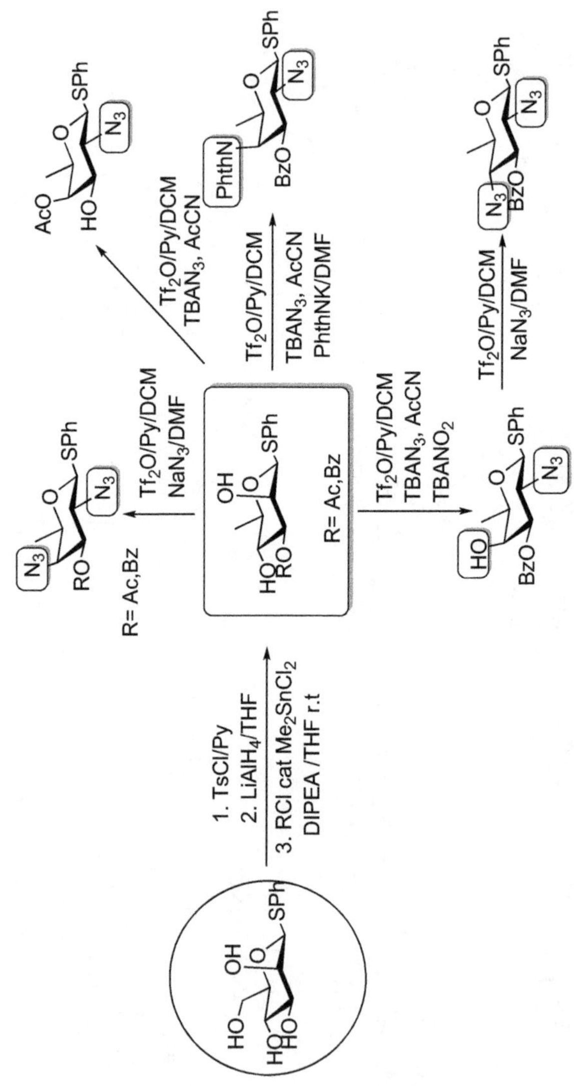

Scheme 10 Synthesis of orthogonally protected D-glucosamine building blocks.

Yoda and coworkers[18] described a stereodivergent synthesis of newly isolated[19] furanodictines A and B belonging to the 3,6-anhydro-amino sugars family. The synthetic approach utilizes D-glucuronolactone as a starting template (Scheme 11). The conventional isopropylidenation at C1 and C-2 followed by benzylation at C-5 produces fully protected lactone. In the next step a C-2 lactone function is reduced and –OH functionality subsequently transformed into a methylene group. Then, the isopropylidene group at C1 and C-2 is opened with methanolic solution and DOWEX 50W X-8 treatment to produce methyl furanoside. Subsequent treatment with triflic anhydride in dichloromethane catalyzed by pyridine, followed by SN_2 nucleophilic displacement with sodium azide in dimethylformamide solution furnished crystalline C-2 azide. Conventional reductions of the azido function with triphenylphosphine in aqueous mixture of tetrahydrofuran and dichloromethane solution, followed by acetylation with acetic anhydride in dichloromethane solution produces 2-acetamido intermediate in good yield. Hydrolysis of methyl furanoside with 70% acetic acid and treatment with *tert*-butyldimethylsilyl iodide followed by debenzylation and subsequent reaction with isovaleric acid under base catalysis with dimethylaminopyridine in dichloromethane solution produced, after final removal of *tert*-butyldimethylsilyl group, furanodictine B in moderate yield.

4 Synthetic strategies for the preparation of C-3 amino sugars

Herein we describe selected examples of new stereoselective synthetic methodologies for this class of amino sugars that have been published in the last decade.

Reissig and Pfrengle[20] have developed an attractive strategy for the synthesis of 3-amino-2,6-dideoxy sugars. The required starting template 3,6-dihydro-2*H*-1,2-oxazine was conveniently produced by the addition of lithiated alkoxyallenes (prepared *in situ*) to D-glyceraldehyde-derived nitrone.[21]

The stereo-divergent approach produced four stereoisomeric 1,2-oxazines. Interestingly, when free nitrone was reacted, exclusively the *syn*-isomer was formed. However, the anti-isomer was formed as the only product when nitrone was pre-complexed with diethylaluminum chloride. The general concept of employing 1,2-oxazines to synthesizing other amino sugars was extensively reviewed.[22,23] It is worth mentioning that the 1,2-oxazines protocol is also applicable to the synthesis of C-2 branched 4-aminosugars (Scheme 12).

Nicolaou and coworkers have reported an expeditious IBX (2-iodoxybenzoic acid, 1-hydroxy-1λ^3,2-benziodoxol-3(1*H*)-one 1-oxide) mediated synthesis of *l*-vancosamine[24] starting from vinyl iodide and protected aldehyde *via* an intramolecular Kishi–Nozaki coupling reaction. The coupling product, as a mixture of alcohols, was immediately oxidized with DMP (Dess-Martin periodinane). The oxidation was followed by a sodium borohydride/cerium chloride reduction to produce a protected alcohol intermediate. In the next step, the required alcohol was reacted

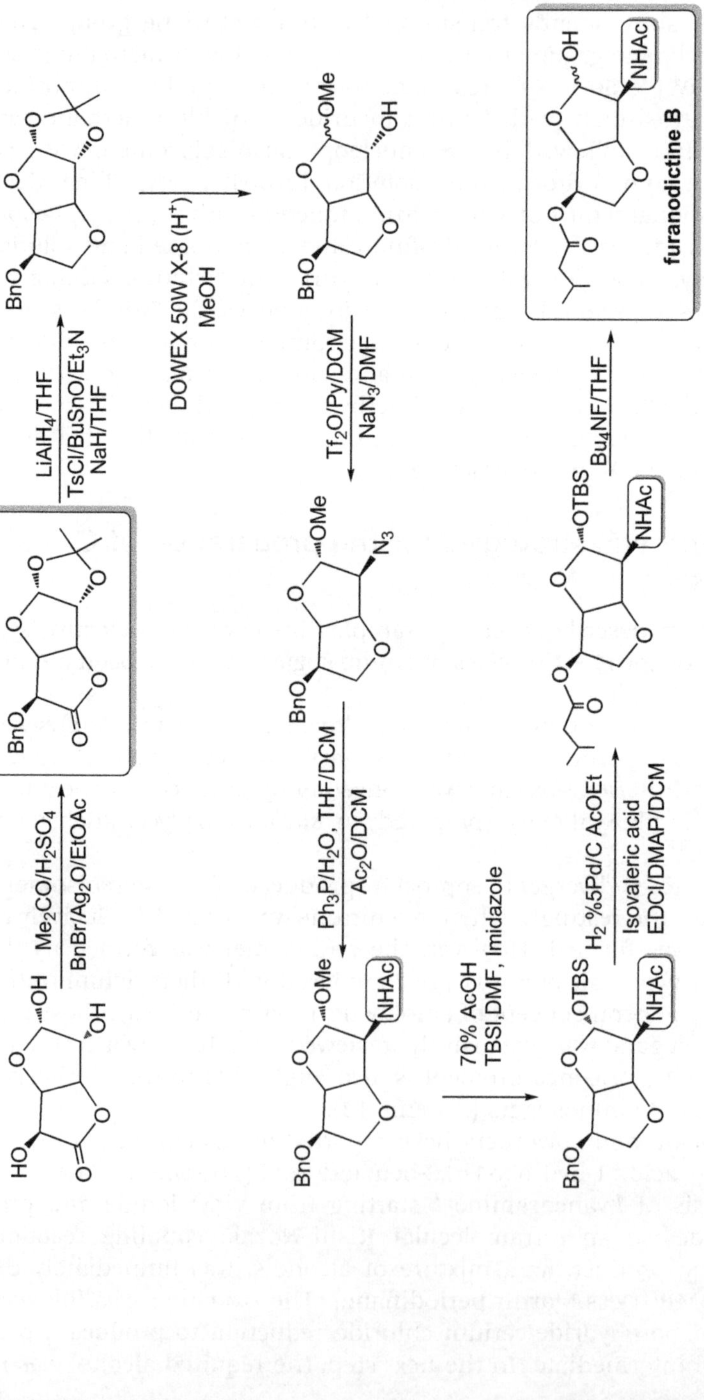

Scheme 11 Yoda's stereoselective synthesis of furanodictine B.

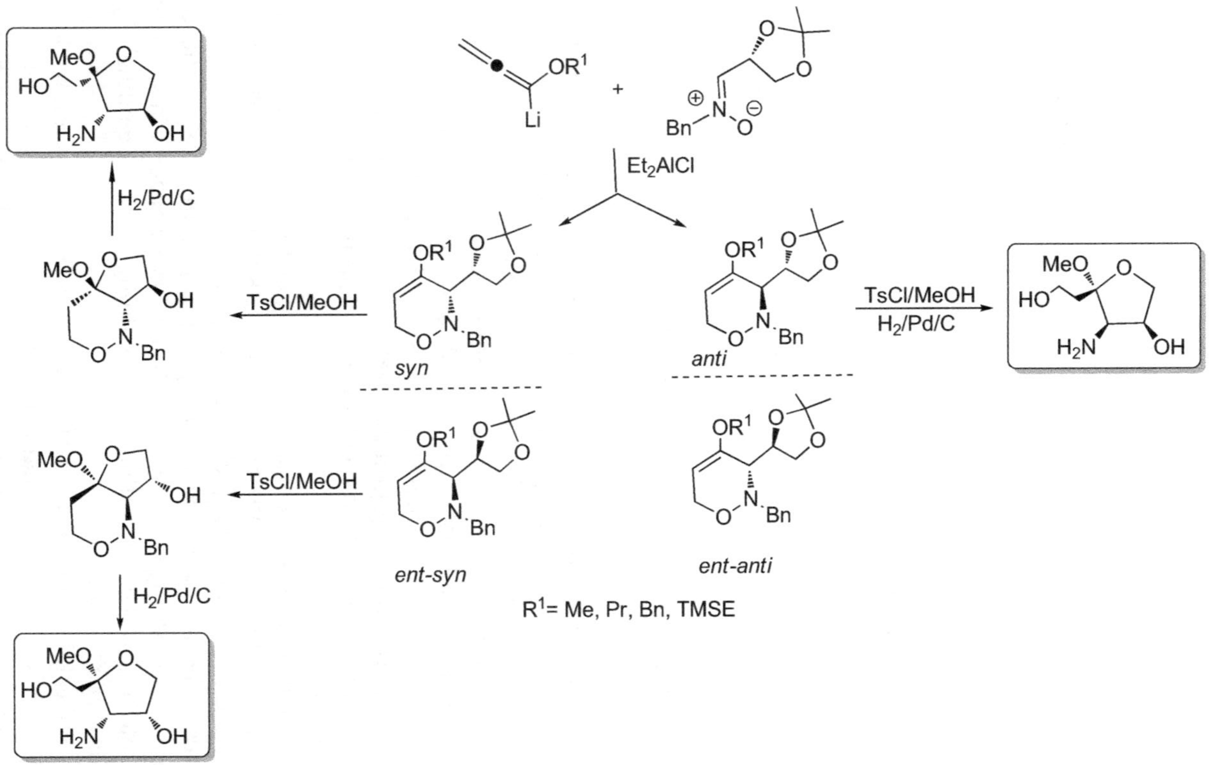

Scheme 12 Pfrengle's synthesis of 3-amino-2,6-dideoxy sugars.

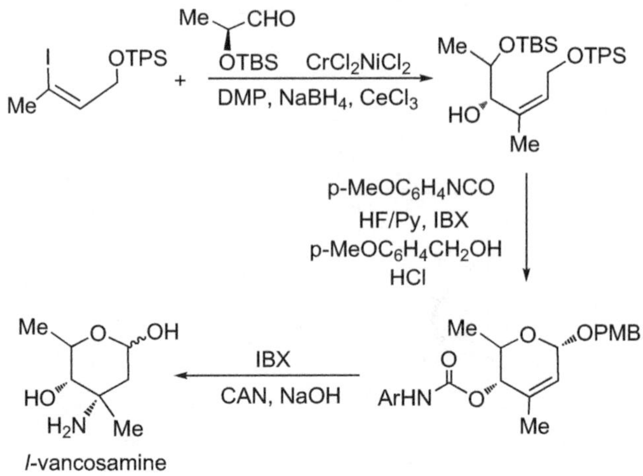

Scheme 13 Nicolaou's synthesis of *l*-vancosamine.

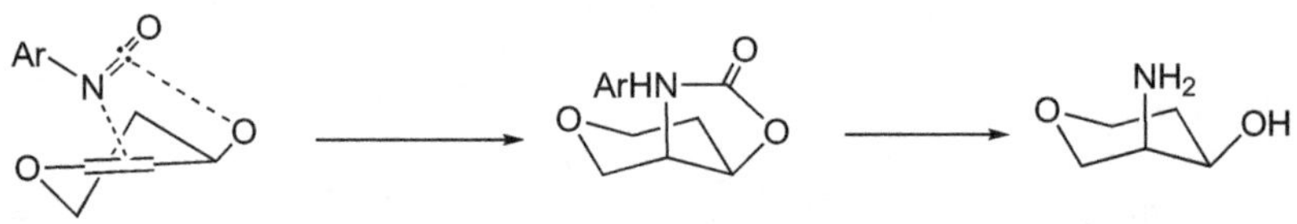

Scheme 14 Nicolaou's general concept for the IBX-mediated construction of amino sugars.

with *p*-methoxyphenyl isocyanate in the presence of catalytic amounts of DBU and the product was subjected to the total deprotection with HF/pyridine, which was followed by a selective primary alcohol oxidation with IBX. Following the protection of the anomeric hydroxyl moiety by *p*-methoxybenzyl alcohol (PMB) the cyclization intermediate was produced in 47% overall yield. The final steps of IBX-mediated cyclization and the removal of both protecting groups (pMP and PMB) followed by a base hydrolysis culminate the stereo-controlled approach to *l*-vancosamine (Scheme 13).

This stereo-controlled approach consisting of 10 steps, which gives an overall yield of 13%, constitutes probably the shortest synthetic route to this important amino sugar (Scheme 13).

Interestingly, the Nicolaou's concept of employing IBX to the construction of amino sugars as depicted below is very general in scope and applicable to a variety of carbohydrate building blocks including allylic alcohols (Scheme 14).

The applicability of this ingenious general strategy to the amino sugars construction with other functionalized carbohydrate templates was demonstrated further by using various carbohydrate-derived allylic alcohols. A representative example is shown below (Scheme 15).

Overall, 11 examples[24] of carbohydrate-derived allylic alcohols utilized in the synthesis of amino sugar precursors were reported.

Friestad and coworkers[25] have reported an original asymmetric synthesis of *N*-trifluoroacetyl-L-daunosamine from achiral precursors.

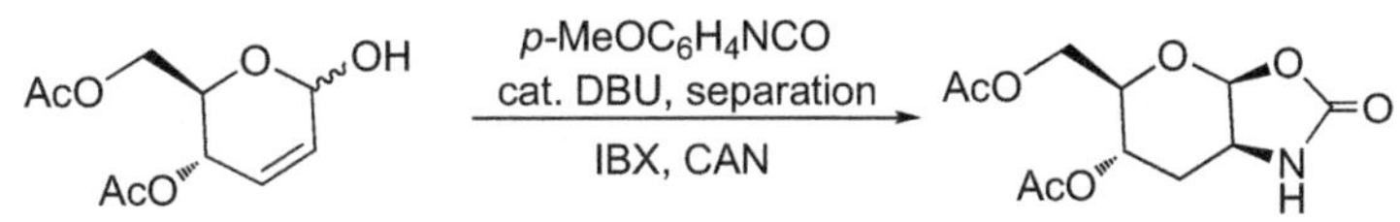

Scheme 15 Nicolaou's IBX-mediated synthesis of amino-sugar templates from allylic alcohols and aryl isocyanate.

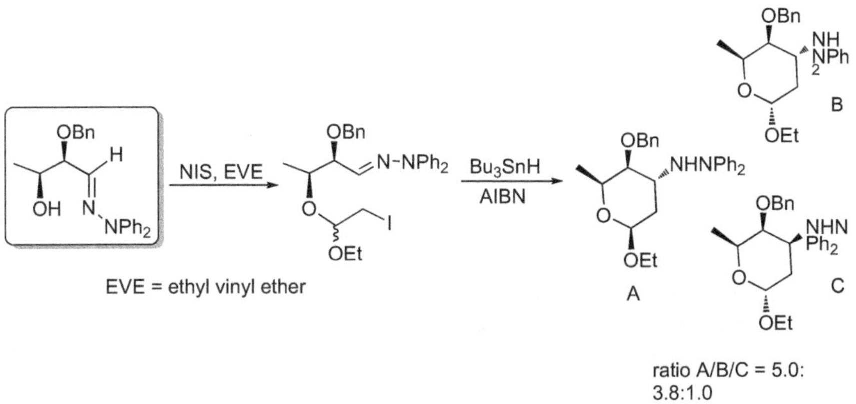

Scheme 16 Friestad's synthesis of *N*-trifluoroacetyl-L-daunosamine.

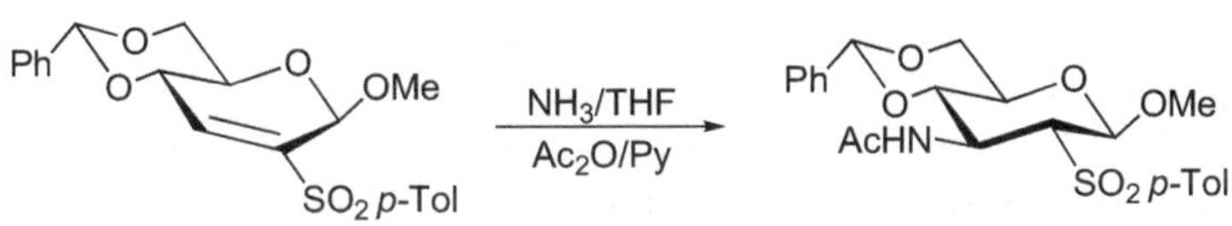

Scheme 17 Pathak's synthesis of 3-aminosugars.

The approach takes advantage of a haloacetal cyclization using an α,β-dioxygenated imino compound as the radical acceptor. The configuration of the isolated products was determined through analysis of the coupling constants among hydrogens at C1–C3 and was consistent with the value characteristic of the equatorial substituent at the anomeric carbon atom *i.e.* of L-3-*epi*-daunosamine (Scheme 16). The crucial step leading to optically active product was Sharpless asymmetric dihydroxylation of crotonaldehyde derivative.

Pathak and Bhattacharya[26,27] have reported an interesting strategy employing vinyl sulfone synthons modified at C-2. A simple addition of ammonia to a double bond, followed by a conventional acetylation of amino functionality at C-3 constitutes a new route to 3-amino-sugars. The approach can also be extended to branched-chain amino-sugars. The general approach is depicted in Scheme 17.

5 Novel synthetic strategies for the preparation of C-4 amino sugars

4-Amino-4-deoxy sugars are a very important class of amino sugars mainly due to the fact that the position 4 is a primary object of coupling

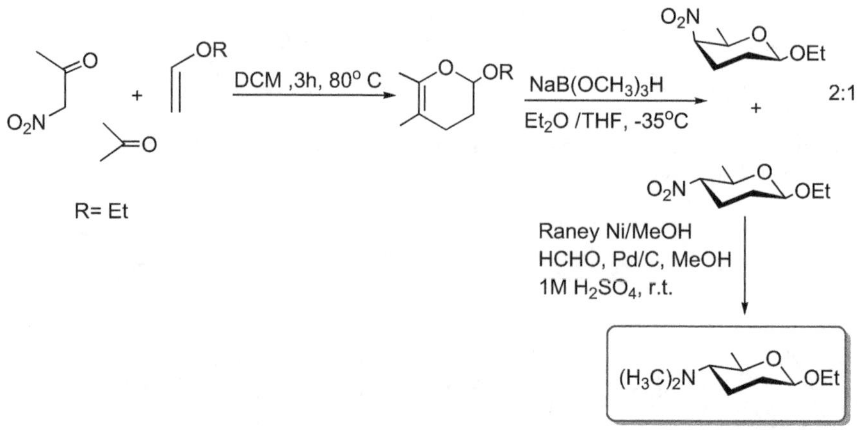

Scheme 18 Tietze's synthesis of the 4-amino sugar (+)-D-forosamine.

with other sugar units. Also, the –NH amino bridge specifically in this position creates an ideal stereo-geometry for the receptor affinity in many biological interactions with proteins. Furthermore, such amino sugars play an active role in many enzyme inhibitors and are components of potential glycomimetics as drug candidates. Amongst the newly reported synthetic methodologies, a three component domino-Knoevenagel-hetero-Diels–Alder reaction, as reported by Tietze and coworkers,[28] was used in the stereoselective approach to (+)-D-forosamine. The methodology is depicted in Scheme 18.

Schmid and coworkers[29] described a synthetic route to fluorinated analogues of tomosamine and 4-amino-4-deoxyarabinose through a stereoselective titanium-mediated aldol addition. This interesting and versatile strategy required a special two-carbon chain elongation of serine- and threonine-derived aldehydes with a chiral fluoroacetyloxazolidinone.

The stereoselective aldol addition was capable of generating 4-acetylamino-2-fluoropentoses and the corresponding 6-deoxyhexoses in overall yields of 16–23% (Scheme 19).

6 Synthetic strategies for the preparation of C 5-amino sugars

The importance of the presence of the amino group at C-5 is well established. In particular, certain valuable classes of nucleosides contain the amino group at the position C-5. One of those important categories of 5-amino-5-deoxy sugars are 5-amino sugar acids. Recently, they were the target of an interesting synthesis described by the Varela's group.[30] The approach, depicted in Scheme 20, uses a universal 2,6-diol template as a starting material. The conventional tosylation was followed by nucleophilic S_N2 displacement of C-6 tosylate with sodium azide in dimethylformamide (DMF) solution. Next, deprotection of the anomeric functionality produced an intermediate with the azide group at C-6 in good yield. The azide template was converted into lactone, which undergoes base hydrolysis caused by a methanolic solution of potassium

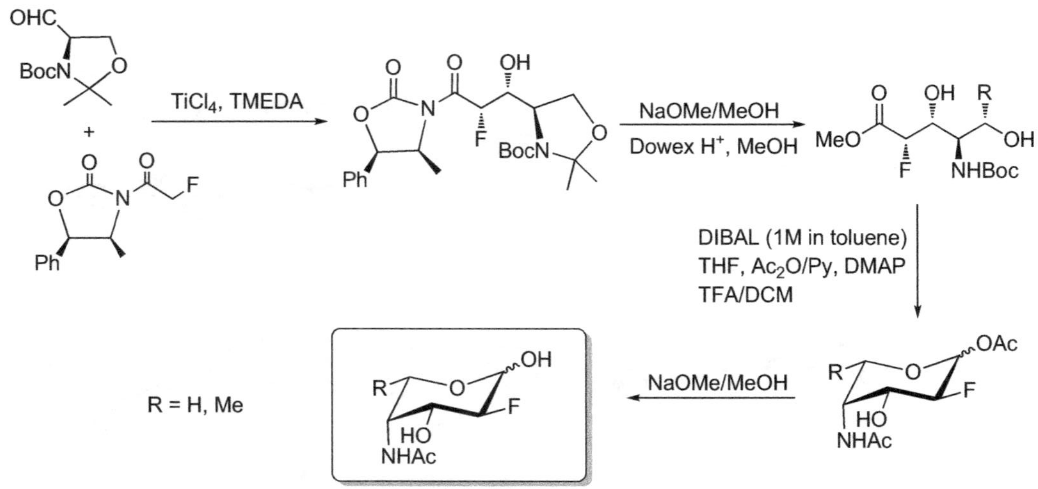

Scheme 19 Schmid's synthesis of 4-amino-4-deoxy sugars.

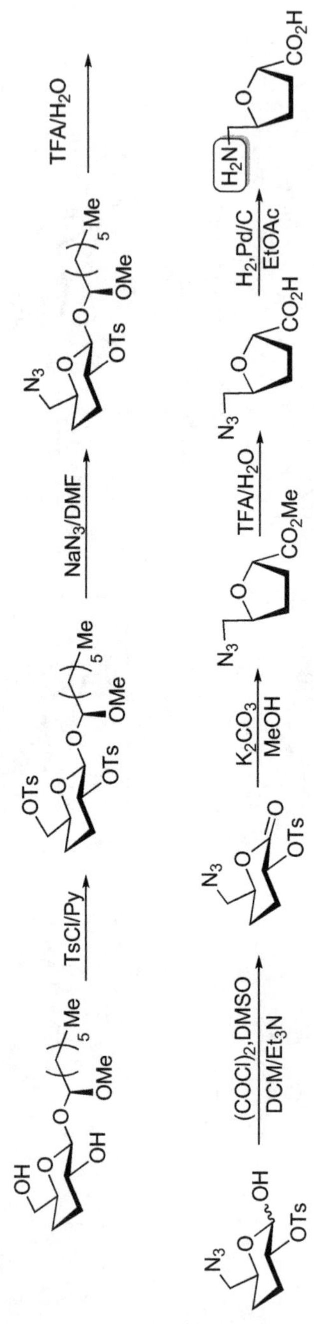

Scheme 20 Varela's synthesis of 5-amino-5-deoxy sugars.

carbonate with the formation of a methyl ester. The final steps of methyl ester hydrolysis with trifluoroacetic acid and azide hydrogenation catalyzed by palladium on carbon (Pd/C) furnished the 5-amino sugar acid in good yield.

7 Synthetic strategies for the preparation 6-amino-6-deoxy sugars

The amino group at the position C-6 of hexoses is linked to a primary carbon atom. Thus, the easy accessible C-6 amino group can be further functionalized. In aminoglycoside antibiotics, a (primary) C-6 functionalization is the key factor to improve their efficacy against methicillin-resistant *Staphylococcus aureus.* Sasaki and coworkers[31] have reported an efficient and regioselective amination strategy at C-4 and C-6 positions. Such functionalization should lead to the increase in the antibacterial activity. Indeed, the newly synthesized 6″-*N*-alkanoylamino functionalized analogs of arbekacin showed an excellent antibacterial activity. It was suggested to result from the inhibition of the bifunctional aminoglycoside-modifying enzyme, that causes *Staphylococcus aureus* resistance to arbekacin. Thus, introducing the amino functionality into position C-6 may offer inhibitors of resistance causing specific enzymes and thus, an increased antibacterial activity. The synthetic strategy of the Sasaki's[31] regioselective amination at C-6 is shown on Scheme 21.

Many interesting aspects of the chemistry of aminoglycoside antibiotics have been recently reviewed by the Garneau-Tsodikowa group.[32] It is worth noting that aminoglycosides and aminoglycoside derivatives/conjugates are capable of targeting many steps in the HIV life cycle, and the development of derivatives showing the specificity towards a particular target has become an ultimate goal of many research efforts during the last few years.

One novel approach reported by Bianchini and coworkers[33] is exploring a new strategy of incorporation the nitrogen atom at the C-6 position *via* *N*-benzyl intermediates. The synthetic route, as depicted in Scheme 22, starts from primary tosylates, which were converted into *N*-benzylamino intermediates and then debenzylated *via* a standard hydrogenation with gaseous hydrogen in the presence of the Pd/C catalyst and acetic acid in acetonitrile solution.

Yenil and coworkers[34] used 1,2-*O*-trichloroethylidene furanoses as convenient precursors for the synthesis of 6-amino sugars, also applied to 5-amino sugars, as shown in Scheme 23. Selective tosylation at C-6 (and C-5 for pentoses) followed by the displacement of primary tosylate with azide, reduction with triphenylphosphine and conventional acetylation gave the target compounds. The sugars with the amino group at the primary position were synthesized in 91% and 98% overall yield as depicted in Scheme 23.

Varela and coworkers[35] reported the enantiospecific synthesis of the extremely important 2,6-diamino sugar component of the sanamycin antibiotic. The synthetic approach starts with an easily available allylic alcohol precursor which reacts with triflic anhydride in dichloromethane

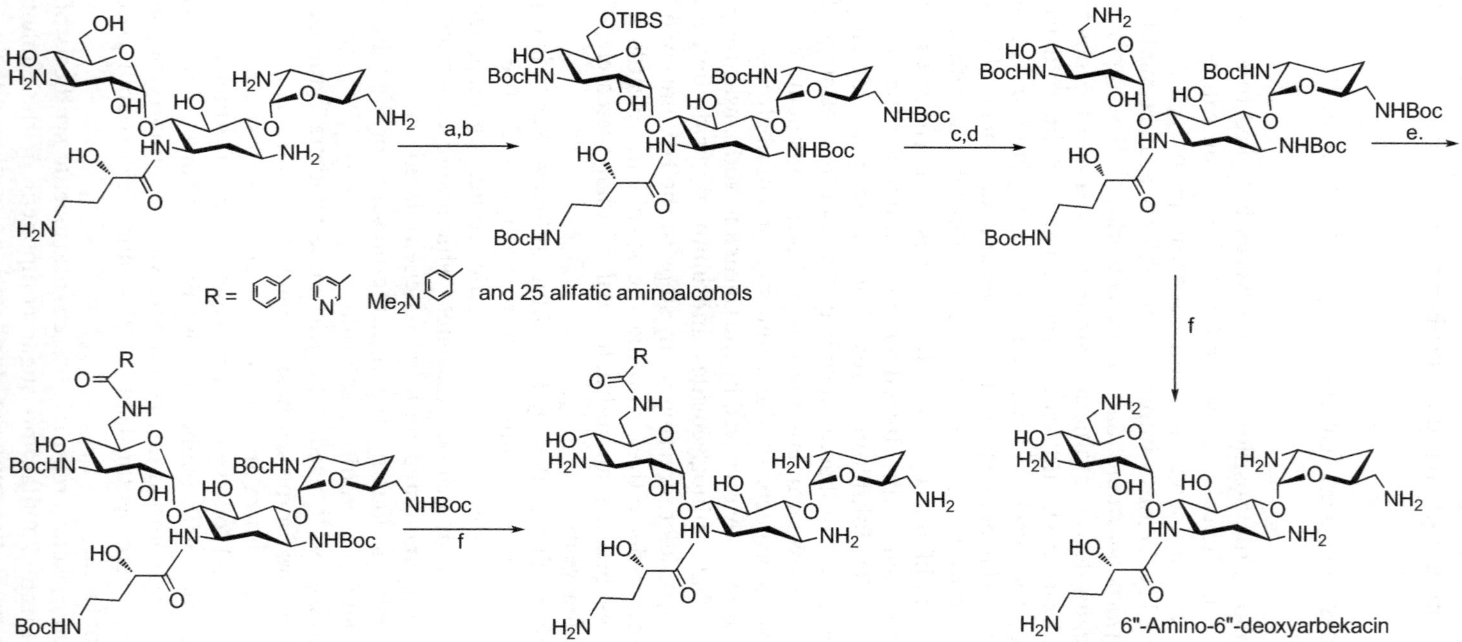

Reagents: (a). Boc_2O, Et_3N, H_2O, dioxane, rt; (b).TIBSCl, Py, rt; (c). NaN_3. DMF, 100° C; (d). PPh_3, H_2O, THF, 50° C; (e). RCO_2H, DMT-MM, or RCO-OSu; (f). aqueous TFA, 0° C.

Scheme 21 Sasaki's regioselective amination strategy in the synthesis of 4,6-diaminosugars.

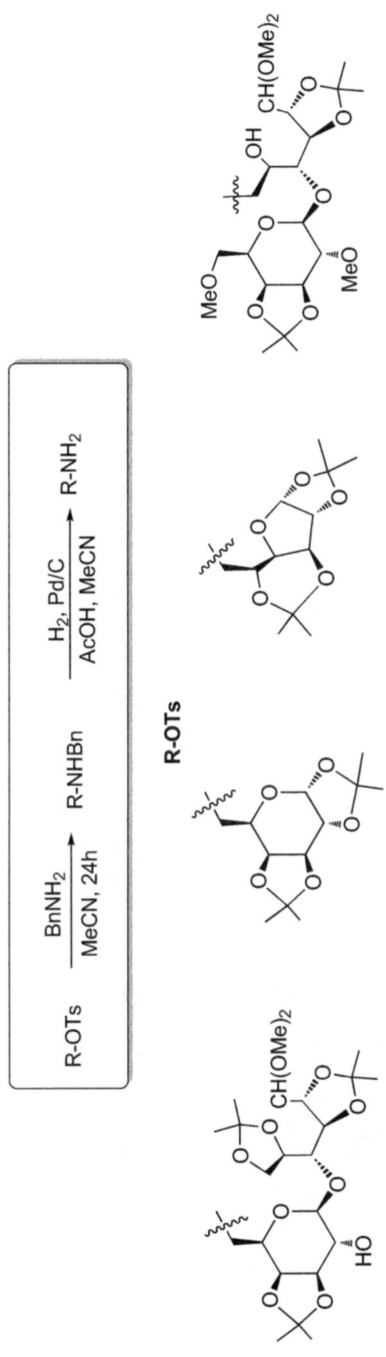

Scheme 22 Bianchini's synthesis of 6-amino sugars.

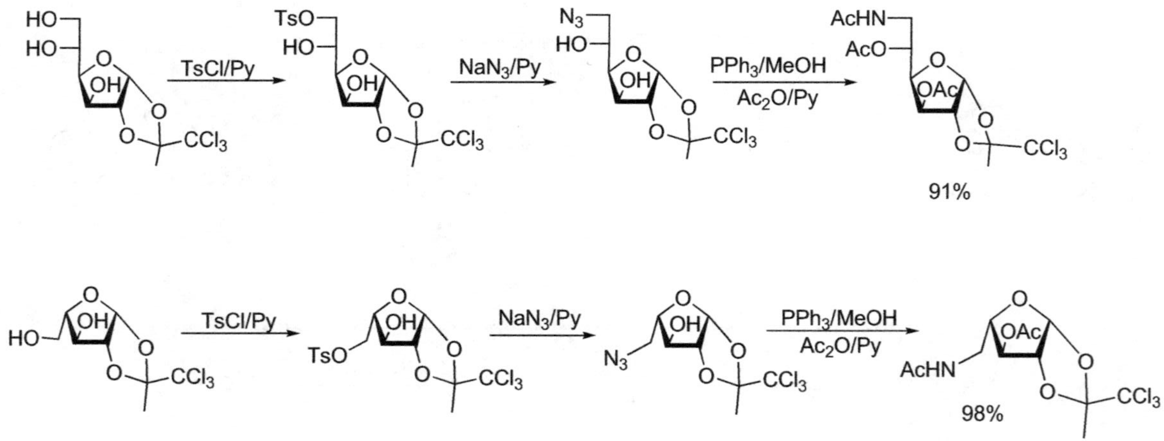

Scheme 23 Yenill's synthetic method for sugar primary position amination.

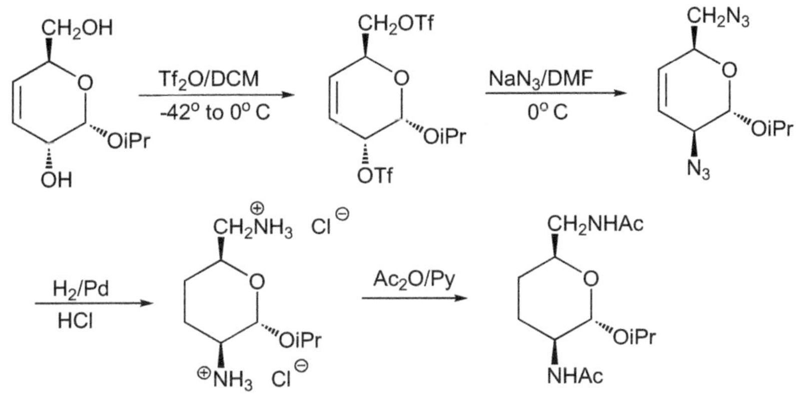

Scheme 24 Varela's enantiospecific synthesis of 2,6-diamino sugars.

to produce 2,6-ditriflate (Scheme 24). Subsequent nucleophilic S_N2 displacement of the triflate group by the azide anion in dimethylformamide, followed by the reduction with hydrogen using the Pd catalyst and treatment with hydrochloric acid afforded the crystalline hydrochloride in a good yield. Conventional acetylation with acetic anhydride in pyridine afforded the crystalline isopropyl 2,6-*N,N*-acetamido-2,3,4,6-tetradeoxy-α-D-*threo*-hexopyranoside in 73% yield.

8 Synthetic strategies for the preparation of branched amino sugars

This category of amino sugars does not fit into the classification followed and based on the position of the amino group. This group can be located inside the branch and the branch may be linked to any sugar ring position. One illustrative example of such compound synthesis will be presented in this section, based on 1,3-dipolar cycloaddition of nitrones to carbohydrate enones. Sinou and coworkers[36] described a cycloaddition of methylideneaniline *N*-oxide (generated *in situ* from phenylhydroxylamine and formaldehyde) to sugar enones. Interestingly, addition of the intermediate nitrone to a C2–C3 double bond generated the *lyxo* configuration and the phenylisoxazolidine derivatives were obtained in 52–62% yield. Hydrogenation of the phenylisoxazolidine derivative using PtO_2 as a catalyst under 1,5 atmosphere of hydrogen resulted in a simultaneous cleavage of the N–O bond of the isoxazolidine ring system and the stereoselective reduction of the carbonyl group at C-4. Upon further acetylation 3-deoxy-*N*-(phenylamino)methyl-α-D-talopyranosides were obtained in moderate yields. The synthetic route is depicted in Scheme 25.

9 Conclusions and future prospects

This contribution has reviewed recent literature reports on synthetic approaches to amino sugars. The synthetic products may act as potential templates for the synthesis of amino sugar antibiotics, glycoconjugates

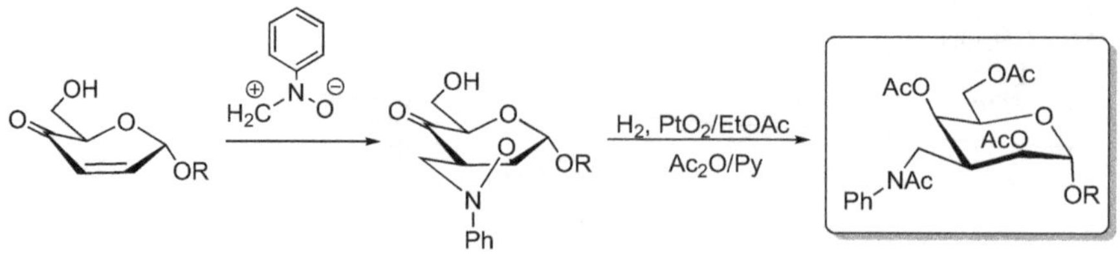

R= Et-, cyclopentyl, cyclohexyl, cyclopropylmethyl

Scheme 25 Sinou's synthesis of branch-chain amino sugars *via* phenylisoxazoline route.

and other biologically important molecules. As the scope of modern synthesis evolves with addition of new techniques and a seemingly unlimited access to large amounts of various amino derivatives, simpler and more economic avenues to a library of amino sugars with diversified configurations must be developed. This important aspect of the synthetic toolbox is steadily growing and prompting an interest of multi-disciplinary teams. Also, in recent years we observe an unprecedented growth in isolating, determining the structure and synthesizing various saccharides including amino sugars. There are several reasons for this trend. They include:

• Amino sugars are components of a plurality of important antibiotics. Sadly, for many years the research in the area of antibiotics was neglected. It seems that we may be coming back to our senses.

• After developing satisfactory methods for sequencing and synthesizing nucleic acids, we realize that we need similar tools for oligosaccharides. While it seemed tremendously difficult at first, there are good reasons to be optimistic. There are many novel methods of synthesizing oligosaccharides comprising amino sugar residues. This is tremendously important since we learn more and more of novel conjugates containing saccharides and their function in living organisms.

• Various glycoconjugates are finding new applications as medicines other than antibiotics.

All these reasons will force the decision makers to spend more money on research and development in areas of chemistry, biology, chemical biology and biochemistry related to sugars in general and amino sugars in particular. Probably, the most important argument derives from the observation that amino sugars play an important role as components of successful cancer vaccines. It is a spectacular reminder that we must understand better this group of compounds. We expect a significant acceleration in all areas of amino sugar R&D.

References

1 S. Knauer, B. Kranke, L. Krause and H. Kunz, *Curr Org. Chem.*, 2004, **23**, 1739.
2 A. F. G. Bongat and A. V. Demchenko, *Carbohydr. Res.*, 2007, **342**, 374.
3 A. Dondoni and A. Marra, *Chem. Rev.*, 2000, **100**, 4395.
4 S. A. W. Grunner, E. Locardi, E. Lohof and H. Kessler, *Chem. Rev.*, 2002, **102**, 491.
5 B. Ernst and J. L. Magnani, *Nat. Rev. Drug Discovery*, 2009, 661.

6 M. Emmadi and S. S. Kulkarni, *Nat Prod. Rep.*, 2014, **31**, 870.
7 K. C. Nicolaou, S. A. Snyder, A. Z. Nalbandian and D. A. Longbottom, *J. Am. Chem. Soc.*, 2004, **126**, 6234.
8 C. P. R. Hackenberger, M. K. O'Reilly and B. Imperiali, *J. Org. Chem.*, 2005, **70**(9), 3574.
9 L. M. Likhosherstov, O. S. Novikova and V. N. Shibaev, *Dokl. Chem.*, 2002, **383**, 89.
10 A. D. Dorsey, J. E. Barbarow and D. Trauner, *Org. Lett.*, 2003, **18**, 3237.
11 B. M. Fusaro, V. Chagnault and D. Postel, *Tetrahedron*, 2013, **69**, 542.
12 L. Colombeau, T. Traore, P. Compai and O. R. Martin, *J. Org Chem.*, 2008, **73**, 8647.
13 W. Schonemann, E. Gallienne, P. Compain, K. Ikeda, N. Asano and O. R. Martin, *Bioorg. Med. Chem. Lett.*, 2010, **18**, 2645.
14 F. Ding, R. William and X.-W. Liu, *J. Org. Chem.*, 2013, **78**, 1293.
15 K. Larsen, C. E. Olsen and M. S. Motavia, *Carbohydr. Res.*, 2008, **343**, 383.
16 Y. Cai, C.-C. Ling and D. R. Bundle, *J. Org Chem.*, 2009, **74**, 580.
17 M. Emmadi and S. S. Kulkarni, *Nat. Protoc.*, 2013, **8**, 1870.
18 D. Matsuura, T. Mitsui, T. Sengoku, M. Takahashi and H. Yoda, *Tetrahedron*, 2008, **64**, 11686.
19 H. Kikuchi, Y. Saito, J. Komija, S. Honma, N. Nkahata, A. Ito and Y. Oshima, *J. Org. Chem.*, 2001, **66**, 6982.
20 B. Bressel, B. Egart, A. Al-harrasi, R. Pulz, H.-U. Reissig and I. Brudgam, *Eur. J. Org. Chem.*, 2008, 467.
21 F. Pfrengle, D. Lentz and H.-U. Reissig, *Angew. Chem., Int Ed.*, 2009, **48**, 3165.
22 B. Bressel and H.-U. Reissig, *Org. Lett.*, 2009, **45**, 7417.
23 F. Pfrengle and H.-U. Reissig, *Chem. Soc. Rev.*, 2010, **39**, 549.
24 K. C. Nicolaou, P. S. Baran, Y.-L. Zhong and J. A. Vega, *Angew. Chem. Int. Ed.*, 2000, **39**, 2525.
25 G. K. Friestad, T. Jiang and G. M. Fioroni, *Tetrahedron*, 2008, **64**, 11549.
26 B. Ravindran and T. Pathak, *Indian J. Chem.*, 2001, **40B**, 1114.
27 T. Pathak and R. Bhattacharya, *Carbohydr. Res.*, 2008, **342**, 1980.
28 L. F. Tietze, N. Bohnke and S. Dietz, *Org. Lett.*, 2009, **11**, 2948.
29 C. Albler and W. Schmid, *Eur J.Org. Chem.*, 2014, 2451.
30 M. L. Uhrig, M. J. Simirgiotis and O. Varela, *Tetrahedron: Asymmetry*, 2010, **21**, 2435.
31 K. Sasaki, Y. Kobayashi, T. Kurihara, Y. Yamashita, Y. Takahashi, T. Mijake and Y. Akamatsu, *J. Antibiot.*, 2015, **68**, 741.
32 J. L. Houghton, K. D. Green, W. Chen and S. Garneau-Tsodikova, *ChemBioChem*, 2010, **11**, 880.
33 M. Corsi, M. Bonnanni, G. Catelani, F. D'Andrea, T. Gragnani and R. Bianchini, *Int. J. Org Chem.*, 2013, **3**, 41.
34 N. Yenil, E. Ay, K. Ay, M. Oskay and J. Maddaluno, *Carbohydr. Res.*, 2010, **345**, 1617.
35 C. A. Iriarte Capaccio and O. Varela, *Tetrahedron: Asymmetry*, 2000, **11**, 4945.
36 J. R. de Freitas Filho, R. M. Srivastava, W. J. P. da Silva, L. Cottier and D. Sinou, *Carbohydr. Res.*, 2003, **338**, 673.